TRENDS IN MATHEMATICS

Trends in Mathematics is a series devoted to the publication of volumes arising from conferences and lecture series focusing on a particular topic from any area of mathematics. Its aim is to make current developments available to the community as rapidly as possible without compromise to quality and to archive these for reference.

Proposals for volumes can be sent to the Mathematics Editor at either

Birkhäuser Verlag
P.O. Box 133
CH-4010 Basel
Switzerland

or

Birkhäuser Boston Inc.
675 Massachusetts Avenue
Cambridge, MA 02139
USA

Material submitted for publication must be screened and prepared as follows:

All contributions should undergo a reviewing process similar to that carried out by journals and be checked for correct use of language which, as a rule, is English. Articles without proofs, or which do not contain any significantly new results, should be rejected. High quality survey papers, however, are welcome.

We expect the organizers to deliver manuscripts in a form that is essentially ready for direct reproduction. Any version of TeX is acceptable, but the entire collection of files must be in one particular dialect of TeX and unified according to simple instructions available from Birkhäuser.

Furthermore, in order to guarantee the timely appearance of the proceedings it is essential that the final version of the entire material be submitted no later than one year after the conference. The total number of pages should not exceed 350. The first-mentioned author of each article will receive 25 free offprints. To the participants of the congress the book will be offered at a special rate.

Graph Theory in Paris

Proceedings of a Conference in Memory of Claude Berge

Adrian Bondy
Jean Fonlupt
Jean-Luc Fouquet
Jean-Claude Fournier
Jorge L. Ramírez Alfonsín

Editors

Editors:

Adrian Bondy
Institut Camille Jordan
Université Claude Bernard Lyon 1
43 boulevard du 11 novembre 1918
69622 Villeurbanne cedex
France
e-mail: jabondy@univ-lyon1.fr

Jean-Claude Fournier
Equipe Combinatoire et Optimisation
Case 189
Université Pierre et Marie Curie, Paris 6
75252 Paris Cedex 05
France
e-mail: Fournier@math.jussieu.fr

Jean Fonlupt
Equipe Combinatoire et Optimisation
Case 189
Université Pierre et Marie Curie, Paris 6
75252 Paris Cedex 05
France
e-mail: Jean.Fonlupt@math.jussieu.fr

Jorge L. Ramírez Alfonsín
Equipe Combinatoire et Optimisation
Case 189
Université Pierre et Marie Curie, Paris 6
75252 Paris Cedex 05
France
e-mail: ramirez@math.jussieu.fr

Jean-Luc Fouquet
Laboratoire d'Informatique Fondamentale
d'Orléans
Bâtiment IIIA
Rue Léonard de Vinci / B.P. 6759
45067 Orléans Cedex 2
France
e.mail: fouquet@univ-lemans.fr

2000 Mathematical Subject Classification 05C, 05D, 05E

A CIP catalogue record for this book is available from the
Library of Congress, Washington D.C., USA

Bibliographic information published by Die Deutsche Bibliothek
Die Deutsche Bibliothek lists this publication in the Deutsche Nationalbibliografie; detailed
bibliographic data is available in the Internet at http://dnb.ddb.de

ISBN 3-7643-7228-1 Birkhäuser Verlag, Basel – Boston – Berlin

This work is subject to copyright. All rights are reserved, whether the whole or part of the
material is concerned, specifically the rights of translation, reprinting, re-use of illustrations,
recitation, broadcasting, reproduction on microfilms or in other ways, and storage in data
banks. For any kind of use permission of the copyright owner must be obtained.

© 2007 Birkhäuser Verlag, P.O. Box 133, CH-4010 Basel, Switzerland
Part of Springer Science+Business Media
Printed on acid-free paper produced from chlorine-free pulp. TCF ∞
Printed in Germany
ISBN-10: 3-7643-7228-1
ISBN-13: 976-3-7643-7228-6

e-ISBN-10: 3-7643-7400-4
e-ISBN-13: 978-3-7643-7400-6

9 8 7 6 5 4 3 2 1

www.birkhauser.ch

Contents

Preface vii

B. Toft
    Claude Berge – Sculptor of Graph Theory 1

M. Abreu and S.C. Locke
    k-path-connectivity and mk-generation: an Upper Bound on m 11

M. Aouchiche and P. Hansen
    Automated Results and Conjectures on Average Distance
    in Graphs 21

E. Birmelé, J.A. Bondy and B.A. Reed
    Brambles, Prisms and Grids 37

Y. Björnsson, R. Hayward, M. Johanson and J. van Rijswijck
    Dead Cell Analysis in Hex and the Shannon Game 45

M. Blidia, M. Chellali and O. Favaron
    Ratios of Some Domination Parameters in Graphs and
    Claw-free Graphs 61

A. Bonisoli and D. Cariolaro
    Excessive Factorizations of Regular Graphs 73

M. Burlet, F. Maffray and N. Trotignon
    Odd Pairs of Cliques 85

K. Cameron, E.M. Eschen, C.T. Hoàng and R. Sritharan
    Recognition of Perfect Circular-arc Graphs 97

M. DeVos, J. Nešetřil and A. Raspaud
    On Edge-maps whose Inverse Preserves Flows or Tensions 109

R.J. Faudree, R.J. Gould and M.S. Jacobson
    On the Extremal Number of Edges in 2-Factor
    Hamiltonian Graphs 139

T. Feder, P. Hell and W. Hochstättler
    Generalized Colourings (Matrix Partitions) of Cographs 149

Z. Fekete and L. Szegő
    A Note on $[k,l]$-sparse Graphs 169

C.M.H. de Figueiredo, F. Maffray and C.R.V. Maciel
    Even Pairs in Bull-reducible Graphs 179

H. Galeana-Sánchez and R. Rojas-Monroy
    Kernels in Orientations of Pretransitive Orientable Graphs 197

J. Grytczuk
    Nonrepetitive Graph Coloring 209

B.L. Hartnell
    A Characterization of the 1-well-covered Graphs with no 4-cycles 219

A.J.W. Hilton and C.L. Spencer
    A Graph-theoretical Generalization of Berge's Analogue of
    the Erdős-Ko-Rado Theorem 225

V.E. Levit and E. Mandrescu
    Independence Polynomials and the Unimodality Conjecture for
    Very Well-covered, Quasi-regularizable, and Perfect Graphs 243

D. Marx
    Precoloring Extension on Chordal Graphs 255

Y. Matsui and T. Uno
    On the Enumeration of Bipartite Minimum Edge Colorings 271

B. Mohar
    Kempe Equivalence of Colorings 287

M. Montassier
    Acyclic 4-choosability of Planar Graphs with Girth at Least 5 299

J. Morris
    Automorphism Groups of Circulant Graphs – a Survey 311

G. Pap
    Hypo-matchings in Directed Graphs 327

B. Randerath and I. Schiermeyer
    On Reed's Conjecture about ω, Δ and χ 339

A. Recski and J. Szabó
    On the Generalization of the Matroid Parity Problem 347

I.P.F. da Silva
    Reconstruction of a Rank 3 Oriented Matroids from
    its Rank 2 Signed Circuits 355

A.K. Wagler
    The Normal Graph Conjecture is True for Circulants 365

S. Zhou
    Two-arc Transitive Near-polygonal Graphs 375

U.S.R. Murty (Editor)
    Open Problems 381

Preface

Following the death of Claude Berge in June 2002, the Equipe Combinatoire, the group founded by Berge in 1975 under the aegis of the C.N.R.S. and in liaison with the Université Pierre et Marie Curie, decided to organise a conference on graph theory in his memory. This meeting, GT04, took place in July 2004. It was the first international conference on graph theory to be held in the Paris region since the memorable meeting in Orsay in 1976, which coincided with Claude's fiftieth birthday. The conference was held in the heart of the Latin Quarter, on one of the campuses of the Université Pierre et Marie Curie, the Couvent des Cordeliers (the former site of a Franciscan convent). Our aim was not only to celebrate the life and achievements of Claude Berge, but also to organise a conference in the line of continuity of the international meetings on graph theory and related topics which had been held successfully in Marseille-Luminy at five-year intervals since 1981.

GT04 brought together many prominent specialists on topics upon which Claude Berge's work has had a major impact, such as perfect graphs and matching theory. The meeting attracted over two hundred graph-theorists, roughly half of whom contributed to the scientific program. Plenary talks were presented by Maria Chudnovsky, Vašek Chvátal, Gérard Cornuéjols, András Frank, Pavol Hell, László Lovász, Jaroslav Nešetřil, Paul Seymour, Carsten Thomassen and Bjarne Toft.

Generous support for the conference was provided by the Université Pierre et Marie Curie, CNRS (Centre National de la Recherche Scientifique), INRIA (Institut National de Recherche en Informatique et en Automatique), the European network DONET (Discrete Optimization NETwork), the Délégation Générale pour l'Armement, France Télécom, ILOG and Schlumberger.

This volume includes contributions from many of the participants. All papers were refereed, and we are pleased to thank those colleagues who assisted us in this task. A short section of open problems presented during the meeting and edited by Rama Murty concludes the book.

The Editors

Adrian Bondy
Jean Fonlupt
Jean-Luc Fouquet
Jean-Claude Fournier
Jorge Ramírez Alfonsín

Claude Berge

Claude Berge – Sculptor of Graph Theory

Bjarne Toft

> **Abstract.** Claude Berge fashioned graph theory into an integrated and significant part of modern mathematics. As was clear to all who met him, he was a multifaceted person, whose achievements, however varied they might seem at first glance, were interconnected in many ways.

1. Introduction

My purpose here is to present an account of some of Claude Berge's activities and achievements, mainly with regard to his role as a graph theorist. The information upon which I draw is mostly available in published sources. But the account is also personal, in that I shall include some of my own experiences and impressions. As a doctoral student in July 1969, I attended the *Colloquium on Combinatorial Theory and its Applications* in Balatonfüred at Lake Balaton. For me, this was like a dream, with its unique Hungarian charm and hospitality, the presence of many young people and of famous mathematicians like Berge, Erdős, Rényi, Rota, Turán and van der Waerden, to mention just a few. This was my first encounter with Berge, and I admired his French intellectual style, if at a distance. I also learned from him – at this meeting Berge emphasized the importance of hypergraphs, then still something of a novelty. Again in Hungary, at the meeting in Keszthely in June 1973 to celebrate Erdős' 60th birthday, I got to know Berge better. I gave my first conference lecture there, in the same afternoon session as Berge – I was nervous talking about hypergraph colouring with Berge sitting in the front row. But Berge was gracious and reassuring – he was without pretention, invariably putting those in his company at ease. As regards his mathematics, too, Berge had a distinctive manner, attempting always to combine the general with the concrete, and to see things in a general mathematical framework. He introduced hypergraphs not merely to generalize, but also to unify and simplify.

2. The late fifties and early sixties

The period around 1960 seems to have been particularly important and fruitful for Berge. Through the book *Théorie des graphes et ses applications* [2] he had established a mathematical name for himself. In 1959 he attended the first graph theory conference ever in Dobogokő, Hungary, and met the Hungarian graph theorists. He published a survey paper on graph colouring [4]. It introduced the ideas that soon led to perfect graphs. In March 1960 he talked about this at a meeting in Halle in East Germany [6]. In November of the same year he was one of the ten founding members of the OuLiPo (*Ouvroir de Littérature Potentiel*). And in 1961, with his friend and colleague Marco Schützenberger, he initiated the *Séminaire sur les problèmes combinatoires de l'Université de Paris* (which later became the *Equipe combinatoire du CNRS*). At the same time Berge achieved success as a sculptor [7].

3. Games, graphs, topology

Games were a passion of Claude Berge throughout his life, whether playing them – as in favorites such as chess, backgammon and hex – or exploring more theoretical aspects. This passion governed his interests in mathematics. He began writing on game theory as early as 1951, spent a year at the Institute of Advanced Study at Princeton in 1957, and the same year produced his first major book *Théorie générale des jeux à n personnes* [1]. Here, one not only comes across names such as von Neumann and Nash, as one would expect, but also names like König, Ore and Richardson. Indeed, the book contains much graph theory, namely the graph theory useful for game theory. It also contains much topology, namely the topology of relevance to game theory. Thus, it was natural that Berge quickly followed up on this work with two larger volumes, *Théorie des graphes et ses applications* [2] and *Espaces topologiques, fonctions multivoques* [3]. *Théorie des graphes et ses applications* [2] is a master piece, with its unique blend of general theory, theorems – easy and difficult, proofs, examples, applications, diagrams. It is a personal manifesto of graph theory, rather than a complete description, as attempted in the book by König [31]. It would be an interesting project to compare the first two earlier books on graph theory, by Sainte-Laguë [34] and König [31] respectively, with the book by Berge [2]. It is clear that Berge's book is more leisurely and playful than König's, in particular. It is governed by the taste of Berge and might well be subtitled 'seduction into graph theory' (to use the words of Rota from the preface to the English translation of [13]). Among the main topics in [2] are factorization, matchings and alternating paths. Here Berge relies on the fundamental paper of Gallai [25]. Tibor Gallai is one of the greatest graph theorists – he is to some degree overlooked – but not by Berge. Gallai was among the first to emphasize min-max theorems and LP-duality in combinatorics. In [26] one finds for the first time in writing the result (in generalized form) that the complement of a bipartite graph is perfect, attributed by Gallai to König and dated to 1932. But also [2] contains a theorem characterizing the size of a maximum independent set of vertices in a bipartite graph, which is easily seen to be equivalent to the fact that the complement of a bipartite graph is perfect. To notice this non trivial, yet simple, result seems to me to be a major step in the direction of the perfect graph conjectures. The 1959 book *Espaces topologiques, fonctions multivoques* [3] deals with general topology, focussing on what is useful in game theory, optimization theory and combinatorics. It includes a theory of multivalued functions, as stated in the title. And as Berge explains, when combinatorial properties of these functions are studied, it may be called a theory of oriented graphs. One of the theorems of [3] is known as *Berge's Maximum Theorem*. It deals with multivalued continuous mappings. It is very useful in economics and well known among economists. Also here, Berge manages to focus on the essential and useful. At *The History of Economic Thought* Website (http://cepa.newschool.edu/het/) the topic *Continuity and all that* is divided into four sections. Two of these deal with Berge's theory. They are called *Upper and lower semicontinuity of correspondences* and *Berge's Theorem*. In the 1960's two more books by Berge appeared, namely *Programmes, jeux et réseaux de transport* [8] and *Principes de combinatoire* [13]. In the preface to the English version of [13], which came out in 1971, Gian-Carlo Rota said:

> Two Frenchmen have played a major role in the renaissance of Combinatorics: Berge and Schützenberger. Berge has been the more prolific writer, and his books have carried the word farther and more effectively than anyone anywhere. I recall the pleasure of reading the disparate examples in his first book, which made it impossible to forget the material. Soon after that reading, I would be one of the many who unknotted themselves from the tentacles of the Continuum and joined the then Rebel Army of the Discrete. What are newed pleasure is it to again read Berge in the present book!

Both books [2] and [3] are now classics, and can still be purchased (in English) as new Dover Paperbacks. In 1970 Berge helped to give graph theory a new aspect by extending it to hypergraphs in the book *Graphes et hypergraphes* [15]. The purpose was to generalize, unify and simplify. The term hypergraph was coined by Berge, following a remark by Jean-Marie Pla, who had used the word hyperedge in a seminar. In 1978 Berge enriched the field once more with his lecture notes *Fractional Graph Theory* [17]. The purpose was again the same – and conjectures changed into elegant theorems in their fractional versions. The 1970 book was later split into two and appeared in the most recent versions as *Graphs* [20] and *Hypergraphs* [19]. In addition to his books Berge edited many collections of papers, some of which have been very influential, such as *Hypergraph Seminar* [16] and *Topics on Perfect Graphs* [18].

4. Perfect graphs

In 1960 Berge wrote a survey paper [4] on graph colouring, a topic not treated in depth in [2]. The paper was reviewed in *Mathematical Reviews* (MR 21, 1608) by Gabriel Andrew Dirac. Berge moved here into an area that Dirac knew like the back of his hand – and where Dirac had thought a lot about how to best prove and present results. Also Dirac, with his Hungarian background, did graph theory in the style of König, and he always maintained that König's book was the best source for learning graph theory. So Dirac was not particularly fond of Berge's more leisurely style, and his review of [4] was quite critical. Seen from today's perspective it seems too harsh. Here the first hints of perfect graphs appeared. As Berge wrote (my translation):

> *We shall here determine certain categories of graphs for which the chromatic number equals the clique number.*

In 1976 I attended the conference in Orsay to celebrate Berge's 50th birthday. I asked him about his reaction to Dirac's negative review. He said that he was just surprised and that he was on good terms with Dirac both before and after 1960. This was his nature – he took no offence. And Dirac invited Berge to visit Aarhus as one of the first graph theory guests after his appointment in Denmark in 1970. Dirac did his utmost to please Berge and make the visit a success, as I witnessed with my own eyes. I can still see the backs of Berge and Dirac, under an umbrella, disappearing in the fog and rain down Ny Munkegade in Aarhus. In 1959 in Dobogokő, Berge met Gallai, who told him about his work about graphs in which every odd cycle has two non-crossing chords, to be published in [27]. Berge saw immediately the importance of Gallai's work and included some of it in [4]. Berge called a graph in which every cycle has a chord a 'Gallai graph' (the terminology commonly used now is 'chordal graph' or 'rigid circuit graph') and proved the new result (in today's terminology) that such a graph is perfect, and he also included a proof of the theorem of Hajnal and Surányi that their complements are perfect – this they had presented in Dobogokő [29]. Berge called line-graphs of bipartite multigraphs 'pseudo-Gallai graphs'. Such graphs, and also their complements, are shown to be perfect. One property of these graphs is that all odd cycles of length at least 5 have chords. Berge remarked that, to establish perfectness, it is not enough to require only that all odd cycles of length at least five have chords, as the complement of the 7-cycle shows. He attributed this observation to A. Ghouila-Houri. Berge lectured about all these ideas at the colloquium on graph theory in Halle in East Germany, in March 1960, and wrote an extended abstract [6]. Here he defined a 'Gallai graph' as in [4] and a 'semi-Gallai graph' as one in which each odd cycle of length at least five has a chord. Berge mentions the result of Hajnal and Surányi [29] that the complement of a Gallai graph is perfect and his own result that a Gallai graph itself is perfect. Moreover he notices that bipartite graphs, line-graphs of bipartite graphs and a class of Shannon graphs are perfect semi-Gallai graphs. At the end he says that it would seem natural to conjecture that all semi-Gallai graphs are perfect, but he then again exhibits the Ghouila-Houri counterexample (the complement of the 7-cycle). Berge does not mention complements of line-graphs of bipartite graphs nor complements of bipartite graphs in the Halle abstract. The abstract [6] is in German. Berge had given it the title (English translation) *Colouring of Gallai and semi-Gallai graphs*. The referee apparently asked Gallai if this was appropriate, and Gallai in his modest style answered that there was a misunderstanding and that he had never looked thoroughly at these classes. So Dirac, involved with the editing, on his way to take up a professorship at Ilmenau in East Germany, suggested the change of title to the one the paper ended up with. In [21] Berge mentions that the strong perfect graph conjecture was stated in Halle in March 1960 at the end of his lecture in the form, that if a graph and its complement are both semi-Gallai graphs (such graphs are now called Berge graphs) then the clique number and chromatic number are equal. It is clear from the abstract that this was certainly a natural question with which to end the lecture, but it also seems clear that the focus was not yet on the conjecture. In 1961 Berge spent the summer at a symposium on combinatorial theory at the RAND Cooperation in Santa Monica in California. There he presented his results, including some new ones on unimodular graphs, and he had many fruitful discussions, among others with Alan J. Hoffman. It seems likely that the whole English terminology, as we know it, was created here. On his return to Paris Berge wrote an English version of the theory of perfect graphs, and he sent it to Hoffman for comments. This manuscript, with some improvements suggested by Hoffman, Gilmore and McAndrew, appeared as *Some classes of perfect graphs* ([9], [12] and [14]). The 1963 paper [9] consists of lecture notes from the Indian Statistical Institute in Calcutta, which Berge visited in March and April 1963. Berge himself had at some point forgotten the existence of these published notes – they are not mentioned in the preprint [21], where he had some difficulties explaining why these important ideas from 1960 had to wait so long to get published. So he was pleased when I sent him a copy of [9]. I discovered [9] in the fine library of the University of Regina, Canada, in 1993. The published lecture notes are however not rare and are present in several libraries around the world. So my bet on the first publication using the term perfect graph and mentioning explicitly the perfect graph conjectures is Berge's paper [9] from 1963. This paper contains the whole basic theory of perfect graphs in the still commonly used terminology. Now, 40 years later the strong perfect graph conjecture has finally been proved [24]. In addition to [6], an abstract [10] from a meeting in Japan in September 1963 is often mentioned as an original source for perfect graphs. However the abstract is very short (nine lines) and mentions neither perfect graphs nor the perfect graph conjectures. At the meeting in Japan Berge did however distribute a manuscript. Judging from titles and from [21] this was the manuscript later to be presented and discussed at Ravello, Italy, in June 1964, and published in 1966 [11]. This paper (in French) contains the strong perfect graph conjecture and acknowledges Paul C. Gilmore's influence.

5. Problems

Berge was influential as an author of books, conference lecturer, thesis director and seminar organizer – the weekly seminar held at the *Maison des sciences de l'homme* on boulevard Raspail was legendary – but he was less so as a problem poser. His problems are relatively few and sometimes seem accidental. The book [2] has an appendix with fourteen unsolved problems. It might be interesting to examine these more closely with modern eyes. There are exceptions to this. In particular the perfect graph conjectures have had a huge impact. There are at least two other widely known circles of problems due to Berge: *The Berge Path Partition Conjectures* and *Berge's Hypergraph Edge Colouring Problems*. See [23] or [30] for a more detailed description. Finally, Berge edited for a period (1960–64) a magazine column [5] of brain teasers. It might also be interesting to take a closer look at these (unfortunately I did not have access to this material).

6. OuLiPo

The OuLiPo is a group of French writers and intellectuals, who experiment with literature. When one writes literature, poetry or music, one imposes on oneself certain restrictions. The main idea of this workshop for potential literature is to make these restrictions of a more precise mathematical nature (as Schönberg did in music and Lewis Carroll did in some of his writings). Martin Gardner wrote two columns on OuLiPo in *Scientific American* in the late 1970's, later to be completely rewritten and included in the book [28]. He said that

> *the most sophisticated and amusing examples of literary word play have been produced by the whimsical, slightly mad French group called the Oulipo.*

There are some books in English about and by the group, among them [32] and [33]. In both these books Berge is prominently featured. Berge was active in OuLiPo and wrote several articles, responsible as he was for "combinatory analysis". His most well-known OuLiPo work is the short story *Who killed the Duke of Densmore?* [22], which is a classical crime story, where the solution however requires knowledge of the Theorem of Hajós, characterizing interval graphs (Berge had heard György Hajós lecture about this theorem in Halle in March 1960). With this theorem it is possible to see that the set of events cannot have taken place as described by the participants, because the overlap graph of the events as described is not an interval graph. So at least one of the participants must be lying. But only the removal of one particular vertex of the corresponding overlap graph changes this into an interval graph, so this reveals the culprit. Berge pays tribute to the author Lewis Carroll and also to Carroll's alter ego, the mathematician C.L. Dodgson, both of whom are represented in the Duke's library. In an interval graph the sequence of events cannot be fully determined since a suitable sequential ordering of the intervals may be reversed. One of the persons in the short story contemplates the possibility of writing a novel with a set of events corresponding to an interval graph, where the two possible orderings in time would give two different solutions to the plot. Maybe Berge himself tried to create such an interesting (possible?) sequence of events? Berge spoke to Adrian Bondy of his wish to write a detective story in which the reader is the murderer, or the author, or the publisher... In [33] there are other interesting contributions, for example the ultimate lipogram, a book where not only one letter, but all letters have been avoided (where however the list of contents, footnotes, index, errata list, foreword and afterword, not being part of the text of the book itself, do use letters!), and a paper by the famous author Raymond Queneau (who had attended Berge's graph theory seminars around 1960), based on Hilbert's axioms for plane geometry, where point has been replaced by word and line by sentence, thus providing a foundation for literature (Hilbert told us that the words point and line are undefined and may be called anything). The OuLiPo group surely had/have a lot of fun, and their meetings, which take place inpublic, are packed. This was in particular so for their meeting in Berge's memory, at which he was officially 'excused' for his absence.

7. Sculpture

In our modern everyday life we are surrounded and bombarded by (too) beautiful, flawless pictures, sculptures and designs. In this stream Claude Berge's sculptures catch our attention, with their authenticity and honesty. They are not pretending to be more than they are. Berge catches again something general and essential, as he did in his mathematics. The sculptures may at first seem just funny, and they certainly have a humorous side. But they have strong personalities in their unique style – you come to like them as you keep looking at them – whether one could live with them if they came alive is another matter! The book *Sculptures multipètres* [7] gives a good impression of Berge's early sculptures, made partly from stones he found in the Seine. It was prefaced by Philippe Soupault, a well-known surrealist writer.

8. Conclusion

Claude Berge's greatest scientific achievement is that he gave graph theory a place in mathematics at large by revealing and emphasizing its connections to set theory, topology, game theory, operations research, mathematical programming, economics and other applications. The influence came mainly through his books, but also via his lectures, discussions at conferences and seminars, and of course very strongly through his many students. He generalized, unified, simplified, and combined in a unique way the general and the concrete. He will for a long time remain an inspirational force. So let us continue to play games and enjoy graph theory Claude Berge style!

9. Acknowledgment

I am indebted to a large number of people who helped me prepare my lecture at the GT04 meeting in Paris in July 2004 on which this paper is based. In particular I wish to thank Jorge Ramírez Alfonsín, Adrian Bondy, Kathie Cameron, Jack Edmonds, Michel LasVergnas, Douglas Rogers and Gert Sabidussi.

References

[1] C. Berge, *Théorie générale des jeux à n personnes*, Memorial des sciences mathématiques **138**, Gauthier-Villars (Paris) 1957.

[2] C. Berge, *Théorie des graphes et ses applications*, Dunod (Paris) 1958. Translation: *The Theory of Graphs and its Applications*, Meuthen (London) and Wiley (New York)1962. Republished as: *The Theory of Graphs*, Dover (New York) 2001.

[3] C. Berge, *Espaces topologiques, fonctions multivoques*, Dunod (Paris) 1959 and 1962. Translation: *Topological Spaces: including a Treatment of Multi-valued Functions, Vector Spaces and Convexity*, Oliver and Boyd (Edinburgh and London) 1963. Republished as: *Topological Spaces*, Dover (New York) 1997.

[4] C. Berge, Les problèmes de coloration en théorie des graphes, *Publ. Inst. Statist. Univ. Paris* **9** (1960), 123–160.

[5] C. Berge, Problèmes plaisants et delectables, Column in *Le Journal de l'A.F.I.R.O.*, 1960–64.

[6] C. Berge, Färbung von Graphen, deren sämtliche bzw. deren ungerade Kreise starr sind, *Wissenschaftliche Zeitschrift der Martin-Luther-Universität Halle-Wittenberg*, Math.-Nat. **X/1** (1960), 114–115.

[7] C. Berge, *Sculptures multipètres*, présenté par Philippe Soupault, 1000 exemplaires numérotés, sur les presses de Lanord, imprimeur à Paris, 1962.

[8] C. Berge and A. Ghouila-Houri, *Programmes, jeux et réseaux de transport*, Dunod 1962. Translation: *Programming, games and transportation networks*, Methuen and Wiley 1965.

[9] C. Berge, Some classes of perfect graphs, in: *Six Papers on Graph Theory*, Indian Statistical Institute, Calcutta, 1963, 1–21.

[10] C. Berge, Sur une conjecture relative au problème des codes optimaux de Shannon, in: *Union Radio Scientifique Internationale, XIVe Assemblée Générale, Tokyo Sept. 9–20, 1963*,Volume **XIII-6**, *Ondes et Circuits Radio électriques*, U.R.S.I. Bruxelles 1963, 317–318.

[11] C. Berge, Une application de la Théorie des Graphes à un Problème de Codage, in: *Automata Theory* (ed. E.R. Caianiello), Academic Press 1966, 25–34.

[12] C. Berge, Some classes of perfect graphs, in: *Graph Theory and Theoretical Physics* (ed. F. Harary), Academic Press 1967, 155–165.

[13] C. Berge, *Principes de combinatoire*, Dunod 1968. Translation: *Principles of Combinatorics*, Academic Press 1971.

[14] C. Berge, Some classes of perfect graphs, in:*Combinatorial Mathematics and its Applications, Proceedings Conf. Univ. North Carolina, Chapel Hill 1967*, Univ. North Carolina Press 1969, 539–552.

[15] C. Berge, *Graphes et hypergraphes*, Dunod 1970. Translation: *Graphs and Hypergraphs*, North-Holland 1973.

[16] *Hypergraph Seminar*, Proceedings First Working Sem. Ohio State Univ., Columbus Ohio 1972 (ed. C. Berge and D. Ray-Chaudhuri), *Lecture Notes in Mathematics* **411**, Springer-Verlag 1974.

[17] C. Berge, *Fractional Graph Theory*, Indian Statistical Institute Lecture Notes No. **1**, The MacMillan Company of India 1978.

[18] *Topics on Perfect Graphs* (edited by C. Berge and V. Chvátal), *North-Holland Mathematical Studies* **88**, *Annals of Discrete Math.* **21**, North-Holland 1984.

[19] C. Berge, *Hypergraphs*, North-Holland 1989.

[20] C. Berge, *Graphs*, North-Holland 1991.

[21] C. Berge, *The history of perfect graphs*, Equipe Combinatoire **94/02**, Mars 1994, preprint, 8 pages.

[22] C. Berge, *Qui a tué le duc de Densmore?* Bibliotèhques Oulipienne No **67**, limited edition of 150 copies, 1994, English translation: *Who killed the Duke of Densmore?* in [33].

[23] C. Berge, Motivation and history of some of my conjectures, *Discrete Math.* **165/166** (1997), 61–70.

[24] M. Chudnovsky, N. Robertson, P. Seymour and R. Thomas, The strong perfect graph theorem, *Annals of Mathematics*, to appear.

[25] T. Gallai, On factorization of graphs, *Acta Math.Acad. Sci. Hung.* **I** (1950), 133–153.

[26] T. Gallai, Maximum-Minimum Sätze über Graphen, *Acta Math. Acad. Sci. Hung.* **IX** (1958), 395–434.

[27] T. Gallai, Graphen mit triangulierbaren ungeraden Vielecken, *Publ. Math. Inst. Hung. Acad. Sci.* **7**, (1962), 3–36.

[28] M. Gardner, *Penrose Tiles to Trapdoor Ciphers and the Return of Dr. Matrix*, Freeman 1989, updated, revised and republished by the Mathematical Association of America 1997.

[29] A. Hajnal and J. Surányi, Über die Auflösung von Graphen in vollständige Teilgraphen, *Ann.Univ. Budapest* **1** (1958), 53–57.

[30] T.R. Jensen and B. Toft, *Graph Coloring Problems*, Wiley Interscience 1995.

[31] D. König, *Theorie der endlichen und unendlichen Graphen*, Teubner, Leipzig 1936.

[32] *OULIPO a primer of potential literature* (W.F. Motte ed.), University of Nebraska Press 1986, and Dalkey ArchivePress, Illinois State University 1998.

[33] *OULIPO Laboratory.* Papers by Raymond Queneaux, Italo Calvino, Paul Fournel, Claude Berge, Jaques Jouet and Harry Mathews, Atlas Anti-classics 1995.

[34] M.A. Sainte-Laguë, *Les réseaux (ou graphes)*, *Mémorial des sciences mathématiques*, Fascicule **XVIII**, Gauthier-Villars (Paris) 1926.

Bjarne Toft
Mathematics and Computer Science
University of SouthernDenmark
Campusvej 55
DK-5230 Odense M, Denmark
e-mail: `btoft@imada.sdu.dk`

k-path-connectivity and mk-generation: an Upper Bound on m

M. Abreu and S.C. Locke

> **Abstract.** We consider simple connected graphs for which there is a path of length at least k between every pair of distinct vertices. We wish to show that in these graphs the cycle space over $\mathbb{Z}_2$ is generated by the cycles of length at least mk, where $m = 1$ for $3 \leq k \leq 6$, $m = 6/7$ for $k = 7$, $m \geq 1/2$ for $k \geq 8$ and $m \leq 3/4 + o(1)$ for large k.
>
> **Keywords.** k-path-connectivity, cycle space, k-generation.

1. Introduction

For basic graph-theoretic terms, we refer the reader to Bondy and Murty [5]. All graphs considered are simple (without loops or multiple edges). For a graph G, we use $V(G)$ for the vertex set of G, $E(G)$ for the edge set, and $\varepsilon(G) = |E(G)|$. For a set $X \subseteq V(G)$, $G[X]$ denotes the subgraph of G induced by X. For a path P, the length of P is $\varepsilon(P)$. If P has ends x and y, we call P an (x, y)-path. For $u, v \in V(P)$ with u preceding v on P, $P[u, v]$ denotes the subpath of P from u to v. For a positive integer k, an $(x, y : k)$-path is an (x, y)-path of length at least k. A simple connected graph is k-path-connected if between every pair of distinct vertices, there is an $(x, y : k)$-path. It is easy to see that every maximal 2-connected subgraph of a k-path-connected graph is itself k-path-connected, and we may therefore restrict our study to graphs which are 2-connected and k-path-connected. Given a subgraph H of G, $d_H(x, y)$ will denote the distance in H between x and y (i.e., the length of the shortest (x, y)-path in H). Recall that $\kappa(G)$ is the (vertex) connectivity of G and that $N(x)$ is the neighborhood of a vertex $x \in V(G)$.

A cycle is a connected, 2-regular graph. For a cycle C, the length of C is $\varepsilon(C)$. We use the term k^+-cycle to refer to a cycle of length at least k. Given $x, y \in V(C)$, if $d_C(x, y) = \max\{d_C(x', y') : x', y' \in V(C)\}$ then x and y are said to be *antipodal*

The authors would like to thank the 'Equipe Combinatoire' Paris 6 and CNRS for the generous support to attend GT04 conference.

vertices of C. If $\varepsilon(C)$ is even there exists a unique antipodal for each vertex of C and an edge joining two antipodal vertices is called a *diameter* of C. If $\varepsilon(C)$ is odd there exist exactly two antipodals for each vertex of C and an edge joining two antipodal vertices is called a *near-diameter* of C. We use the standard notation $(v_1, \ldots, v_t)$ for the cycle C with edges $v_i v_{i+1}$ for $i = 1, \ldots, t$ and the edge $v_t v_1$. This notation induces a natural orientation for the cycle from which we may denote by $C[v_i, v_j]$ the path on C with edges $v_s v_{s+1}$ for $s = i, i+1, \ldots, j-1 \mod \varepsilon(C)$; and by $C^{-}[v_i, v_j]$ the path on C with edges $v_s v_{s-1}$ for $s = i, i-1, \ldots, j+1 \mod \varepsilon(C)$. Let $\{x, y\}$ and $\{x', y'\}$ be two pairs of all distinct vertices of C. Then $\{x', y'\}$ is said to be a *separating* pair for the pair $\{x, y\}$ if the vertices appear on C in the order $xx'yy'$ or $xy'yx'$. A *chord* is an edge joining two non consecutive vertices of a cycle. Two chords are said to be crossing if the end vertices of one is a separating pair for the end vertices of the other. The *circumference* of a graph is the length of its longest cycle.

The *cycle space*, $Z(G)$, of a graph G is the vector space of edge sets of Eulerian subgraphs of G. A graph G is *k-generated* if the cycle space of G over $\mathbb{Z}_2$ is generated by the cycles of length at least k. A 2-connected graph G is a *k-generator* if it is both k-generated and $(k-1)$-path-connected. In [8] it was established that any 2-connected graph which contains a k-generator must itself be a k-generator.

The relation between long paths, cycle space, k-path-connectivity and k-generation of a graph has been studied by several authors. In particular, Bondy [4] conjectured that *if G is a 3-connected graph with minimum degree at least d and at least $2d$ vertices, then every cycle of G can be written as the symmetric difference of an odd number of cycles, each of whose lengths are at least $2d - 1$* and Hartman [6] proved that *if G is a 2-connected graph with minimum degree d, where G is not K_{d+1} if d is odd, then the cycles of length at least $d+1$ generate the cycle space of G*. Locke [8, 9] partially proved Bondy's conjecture and gave ideas to extend the results presented. Furthermore, Locke [7, 8] gave another proof of Hartman's theorem and together with Barovich [3] generalized that result by considering fields other than $\mathbb{Z}_2$. Locke and Teng in [10] give some results on odd sums of long cycles in 2-connected graphs.

The families of graphs studied by Locke in [7, 8] turned out to be k-path-connected and $(k+1)$-generated. So in [9] he conjectured:

Conjecture 1.1. *For some constant m, $0 < m \leq 1$, every k-path-connected graph is mk-generated.*

From [2] we recall that a k-path-connected graph G (other than K_1) must have a cycle of length at least $k+1$ in each block. Thus, G is t-generated for $t \leq \lfloor \frac{k+3}{2} \rfloor$. This immediately improves the lower bound, so every k-path-connected graph is $\lfloor \frac{k+3}{2} \rfloor$-generated, for $k \geq 1$. While noting that any $(2k-3)^+$-cycle is a k-generator, implies that we only need to study k-path-connected graphs which contain cycles of length less than or equal to $2k - 4$.

Locke in [9] proved that $m = 1$ for $3 \leq k \leq 5$ and Abreu, Labbate, Locke in [1] proved the following result:

Theorem 1.2. *Let G be a 2-connected, 6-path-connected graph with $|V(G)| \geq 9$ and minimum degree at least 3. Then G is 6-generated.*

In the next section we complete the proof of the following:

Theorem 1.3. *Let G be a 2-connected, 6-path-connected graph. Then G is 6-generated.*

The dodecahedron is an 18-path-connected graph but is only 17-generated [9], so it will not be possible to prove in general that a k-path-connected graph is k-generated.

However, in Section 3 we present a family of graphs that is $(4a + 3)$-path-connected and $(3a+3)$-generated but not $(3a+4)$-generated for $a \geq 1$. This family allows us to prove

Theorem 1.4. *Let G be a 2-connected, k-path-connected graph, and $k \geq 7$. Then G is mk-generated where*

(i) $m = 6/7$ for $k = 7$, and
(ii) $m \leq 3/4 + o(1)$ for large k.

2. 6-path-connected graphs

We first recall some results from [1].

Theorem 2.1. *Let G be a 2-connected, $(k-1)$-path-connected graph with a cycle C of length $2k - 4$, then G is k-generated if one of the following holds:*

(1) *G is at least 3-connected,*
(2) *There are no diameters of C in $E(G)$,*
(3) *There is exactly one diameter of C in $E(G)$,*
(4) *There are at least three diameters of C in $E(G)$.*

Theorem 2.2. *Let G be a 2-connected, $(k-1)$-path-connected graph with a cycle C of length $2k - 5$, then G is k-generated if one of the following holds:*

(1) *There are no near-diameters of C in $E(G)$,*
(2) *There is exactly one near-diameter of C in $E(G)$,*
(3) *There are at least three pairwise crossing near-diameters of C in $E(G)$.*

Now we present a couple of new results on 2-connected, $(k-1)$-path-connected graphs with a cycle of length at least $(2k - 4)$.

Lemma 2.3. *Let G be a 2-connected, $(k-1)$-path-connected graph with a cycle $C = (v_1, v_2, \ldots, v_{2k-4})$ of length $(2k-4)$. If there is a diameter $v_1 v_{k-1}$ and a (v_i, v_j)-path P, internally disjoint from C where $v_1 v_{k-1}$ separates $\{v_i, v_j\}$, then either $|i - j| = k - 2$ and P is a diameter or G is a k-generator.*

Proof. If $|i-j| = k-2$ and P is a diameter, there is nothing to prove. We may assume that $j > i$. Let $C_1 = C[v_1, v_i] \cup P \cup C[v_j, v_1]$ and $C_2 = C[v_i, v_j] \cup P$. $\varepsilon(C_1) + \varepsilon(C_2) = \varepsilon(C) + 2\varepsilon(P) = 2k - 4 + 2\varepsilon(P)$. If $\varepsilon(P) > 1$, then $\varepsilon(C_1) + \varepsilon(C_2) \geq 2k$, so $\max\{\varepsilon(C_1), \varepsilon(C_2)\} \geq k$. If $\varepsilon(P) = 1$, then $\varepsilon(C_1) + \varepsilon(C_2) = 2(k-1)$ and either $\varepsilon(C_1) = \varepsilon(C_2) = k-1$, in which case $|i-j| = k-2$ and P is a diameter, or $\max\{\varepsilon(C_1), \varepsilon(C_2)\} \geq k$. In both cases in which P is not a diameter, we can conclude that there is $C_3 \in \{C_1, C_2\}$ such that $\varepsilon(C_3) \geq k$. Therefore $C \cup P$ is k-generated.

Let $C_4 = C[v_1, v_i] \cup P \cup C^{-}[v_j, v_{k-1}] \cup v_{k-1}v_1$ and $C_5 = C^{-}[v_1, v_j] \cup P \cup C[v_i, v_{k-1}] \cup v_{k-1}v_1$. $\varepsilon(C_4) + \varepsilon(C_5) = \varepsilon(C) + 2\varepsilon(P) + 2 = 2k - 4 + 2 + 2\varepsilon(P)$. Since $\varepsilon(P) \geq 1$, $\max\{\varepsilon(C_4), \varepsilon(C_5)\} \geq k$ and therefore, there is $C_6 \in \{C_4, C_5\}$ such that $\varepsilon(C_6) \geq k$. Hence $C \cup P \cup \{v_1 v_{k-1}\}$ is k-generated.

Now we need to prove that if P is not a diameter, $H = C \cup P \cup \{v_1 v_{k-1}\}$ is $(k-1)$-path-connected. For $x, y \in V(H) - V(C)$, with $x \neq y$, there are $x', y' \in V(C)$ such that there is an (x, x')-path P_1 in H and an a (y, y')-path P_2 in H with $\varepsilon(C[x', y']) \geq k - 2$ or $\varepsilon(C^{-}[x', y']) \geq k - 2$, and with P_1 disjoint from P_2. Therefore, either $\varepsilon(P_1 \cup C[x', y'] \cup P_2) \geq k$ or $\varepsilon(P_1 \cup C^{-}[x', y'] \cup P_2) \geq k$. Thus, there is an $(x, y : k)$-path in H.

For $x \in V(H) - V(C)$ and $y \in V(C)$, there is an $x' \in V(C)$ and an (x, x')-path P_3 with either $\varepsilon(P_3 \cup C[x', y]) \geq k - 2$ or $\varepsilon(P_3 \cup C^{-}[x', y]) \geq k - 2$, which gives us an $(x, y : k-1)$-path in H.

For $x, y \in V(C)$ with x and y not antipodal on C, there already is an $(x, y : k-1)$-path in $C \subseteq H$. Hence, we need only consider the case in which x and y are antipodal. If $\{x, y\} \neq \{v_1, v_{k-1}\}$, we may assume, without loss of generality, that $v_1 x v_{k-1} y$ appear in this order on C and either $C^{-}[x, v_1] \cup v_1 v_{k-1} \cup C[v_{k-1}, y] \geq k-1$ or $C[x, v_{k-1}] \cup v_1 v_{k-1} \cup C^{-}[v_1, y] \geq k-1$. Hence there is an $(x, y : k-1)$-path in $C \cup P \cup \{v_1 v_{k-1}\}$.

If $\{x, y\} = \{v_1, v_{k-1}\}$ we may assume, without loss of generality, that $x = v_1$ and $y = v_{k-1}$ and then either $C[v_1, v_i] \cup P \cup C^{-}[v_j, v_{k-1}] \geq k-1$ or $C^{-}[v_1, v_j] \cup P \cup C[v_i, v_{k-1}] \geq k-1$. Giving an $(x, y : k-1)$-path in H.

Therefore when P is not a diameter, $C \cup P \cup \{v_1 v_{k-1}\}$ is $(k-1)$-path-connected and k-generated, hence a k-generator. □

Lemma 2.4. *Let G be a 2-connected, k-path-connected graph with a cycle $C = (v_1, v_2, \ldots, v_{2k-4})$ of length $2k-4$. If G contains two consecutive diameters of C, then G is a k-generator.*

Proof. Without loss of generality, suppose these two consecutive diameters are $v_1 v_{k-1}$ and $v_2 v_k$. Suppose there is a (v_i, v_j)-path P crossing $\{v_2, v_{k-1}\}$. Thus P crosses at least one of $v_1 v_{k-1}$ or $v_2 v_k$. By the previous lemma, $\varepsilon(P) = 1$, and the unique edge e of P is a diameter of C. But then, $e \notin \{v_1 v_{k-1}, v_2 v_k\}$, and $\{e, v_1 v_{k-1}, v_2 v_k\}$ is a set of three diameters of G, which by Theorem 2.1 implies that G is a k-generator.

Hence, we may assume that $G - \{v_2, v_{k-1}\}$ is disconnected. Since G is k-path-connected, there is a $(v_2, v_{k-1} : k)$-path Q in G. If $V(Q) \cap \{v_3, \ldots, v_{k-2}\} = \emptyset$, then $Q \cup C[v_2, v_{k-1}]$ is a $(2k-3)^+$-cycle, hence a k-generator. However if $V(Q) \cap \{v_3, \ldots, v_{k-2}\} \neq \emptyset$, then $V(Q) \cap \{v_k, \ldots, v_{2k-4}, v_1\} = \emptyset$, and $Q \cup C[v_{k-1}, v_2]$ is a $(2k-3)^+$-cycle, hence a k-generator. Thus, G is a k-generator. $\square$

We now return to the discussion of a 2-connected, 6-path-connected graph G which is not a 6-generator. If G contains a 9^+-cycle, then G is a 6-generator. We may therefore assume G has no 9^+-cycle.

Suppose G has an 8-cycle $C = (v_1, v_2, \ldots, v_8)$. By Theorem 2.1, we may assume that G has exactly two diameters, and by Lemma 2.4 we may assume, without loss of generality, that these diameters are v_1v_5 and v_3v_7. From Lemma 2.3, no chord, which is not itself a diameter, can cross a diameter. Also, by Lemma 2.3, there can be no (v_i, v_j)-path P internally-disjoint from C if $\{v_i, v_j\}$ separates $\{v_1, v_5\}$ or if $\{v_i, v_j\}$ separates $\{v_3, v_7\}$.

Suppose that there is an edge xy in $G - V(C)$. Then, we can find a (v_i, v_j)-path P containing xy, and internally disjoint from C, with $v_i \neq v_j$. Since P cannot cross either diameter, without loss of generality, $\{v_i, v_j\} \subseteq \{v_1, v_2, v_3\}$. But then, $P \cup C$ contains a 9^+-cycle, contradicting our hypotheses about G. Hence, $\varepsilon(G - V(C)) = 0$. Suppose there is a vertex $x \in V(G) - V(C)$. Then, we can find a (v_i, v_j)-path P containing x, and internally disjoint from C, with $v_i \neq v_j$. Since P cannot cross either diameter, $\{v_i, v_j\} \subseteq \{v_{2m+1}, v_{2m+2}, v_{2m+3}\}$, for some $m \in \{0, 1, 2, 3\}$, subscripts modulo 8. Since G can have no cycle of length exceeding 8, $\{v_i, v_j\} = \{v_{2m+1}, v_{2m+3}\}$, and $\varepsilon(P) = 2$. Note that $N(x) \subseteq V(C)$, and thus $N(x) \subseteq \{v_1, v_3, v_5, v_7\}$. If $N(x) \neq \{v_{2m+1}, v_{2m+3}\}$, then (reversing the direction along the cycle, if necessary) we may assume that $xv_{2m+5} \in E(G)$, and $v_{2m+1}xv_{2m+5}$ is a path of length exceeding one and crossing the diameter $v_{2m+3}v_{2m+7}$, in contradiction to our assumption. Thus, $|N(x)| = 2$.

We now have a characterization of G. The graph G is an eight cycle $C = (v_1, v_2, \ldots, v_8)$, together with two crossing diameters, v_1v_5 and v_3v_7 and possibly some vertices of degree two joined to $\{v_{2m+1}, v_{2m+3}\}$, for various choices of m, and also possibly some chords $\{v_{2m+1}v_{2m+3}\}$, again for various choices of m. But, this graph has no (v_1, v_5)-path of length at least 6. This violates our assumptions about G.

This concludes all of the cases in which the circumference of G is 8, leaving only the cases where the circumference of G is 7, since a $(k-1)$-path-connected graph must contain a k^+-cycle. Suppose that G has a vertex v of degree 2, with $N(v) = \{x, y\}$. There is an $(x, y : 6)$-path P in G, and $P \cup xvy$ is an 8^+-cycle. Therefore, we may assume that $\delta(G) \geq 3$.

Let $C = (v_1, v_2, \ldots, v_7)$ be a 7-cycle in G and let $R_m = \{v \in V(G) : d(v, V(C)) = m\}$ be the set of vertices of G at distance exactly m from the cycle C. In previous work [1], we showed that $|R_2| = 0$ (or G contains an 8^+-cycle) and that $\varepsilon(G[R_1]) = 0$, (or G contains an 8^+-cycle). But then, for any $v \in R_1$, $H = C \cup \{v\} \cup \{vw : w \in N(v)\}$ is a 6-generator. Hence, $|R_1| = 0$, and $V(G) = V(C)$.

What remains is to check the Hamiltonian graphs on 7 vertices to see that any of which are 6-path-connected are also 6-generated. In order to do this we first prove some general results.

Lemma 2.5. *Let G be a graph, C a cycle of length $2k-5$ in G and S a set of chords of C in G. Then $C+S$ is $(k-1)$-path-connected if and only if for every antipodal pair of vertices $\{v_i, v_{i+k-3}\}$ there is a chord $v_s v_t \in S$ separating the pair $\{v_i, v_{i+k-3}\}$.*

Proof. Given a pair $\{v_i, v_{i+k-3}\}$, if there is chord $v_s v_t$ that separates it, without loss of generality we may assume that the vertices appear in the order $v_i v_s v_{i+k-3} v_t$ on C. Then the paths $P_1 = C[v_i, v_s] \cup v_s v_t \cup C^-[v_t, v_{i+k-3}]$ and $P_2 = C^-[v_i, v_t] \cup v_t v_s \cup C[v_s, v_{i+k-3}]$ satisfy $\varepsilon(P_1) + \varepsilon(P_2) = \varepsilon(C) + 2 = 2k-3$, implying that $\max\{\varepsilon(P_1), \varepsilon(P_2)\} \geq k-1$. Therefore there is path $P_3 \in \{\varepsilon(P_1), \varepsilon(P_2)\}$ which is a $(v_i, v_{i+k-3} : k-1)$-path.

For the converse, let $\{v_i, v_{i+k-3}\}$ be a pair of vertices for which there is no chord in S separating it. This implies that $C+S-\{v_i, v_{i+k-3}\}$ is disconnected with exactly two components A_1 and A_2. For $i=1,2$, let $H_i = A_i \cap (C+S)$. Then the longest path in H_i for $i=1,2$ has length $k-2$. Therefore $C+S$ is not $(k-1)$-path-connected. $\square$

Let H be a 2-connected graph and $s,t \in V(H)$. The graph H is said to be $\{s,t\}$-*near-$(k-1)$-path-connected* if $k \geq 3$, there is an $(s,t: k-2)$-path in H, and for every pair of distinct vertices $x,y \in V(H)$ with $\{x,y\} \neq \{s,t\}$, there is an $(x,y: k-1)$-path in H. If H is also k-generated, then it is said to be an $\{s,t\}$-*near-k-generator*. In [1] the following lemma was proved.

Lemma 2.6. *Let G be a 2-connected, $(k-1)$-path-connected graph that contains an $\{s,t\}$-near-k-generator H. Then, G contains a k-generator (and then G is a k-generator).*

Remark 2.7. In a cycle $C = (v_1, \ldots, v_{2k-5})$ of length $2k-5$ there are $(x,y: k-1)$-paths among all pairs of non-antipodal vertices $x,y \in C$. A near-diameter $e = v_i v_{i+k-3}$ separates every pair of antipodal vertices $\{v_j, v_{j+k-3}\}$ except for the pairs $p_1 = \{v_i, v_{i+k-2}\}$ and $p_2 = \{v_{i-1}, v_{i+k-3}\}$. Then any chord e' which is different from the near-diameters $e_1 = v_i v_{i+k-2}$ and $e_2 = v_{i-1} v_{i+k-3}$ and which crosses e, also crosses at least one of e_1, e_2. Therefore $C + e + e'$ is either a k-generator or a p_i-near-k-generator for some $i=1,2$.

Remark 2.8. In particular if $C = (v_1, \ldots, v_7)$ is a 7-cycle and if a 2-chord crosses a 3-chord (near-diameter) in C, we have a near-6-generator. To prove this, suppose without loss of generality that the 2-chord is the edge $e = v_2 v_7$ and the 3-chord is the edge $e' = v_1 v_4$. The cycles $C_1 = (v_1, v_2, v_3, v_4, v_5, v_6, v_7)$, $C_2 = (v_2, v_3, v_4, v_5, v_6, v_7)$ and $C_3 = (v_1, v_2, v_7, v_6, v_5, v_4)$ generate the cycle space of $C + e + e'$, so it is 6-generated and by Remark 2.7 it is near-5-path-connected.

Let G be 6-path-connected Hamiltonian graph on 7 vertices and let $C = (v_1, \ldots, v_7)$ be a 7-cycle in G. The possible near-diameters of C are of the form

$e_j = v_j v_{j+3}$, subscripts modulo 7. We record the vector $F = (f_1, f_2, \ldots, f_7)$, where $f_j = 1$ if $e_j \in E(G)$ and $f_j = 0$ if $e_j \notin E(G)$. We list a representative of the equivalence class $[F]$ of F under the dihedral group D_7, acting on C. We write $C + F$ for the graph whose vertex set is $\{v_1, v_2, \ldots, v_7\}$ and whose edges are the edges of C, together with the edges $\{e_j : f_j = 1\}$. Those patterns which force three pairwise crossing near-diameters are marked with the symbol $\sqrt{}$, as are those with zero or one near-diameters (hence covered by Theorem 2.2). For each of the remaining patterns we study the addition of chords which are not near-diameters to $C + F$ and in each case we get either a 6-generator or a graph which is not 6-path-connected. Each pattern is marked with the corresponding observation below in which it is studied.

F	$\|[F]\|$		F	$\|[F]\|$	
$(0,0,0,0,0,0,0)$	1	$\sqrt{}$	$(1,1,1,1,0,0,0)$	7	$\sqrt{}$
$(1,0,0,0,0,0,0)$	7	$\sqrt{}$	$(1,1,1,0,1,0,0)$	14	$\sqrt{}$
$(1,1,0,0,0,0,0)$	7	Obs. 2.9	$(1,1,0,1,1,0,0)$	7	Obs. 2.12
$(1,0,1,0,0,0,0)$	7	Obs. 2.10	$(1,1,0,1,0,1,0)$	7	Obs. 2.12
$(1,0,0,1,0,0,0)$	7	Obs. 2.11	$(1,1,1,1,1,0,0)$	7	$\sqrt{}$
$(1,1,1,0,0,0,0)$	7	$\sqrt{}$	$(1,1,1,1,0,1,0)$	7	$\sqrt{}$
$(1,1,0,1,0,0,0)$	14	Obs. 2.12	$(1,1,1,0,1,1,0)$	7	$\sqrt{}$
$(1,1,0,0,1,0,0)$	7	Obs. 2.9	$(1,1,1,1,1,1,0)$	7	$\sqrt{}$
$(1,0,1,0,1,0,0)$	7	Obs. 2.10	$(1,1,1,1,1,1,1)$	1	$\sqrt{}$

Observation 2.9. Let $F_1 = (1,1,0,0,0,0,0)$. By Remark 2.8 and Lemma 2.5 the graphs $C + F_1 + v_2 v_7$, $C + F_1 + v_4 v_6$, $C + F_1 + v_1 v_3$ and $C + F_1 + v_3 v_5$ are 6-generators or near-6-generators. On the other hand for $F_2 = (1,1,0,0,1,0,0)$, by Lemma 2.5 $C + F_2 + \{v_2 v_4, v_5 v_7, v_1 v_6\}$ is not 5-path-connected, since there is no $(v_1, v_5 : 5)$-path, therefore neither is $C + F_1 + \{v_2 v_4, v_5 v_7, v_1 v_6\}$.

Observation 2.10. Let $F_3 = (1,0,1,0,0,0,0)$. By Remark 2.8 and Lemma 2.5 the graphs $C + F_3 + v_2 v_7$, $C + F_3 + v_5 v_7$, $C + F_3 + v_2 v_4$ and $C + F_3 + v_3 v_5$ are 6-generators. On the other hand $C + F_3 + \{v_1 v_3, v_4 v_6, v_1 v_6\}$ is not 6-path-connected. For $F_4 = (1,0,1,0,1,0,0)$ we have that $C + F_4 + \{v_1 v_3, v_1 v_6\}$ is not 6-path-connected, since there is no $(v_3, v_6 : 6)$-path, while by Remark 2.8 $C + F_4 + v_4 v_6$ is a near-6-generator.

Observation 2.11. Let $F_5 = (1,0,0,1,0,0,0)$. By Remark 2.8 and Lemma 2.5 the graphs $C + F_5 + v_1 v_6$, $C + F_5 + v_2 v_7$ and $C + F_5 + v_3 v_5$ are 6-generators or near-6-generators. On the other hand, by Lemma 2.5 $C + F_5 + \{v_1 v_3, v_2 v_4, v_4 v_6, v_5 v_7\}$ is not 5-path-connected, since there is no $(v_1, v_4 : 5)$-path.

Observation 2.12. Let $F_6 = (1,1,0,1,0,0,0)$ and note that $C + F_6$ is contained in $C + F_0 + v_4 v_7$ where $F_0 = (1,0,0,0,0,0,0)$. Therefore $C + F_6$ is a 6-generator, as well as $C + F_7$ and $C + F_8$ with $F_7 = (1,0,1,1,0,0)$ and $F_8 = (1,1,0,1,0,1,0)$, which contain $C + F_6$.

This completes the proof of Theorem 1.3 $\qquad\square$

3. The $M^+_{r,s,t}$ family

Let $X = \{u_1, u_2, u_3, u_4\}$ be the vertices of a copy of K_4. We replace each edge $u_i u_j$ by a path P_{ij}, where $\varepsilon(P_{13}) = \varepsilon(P_{24}) = r$, $\varepsilon(P_{12}) = \varepsilon(P_{34}) = s$, $\varepsilon(P_{14}) = \varepsilon(P_{23}) = t$, with $1 \leq r < s < t$. The resulting graph, $M_{r,s,t}$, has only seven cycles. Four of these cycles are of length $r+s+t$ and the other three have lengths: $2r+2s$, $2r+2t$ and $2s+2t$. Any basis for the cycle space uses at least one cycle of length $r+s+t$, and thus $M_{r,s,t}$ is $(r+s+t)$-generated but not $(r+s+t+1)$-generated. The cycle $P_{12} \cup P_{23} \cup P_{34} \cup P_{14}$ contains a (u_i, u_{i+1})-path of length at least $2s+t$, for $i = 1, 2, 3, 4$ (with $u_5 = u_1$). The cycle $P_{13} \cup P_{23} \cup P_{24} \cup P_{14}$ contains a (u_1, u_3)-path of length at least $2s+t$, and a (u_2, u_4)-path of length at least $2s+t$. The graph $M_{r,s,t}$ is at most $(\min\{2s+t, 2t+r\})$-path-connected. However, in general, the path-connectivity of $M_{r,s,t}$ is strictly less than this. For example, an $M_{1,a+1,2a+1}$ can't be more than $(4a+2)$-path-connected since for $x \in V(P_{23})$ at distance a from u_2 on P_{23}, the longest (x, u_4)-path in $M_{1,a+1,2a+1}$ has length $4a+2$.

We would like to modify $M_{r,s,t}$ to yield a graph with higher path-connectivity but which is not $(r+s+t+1)$-generated. Let $M^+_{r,s,t} = M_{r,s,t} \cup Q_{14} \cup Q_{23}$, where Q_{14} is a (u_1, u_4)-path and Q_{23} is a (u_2, u_3)-path, each of length t, internally disjoint from each other and from $M_{r,s,t}$. The graph $M^+_{r,s,t}$ has only 19 cycles. Eight of these cycles have length $r+s+t$, two have length $2t$, one has length $2r+2s$, four have length $2t+2r$ and four have length $2t+2s$. Any basis for the cycle space uses at least one cycle of length $r+s+t$, and thus $M^+_{r,s,t}$ is $(r+s+t)$-generated but not $(r+s+t+1)$-generated.

Lemma 3.1. *For $a \geq 1$, the graph $M^+_{1,a+1,2a+1}$ is $(4a+3)$-path-connected.*

Proof. For distinct vertices $x, y \in X$. As before we realize that we can always find an $(x, y : 4a+3)$-path in $M^+_{1,a+1,2a+1}$.

For $x \in V(P_{ij})$ we have $(x, u_i : 4a+3)$-paths and $(x, u_j : 4a+3)$-paths making use of the $(u_i, u_j : 4a+3)$-path found in the previous case, together with a segment of P_{ij}. Similarly for $x \in V(Q_{ij})$.

For $x \in V(M^+_{1,a+1,2a+1}) - X$ but $x \notin V(P_{ij}) \cup V(Q_{ij})$ there are $(x, u_i : 4a+3)$-paths and $(x, u_j : 4a+3)$-paths using two paths of length $2a+1$. For example, for $x \in V(P_{23})$, $P_{23}[x, u_3] \cup Q^-_{23} \cup u_2 u_4 \cup P^-_{14}$ is an $(x, u_1 : 4a+3)$-path.

For distinct vertices $x, y \in V(M^+_{1,a+1,2a+1}) - X$. If $x, y \in V(P_{ij})$ we may assume without loss of generality that the vertices appear in the order u_i, x, y, u_j in $M^+_{1,a+1,2a+1}$. Consider the path $P_1 = P_{ij}[u_i, x] \cup P_{ik} \cup P_{kl} \cup P_{lj} \cup P^-_{ij}[u_j, y]$, where P_{kl} is a path of length $a+1$, and P_{ik} and P_{lj} are paths of length $2a+1$. Then P_1 is an $(x, y : 4a+3)$-path. Similarly for $x, y \in V(Q_{ij})$.

If $x \in V(P_{ij})$, $y \notin V(P_{ij})$ and $\varepsilon(P_{ij}) = a+1$. Then if $y \in V(P_{kl})$ with $\varepsilon(P_{kl}) = a+1$, let $P_2 = P^-_{ij}[u_i, x] \cup P_{ik} \cup u_k u_l \cup P_{jl} \cup P^-_{kl}[y, u_l]$ where P_{ik} and P_{jl} are paths of length $2a+1$. Then P_2 is an $(x, y : 4a+3)$-path. If $y \in V(P_{ik})$ with $\varepsilon(P_{ik}) = 2a+1$, let $P_3 = P_{ij}[x, u_j] \cup P_{jl} \cup u_l u_i \cup Q_{ik} \cup P^-_{ik}[u_k, y]$ where P_{jl} is a path of length $2a+1$. Then P_3 is an $(x, y : 4a+3)$-path. Similarly if $y \in V(Q_{ik})$ or $y \in V(Q_{jl})$ or $y \in V(P_{jl})$ with $\varepsilon(P_{jl}) = 2a+1$.

If $x \in V(P_{ij})$, $y \notin V(P_{ij})$ and $\varepsilon(P_{ij}) = 2a+1$. Then if $y \in V(Q_{ij})$, let $P_4 = P_{ij}^-[u_i, x] \cup P_{ik} \cup P_{kl} \cup P_{jl}^- \cup Q_{ij}^-[u_j, y]$ where P_{ik} and P_{jl} are paths of length $a+1$ and P_{kl} is a path of length $2a+1$. Then P_4 is an $(x, y : 4a+3)$-path. Similarly if $y \in V(P_{kl})$ or $y \in V(Q_{kl})$. Analogously if $x \in V(Q_{ij})$ and $y \notin V(Q_{ij})$. □

Proof of Theorem 1.4. (i) For $a = 1$, $M_{1,a+1,2a+1}^+$ is a 7-path-connected graph which is only 6-generated, hence $m = 6/7$ for $k = 7$.

(ii) We have found a family of graphs which are $(4a+3)$-path-connected and $(3a+3)$-generated but not $(3a+4)$-generated. Since these graphs are also $(4a+2)$, $(4a+1)$ and $(4a)$-path-connected, we have, for $0 \leq b \leq 3$, graphs which are $(4a+b)$-path-connected and $(3a+3)$-generated but not $(3a+4)$-generated. Now, $\frac{3a+3}{4a+b}$ has a limit of approximately $\frac{3}{4}$, and approaching $\frac{3}{4}$ as a increases, hence $m \leq 3/4 + o(1)$ for large k. □

We may now conclude that for $k \geq 8$ every k-path-connected graph is mk-generated for some constant m, with $\frac{1}{2} \leq m \leq \frac{3}{4}$.

References

[1] M. Abreu, D. Labbate and S.C. Locke, 6-path-connectivity and 6-generation, Discrete Math. **301** (2005), no. 1, 20–27.

[2] S.C. Locke, Personal communication with M. Abreu (2000).

[3] M. Barovich and S.C. Locke, The cycle space of a 3-connected Hamiltonian graph. Discrete Math. **220**, (2000), 13–33.

[4] J.A. Bondy, Personal communication with S.C. Locke (1979).

[5] J.A. Bondy and U.S.R. Murty, *Graph Theory with Applications*, Elsevier, North-Holland, 1976.

[6] I.B.-A. Hartman, Long cycles generate the cycle space of a graph, Europ. J. Combin. **4** (1983), 237–246.

[7] S.C. Locke, A basis for the cycle space of a 2-connected graph, Europ. J. Combin. **6** (1985), 253–256.

[8] S.C. Locke, A basis for the cycle space of a 3-connected graph, Cycles in Graphs, Annals of Discrete Math. **27** (1985), 381–397.

[9] S.C. Locke, Long Paths and the Cycle Space of a Graph, Ars Combin. **33**, (1992), 77–85.

[10] S.C. Locke and C. Teng, Odd sums of long cycles in 2-connected graphs, Congressus Numerantium **159**, (2002), 19–30.

M. Abreu
Dipartimento di Matematica, Università degli Studi della Basilicata, Potenza, Italy
e-mail: abreu@unibas.it

S.C. Locke
Department of Mathematical Sciences, Florida Atlantic University, Florida, USA
e-mail: LockeS@fau.edu

Automated Results and Conjectures on Average Distance in Graphs

Mustapha Aouchiche and Pierre Hansen

Abstract. Using the *AutoGraphiX* 2 system, a systematic study is made on generation and proof of relations of the form
$$\underline{b}_n \leq \bar{l} \oplus i \leq \bar{b}_n$$
where $\bar{l}$ denotes the average distance between distinct vertices of a connected graph G, i one of the invariants: diameter, radius, girth, maximum, average and minimum degree, $\underline{b}_n$ and $\bar{b}_n$ are best possible lower and upper bounds, functions of the order n of G and $\oplus \in \{-, +, \times, /\}$. In 24 out of 48 cases simple bounds are obtained and proved by the system. In 21 more cases, the system provides bounds, 16 of which are proved by hand.

Mathematics Subject Classification (2000). Primary 05C35; Secondary 05C12.

Keywords. Graph, Invariant, AGX, Conjecture, Average Distance.

1. Introduction

Classical books on graph theory, such as Berge's *Graphs and Hypergraphs* [4], present many lower and upper bounds on graph invariants (*i.e.*, numerical functions of graphs which do not depend on the numbering of vertices or edges) in terms of the graph's number n of vertices and/or m of edges. So it appears to be a naturel challenge to see if such bounds can be discovered automatically [8] by some computer system, (and if not, if such a system can provide substantial help, *e.g.*, by discovering extremal graphs for a given expression).

Recently, using the AutoGraphiX 2 (AGX 2) software [1, 5, 6], a systematic study has been performed [2] on automated generation of bounds of the following form:
$$\underline{b}_n \leq i_1 \oplus i_2 \leq \bar{b}_n$$
where $\underline{b}_n$ and $\bar{b}_n$ are expressions depending only on the order n of the graphs under study, i_1 and i_2 are graph invariants and $\oplus$ belongs to $\{+, -, \times, /\}$. Moreover, it is requested that the bound $\underline{b}_n$ and $\bar{b}_n$ be best possible in the strong sense that

for all n (except very small values, due to border effects) there exists a graph for which the bound is tight. The proposed form generalizes formulae of the well-known Nordhaus-Gaddum [12] form, in that i_1 and i_2 are independent invariants instead of the same one in G and its complement $\overline{G}$ and that the operations $-$ and $/$ are considered in addition to $+$ and $\times$.

In the present paper we report in detail on results of the comparison of average distance in graphs with six other invariants: diameter, radius, girth, maximum, average and minimum degree.

These results fall into the following categories:

(a) *Automated complete results*: structural conjectures on the family of extremal graphs, algebraic expression of the bound, automated proof of this bound's validity and tightness (it turns out that such results are frequently obtained in a simple way; they are therefore referred to as *observations*);

(b) *Automated complete conjectures*: structural conjectures and algebraic relations obtained as above, but without automated proof. Some conjectures are proved by hand (and referred to as *propositions*), others remain open;

(c) *Semi-automated conjectures*: structural conjectures obtained automatically, but algebraic relations derived from them by hand; of those some are proved and some remain open;

(d) *Automated structural conjectures*, for which algebraic expressions have not been found (or do not exist);

(e) *No results*, as the (presumably) extremal graphs do not present any regularity.

In order to enable an informed evaluation of the results obtained, they are all presented. Simple ones are briefly listed. Their main interest is that they can enrich the database of relations used in the automated proofs. Other results are given with full proofs or with indications about how to prove them if a previous proof technique carries over.

The paper is organized as follow: each of the next six sections presents a comparison of average distance with diameter, radius, girth, maximum, average and minimum degree respectively. Brief conclusions are given in the last section. Observations made by AGX 2 are collected in the Appendix.

2. The diameter

The diameter D of a graph $G = (V, E)$ is defined by $D = \max\{d(u,v), u, v \in V\}$, where $d(u,v)$ is the distance between u and v in G. A *diametric* path in G is a path between two vertices u and v such that $d(u,v) = D$.

Automated results obtained by AGX 2, in 6 cases out of 8, when comparing the average distance $\bar{l}$ and the diameter D are given in Table 2 of the Appendix.

The following proposition was obtained automatically by AGX 2 and then proved by hand.

Proposition 2.1. *For any connected graph on at least 3 vertices,*

$$D - \bar{l} \leq \frac{2n}{3} - \frac{4}{3}.$$

The bound is attained if and only if the graph is a path.

Proof. Let G be a connected graph of diameter D and average distance $\bar{l}$, and H a subgraph of G induced by a diametric path. Let

$$\sigma = \sum_{u,v \in V} d(u,v) \quad \text{and} \quad \sigma_H = \sum_{u,v \in V(H)} d(u,v),$$

where $V(H)$ is the set of vertices of H. It is easy to see that

$$\sigma \geq \sigma(H) = D \cdot (D+1) \cdot (D+2)/6$$

and

$$\bar{l} \geq D \cdot (D+1) \cdot (D+2)/(3n(n-1)).$$

Thus

$$\begin{aligned}
D - \bar{l} &\leq D - \frac{D \cdot (D+1) \cdot (D+2)}{3n(n-1)} \\
&\leq \frac{3n(n-1) \cdot D - D \cdot (D+1) \cdot (D+2)}{3n(n-1)} \\
&\leq \frac{-D^3 - 3D^2 + (3n(n-1) - 2) \cdot D}{3n(n-1)}.
\end{aligned}$$

Easy algebraic manipulations show that this last expression is an increasing function of D. It thus reaches its maximum if and only if $D = n - 1$, i.e., if G is a path. □

Before stating the next conjecture, let us define the family of graphs called bugs [10]. A *bug* Bug_{p,k_1,k_2} is a graph obtained from a complete graph K_p by deleting an edge uv and attaching paths P_{k_1} and P_{k_2} at u and v, respectively. A bug is *balanced* if $|k_1 - k_2| \leq 1$. (In a bug, $n = p + k_1 + k_2$ and $m = \frac{p(p-1)}{2} + k_1 + k_2 - 1$).

Conjecture 2.2. *Among all connected graphs on at least 3 vertices, $D/\bar{l}$ is maximum for a balanced bug.*

3. The radius

The eccentricity of a vertex v in a graph $G = (V, E)$ is defined by $\text{ecc}(v) = \max\{d(u,v), u \in V\}$, where $d(u,v)$ is the distance between u and v in G. The radius of G is the minimum of its eccentricities, i.e., $r = \min\{\text{ecc}(v), v \in V\}$.

Automated results obtained by AGX 2, in 4 cases out of 8, when comparing the average distance $\bar{l}$ and the radius r are given in Table 2 of the Appendix.

The following proposition was obtained as a conjecture using $AGX\ 2$ in automated mode, and then proved by hand.

Proposition 3.1. *For any connected graph on at least 3 vertices,*
$$\bar{l}/r \leq 2 - \frac{2}{n}.$$
The bound is attained if and only if the graph is a star.

Proof. If G is a connected graph of radius r, it contains a spanning tree T of the same radius $r(T) = r$. It is obvious that $\bar{l}(T) \geq \bar{l}$, where $\bar{l}(T)$ and $\bar{l}$ denote the average distance in T and G respectively, with equality if and only if $G \equiv T$. So $\bar{l}/r$ is maximum for a tree and we can assume that G is a tree.

Let m_i denote the number of vertex pairs in G at distance i, for $i = 1, \ldots, D$ where D is the diameter of G. We have:
$$\begin{aligned}
\bar{l} &= 2 \cdot (m_1 + 2m_2 + 3m_3 + \cdots + Dm_D)/(n(n-1)) \\
\bar{l} &\leq (2n - 2 + D((n(n-1)) - 2n + 2))/(n(n-1)) \\
\bar{l} &\leq D - 2(D-1)/n.
\end{aligned}$$
Then we obtain:
$$\bar{l}/r \leq \frac{D}{r} - \frac{(D-1)}{r} \cdot \frac{2}{n}.$$
Since G is assumed to be a tree, we have [4] $D = 2r$ or $D = 2r - 1$, thus
$$\bar{l}/r \leq 2 - \frac{4}{n} + \frac{2}{rn}$$
which is largest for $r = 1$ and the bound follows.

Now, let G be a tree such that: $\bar{l}/r = 2 - \frac{2}{n}$. Because of
$$\bar{l}/r = 2 - 2/n \leq D/r - (D-1)/r \cdot 2/n \leq 2 - 2/n$$
necessarily
$$D/r - (D-1)/r \cdot 2/n = 2 - 2/n$$
which implies $D/r = 2$ and $D - 1 = r$, i.e., $D = 2$ and $r = 1$. The star is the unique tree satisfying these conditions. □

Before stating the conjectures about $\bar{l} - r$ and $\bar{l}/r$, let us define the family of graphs called bags [10]. A *bag* $\text{Bag}_{p,k}$ is a graph obtained from a complete graph K_p by replacing an edge uv with a path P_k (as P_k has $k - 2$ internal vertices, for bags $n = p + k - 2$ and $m = \frac{p(p-1)}{2} + k - 2$). A bag is *odd* if k is odd, otherwise it is *even*.

Conjecture 3.2. *For given $n \geq 3$, among all connected graphs on n vertices,*
$$\bar{l} - r \geq \begin{cases} \frac{-n(n-2)}{4(n-1)} & \text{if } n \text{ is even,} \\ \frac{8-(n-1)^3}{4n(n-1)} & \text{if } n \text{ is odd.} \end{cases}$$
The bound is attained for a cycle if n is even and for a bag $\text{Bag}_{4,n-2}$ if n is odd.

Conjecture 3.3. *For given $n \geq 3$, among all connected graphs on n vertices, $\bar{l}/r$ is minimum for bags.*

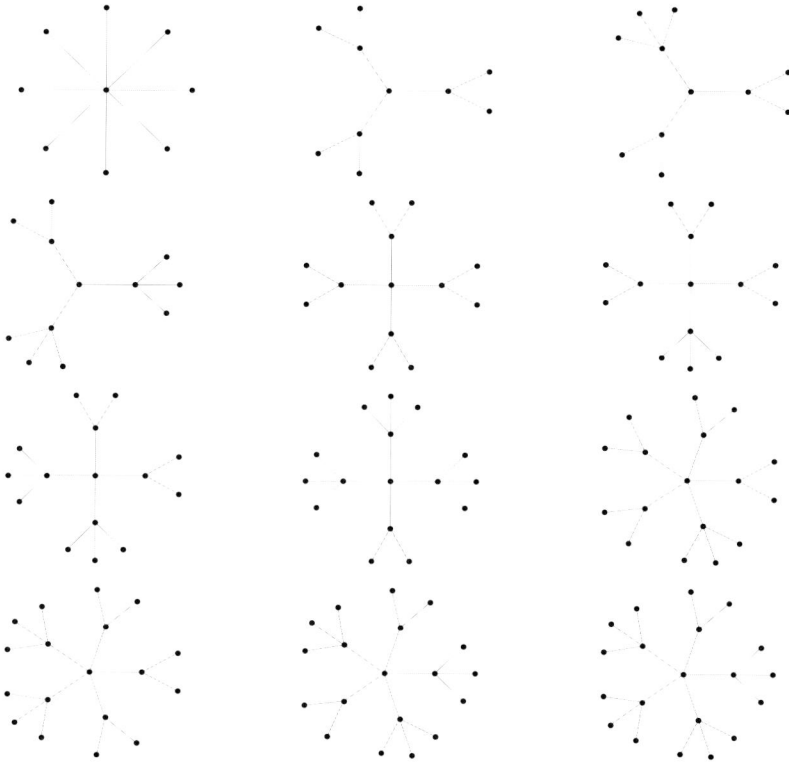

FIGURE 1. Graphs obtained by $AGX\,2$ when maximizing $\bar{l} - r$.

When studying the maximum of $\bar{l} - r$, $AGX\,2$ obtained the trees represented in Figure 1. They are not sufficiently regular to derive a conjecture.

4. The girth

The girth of a connected graph G is the length of its smallest cycle. If G contains at least a cycle, $3 \leq g \leq n$, and if not $g = \infty$.

Automated results obtained by AGX 2, in 2 cases out of 8, when comparing the average distance $\bar{l}$ and the girth g are given in Table 2 of the Appendix.

Before stating the next result, we need to define the Soltés graph [13]. Let u be an isolated vertex or one end vertex of a path. Let us join u with at least one vertex of a complete graph. The graph so obtained is the Soltés graph $PK_{n,m}$, also called the *path-complete* graph, where n is its order and m its size. There is exactly one $PK_{n,m}$ for given n and m such that $1 \leq n - 1 \leq m \leq \frac{n(n-1)}{2}$. For given n

and m, $PK_{n,m}$ maximizes (non uniquely) the diameter D [11] and (uniquely) the average distance $\bar{l}$ [13].

Proposition 4.1. *For any connected graph with a finite girth g,*

$$\left.\begin{array}{ll} \text{if } n \text{ is even} & \frac{n^2}{4(n-1)} - n \\ \text{if } n \text{ is odd} & \frac{n+1}{4} - n \end{array}\right\} \leq \bar{l} - g \leq \frac{(n+1)(n^2 - 10n + 12)}{3n(n-1)}.$$

The lower bound is attained if and only if the graph is a cycle, and the upper bound if and only if the graph is a triangle with an appended path.

Proof. The lower bound: Let G be a graph of girth g and H a subgraph of G induced by a smallest cycle in G. Let σ and $\sigma(H)$ denote the sum of all distances in G and H respectively. We have $\sigma \geq \sigma(H)$, so

$$\bar{l} - g \geq \frac{2\sigma(H)}{n(n-1)} - g.$$

If we denote the right-hand side of this last expression by $f(g)$, we have

$$f(g) = \begin{cases} \frac{g^3 - 4(n-1/2)^2 g}{4n(n-1)} & \text{if } g \text{ is odd,} \\ \frac{g^3 - 4n(n-1)g}{4n(n-1)} & \text{if } g \text{ is even.} \end{cases}$$

For $3 \leq g \leq n$, f is maximum if and only if $g = n$, and we have:

$$f(n) = \begin{cases} n^2/(4(n-1)) - n & \text{if } n \text{ is odd,} \\ (n+1)/4 - n & \text{if } n \text{ is even.} \end{cases}$$

The cycle is the only graph for which $g = n$.

The upper bound: If G is a connected graph on n vertices and m edges with a finite girth g, we have $m \geq n$. Moreover, the deletion of an edge increases the average distance in a graph, so

$$\max_{m \geq n}(\bar{l} - g) = \max_{m=n}(\bar{l} - g) \leq \max_{m=n} \bar{l} - 3.$$

According to [13] and subject to $m = n$, $\bar{l}$ is maximum for a $PK_{n,n}$ graph, which is a triangle with an appended path. Then easy computations lead to the formula. $\square$

Remark 4.2. As the average distance of a graph is at most $(n+1)/3$, which is reached for a path on n vertices, and the girth is at least 3, a trivial upper bound on $\bar{l} - g$ is $a_n = (n-8)/3$. The bound is not sharp. However the difference between a_n and the upper bound, say b_n, given in Proposition 4.1 is very small and asymptotically null. Indeed

$$a_n - b_n = \frac{n-8}{3} - \frac{(n+1)(n^2 - 10n + 12)}{3n(n-1)} = \frac{4}{n} - \frac{2}{n-1}.$$

Proposition 4.3. *For any connected graph with a finite girth g,*

$$\bar{l} + g \leq \begin{cases} \frac{n^2}{4(n-1)} + n & \text{if } n \text{ is even,} \\ \frac{5n+1}{4} & \text{if } n \text{ is odd.} \end{cases}$$

The bound is attained if and only if the graph is a cycle.

Proof. As the deletion of an edge increases the average distance and G must contain a cycle, the maximum of $\bar{l}+g$ is attained for a unicyclic graph. Let G be such a graph. If $g = n$, G is a cycle and we have the bound. Let us assume that $g < n$ and consider a vertex u of degree at least 3 on the unique cycle of G (such a vertex exists because G is unicyclic and not a cycle). Let v and w be neighbors of u such that v is on the cycle and w is not. Consider the graph H constructed from G by deleting the edge uv and adding the edge vw. It is clear that $g(H) = g + 1$, where $g(H)$ denotes the girth of H. It is easy to see that for any pair (x, y) of vertices

$$d(x, y) - 1 \leq d_H(x, y) \leq d(x, y) + 1.$$

where d and d_H are the distance functions in G and H respectively, with the first inequality strict for at least the $n - 1$ common pairs of adjacent vertices between G and H. This implies

$$\bar{l}(G) < \bar{l}(H) + 1.$$

Therefore

$$\bar{l}(G) + g(G) < \bar{l}(H) + g(G) + 1 = \bar{l}(H) + g(H).$$

Iterating this operation leads to a cycle C_n. So, for any unicyclic graph G

$$\bar{l}(G) + g(G) \leq \bar{l}(C_n) + g(C_n),$$

with equality if and only if $G \equiv C_n$. The bounds are then easily checked. □

Proposition 4.4. *For any connected graph with a finite girth g,*

$$\frac{\bar{l}}{g} \leq \frac{n^3 - 7n + 12}{9n(n-1)}.$$

The bound is attained if and only if the graph is a triangle with an appended path.

Proof. This proposition can be proved exactly as the upper bound of $\bar{l} - g$. □

Conjecture 4.5. *For any connected graph with a finite girth g,*

$$\frac{\bar{l}}{g} \geq \begin{cases} \frac{n}{4(n-1)} & \text{if } n \text{ is even,} \\ \frac{n+1}{4n} & \text{if } n \text{ is odd.} \end{cases}$$

The bound is attained if and only if the graph is a cycle.

Conjecture 4.6. *For any connected graph with a finite girth g,*

$$\bar{l} \cdot g \leq \begin{cases} \frac{n^3}{4(n-1)} & \text{if } n \text{ is even,} \\ \frac{n^2+n}{4} & \text{if } n \text{ is odd.} \end{cases}$$

The bound is attained if and only if the graph is a cycle.

5. The maximum degree

Automated results obtained by AGX 2, in 4 cases out of 8, when comparing the average distance $\bar{l}$ and the maximum degree Δ are given in Table 2 of the Appendix. The following propositions were first obtained automatically as conjectures and then proved by hand.

Proposition 5.1. *For any connected graph on at least 3 vertices,*
$$\Delta + \bar{l} \leq n + 1 - \frac{2}{n}.$$
The bound is attained if and only if the graph is a star.

Proof. Recall that a *comet* is a star with a path appended at one of its pending vertices.

(i) *Any graph G maximizing $\Delta + \bar{l}$ must be a tree, with a vertex u such that $d(u) = \Delta$.* Indeed assume not; then G must contain a cycle, at least one edge of which is not incident to u. Removing this edge keeps Δ unchanged and strictly increases $\bar{l}$, a contradiction.

(ii) *The average distance in a tree with maximum degree Δ is maximized by a comet.* Let T be a tree with a vertex u such that $d(u) = \Delta$. Let v be a vertex of degree $d(v) \geq 3$ (if any) farthest from u and not equal to u. Let r and s be pending vertices on two paths from u to r through v and from u to s through v in T. Denote by P_1 (resp. P_2) the path joining v to r (resp. v to s) in T. Let $l(P_1)$ (resp. $l(P_2)$) denote the length of P_1 (resp. P_2). Let P denote the path from r to s in T, with a length $l(P_1) + l(P_2)$.

(ii-a) Consider then the following transformation on T: disconnect path P_1 at v and reconnect it at s, thus obtaining a path P' from v to r with a length $l(P') = l(P)$. Let us show this increases the total distance and hence the average distance, between pairs of distinct vertices of T. Let $d(V_1)$ denote the sum of all distances between vertices of $V_1 \subset V$. Let $V_1 = V(P) = V(P')$. Then $d(V_1)$ is unchanged as well as $d(V \backslash V_1)$. Let $d(V_1, V \backslash V_1)$ denote the sum of all distances between a vertex of V_1 and one of $V \backslash V_1$ in T. Then the transformation increases $d(V_1, V \backslash V_1)$ by $|V \backslash V_1| \cdot l(P_1) > 0$ and $d(V) = d(V_1) + d(V \backslash V_1) + d(V_1, V \backslash V_1)$ increases.

Iterating the transformation yields a tree T' homeomorphic to a star (a *daddy-long-legs*).

(ii-b) Consider the following transformation on T'. Let P_3 denote a longest path from u to a pending vertex r in T', and P_4 the shortest path from a pending vertex $s \neq r$ in T' to its closest neighbor w of u, with $l(P_4) \geq 1$ (if any). Disconnect the path P_4 at w and connect it at r, thus obtaining a tree T'' with the same maximum degree as T'. Let us show this increases the total distance between pairs of distinct vertices of T'. Let P'' denote the path from r to s in T' and P''' the

TABLE 1. Optimal values for Δ.

n	12	13	14	15	16	17	18	19	20	21	22	23	24
Δ	7	8	8	9	9	9	10	10	10	11	11	12	12
n	25	26	27	28	29	30	31	32	33	34	35	36	37
Δ	13	13	13	14	14	15	15	15	16	16	17	17	17
n	38	39	40	41	42	43	44	45	46	47	48	49	50
Δ	18	18	19	19	20	20	20	21	21	22	22	23	23
n	51	52	53	54	55	56	57	58	59	60	61	62	63
Δ	23	24	24	25	25	25	26	26	27	27	28	28	28

path from w to s in T''. Clearly $l(P'') = l(P''')$ and then noting by V_2 the set of vertices of these paths, $d(V_2)$ and $d(V \setminus V_2)$ are unchanged. Moreover $d(V_2, V \setminus V_2)$ is increased by $|V \setminus V_1| \cdot l(P_3) \cdot l(P_4) > 0$. Iterating the transformation yields a comet T'''.

(iii) To prove the result, we consider a third transformation: let r denote the vertex farthest from u in the comet T''. Denote by P_5 the path from u to r in T'' and assume $l(P_5) \geq 2$. Let w be the neighbor of r in T''. Disconnect the edge (w, r) at w and reconnect it at u thus obtaining a comet T''' with a maximum degree $\Delta + 1$. Let P_6 denote the path from r to w in this comet. Clearly, $l(P_5) = l(P_6)$ and thus denote by V_3 the set of vertices of these paths in T'' and T''', $d(V_3)$ and $d(V \setminus V_3)$ are unchanged. In $d(V_3, V \setminus V_3)$, only the distances between r and the $\Delta - 1$ pending neighbors of u, change. For every u' pending neighbor of u, the distance $d(r, u')$ decreases by exactly $n - \Delta - 1$. Then the average distance (from T'' to T''') decreases by exactly $\frac{2(\Delta-1)(n-\Delta-1)}{n(n-1)} < 1$. Thus $\bar{l} + \Delta$ increases strictly.

Iterating the transformation yields a star T''', for which $\Delta = n - 1$ and $\bar{l} = 2 - \frac{2}{n}$, which gives the desired bound. $\square$

Proposition 5.2. *Among all graphs on at least 3 vertices $\Delta \cdot \bar{l}$ is maximum for a star if $n \leq 11$, and for a comet if $n \geq 12$ for which the maximum degree (for $12 \leq n \leq 63$) is given in Table 1.*

Proof. Correctness of the structural result follows from the fact that the extremal graph for a given Δ is a comet as shown in the proof of Proposition 5.1. Due to the apparent irregularity of optimal values of Δ, no algebraic conjecture could be found.

To prove that each value of Δ given in Table 1 is optimal for the corresponding order n, it is sufficient to show that it maximizes the function

$$f_n(\Delta) = \Delta \cdot \bar{l}(\Delta, n)$$
$$= \Delta \cdot \frac{(n - \Delta + 1)(n - \Delta + 2)(n - 4\Delta - 3) + 12(\Delta - 1)(\Delta - 2)}{3n(n-1)},$$

where $\bar{l}(\Delta, n)$ denotes the average distance in a comet of order n and maximum degree Δ.

For example if $n = 12$, we have
$$f_{12}(\Delta) = \frac{-4\Delta^4 + 123\Delta^3 - 989\Delta^2 + 1650\Delta}{396}.$$
which is maximum for $\Delta = 7$. □

When studying the minimum of $\Delta + \bar{l}$ and of $\Delta \cdot \bar{l}$, AGX 2 obtained graphs with no apparent regularity, thus no conjecture was derived.

6. The average degree

Automated results obtained by AGX 2, in 4 cases out of 8, when comparing the average distance $\bar{l}$ and the average degree $\bar{d}$ are given in Table 2 of the Appendix. The following propositions were first obtained automatically as conjectures and then proved by hand.

Proposition 6.1. *For any connected graph on at least 3 vertices,*
$$4 - \frac{4}{n} \leq \bar{l} + \bar{d} \leq n.$$
The lower bound is attained if and only if the graph is a star, and the upper bound if and only if the graph is complete.

Proof. **The lower bound:**
Let $\sigma(r, m)$ be the sum of all distances in a graph G of (fixed) order n, size m and radius r. If $r = 1$, we have
$$\sigma(1, m) = m + 2(n(n-1)/2 - m) = n(n-1) - m.$$
Therefore,
$$\bar{l} + \bar{d} = 2(n(n-1) - m)/(n(n-1)) + 2m/n = 2 + m(2n-4)/(n(n-1)).$$
This last expression is minimum if and only if $m = n - 1$, i.e., when the graph is a star.

Consider now, the case where $r > 1$, and recall that m_i is the number of vertex-pairs at distance i in G,
$$\sigma(r, m) = m + 2m_2 + 3m_3 + \cdots \geq m + 2(m_2 + m_3 + \cdots)$$
$$\sigma(r, m) > m + 2(n(n-1)/2 - m) = \sigma(1, m).$$
Therefore, we have:
$$\bar{l} + \bar{d} = \frac{2\sigma(r, m)}{n(n-1)} + \frac{2m}{n} \geq \frac{2\sigma(1, m)}{n(n-1)} + \frac{2m}{n} \geq 4 - \frac{4}{n}.$$
This proves the lower bound and characterizes the associated graph.

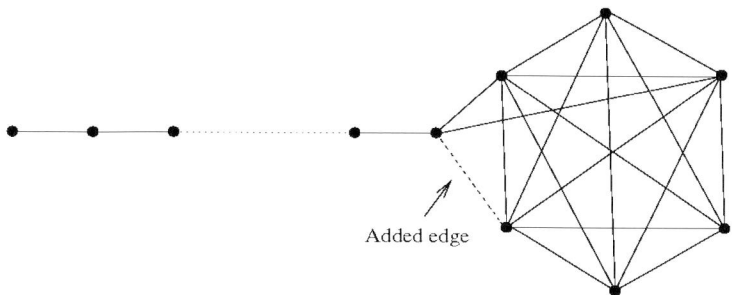

FIGURE 2. Obtaining $PK_{n,m+1}$ from $PK_{n,m}$.

The upper bound:
According to [13], for any connected graph G on n vertices and m edges, we have,

$$\bar{l}(G) \leq \bar{l}(PK_{n,m})$$

where $PK_{n,m}$ is the path-complete graph on n vertices and m edges. So

$$\bar{l}(G) + \bar{d}(G) \leq \bar{l}(PK_{n,m}) + \frac{2m}{n}.$$

So to prove the upper bound, we have to show that for any size m,

$$\bar{l}(PK_{n,m}) + \frac{2m}{n} \leq \bar{l}(PK_{n,m+1}) + \frac{2(m+1)}{n}$$

To obtain $PK_{n,m+1}$ from $PK_{n,m}$, we add an edge between any vertex from the clique of $PK_{n,m}$ and the nearest end vertex of its path (see Figure 2).

It is easy to see that by adding an edge to $PK_{n,m}$ the average distance decreases by at most $2(n-2)/(n(n-1))$, and the average degree increases by exactly $2/n$, then

$$\bar{l}(PK_{n,m+1}) - \bar{l}(PK_{n,m}) \geq -\frac{2(n-2)}{n(n-1)} + \frac{2}{n} = \frac{2}{n(n-1)} > 0.$$

Iterating the operation leads to the complete graph K_n, thus

$$\bar{l}(G) + \bar{d}(G) \leq \bar{l}(K_n) + \bar{d}(K_n) = n.$$

with equality if and only if $G \equiv K_n$. $\square$

Proposition 6.2. *For any connected graph on at least 3 vertices,*

$$\bar{l} \cdot \bar{d} \geq 4\left(\frac{n-1}{n}\right)^2.$$

The bound is attained if and only if the graph is a star.

Proof. As shown in the proof of Proposition 6.1, for given order n and size m the sum of all distances in a graph is minimum for radius $r = 1$, so

$$\bar{l} \cdot \bar{d} \geq \frac{2n(n-1) - 2m}{n(n-1)} \cdot \frac{2m}{n} = \frac{4n(n-1)m - 4m^2}{n^2(n-1)}.$$

This last expression is minimum only for $m = n - 1$, i.e., for a star. □

For the upper bound, $AGX\ 2$ leads to a structural conjecture, proved by hand in the following proposition.

Proposition 6.3. *Among all graphs on at least 3 vertices, $\bar{l} \cdot \bar{d}$ is maximum for a complete graph if $n \leq 9$ and for a Soltés graph if $n \geq 10$.*

Proof. This follows immediately from the result of *Soltés* [13]. □

Some extremal graphs are presented in Figure 3; no algebraic relation has (as yet) been obtained.

7. The minimum degree

Automated results obtained by AGX 2, in 4 cases out of 8, when comparing the average distance $\bar{l}$ and the minimum degree δ are given in Table 2 of the Appendix.

The following two propositions were obtained by $AGX\ 2$ as conjectures and then proved by hand.

Proposition 7.1. *For any connected graph on at least 3 vertices,*

$$\frac{2n^2 - 4}{n(n-1)} \leq \bar{l} + \delta \leq n.$$

The lower bound is obtained if and only if the graph is a clique on $n-1$ vertices with a pending edge. The upper bound is obtained if and only if the graph is complete.

Proof. The lower bound: let

$$F(m, \delta, D) = \bar{l} + \delta$$

for a graph of (fixed) order n, size m, minimum degree δ and diameter D. If $D = 1$, necessarily, $m = n(n-1)/2$, $\delta = n - 1$ and $F(m, \delta, 1) = n$.

If $D \geq 2$:

$$F(m, \delta, D) = 2(m + 2m_2 + 3m_3 + \cdots + Dm_D)/(n(n-1)) + \delta$$

$$F(m, \delta, D) \geq 2(m + 2(m_2 + m_3 + \cdots + m_D))/(n(n-1)) + \delta$$

$$F(m, \delta, D) \geq 2(m + 2(\frac{n(n-1)}{2} - m))/(n(n-1)) + \delta = F(m, \delta, 2).$$

The function $F(m, \delta, 2)$ is decreasing with respect to m, so it is minimum for the largest possible value of m, that is $m = \frac{(n-1)(n-2)}{2} + \delta$, thus

$$F(m, \delta, 2) \geq F(\frac{(n-1)(n-2)}{2} + \delta, \delta, 2) = \frac{(1+\delta)n^2 + (1-\delta)n - 2(1+\delta)}{n(n-1)}.$$

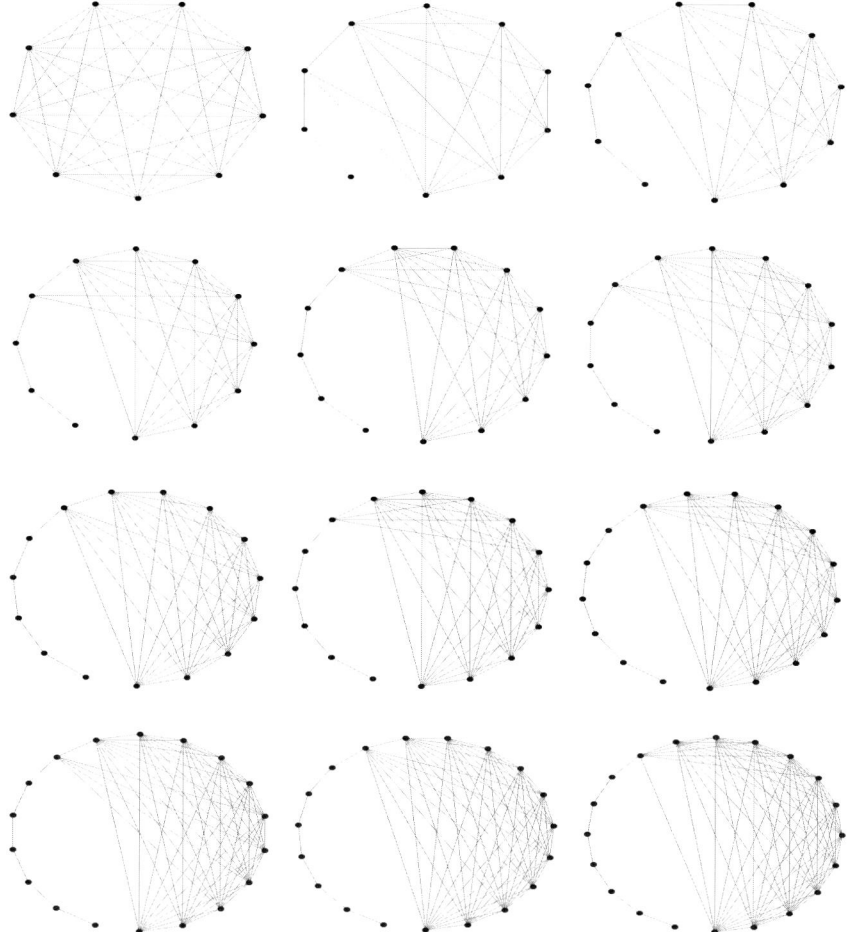

FIGURE 3. Graphs obtained by $AGX\,2$ when maximizing $\bar{l}\cdot\bar{d}$.

To be done, it is sufficient to observe that
$$F(\frac{(n-1)(n-2)}{2}+\delta,\delta,2) \geq F(\frac{(n-1)(n-2)}{2}+1,1,2) = \frac{2n^2-4}{n(n-1)}.$$
and that the unique graph of order n, size $(n-1)(n-2)/2+1$, diameter 2 and minimum degree 1 is a clique on $n-1$ vertices with a pending edge.

The upper bound is a consequence of Proposition 6.1. □

The upper bound of the next proposition, obtained by $AGX\,2$ in automated mode, improves upon conjecture 127 of Graffiti [14], *i.e.*, $\bar{l}\delta \leq n$, which was proved in [3] as a corollary of a stronger result.

Proposition 7.2. *For any connected graph on at least 3 vertices,*
$$\frac{n^2+n-4}{n(n-1)} \leq \bar{l}\cdot\delta \leq n-1.$$
The lower bound is obtained if and only if the graph is a clique on $n-1$ vertices with a pending edge. The upper bound is obtained if and only if the graph is complete.

Proof. The lower bound can be proved exactly as the lower bound of Proposition 7.1.

For the upper bound, according to Beezer *et al.* [3],
$$\bar{l} \leq \frac{(n+1)n(n-1)-2m}{(\delta+1)n(n-1)} = \frac{n+1}{\delta+1} - \frac{2m}{n}\cdot\frac{1}{(\delta+1)(n-1)}.$$
If we substitute $\frac{2m}{n}$ by $\bar{d}$, multiply by δ, and use the fact that $\delta \leq \bar{d}$, we get
$$\bar{l}\cdot\delta \leq (n+1-\frac{\bar{d}}{n-1})\frac{\delta}{\delta+1} \leq (n+1-\frac{\delta}{n-1})\frac{\delta}{\delta+1}.$$
The last expression is maximum if and only if $\delta = n-1$, *i.e.*, when the graph is complete. (This proof was obtained by the first author in April 2004; another proof was obtained independently by Smith at about the same time.) □

8. Conclusion

A systematic comparison of average distance with other graph invariants, *i.e.*, diameter, radius, girth, maximum, average and minimum degree, has been made with the system *AutoGraphiX* 2. In each case, eight bounds have been sought for, *i.e.*, lower and upper bounds for an expression of the form $\bar{l} \oplus i$ where $\oplus$ belongs to $\{+, -, \times, /\}$. In 24 of 48 cases, simple best possible bounds could be found and proved automatically. In 21 more cases, conjectures were obtained either as (*i*) both a structural result, *i.e.*, description of extremal graphs, and an algebraic relation (17 cases) or (*ii*) a structural result from which no relation was, as yet, obtained (4 cases). Of these 21 conjectures 16 are proved in this paper and the remaining ones are open. Finally, no conjectures were obtained in 3 cases.

It thus appears that it is possible to obtain automatically best possible bounds for expressions of 2 graph invariants, which range from simple observations to more substantial results and open questions. It is planned to do similar work for a variety of graph invariants, as well as to explore other, more general or different forms of conjectures [9] with the system *AutoGraphiX* in the near future.

Appendix: Results proved automatically by AGX 2

Table 2 contains all observations obtained and proved automatically by AGX 2.

TABLE 2. Automated results obtained by AGX 2.

Lower bound		$i_1 \oplus i_2$		Upper bound
2	$\leq$	$D + \bar{l}$	$\leq$	$(4n-2)/3$
2	$\leq$	$D \cdot \bar{l}$	$\leq$	$(n^2 - 1)/3$
0	$\leq$	$D - \bar{l}$		
1	$\leq$	$D / \bar{l}$		
2	$\leq$	$r + \bar{l}$	$\leq$	$\lfloor n/2 \rfloor + (n+1)/3$
1	$\leq$	$r \cdot \bar{l}$	$\leq$	$\lfloor n/2 \rfloor \cdot (n+1)/3$
4	$\leq$	$g + \bar{l}$		
3	$\leq$	$g \cdot \bar{l}$		
$2 - n$	$\leq$	$\bar{l} - \Delta$	$\leq$	$(n-5)/3$
$1/(n-1)$	$\leq$	$\bar{l} / \Delta$	$\leq$	$(n+1)/6$
$2 - n$	$\leq$	$\bar{l} - \bar{d}$	$\leq$	$(n-5)/3 + 2/n$
$1/(n-1)$	$\leq$	$\bar{l} / \bar{d}$	$\leq$	$(n^2 + n)/(6n - 6)$
$2 - n$	$\leq$	$\bar{l} - \delta$	$\leq$	$(n-2)/3$
$\frac{1}{n-1}$	$\leq$	$\bar{l} / \delta$	$\leq$	$(n+1)/3$

All the lower bounds are obtained for a complete graph K_n and all the upper bounds are obtained for a path P_n. Note that the results in third and fourth line are equivalent.

Acknowledgment

We are grateful to an anonymous referee for a careful reading of our paper, and several remarks. We also thank Gilles Caporossi and Dragan Stevanović for useful discussions about AGX 2 and its results.

References

[1] M. Aouchiche, J.-M. Bonnefoy, A. Fidahoussen, G. Caporossi, P. Hansen, L. Hiesse, J. Lacheré, A. Monhait, *Variable Neighborhood Search for Extremal Graphs. 14. The AutoGraphiX 2 System.* In L. Liberti and N. Maculan (editors). *Global Optimization: From Theory to Implementation,* Springer (2006), 231–310.

[2] M. Aouchiche, G. Caporossi, P. Hansen, *Automated Comparison of Graph Invariants.* Les Cahiers du GERAD, G–2005–40, Technical report, HEC Montréal.

[3] R.A. Beezer, J.E. Riegsecker, B.A. Smith, *Using Minimum Degree to Bound Average Distance.* Discrete Math. **226**(1–3) (2001), 365–371.

[4] C. Berge, *Graphs and Hypergraphs.* North-Holland, Amsterdam, 1976.

[5] G. Caporossi, P. Hansen, *Variable Neighborhood Search for Extremal Graphs. I. The AutoGraphiX System.* Discrete Math. **212**(1–2) (2000), 29–44.

[6] G. Caporossi, P. Hansen, *Variable Neighborhood Search for Extremal Graphs. V. Three Ways to Automate Finding Conjectures.* Discrete Math. **276**(1–3) (2004), 81–94.

[7] S. Fajtlowicz, P. Fowler, P. Hansen, M. Janowitz, F. Roberts, (editors). *Graphs and Discovery.* DIMACS Series in Discrete Mathematics and Theoretical Computer Science Vol. 69, AMS, 2005.

[8] P. Hansen, *How Far Is, Should and Could Be Conjecture-Making in Graph Theory an Automated Process?* In S. Fajtlowicz, P. Fowler, P. Hansen, M. Janowitz, F. Roberts, (editors). *Graphs and Discovery.* DIMACS Series in Discrete Mathematics and Theoretical Computer Science, AMS, 2005, 189–230.

[9] P. Hansen, M. Aouchiche, G. Caporossi, H, Mélot, D. Stevanović, *What Forms Do Interesting Conjectures Have in Graph Theory?* In S. Fajtlowicz, P. Fowler, P. Hansen, M. Janowitz, F. Roberts, (editors). *Graphs and Discovery.* DIMACS Series in Discrete Mathematics and Theoretical Computer Science, AMS, 2005, 231–252.

[10] P. Hansen and D. Stevanović, *On bags and bugs.* Electronic Notes in Discrete Mathematics **19**, (1 June 2005), 111–116.

[11] F. Harary, *The Maximum Connectivity of a Graph.* Proc. Nat. Acad. Sci. U.S. **48** (1962), 1142–1146.

[12] E.A. Nordhaus, J.W. Gaddum, *On Complementary Graphs.* Amer. Math. Monthly **63** (1956), 175–177.

[13] L. Soltés, *Transmission in Graphs: a Bound and Vertex Removing.* Math. Slovaca **41**(1) (1991), 11–16.

[14] *Written on the Wall.* Electronic file available from http://math.uh.edu/∼clarson/, 1999.

Mustapha Aouchiche
GERAD and Département de mathématiques et de génie industriel
École Polytechnique de Montréal
C.P. 6079, Succ. Centre-ville
Montréal (Québec) Canada H3C 3A7
e-mail: `mustapha.aouchiche@gerad.ca`

Pierre Hansen
GERAD and Service de l'enseignement des méthodes quantitatives de gestion
HEC Montréal
3000, chemin de la Côte-Sainte-Catherine
Montréal (Québec) Canada H3T 2A7
e-mail: `pierre.hansen@gerad.ca`

Brambles, Prisms and Grids

E. Birmelé, J.A. Bondy and B.A. Reed

Abstract. The Cartesian product $C_k \times K_2$ of a circuit of length k with K_2 is called a *k-prism*. It is well known that graphs not having the k-prism as a minor have their tree-width bounded by an exponential function of k. Using brambles and their well-studied relation to tree-width, we show that they have in fact tree-width $\mathcal{O}(k^2)$. As a consequence, we obtain new bounds on the tree-width of graphs having no small grid as a minor.

Mathematics Subject Classification (2000). 05C83.

Keywords. Bramble, minor, tree-width.

1. Introduction

A *bramble* in a graph G (see [5]) is a set $\mathcal{B}$ of connected subgraphs every two of which *touch*, that is, either intersect or are joined by an edge. A *transversal* of a bramble $\mathcal{B}$ is a set of vertices which meets each element of $\mathcal{B}$. The minimum size of a transversal is the *order* of the bramble, and the maximum order of a bramble in a graph G is the *bramble number* of G, denoted $\mathrm{BN}(G)$. This parameter is of interest because of the duality theorem of Seymour and Thomas [7] which links it to the tree-width $\mathrm{TW}(G)$:

Theorem. $$\mathrm{BN}(G) = \mathrm{TW}(G) + 1.$$

If xy is an edge of a graph G, we shall denote by G_{xy} the graph obtained by contracting xy to a vertex $x * y$; thus $V(G_{xy}) = (V(G) \setminus \{x, y\}) \cup \{x * y\}$ and $E(G_{xy}) = E(G - \{x, y\}) \cup \{(x * y)z : xz \text{ or } yz \in E(G)\}$. A *minor* of G is a graph obtained from G by a sequence of edge deletions, edge contractions and vertex deletions. If H is a minor of G, we say that G has an H-*minor*.

Brambles are complicated objects which are hard to grasp. A graph which has a K_n-minor contains a bramble of order n consisting of the preimages (under contraction) of the vertices of the minor. But the converse is far from true. Consider, for example, the $(k \times k)$-grid, the graph F_k on $\{1, \ldots, k\}^2$ with edge set

$$\{(i,j)(i',j') : |i - i'| + |j - j'| = 1\}.$$

FIGURE 1. The (6×6)-grid

The grid F_6 is shown in Figure 1. For $1 \leq i \leq k$, the path R_i induced by the vertices of first coordinate i is called the ith *row* of the grid and the path C_i induced by the vertices of second coordinate i is called the ith *column* of the grid.

Being planar, F_k has no K_5-minor. However, its bramble number is at least k. The set $\mathcal{B} := \{B_{i,j} : 1 \leq i, j \leq k\}$, where $B_{i,j} := R_i \cup C_j$, is a bramble; as each row intersects each column, every $B_{i,j}$ is connected and any two of these subgraphs intersect. However, any set X of at most $k-1$ vertices of F_k misses at least one row, say R_i, and at least one column, say C_j. So $X \cap V(B_{i,j}) = \emptyset$ and X is not a transversal of $\mathcal{B}$. This shows that

$$BN(F_k) \geq k.$$

If $xy \in E(G)$ and $\mathcal{B}$ is a bramble of G_{xy}, we obtain a bramble of the same order in G by replacing $x * y$ by $\{x, y\}$ in each element of $\mathcal{B}$ which contains it. So $\mathrm{BN}(G_{xy}) \leq \mathrm{BN}(G)$.

Deleting edges or vertices also decreases the bramble number. Thus, if we define $h(G)$ as the greatest integer k such that G has an F_k-minor, we obtain the following generalization of the above inequality for grids:

Theorem. $\qquad \mathrm{BN}(G) \geq h(G).$

In [6], Robertson and Seymour obtained an upper bound for $\mathrm{BN}(G)$ in terms of $h(G)$:

Theorem. $\qquad \mathrm{BN}(G) \leq 2^{20h(G)^5}.$

The authors of [3] gave a somewhat simpler proof of the Robertson-Seymour theorem, but their bound is still greater. These bounds on $\mathrm{BN}(G)$ seem to be far from sharp. Indeed, it may well be that $\mathrm{BN}(G)$, and therefore also $\mathrm{TW}(G)$, is bounded above by a polynomial in $h(G)$.

If $h(G) = 1$, then G contains no circuit of length four or more, so each block is an edge or a triangle, and $\mathrm{BN}(G) \leq 3$. (Indeed, any graph with no K_4-minor has tree-width at most two – see, for example, Diestel [2].)

Recently, the authors [1] proved that if $h(G) = 2$ then $\mathrm{BN}(G) \leq 8$. This bound is tight, as shown by the complete graph K_8.

In this paper, using different techniques, we show that if $h(G) = 3$ then $\mathrm{BN}(G) \leq 7263$. We obtain this bound as a consequence of a stronger result which is of independent interest.

The Cartesian product $C_k \times K_2$ of a k-circuit with K_2 is called a *k-prism* (see Figure 2).

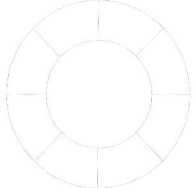

FIGURE 2. The 8-prism

Theorem 1.1. *If G contains no k-prism as a minor,*
$$\mathrm{BN}(G) \leq 60k^2 - 120k + 63.$$

We note that the (4×4)-grid F_4 is a minor of the 12-prism, so Theorem 1.1 implies that if $h(G) = 3$ then $\mathrm{BN}(G) \leq 7263$.

To prove Theorem 1.1, we first prove:

Theorem 1.2. *If G does not contain two disjoint circuits linked by l disjoint paths, $\mathrm{BN}(G) \leq 60l - 57$.*

(By *disjoint*, we shall always mean *vertex-disjoint*.)

As we are about to show, Theorem 1.2 (combined with a well-known result of Erdős and Szekeres [4] on monotone subsequences of integer sequences) implies Theorem 1.1. We outline our proof of Theorem 1.2 in Section 2, and fill in the details in Sections 3 and 4.

Proposition 1.3 (P. Erdős and G. Szekeres). *A sequence of $(k-1)(l-1)+1$ distinct integers has either an increasing subsequence of length k or a decreasing subsequence of length l.*

Let G be a graph such that $\mathrm{BN}(G) \geq 60k^2 - 120k + 64$. By Theorem 1.2, G contains two disjoint circuits C_1 and C_2 linked by $k^2 - 2k + 2$ disjoint paths. By Lemma 1.3, there are k paths among them whose endpoints appear in the same order on C_1 and C_2 (the increasing and decreasing aspect in Lemma 1.3 corresponds here to the two possible orientations of C_2). Thus G has the k-prism as a minor.

2. Well-attached circuits

Given a bramble $\mathcal{B}$ of order k in a graph G, and a vertex-cut X of G with $|X| < k$, there is a bramble element disjoint from X and hence a component of $G \setminus X$ which contains an element of $\mathcal{B}$. Because every pair of elements in a bramble touch, this component is unique. We call it the *big component* of $G \setminus X$ (with respect to $\mathcal{B}$) and denote its vertex set by X^*.

Lemma 2.1. *Let $\mathcal{B}$ be a bramble and S a minimum transversal of $\mathcal{B}$. Then, for all X with $|X| < |S|$, $(S \cap X^*) \cup X$ is a transversal of $\mathcal{B}$. In particular, $|S \setminus X^*| \leq |X|$.*

Proof. As X^* contains an element of $\mathcal{B}$, every element of $\mathcal{B}$ either is contained in $G[X^*]$ or else includes a vertex of the cut X. It follows that $(S \cap X^*) \cup X$ is a transversal of $\mathcal{B}$. □

Let $\mathcal{B}$ be a bramble of order at least k, and let l be an integer, $l \leq k$. We say that a subgraph of G is *l-attached* to $\mathcal{B}$ if it intersects $G[X^*]$ for all X with $|X| < l$.

Lemma 2.2. *Suppose that G contains a bramble $\mathcal{B}$ of order at least l, and also two disjoint subgraphs, G_1 and G_2, each of which is l-attached to $\mathcal{B}$. Then there are l disjoint paths linking G_1 and G_2 in G.*

Proof. If G_1 and G_2 are not linked by l disjoint paths, there exists a vertex cut X of G with $|X| < l$ separating G_1 and G_2. But now either G_1 or G_2 fails to intersect $G[X^*]$. □

We shall show that every bramble of order at least three has an l-attached circuit. Before doing so, we establish an analogous result for paths.

Theorem 2.3. *For every bramble $\mathcal{B}$ of order l, there exists a path P meeting every element of $\mathcal{B}$. In particular, the path P is l-attached to $\mathcal{B}$.*

Proof. For notational convenience, we set $\mathcal{B} = \mathcal{B}_0$. For an xy-path P, we denote by $[xPy[$ the path $P \setminus y$.

Let B_0 be an element of $\mathcal{B}_0$ and $x_0 \in B_0$. Set $\mathcal{B}_1 = \mathcal{B}_0 \setminus B_0$.
While $\mathcal{B}_{i+1} \neq \emptyset$

- we choose $B_{i+1} \in \mathcal{B}_{i+1}$
- we define P_i as a shortest path linking x_i to a vertex $x_{i+1} \in B_{i+1}$ such that $[x_i P_i x_{i+1}[\subset B_i$. Such a path exists because $\mathcal{B}$ is a bramble.
- we set $\mathcal{B}_{i+2} = \mathcal{B}_{i+1} \setminus \{B \in \mathcal{B}_{i+1} | B \cap P_i \neq \emptyset\}$

We thus obtain a finite sequence $P_0, \ldots, P_r$ and denote by P the walk $P_1 \cup \cdots \cup P_r$. As $\mathcal{B}_{r+2}$ is empty, it is clear that P meets every element of the bramble.

We claim that P is a path. To prove this, suppose to the contrary that a vertex y appears twice in P. Then there exist two integers $i < j$ such that $y \in [x_i P_i x_{i+1}[$ and $y \in [x_j P_j x_{j+1}[$.

If $j > i+1$, $B_j \in \mathcal{B}_{i+2}$ and thus $B_j \cap P_i = \emptyset$. This contradicts the fact that $y \in V(B_j \cap P_i)$. Thus $j = i+1$. As P_i was chosen as short as possible, $[x_i P_i x_{i+1}[\cap B_{i+1} = \emptyset$, again contradicting the fact that $y \in [x_i P_i x_{i+1}[\cap B_j$.

Therefore P is a path and satisfies the theorem. □

Theorem 2.4. *Let $\mathcal{B}$ be a bramble of order $l \geq 3$. Then there exists a circuit C meeting every element of $\mathcal{B}$. In particular, the circuit C is l-attached to $\mathcal{B}$.*

Proof. Let $P = x_1 x_2 \ldots x_n$ be a shortest path meeting every element of $\mathcal{B}$.

By the minimality of P, there exist two elements $B_1, B_n \in \mathcal{B}$ such that $B_1 \cap P = \{x_1\}$ and $B_n \cap P = \{x_n\}$.

Also, as $\mathcal{B}$ is a bramble, there is an (x_n, x_1)-path Q all of whose vertices belong to B_1 or B_n. The paths P and Q are thus internally-disjoint, and the circuit C obtained by concatenating them meets all the elements of $\mathcal{B}$. □

In Section 4, we shall prove:

Theorem 2.5. *Let $\mathcal{B}$ be a bramble in a graph G. If G contains a circuit which is $(60l-56)$-attached to $\mathcal{B}$, it contains two disjoint circuits, each of which is l-attached to $\mathcal{B}$.*

Combining Lemma 2.2 with Theorems 2.4 and 2.5 yields Theorem 1.2. Thus it remains only to give the proof of Theorem 2.5. A key concept in our proof of this theorem is that of a sun. Suns are defined in the next section, and their relationship to l-attached circuits is shown.

3. Suns

For $X \subseteq V$, an X-*sun* $(C, P_1, \ldots, P_l)$ consists of a circuit C together with l disjoint (C, X)-paths $P_1, \ldots, P_l$, all internally-disjoint from C; note that if $X \cap V(C) \neq \emptyset$, some of these paths will be of length zero. The paths P_i are the *rays* of the sun, and their number, l, its *order* (see Figure 3). The vertex of a ray on C is its *root*.

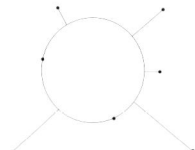

FIGURE 3. A sun of order seven

Suns are of interest because they are easier to handle than l-attached circuits, but are nonetheless closely linked to them, as described in the following lemma.

Lemma 3.1. *Let $\mathcal{B}$ be a bramble of order l and S a minimum transversal of $\mathcal{B}$. If C is a circuit which is l-attached to $\mathcal{B}$, there exist paths $P_1, \ldots, P_l$ such that $(C, P_1, \ldots, P_l)$ is an S-sun of order l.*
Conversely, if $(C, P_1, \ldots, P_l)$ an S-sun of order l, the circuit C is $\lceil \frac{l}{2} \rceil$-attached to $\mathcal{B}$.

Proof. First, let C be a circuit which is l-attached to $\mathcal{B}$. Suppose that it cannot be linked to S by l disjoint paths. Then there is a set X, with $|X| < l$, separating C and S. Since C is l-attached to $\mathcal{B}$, $X^* \cap V(C) \neq \emptyset$ and hence $S \cap X^* = \emptyset$. By Lemma 2.1, this implies that X is a transversal of $\mathcal{B}$. But $|X| < |S|$, a contradiction.

Suppose, now, that $(C, P_1, \ldots, P_l)$ is an S-sun of order l, and that C is not $\lceil \frac{l}{2} \rceil$-attached to $\mathcal{B}$. Then there exists a set X, with $|X| < \lceil \frac{l}{2} \rceil$, such that C includes no vertex of X^* (see Figure 4).

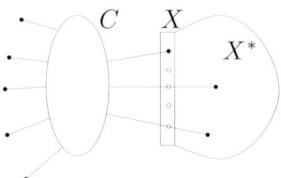

FIGURE 4

But, as $(C, P_1, \ldots, P_l)$ is an S-sun of order l, and as at most $|X|$ of the paths $P_1, \ldots, P_l$ can meet X, $|S \setminus X^*| \geq l - |X| > |X|$, contradicting Lemma 2.1. □

4. Disjoint well-attached circuits

We are now ready to prove:

Theorem 2.5 *Let $\mathcal{B}$ be a bramble in a graph G. If G contains a circuit which is $(60l-56)$-attached to $\mathcal{B}$, it contains two disjoint circuits, each of which is l-attached to $\mathcal{B}$.*

To prove this theorem, we use the suns introduced in the previous section. By virtue of Lemma 3.1, it suffices to show:

Theorem 4.1. *Let $\mathcal{B}$ be a bramble, and S a minimal transversal of $\mathcal{B}$. Suppose that G has an S-sun of order at least $60l - 56$. Then G contains two disjoint circuits, each of which is l-attached to $\mathcal{B}$.*

We shall need the following result, which is a consequence of the minimality of S.

Lemma 4.2. *Let S_1 and S_2 be disjoint subsets of S with $|S_1| = |S_2| = k$. Then there are k disjoint paths linking S_1 and S_2.*

Proof. Suppose the conclusion false. Then there exists a set X with $|X| < k$ which separates S_1 and S_2. Thus one of the two sets, say S_2, is disjoint from X^*, implying that $|X \cup (S \cap X^*)| < |S|$, and thereby contradicting Lemma 2.1. □

Proof of Theorem 4.1. Consider an S-sun of order at least $60l - 56$, with circuit C. Let C_1 and C_2 be disjoint segments of C, each containing the roots of at least $30l - 28$ rays of the sun, and let S_i denote the set of vertices of S reached by the rays rooted in C_i, $i = 1, 2$.

By Lemma 4.2, there exist $30l - 28$ disjoint paths linking S_1 and S_2. Each of these paths shares its endpoint in S_1 with a ray rooted in C_1 and its endpoint in S_2 with a ray rooted in C_2. Concatenating these three paths results in a walk

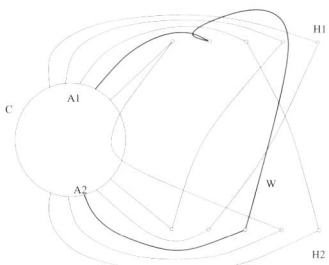

FIGURE 5. Concatenating paths into a walk

from C_1 to C_2 (see Figure 5). Denote by W_i, $1 \leq i \leq 30l - 28$, the walks thereby obtained.

Each W_i contains a path P_i from C_1 to C_2, internally-disjoint from C. Let H be the union of these $30l - 28$ paths. Each vertex of H is in at most one ray and at most one of the paths linking S_1 and S_2, so in at most two of the walks W_i and thus in at most two of the paths P_i. Therefore there exist $15l - 14$ paths which are disjoint and whose edges are edges of the P_i's: if not, there would exist a set X of at most $15l - 15$ vertices separating C_1 and C_2, and so three distinct paths P_i would pass through the same vertex of X. Call these $15l - 14$ paths $Q_1, \ldots, Q_{15l-14}$; note that the Q_i are internally-disjoint from C.

Let $X := \{x_0, \ldots, x_{15(l-1)}\}$ be the set of endpoints of the paths Q_i on C_1, enumerated in the order that they appear on this segment. For $0 \leq i \leq 5$, we set $y_i := x_{3i(l-1)}$. Then, for $0 \leq i \leq 4$, the segment of C_1 between y_i and y_{i+1} contains $3l - 2$ vertices x_j.

The following observation is important:

Claim. *Any circuit containing $3l - 2$ vertices of X is l-attached to $\mathcal{B}$.*

Let D be a circuit containing $3l - 2$ vertices of X. Suppose that D is not l-attached to $\mathcal{B}$. Then there is a set Y with $|Y| \leq l - 1$ such that $V(D) \cap Y^* = \emptyset$. Thus, by Lemma 2.1, $|S \setminus Y^*| \leq l - 1$, so one of the sets S_i, $i = 1, 2$, satisfies $|S_i \setminus Y^*| \leq \lfloor \frac{l-1}{2} \rfloor$. Suppose that this set is S_2. (A similar argument applies to S_1.)

Every vertex in $X \cap V(D)$ is connected to a vertex of S_2 by a segment of the walk W_i to which it belongs, and thus by a path R_i contained in W_i. Moreover, each vertex in G is in at most two such paths R_i, as it is in at most two of the walks W_i. Thus at most $l - 1$ of the paths R_i link X to $S_2 \setminus Y^*$ and at most $2l - 2$ of them link X to $S_2 \cap (Y \cup Y^*)$ by way of Y. This is a contradiction, as there are $3l - 2$ such paths, and the claim is established.

Consider the four paths Q_i starting from the vertices y_1, y_2, y_3 and y_4. By Lemma 1.3, either two of them are parallel, that is, reach C_2 in the same order, or all four of them cross one another, that is, reach C_2 in the opposite order (see Figure 6).

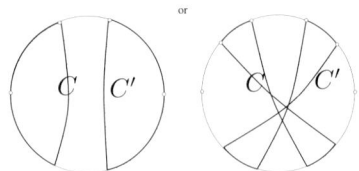

Figure 6. Parallel and crossing paths

In each case, there are two disjoint circuits, each containing at least $3l - 2$ vertices of X, and thus, by the above claim, each l-attached to $\mathcal{B}$. □

References

[1] E. Birmelé, J.A. Bondy and B. Reed, *The tree-width of the (3×3)-grid*, manuscript.

[2] R. Diestel, *Graph Theory*, Second edition. Graduate Texts in Mathematics, **173**. Springer-Verlag, New York, 2000.

[3] R. Diestel, K.Yu. Gorbunov, T.R. Jensen and C. Thomassen, *Highly connected sets and the excluded grid theorem*, J. Combin. Theory Ser.B, **75** (1999), 61–73.

[4] P. Erdős and G. Szekeres, *A combinatorial problem in geometry*, Compositio Math. **2** (1935), 463–470.

[5] B.A. Reed, *Tree width and tangles: a new connectivity measure and some applications*, in Surveys in Combinatorics, London Math. Soc. Lecture Note Ser. **241**, Cambridge Univ. Press, Cambridge, 1997, 87–162.

[6] N. Robertson and P.D. Seymour, *Graph minors V: Excluding a planar graph*, J. Combin. Theory Ser.B, **41** (1986), 92–114.

[7] P.D. Seymour and R. Thomas, *Graph searching and a min-max theorem for tree-width*, J. Combin. Theory Ser. B, **58** (1993), 22–33.

E. Birmelé
Laboratoire MAGE, FST, Université de Haute-Alsace
4, rue des frères Lumière, F-68093 Mulhouse Cedex, France
e-mail: `Etienne.Birmele@uha.fr`

J.A. Bondy
LaPCS, Université Claude-Bernard Lyon 1,
Domaine de Gerland, 50 avenue Tony Garnier, F-69366 Lyon Cedex 07, France
e-mail: `jabondy@moka.ccr.jussieu.fr`

B.A. Reed
School of Computer Science, McGill University,
3480 University, Montreal, Quebec, Canada
e-mail: `breed@cs.mcgill.ca`

Dead Cell Analysis in Hex
and the Shannon Game

Yngvi Björnsson, Ryan Hayward, Michael Johanson
and Jack van Rijswijck

> **Abstract.** In 1981 Claude Berge asked about combinatorial properties that might be used to solve Hex puzzles. In response, we establish properties of dead, or negligible, cells in Hex and the Shannon game.
>
> A cell is dead if the colour of any stone placed there is irrelevant to the theoretical outcome of the game. We show that dead cell recognition is NP-complete for the Shannon game; we also introduce two broader classifications of ignorable cells and present a localized method for recognizing some such cells. We illustrate our methods on Hex puzzles created by Berge.
>
> **Keywords.** Game theory, Hex, Shannon game, dead cell, negligible, induced path.

1. Introduction

Claude Berge, who loved to play Hex, commented in [4] that

> "[it] would be nice to solve some Hex problem by using nontrivial theorems about combinatorial properties of sets (the sets considered are groups of critical [board cells])."

In response, we investigate properties of cells that are dead, or ignorable in a certain sense, in Hex and the Shannon game and show how these properties simplify the solution of Hex puzzles.

In §2 we review Hex and the Shannon game. In §3 we define dead cells and show some basic properties; in particular, dead cell recognition is NP-complete in the Shannon game. In §4 we introduce captured and dominated sets, two broader classes of ignorable cells, and explain how some such sets can be defined in terms of local subgames. In §5 we explain the strategic implications of these results, while in §6 we describe how some such sets can be recognized. In §7 we illustrate our analysis on some Hex puzzles created by Berge.

The authors gratefully acknowledge the research support of NSERC.

2. Hex and the Shannon game

Hex is played on a board containing a rhombic array of hexagonal cells with an equal number of hexagons on each side.[1] Commonly used board sizes are 11×11[2] and 14×14.[3] The two players, Black and White, take turns placing a stone on the board. White's goal is to connect the lower-left and upper-right sides of the board with a chain of white stones; Black's goal is to connect the upper-left and lower-right sides with a black chain. Lest the players forget their goals, each marks their two sides with a pair of extra stones off the board. Figure 1 shows a completed game which White has won.

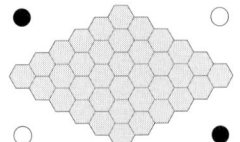 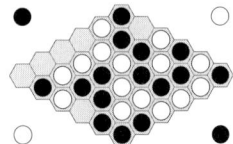

FIGURE 1. An empty 6×6 Hex board (left) and a completed game (right).

For Hex on any board size, there exists a winning strategy for the first player; however, the only known proof is by *reductio ad absurdum* and no explicit general winning strategy is known [9]. Indeed, determining the existence of a winning strategy is PSPACE-complete for Hex [18] and so also for the Shannon game [6]. For more on Hex, see the website by Thomas Maarup [16], the survey by two of the authors of this paper [12], or the book by Cameron Browne [5].

The Shannon game is played on any graph G with two distinguished *terminal vertices*. Shannon originally formulated the game as played by colouring the edges of the graph; in this paper we consider only the vertex colouring game.[4] Further, we restrict our attention to finite graphs. We assume that the colours used are χ_s and χ_c.

The two players of a Shannon game are called *Short* and *Cut*. To start, the terminal vertices are coloured χ_s and all other vertices are uncoloured. Play proceeds with each player in turn colouring any previously uncoloured vertex. Short's goal is to form a monochrome path containing the terminal vertices; Cut's goal is to prevent this. Thus Short needs to create a χ_s-coloured terminal-connecting path while Cut needs to establish a χ_c-coloured terminal-separating cutset.

Hex is a special case of the Shannon game. Each Hex position, or board state, can be represented as a Shannon graph in two dual ways. Figure 2 shows

---

[1] If the side lengths are unequal, the game is trivial due to an easy pairing strategy [9].
[2] This is the size used by Piet Hein, the original inventor of Hex [14].
[3] This is the size preferred by Berge [3, 10].
[4] The edge colouring game, known as the *Shannon switching game*, is actually a special case of the Shannon game since it is equivalent to colouring vertices on the line graph of the original graph. Lehman found a polynomial-time algorithmic solution for the Shannon switching game [15].

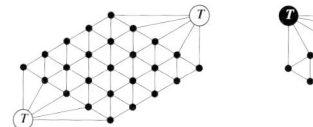

FIGURE 2. The two dual Shannon game representations of the 5×5 Hex board. Short is White on the left and Black on the right. Terminal vertices are labelled 'T'.

the two graphs that correspond to the empty 5×5 Hex board. Neither Hex nor the Shannon game can end in a draw; proofs which hold for Hex are in [1, 8].

For the Shannon game, note that when a vertex is coloured by Cut, it may equivalently be *cut*, or deleted, from the graph. On the other hand, when a vertex is coloured by Short, it may equivalently be *shorted* or contracted by adding edges between all pairs of its neighbors and then deleting the vertex. We refer to the graph that results by shorting and cutting all coloured nonterminal vertices as the *Shannon reduced graph*. For a Shannon board state represented by a graph G, the vertices of the Shannon reduced graph G' are the terminals and uncoloured vertices of G, with two vertices adjacent in G' if and only if they are adjacent or connected by a χ_s-coloured path in G. Short has a winning path in G if and only if the terminals are adjacent in the Shannon reduced graph G'; Cut has a winning cutset in G if and only if the terminals are disconnected in G'. Figure 3 shows a Hex position and the two Shannon reduced graphs that represent it.

In the remainder of this paper we use *node* to refer to both a vertex in a Shannon graph and a cell on a Hex board.

3. Dead nodes

The Shannon game is a special case of a more general class of games known as *division games*, which can be seen as set-colouring games with some function that assigns a winner to each completely coloured set. Following Yamasaki, who gave a theory of division games [21], an element in a particular state of a division game is *regular* if it is never disadvantageous to own it, *misère* if it is never advantageous to own it, and *negligible* if it does not matter who owns it. In the Shannon game,

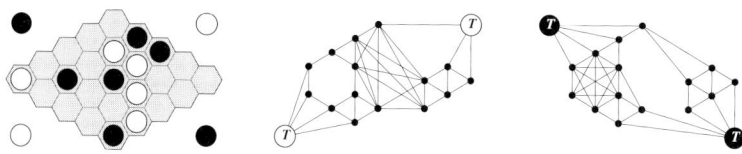

FIGURE 3. A Hex position and its two Shannon reduced graph representations.

all nodes are regular and some nodes may become negligible during the course of the game.

We shall refer to negligible nodes as *dead*. The notion of a dead cell in Hex or other related games was implicitly recognized by Schensted and Titus in their discussion of "worthless triangles" [19] and Beck et al. in the proofs of their opening move results [1, 2].

Formally, define a *board state* (G, T, Ψ) of a Shannon game as a graph G with a set of terminal nodes T and a colouring Ψ of some subset of the nonterminal nodes. For colourings Ψ_1 and Ψ_2, say that Ψ_2 *extends* Ψ_1 if every coloured node in Ψ_1 has the same colour in Ψ_2. A colouring is *complete* if every node is coloured. If a colouring is complete, then the game is over and the winner is known.

Definition 3.1. A nonterminal node v in a board state (G, T, Ψ) is *dead* if, for every complete colouring Ψ^* that extends Ψ, the winner of (G, T, Ψ^*) is independent of the colour of v; v is *live* otherwise.

Note that this definition applies to both coloured and uncoloured nodes.

The reader may enjoy working out which nodes are dead in Figure 3; the answer is shown in Figure 4. Each of the three representations contains exactly one dead node, and it is the same node in each case. As we shall see shortly, this is no coincidence.

Consider a board state (G, T, Ψ) and a set S of uncoloured nodes. When Ψ is extended to a complete colouring by assigning χ_π to all nodes in S and $\chi_{\bar\pi}$, the colour different from χ_π, to all other uncoloured nodes, the resulting colouring is denoted by $\Psi \oplus_\pi S$. With respect to a board state, we say that S is a *short set* if $\Psi \oplus_s S$ has a winning chain for Short. A short set is *minimal* if no proper subset is a short set.

Theorem 3.2. *For a board state of the Shannon game, an uncoloured node is live if and only if some minimal short set contains it.*

Proof. ($\Leftarrow$) Let S be a minimal short set that contains v, and let $\Psi^* = \Psi \oplus_s S$. Short is the winner in (G, T, Ψ^*). If the colour of v is subsequently changed then Cut is the winner, since otherwise $S - v$ would be a short set and S would not be minimal. Therefore v is live.

($\Rightarrow$) Suppose v is live. Then there is an extension of Ψ to a complete colouring Ψ^* in which the colour of v determines the winner for Ψ^*. Let Ψ_s^* and Ψ_c^* be Ψ^* with v repainted with χ_s and χ_c respectively, and let S be the uncoloured nodes of

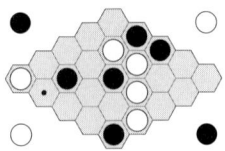

FIGURE 4. The only dead cell for this Hex position.

Ψ which are coloured χ_s in both Ψ_s^* and Ψ_c^*. Since Short wins (G, T, Ψ_s^*), $S \cup \{v\}$ contains a minimal short set S' for (G, T, Ψ). Since Cut wins (G, T, Ψ_c^*), S contains no short set for (G, T, Ψ), so S' is not a subset of S, so S' contains v. □

For a Shannon board state with all non-terminal nodes uncoloured, a set is a short set if and only if it contains a path between the terminals, and a set is a minimal short set if and only if it is an induced, or chordless, inter-terminal path. Thus we have the following.

Corollary 3.3. *In the Shannon game, an uncoloured node is live if and only if it is on some induced inter-terminal path in the Shannon reduced graph.*

A coloured node is live if and only if it is live when uncoloured, so the preceding corollary can also be used to tell whether a coloured node is live.

Recognizing dead nodes simplifies Shannon game analysis, since:

Observation 3.4. *A dead node can be assigned an arbitrary colour or removed from the game without affecting the outcome.*

Thus playing a move at a dead node is equivalent to skipping a move. In the Shannon game no move can be worse than skipping a move, since all nodes are regular, so every move wins if a dead cell represents a winning move. Also, if the game has not ended, then at least one move is to a live node. Hence there is a winning move if and only if there is a live winning move, and so:

Theorem 3.5. *In the Shannon game, a player with a winning strategy has a winning strategy in which every move is to a live node.*

In the interest of streamlining the search for a winning strategy, a player would thus like to be able to recognize dead nodes efficiently. So, how hard is it to recognize dead nodes?

For a vertex subset S of a graph G, the set of all vertices on some induced path between two vertices of S is known as the *monophonic interval* $J(S)$. By the preceding theorem, for a Shannon board state with reduced graph G', the set of live nodes is the monophonic interval $J(T)$ in G', where T consists of the two terminal nodes.

The *induced path pairs* problem is as follows: given a graph and vertex pairs $(a_1, b_1), \ldots, (a_k, b_k)$, is there a vertex induced subpath consisting of k disjoint induced paths joining a_j to b_j? Fellows showed the following [7].

Theorem 3.6 (Fellows). *For $k \geq 2$ the induced path pairs problem is NP-complete.*

For $k = 2$, Marcus Schaefer observed[5] that the induced path pairs problem reduces to the problem of finding a monophonic interval by adding a new vertex x adjacent to b_1 and b_2 and asking whether x is in $J(a_1, a_2)$ Thus we have the following.

Corollary 3.7. *Determining membership of a monophonic interval is NP-complete.*

---
[5] Private communication.

Corollary 3.8. *In the Shannon game, recognizing dead nodes is NP-complete.*

Dead cell recognition might be easier in Hex than in the Shannon game. In particular, in the Shannon reduced graph of a Hex position, dead nodes are often simplicial or, more generally, separated from both terminals by a clique cutset. Such nodes can be efficiently recognized using Whitesides's algorithm for finding clique cutsets [20].

Unfortunately, these observations do not simplify Hex analysis as much as one might hope, since dead cells arise infrequently in typical Hex positions. On the other hand, considering nodes that are at risk of becoming dead seems more useful, as we explain in the next section.

4. Beyond death: captured and dominated sets

We introduce in this section two concepts that generalize node death. Informally, we call a set of uncoloured nodes π-*captured* if player π can "own" the entire set no matter who plays first, and π-*dominated* if π can own it provided π has the first move. These notions evolved from [17, 11, 13].

By the *value* of a position, we mean the outcome assuming perfect play.

Consider for example the two Hex diagrams shown in Figure 5. On the left, the two marked nodes are effectively captured by Black, since a White stone played at either node becomes dead if Black then plays at the other. Thus adding Black stones to these two nodes will not change the value of the game. On the right, the three marked nodes form a dominating set for Black, since a Black move to the node with a large dot captures the two other nodes.

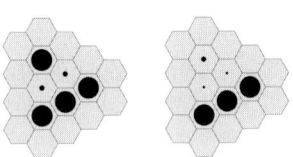

FIGURE 5. A Black-captured set (left) and a Black-dominated set (right).

Before formally defining capture and domination, we first introduce a generalization of the Shannon game.

The *multi-Shannon game* is a Shannon game played on a graph with two or more terminals. At the start of the game, all terminals are χ_s-coloured. Short's goal is to join each pair of connected terminals with some χ_s-coloured path; these paths may intersect. Cut's goal is to separate each pair of nonadjacent terminals with some χ_c-coloured cutset; these cutsets may intersect. This game can end in a draw.

Consider the two graphs of Figure 6. On the left graph, Cut has a second-player winning strategy, meaning that Cut wins even when Short goes first. On the right graph, Short has a second-player winning strategy.

FIGURE 6. Multi-Shannon games with second-player wins for Cut (left) and Short (right).

When such a graph occurs as a particular subgraph in a regular Shannon game, we can simplify the analysis of the game. In a graph, the *neighbourhood* $N(S)$ of a set S of nodes is the set of all nodes not in S but adjacent to some node in S. Let G be a reduced Shannon graph with a set S of nonterminal nodes and let Γ_S be the multi-Shannon game played on the subgraph of G induced by $S \cup N(S)$ with terminals $N(S)$. Then:

Theorem 4.1. *If player π has a second-player winning strategy for Γ_S, then S can be filled in with χ_π stones without changing the Shannon value of Γ.*

Proof. Let $\Gamma = (G, T, \Psi)$ be a board state, and let $\Gamma' = (G, T, \Psi')$ where Ψ' is the extension of Ψ by filling in S with χ_π. The theorem is equivalent to stating that π wins Γ if and only if π wins Γ'.

($\Rightarrow$) If π wins Γ, then π trivially also wins on Γ', since Γ' is formed by giving π a number of "free" moves.

($\Leftarrow$) If π has a winning strategy for Γ' and a second-player winning strategy for Γ_S, then π can adopt the following strategy for Γ. Whenever $\bar\pi$ plays in S, π responds with a move in S according to the winning strategy for Γ_S. Otherwise, π plays a move in $G - S$ according to the winning strategy for Γ'. Let Γ^* be the final board state when all nodes are coloured. There are two cases to distinguish:

- π plays Cut: If Short forms a chordless winning path P in Γ then P cannot intersect S, for otherwise the two nodes $P \cap T$ would be nonadjacent and Short would have at least achieved a draw in Γ_S with the path $P \cap S$. But if P does not intersect S then Short has won Γ', contradicting Cut's winning strategy for Γ'.
- π plays Short: Short forms some winning path P' in Γ'. If this path does not intersect S then it is also a winning path for Short on Γ. If it does intersect S then there is a winning path on Γ that consists of $P' - S$ plus some Short path connecting the two nodes $P \cap S_T$; the latter is guaranteed to exist due to Short's win on Γ_S.

Therefore π wins Γ. □

For example, let S consist of the two marked nodes in the left diagram of Figure 5 and consider the resulting subgame Γ_S; thus $N(S)$ consists of the twelve

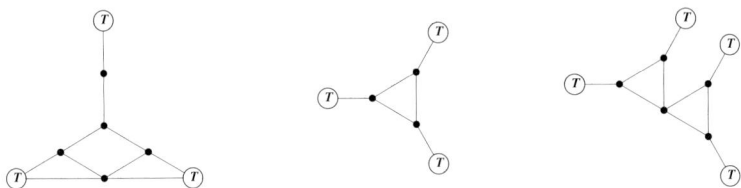

FIGURE 7. Multi-Shannon games with first-player wins for Cut.

nodes forming the diagram boundary. Short has a second-player winning strategy for Γ_S: play at whichever node of S that Cut does not play at. Thus the theorem tells us that for any Hex position containing this diagram the marked nodes can be filled in with black stones without changing value of the position.

In the graph on the left in Figure 7, the game ends in a draw when Short has the first move. If Cut has the first move, then Cut wins by colouring the bottom node. If this graph occurs as a subgraph in a regular Shannon graph, then a Cut move in the bottom node is equivalent to cutting all the nodes in the subgraph, since the remaining part of the subgraph is a second-player win for Cut and can therefore be filled in with χ_c. This leads to the following theorem.

Theorem 4.2. *If player π has a first-player winning strategy for Γ_S with winning move v, then in Γ v is at least as good for π as any other move in S.*

Proof. A π-move to v is equivalent to simultaneous π-moves to all nodes of S. □

We now formally define capture and domination with respect to a Shannon game Γ, a set of uncoloured nodes S, and the local multi-Shannon game Γ_S.

Definition 4.3. *S is π-captured if π has a second-player winning strategy for Γ_S.*

Definition 4.4. *S is π-dominated if π has a first-player winning strategy for Γ_S. For any initial move m in such a strategy, we say that m π-dominates S.*

Based on the usual recursive definitions of wins and losses in games, we have the following.

Observation 4.5. *S is π-captured if and only if S is the empty set or, for each $\bar{\pi}$-move m in S, both (i) $S - m$ is π-dominated and (ii) m is dead if $S - m$ is filled in with χ_π stones.*

Observation 4.6. *S is π-dominated if and only if S is the empty set or there is some π-move m in S such that $S - m$ is π-captured.*

In Observation 4.5, (i) guarantees that π can capture all cells of $S - m$ after an $\bar{\pi}$ move at m, while (ii) guarantees that the $\bar{\pi}$-stone at m would then also be dead; thus π captures all cells of S.

		Short moves first		
		Short wins	draw	Cut wins
Cut moves first	Short wins	Short-captured	–	–
	draw	Short-dominated	indifferent	–
	Cut wins	both dominated	Cut-dominated	Cut-captured

TABLE 1. Possible outcomes of a local position. E.g., for the event "Short wins when both Short and Cut go first", the outcome is "Short-captured".

Note that for any position of a Shannon game, each uncoloured node is dominated by both players and each dead node is captured by both players. Also, note that captured and dominated sets are defined only for uncoloured nodes, whereas dead cells are defined for coloured and uncoloured nodes.

5. Strategic advice

A local subgame Γ_S can be a win for Short or Cut or a draw, depending on whether Short or Cut has the first move. The possible outcomes are listed in Table 1. The boxes marked '–' represent impossible combinations, since having the first move cannot be a disadvantage in the multi-Shannon game.

Suppose player π is considering a move in Γ_S. According to the outcomes in Table 1, the following cases can be distinguished:

- One of the players has a second-player win. Then the set is captured and can be filled in, as per Theorem 4.1. Any π-move in S would be wasted.
- Set S is dominated by π. Then π has a locally winning move in Γ_S. By Theorem 4.2, π can safely play such a move.
- Set S is dominated by $\overline{\pi}$ but not by π. Then the best local move available for π is a local draw, while other moves may be local losses. Since a locally losing move can be followed by a $\overline{\pi}$-move that captures S and kills π's move, π should avoid locally losing moves.
- Set S is not dominated by either player. Any π-move leads to a local draw with locally optimal play. The choice of move depends on which pairs of terminals are favourable for π to connect or disconnect in the global game.

This information can be summarized by the following theorem.

Theorem 5.1. *Let Γ be a multi-Shannon game, and let v and w be moves in a subgame Γ_S defined by a set of nodes S. If v is at least as good as w for player π in Γ_S, then v is at least as good as w for π in Γ.*

Thus, if π is going to move in S, then π should make a move that is optimal in Γ_S. In short, "do not make any local mistakes". Here we consider the act of making no move at all to be optimal in a second-player win game, since it is better than wasting a move.

Once dominated and captured sets have been identified, the strategic advice for player π to move, is:

1. Fill in all captured sets, iterate until no new captured nodes are found.
2. For any π-dominated set, pick one dominating move and ignore the rest.
3. For any $\bar\pi$-dominated set, ignore the locally losing moves.

A special case of Step 3 is a $\bar\pi$-dominated set with two nodes $\{v,w\}$. There, the dominating move for $\bar\pi$, say v, is also the only locally drawing move for π. If π moves in w then $S - w$ is still dominated by $\bar\pi$, since it is a single uncoloured node, and v is dead after $\bar\pi$ moves in w, since S was $\bar\pi$-dominated by w.

Theorems 4.1–5.1 also hold in the more general case where a multi-Shannon game is a subgame of another multi-Shannon game.

6. Recognizing captured and dominated sets

In order for a player to benefit from our results so far, the player should be able to recognize captured and dominated sets efficiently. One way to achieve this is to build a library of local patterns, or subgames, that yield such sets.

Consider a set Q of nodes of a Hex position, where Q may contain Black or White stones. Let S be the uncoloured nodes of Q. Create a new board position by uncolouring all coloured nodes in $N(Q)$. Now let Γ be the reduced Shannon graph of this new position. Any move in S that is suboptimal in the multi-Shannon game Γ_S is also suboptimal in the original Hex position. This allows moves in Q to be analysed while ignoring the rest of the board. Moreover, whenever this same node pattern occurs anywhere else on the board, the local analysis results are the same. We consider such a pattern to be *irreducible* if none of the captured cells or suboptimal moves could have been detected by considering a smaller pattern.

Such a library need not be exhaustive, as there is a trade-off in effort saved by disregarding locally suboptimal moves and effort invested in detecting them. Experimental computer results suggest that in Hex games such sets are almost always reducible to a few base cases that can be derived from simplicial nodes.

Such a derivation is illustrated in Figure 8, which shows irreducible local Hex patterns with two empty cells that can be built up by starting from all cases in which a node is found to be simplicial by considering only its immediate neighborhood in the nonreduced graph. In the five starting patterns shown on the bottom of the figure, the empty cell is dead. Removing a White stone from each pattern yields a pattern with a two-cell White-dominated set; the dominating move is indicated with a dot. Now, pairs of these six patterns are combined to form larger patterns with two empty cells, such that each empty cell White-dominates the other; these two cells form a White-captured set. The eleven larger patterns that can be created in this way are shown as entries of the central array of the chart, with each entry indexed by the two smaller patterns that yield it.

There are no irreducible captured set patterns with three connected empty cells. Figure 9 shows examples of irreducible captured sets with four empty cells.

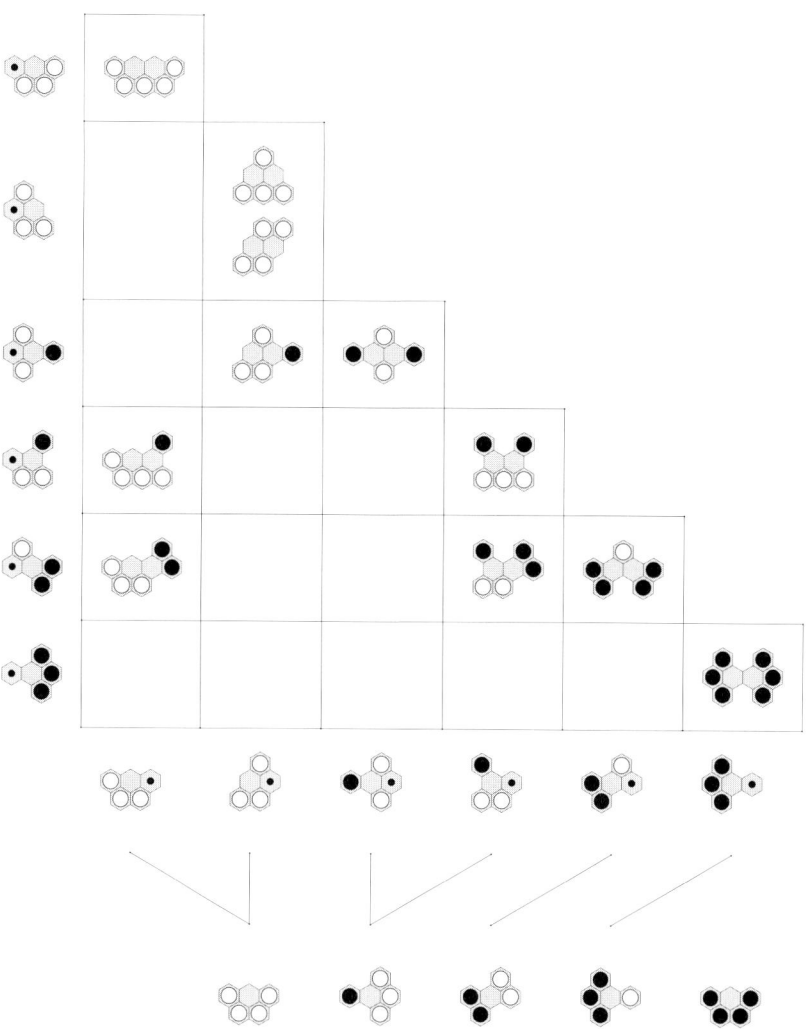

FIGURE 8. Constructing irreducible White-captured sets (empty hexagons of table entry patterns) and White-dominated sets (empty hexagons of row and column index patterns) from the five bottom dead-cell patterns.

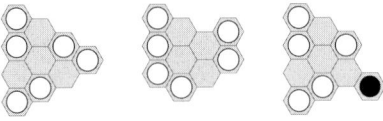

FIGURE 9. Some irreducible White-captured sets.

7. Hex examples

Claude Berge presented five Hex puzzles in his introduction to the game [3][6] (see also [10]). Figures 10–13 show the static analysis of these puzzles according to the notions of captured and dominated sets. Captured cells, including dead cells, have been marked by a large dot. Small dots represent suboptimal moves in locally dominated sets. Table 2 summarizes the statistics for these puzzles.

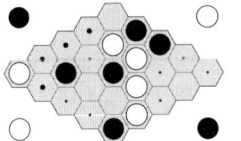 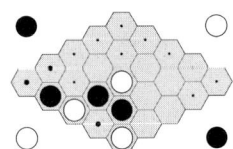

FIGURE 10. Berge's Puzzles 1 and 2. White to play can ignore all but the unmarked empty cells. Large dots are captured; small dots are locally inferior.

puzzle number	available moves	ignored moves	viable moves	percentage ignored
1	15	11	4	73%
2	19	13	6	68%
3	163	65	98	40%
4	145	75	70	52%
5	120	59	61	49%

TABLE 2. Statistics for dead cell analysis of Berge's puzzles.

Acknowledgment

We thank Cameron Browne for the use of his Hex drawing software and Marcus Schaefer for bringing Theorem 3.6 and Corollary 3.7 to our attention.

---

[6]The Puzzle 5 that we include is from an early version of Berge's manuscript; a later version has two additional white stones near the lower-right side.

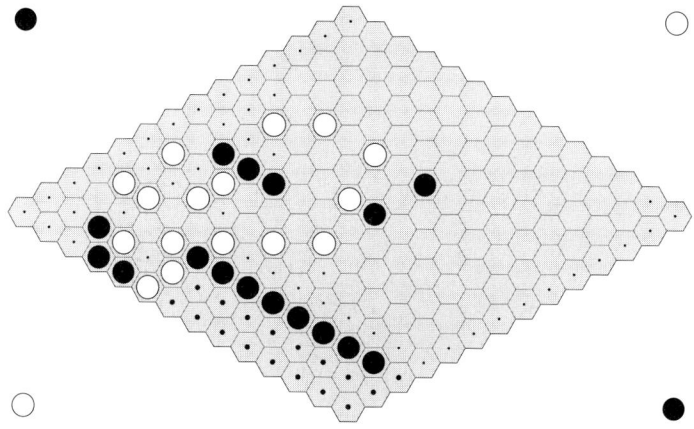

FIGURE 11. Berge's Puzzle 3: Black to play.

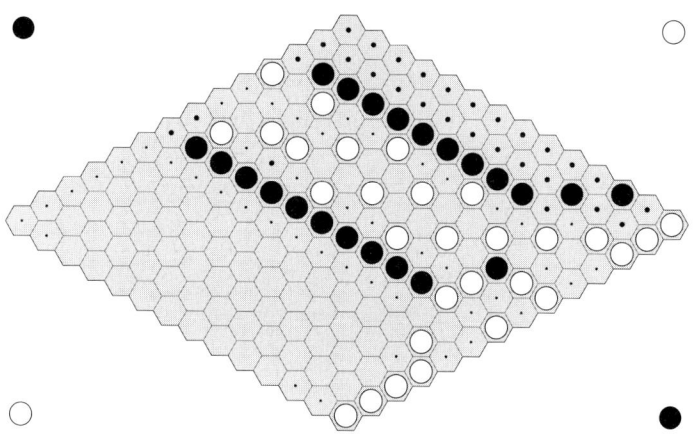

FIGURE 12. Berge's Puzzle 4: Black to play.

References

[1] Anatole Beck, Michael N. Bleicher, Donald W. Crowe, *Excursions into Mathematics* 317–387, Worth, New York, 1969.

[2] Anatole Beck, Michael N. Bleicher, Donald W. Crowe, *Excursions into Mathematics: the Millenium Edition* 496–497, A.K. Peters, Natick, 2000.

[3] Claude Berge, L'Art Subtil du Hex, manuscript, 1977.

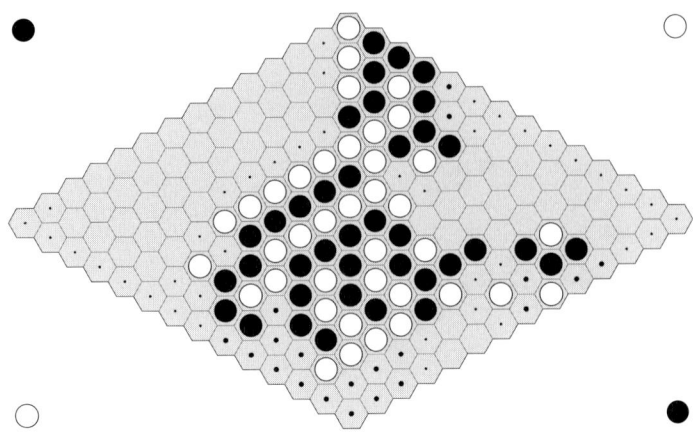

FIGURE 13. Berge's Puzzle 5: White to play.

[4] Claude Berge, Some remarks about a Hex problem, in D. Klarner, ed., *The Mathematical Gardner*, Wadsworth International, Belmont, 1981.

[5] *Cameron Browne, Hex Strategy: Making the Right Connections*, A.K. Peters, Natick, 2000.

[6] Shimon Even and Robert E. Tarjan, A combinatorial problem which is complete in polynomial space, *JACM* **23** 710–719, 1976.

[7] Michael Fellows, The Robertson-Seymour theorems: a survey of applications, *Contemporary Mathematics* **89** 1–18, 1989.

[8] David Gale, The game of Hex and the Brouwer fixed-point theorem, *Amer. Math. Monthly* **86**(10) 818–827, 1979.

[9] Martin Gardner, Mathematical Games, *Scientific American* **197** Jul. 145–150, Aug. 120–127, Oct. 130–138; also in M. Gardner, *The Scientific American Book of Mathematical Puzzles and Diversions*, Simon and Schuster, New York, 1959.

[10] Ryan Hayward, Berge and the Art of Hex, manuscript, www.cs.ualberta.ca/~hayward/publications.html, 2003.

[11] Ryan Hayward, A note on domination in Hex, www.cs.ualberta.ca/~hayward/publications.html, 2003.

[12] Ryan Hayward and Jack van Rijswijck, Hex and Combinatorics, to appear in *Disc. Math.*, www.cs.ualberta.ca/~hayward/publications.html, 2005.

[13] Ryan Hayward, Yngvi Björnsson, Michael Johanson, Morgan Kan, and Jack van Rijswijck, Solving 7 × 7 Hex with domination, fill-in, and virtual connections, *Theoretical Computer Science* **349** 123–139, 2005.

[14] Piet Hein, Vil de laere Polygon?, *Politiken*, Copenhagen, Dec. 26, 1942.

[15] Alfred Lehman, A solution of the Shannon switching game, *J. SIAM* **12** 687–725, 1964.

[16] Thomas Maarup, Hex, http://maarup.net/thomas/hex/, 2005.

[17] Jack van Rijswijck, Computer Hex: are Bees better than Fruitflies? MSc. thesis, U. Alberta, Edmonton, 2000. www.cs.ualberta.ca/~javhar

[18] Stefan Reisch, Hex ist PSPACE-vollständig, *Acta Informatica* **15** 167–191, 1981.

[19] Craige Schensted and Charles Titus, *Mudcrack Y and Poly-Y*, Neo Press, Peaks Island, 1975.

[20] Sue Whitesides, A method for solving certain graph recognition and optimization problems with applications to perfect graphs, in *Topics on Perfect Graphs*, Claude Berge and Vašek Chvátal eds., *Annals of Disc. Math.* **21**, North Holland, Amsterdam, 1984.

[21] Yohei Yamasaki, Theory of division games, *Pub. Res. Inst. Math. Sci.* **14** 337–358, 1978.

Yngvi Björnsson
School of Computer Science
Reykjavik University, Iceland
e-mail: yngvi@ru.is

Ryan Hayward, Michael Johanson and Jack van Rijswijck
Dept. of Comp. Science
University of Alberta, Canada
e-mail: hayward@cs.ualberta.ca
e-mail: mikej@cs.ualberta.ca
e-mail: javhar@cs.ualberta.ca

Ratios of Some Domination Parameters in Graphs and Claw-free Graphs

Mostafa Blidia, Mustapha Chellali and Odile Favaron

Abstract. In the class of all graphs and the class of claw-free graphs, we give exact bounds on all the ratios of two graph parameters among the domination number, the total domination number, the paired domination number, the double domination number and the independence number. We summarize the old and new results in a table and give for each bound examples of extremal families.

Mathematics Subject Classification (2000). 05C.

Keywords. Total, paired, domination, independence, claw-free graphs.

1. Introduction

We consider finite simple graphs G with vertex-set $V(G)$, edge-set $E(G)$ and minimum degree $\delta(G)$ (we use V, E and δ if there is no ambiguity). The neighborhood of a vertex x is $N(x) = \{y \in V(G); xy \in E(G)\}$ and its closed neighborhood is $N[x] = N(x) \cup \{x\}$. If S is a subset of $V(G)$ then $N(S) = \cup_{x \in S} N(x)$, $N[S] = \cup_{x \in S} N[x]$ and the subgraph induced by S in G is denoted $G[S]$. A *dominating set* of G is a subset S of V such that every vertex in $V - S$ has at least one neighbor in S, in other terms $N[S] = V$. The notion of domination is essential in the representation of many problems and since its first formal definition by Berge [2], many variants have been introduced to deal with different kinds of problems (see [16]). We give below some of them. A subset $S \subset V$ is a *total dominating set* [5] if every vertex of V has at least one neighbor in S, in other terms if S is dominating and $G[S]$ has no isolated vertex, or if $N(S) = V$. A dominating set S is a *paired dominating set* [17] if $G[S]$ admits a perfect matching. A dominating set S is an *independent dominating set* if $G[S]$ is independent. A subset $S \subset V$ is a *2-dominating set* [13] if every vertex of $V - S$ has at least two neighbors in S and a *double dominating set* [14] if S is both a 2-dominating set and a total dominating set (in [10] these sets were also introduced under the term 1-total 2-dominating sets). Since some of these sets do not exist if G has isolated vertices, we suppose in the whole paper $\delta(G) \geq 1$. We consider the following parameters associated with the previous definitions: $\gamma_t(G)$ ($\gamma_{pr}(G), \gamma_2(G), \gamma_{\times 2}(G)$ respectively) is the minimum cardinality of a total dominating set (paired dominating set, 2-dominating set, double dominating set respectively) of G. Note that $\gamma_{\times 2}(G)$ was denoted $dd(G)$ in [14] and $\gamma_{1,2}(G)$ in [10]. The minimum and maximum cardinalities of an independent dominating set of G are denoted $i(G)$ and $\beta(G)$. Of course these different notions are not independent and the corresponding parameters satisfy in every graph some obvious and well-known inequalities as

$$\gamma(G) \leq \gamma_t(G) \leq \gamma_{pr}(G), \ \gamma_t(G) \leq \gamma_{\times 2}(G),$$

$$\gamma(G) \leq \gamma_2(G) \leq \gamma_{\times 2}(G), \ \gamma(G) \leq i(G) \leq \beta(G).$$

When two graph parameters satisfy $a \leq b$, one can wonder how large b can be with respect to a. This is often done by studying the difference $b - a$ but the ratio b/a is more informative. For instance both differences $\beta(G) - \gamma(G)$ and $\gamma_t(G) - \gamma(G)$ can be arbitrarily large. But $\gamma_t(G)/\gamma(G)$ cannot since $\gamma_t(G) \leq 2\gamma(G)$ for every G while $\beta(G)/\gamma(G)$ is not bounded. In this paper we are interested in exact upper bounds, if any, on all the ratios $b(G)/a(G)$ where a and b are each of the five parameters $\gamma, \gamma_t, \gamma_{pr}, \gamma_{\times 2}$ and β. Some of these bounds were already known. We determine the other ones and construct families of graphs showing the sharpness of all the bounds.

The class of claw-free graphs, i.e., graphs containing no claw $K_{1,3}$ as an induced subgraph, has particularly interesting properties in domination theory. It is known that every claw-free graph satisfies $\gamma(G) = i(G)$ [1] and $\beta(G) \leq 2i(G)$ [19]. This means that the ratios i/γ and β/i, which can be arbitrarily large in general graphs, are bounded respectively by 1 and 2 in the class of claw-free graphs. We extend these results by looking for upper bounds on the ratios of any pair of the five above mentioned parameters in claw-free graphs. Note that as $\beta(G)$ can be determined in polynomial time in this class [18], the inequalities $\beta(G)/2 \leq \gamma(G) \leq \beta(G)$, $2\beta(G)/3 \leq \gamma_t(G) \leq \gamma_{pr}(G) \leq 2\beta(G)$, $\beta(G) \leq \gamma_{\times 2}(G) \leq 2\beta(G)$ (cf. Table 1) give polynomial approximations for the other four parameters of claw-free graphs.

Other results on domination parameters in claw-free graphs and on the ratio of domination parameters in particular classes of graphs can be found in [3, 4, 6, 7, 8, 9, 10, 11, 12, 14, 15, 17, 19, 20, 21].

2. Some families of extremal graphs

We describe here some families of connected graphs which will provide extremal examples in Section 3. Each family depends on one or two arbitrarily large parameters which proves that the bounds given in the table below cannot be improved even if we suppose each component of the graphs large enough.

Family B_k (Figure 1.a)

The claw-free graph B_k (Figure 1.a) is constructed from a clique K_k of vertex-set $\{x_1, \ldots, x_k\}$ by adding k disjoint triangles $a_i b_i c_i$ and the k edges $a_i x_i$ for $1 \leq i \leq k$. It is easy to check that $\gamma(B_k) = i(B_k) = k$, $\beta(B_k) = k+1$, $\gamma_t(B_k) = \gamma_{pr}(B_k) = 2k$ and $\gamma_2(B_k) = \gamma_{\times 2}(B_k) = 2k+1$.

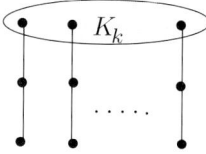
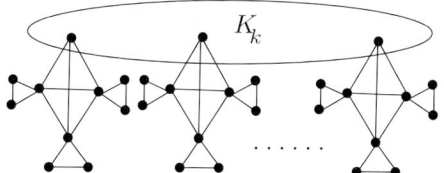

Figure 1.a: N_k　　　　　　　　　Figure 1.b: L_k

Family L_k (Figure 1.b)

The claw-free graph L_k (Figure 1.b) is constructed by attaching at each vertex a of a clique K_k a pendant copy of a graph of vertex set $\{a\} \cup_{1 \leq i \leq 3} \{b_i, c_i, d_i\}$ where $\{a, b_1, b_2, b_3\}$ induces a clique and $\{b_i, c_i, d_i\}$ a triangle for $1 \leq i \leq 3$. The graph L_k satisfies $\gamma_t(L_k) = 3k$ and $\gamma_{pr}(L_k) = 4k$.

Family J_k (Figure 2 for $k = 5$)

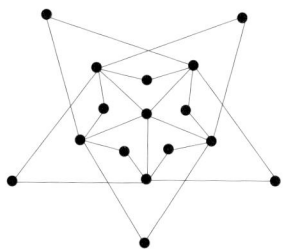

Figure 2: J_5

For $k \geq 4$, the graph J_k is constructed from a star $K_{1,k}$ with center a and leaves $v_1, \ldots, v_k$ by adding $\binom{k}{2}$ new vertices $x_{i,j}$ and the edges $v_i x_{i,j}$, $v_j x_{i,j}$ for $1 \leq i < j \leq k$. For J_k, $\{a, v_1, \ldots, v_{k-1}\}$ is a minimum dominating set which is total. Moreover all the total dominating sets of order k are of this type and are not double. Hence $\gamma(J_k) = \gamma_t(J_k) = k$ and $\gamma_{\times 2}(J_k) \geq k+1$. Since the set $\{a, v_1, \ldots, v_k\}$ is a double dominating set, $\gamma_{\times 2}(J_k) = k+1$. Finally let D be any dominating set of J_k and $\beta(D)$ the order of a maximum independent set of D. If $|D \cap \{v_1, \ldots, v_k\}| \geq k-1$, then $\beta(D) \geq k-1$. Otherwise let $W = \{v_1, \ldots, v_k\} - D$ with $|W| \geq 2$. To dominate J_k, the set D contains the set X of order $|W|(|W|-1)/2$ of all the vertices x_{ij} such that $\{v_i, v_j\} \subset W$. Hence again $\beta(D) \geq |D \cap \{v_1, \ldots, v_k\}| + |X| = k - |W| + |W|(|W|-1)/2 \geq k-1$. Therefore every dominating set of G contains at least $k-1$ independent vertices. This implies that every paired dominating set has order at least $2(k-1)$. Since $\{v_1, x_{1,2}, v_2, x_{2,3}, \ldots, v_{k-1}, x_{k-1,k}\}$ is a paired dominating set, $\gamma_{pr}(J_k) = 2k - 2$. Note the graph J_k contains claws.

Family H_k (Figure 3)

Let H_k be the claw-free graph constructed from k disjoint copies of the graph H of Figure 3 by adding the edges $c_i a_{i+1}$ and $d_i b_{i+1}$ for $1 \le i \le k-1$. For this graph, $\gamma(H_k) = 3k$ and $\gamma_t(H_k) = \gamma_{pr}(H_k) = \gamma_{\times 2}(H_k) = 4k$.

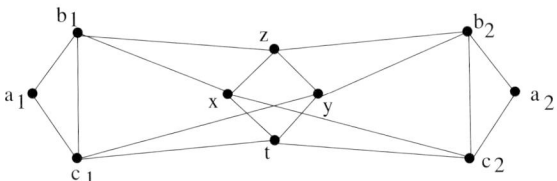

Figure 3: H

Family $M_{k,p}$ (Figure 4.a)

For $p \ge 2$, the graph $M_{k,p}$ is obtained by attaching p pendant edges at each vertex of a clique K_{2k}. Clearly $M_{k,p}$ contains claws and satisfies $\gamma(M_{k,p}) = \gamma_t(M_{k,p}) = \gamma_{pr}(M_{k,p}) = 2k$ and $\gamma_{\times 2}(M_{k,p}) = 2kp$.

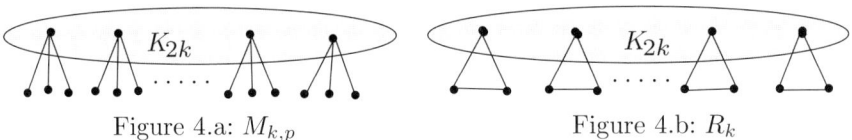

Figure 4.a: $M_{k,p}$ Figure 4.b: R_k

Family R_k (Figure 4.b)

The claw-free graph R_k is obtained by attaching a pendant triangle at each vertex of a clique K_{2k}. This graph satisfies $\gamma_t(R_k) = \gamma_{pr}(R_k) = 2k$ and $\gamma_{\times 2}(R_k) = 4k$.

The last five families could be gathered into one using two parameters. However the notation would be formally heavy and we define them separately to facilitate the reading.

Family Q_k (Figure 5.a)

The claw-free graph Q_k is obtained from a cycle $C_{3k} = a_1 b_1 c_1 a_2 b_2 c_2 \cdots a_k b_k c_k$ by adding $2k$ new vertices, d_i joined to a_i and b_i and e_i joined to b_i and c_i for $1 \le i \le k$. In Q_k, $\{b_1, \ldots, b_k\}$ is a minimum dominating set and $\gamma(Q_k) = k$. Moreover each double dominating set D must contain at least three vertices in each set $\{a_i, b_i, c_i, d_i, e_i\}$ (for if $d_i \notin D$, then $\{a_i, b_i\} \subset D$ and c_i or e_i is in D). Therefore $V(C_{3k})$ is a minimum double dominating set and $\gamma_{\times 2}(Q_k) = 3k$. Similarly, each total or paired dominating set contains at least two vertices in each set $\{a_i, b_i, c_i, d_i, e_i\}$ and $\gamma_t(Q_k) = \gamma_{pr}(Q_k) = 2k$.

Family S_k (Figure 5.b)

The claw-free graph S_k is obtained from a cycle $C_{2k} = a_1 a_2 \cdots a_{2k}$ by adding $2k$ new vertices c_i joined to a_i and a_{i+1} (mod $2k$) for $1 \leq i \leq 2k$. Clearly $\gamma(S_k) = k$ and $\beta(S_k) = 2k$.

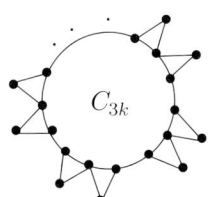
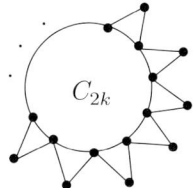
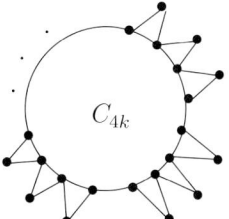

Figure 5a: Q_k Figure 5b: S_k Figure 5c: T_k

Family T_k (Figure 5.c)

The claw-free graph T_k is obtained from a cycle $C_{4k} = a_1 b_1 c_1 d_1 a_2 b_2 c_2 d_2 \cdots a_k b_k c_k d_k$ by adding $3k$ new vertices e_i joined to a_i and b_i, f_i joined to b_i and c_i, g_i joined to c_i and d_i for $1 \leq i \leq k$. Clearly $\beta(T_k) = 3k$. Every dominating set of T_k contains at least two vertices in each set $\{a_i, b_i, c_i, d_i, e_i, f_i, g_i\}$. Since $\{b_1, c_1, \ldots, b_k, c_k\}$ is a paired dominating set of T_k, $\gamma_t(T_k) = \gamma_{pr}(T_k) = 2k$.

Family W_k (Figure 5.d)

The claw-free graph W_k is obtained from a cycle $C_{3k} = a_1 b_1 c_1 a_2 b_2 c_2 \cdots a_k b_k c_k$ by adding k new vertices d_i joined to a_i and b_i for $1 \leq i \leq k$. Clearly $\{c_1, d_1, \ldots, c_k, d_k\}$ is a maximum independent set and $\beta(W_k) = 2k$. Every 2-dominating set contains at least two vertices in each set $\{a_i, b_i, c_i, d_i\}$ and since $\{a_1, b_1, \ldots, a_k, b_k\}$ is a double dominating set which is total, $\gamma_2(W_k) = \gamma_{\times 2}(W_k) = 2k$.

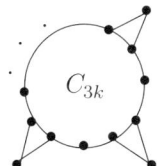
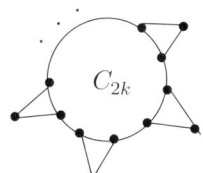

Figure 5d: W_k Figure 5e: Z_k

Family Z_k (Figure 5.e)

The claw-free graph Z_k is obtained from a cycle $C_{2k} = a_1 b_1 a_2 b_2 \cdots a_k b_k$ by adding k new vertices c_i joined to a_i and b_i for $1 \leq i \leq k$. Clearly $\gamma(Z_k) = \beta(Z_k) = k$ and $\gamma_{\times 2}(Z_k) = 2k$.

3. Ratios of different parameters

	γ	γ_t	γ_{pr}	$\gamma_{\times 2}$	β
γ		2 / 2	2 [17] / 2	∞ / 3 [Cor 3.9]	∞ / 2 [19]
γ_t	1 / 1		2 [17] / 4/3 [6]	∞ / 2 [Th 3.3]	∞ / 3/2 [Th 3.4]
γ_{pr}	1 / 1	1 / 1		∞ / 2 [Th 3.3]	∞ / 3/2 [Th 3.4]
$\gamma_{\times 2}$	1 / 3/4 [10]	1 / 1	2 [Cor 3.2] / 1 [4]		∞ / 1 [Th 3.5]
β	1 / 1	2 / 2	2 / 2	2 [Cor 3.7] / 2 [Cor 3.7]	

Table 1

We present the results of this section in a table the entries of which are the five parameters $a_1 = \gamma$, $a_2 = \gamma_t$, $a_3 = \gamma_{pr}$, $a_4 = \gamma_{\times 2}$ and $a_5 = \beta$. The square (i, j) corresponding to the row i and the column j consists of two subsquares respectively containing an upper bound on the ratio a_j/a_i in the class of all graphs for the upper subsquare and in the class of claw-free graphs for the lower one. For instance the two values 2 and 1 contained in the square (4,3) indicate that $\gamma_{pr}(G)/\gamma_{\times 2}(G)$ is at most 2 for every G and at most 1 if G is claw-free. All the bounds are sharp or asymptotically sharp. When a bound is not the direct consequence of the obvious inequalities between the parameters given in the introduction, we refer to the article where it was first established or to a theorem in this paper. The following comments on the squares of the table contain the justification of the bound given in each square and arbitrarily large examples of extremal graphs. Moreover in each extremal example, the domination parameters involved in the corresponding inequality can take arbitrarily large values.

Squares (2,1),(3,1),(3,2),(4,2): γ/γ_t, γ/γ_{pr}, γ_t/γ_{pr}, $\gamma_t/\gamma_{\times 2}$.

Since in every graph with $\delta \geq 1$, $\gamma(G) \leq \gamma_t(G) \leq \gamma_{pr}(G)$ and $\gamma_t(G) \leq \gamma_{\times 2}(G)$, these four ratios are bounded by 1. The bound 1 is sharp even in claw-free graphs. This can be seen for the first three ratios on the corona $G = K_{2k} \circ K_1$ obtained by adding a pendant edge at each vertex of a clique K_{2k}. Indeed $\gamma(G) = \gamma_t(G) = \gamma_{pr}(G) = 2k$. For the fourth ratio, the graphs H_k (Figure 3) show that the bound 1 on $\gamma_t/\gamma_{\times 2}$ is sharp even in claw-free graphs.

Squares (1,2),(1,3),(2,3): γ_t/γ, γ_{pr}/γ, γ_{pr}/γ_t.

As observed in [17], every graph with $\delta \geq 1$ satisfies $\gamma_t(G) \leq \gamma_{pr}(G) \leq 2\gamma(G)$. Hence $\gamma_t(G)/\gamma(G) \leq 2$ and $\gamma_{pr}(G)/\gamma(G) \leq 2$ for all graphs. These two bounds are sharp even for claw-free graphs as shown by the graph B_k of Figure 1.a or the graph Q_k of Figure 5.a.

It is also proved in [17] that every graph G with $\delta \geq 1$ satisfies $\gamma_{pr}(G) \leq 2\gamma_t(G) - 2$. Hence $\gamma_{pr}(G)/\gamma_t(G) < 2$ for all graphs. The subdivided star SS_k obtained by subdividing each edge of a star $K_{1,k}$ by exactly one vertex satisfies $\gamma_t(SS_k) = k+1$ and $\gamma_{pr}(SS_k) = 2k$, which proves that the bound 2 on $\gamma_{pr}(G)/\gamma_t(G)$ is asymptotically sharp. However this example contains claws and the upper bound is lower in the class of claw-free graphs. Brigham and Dutton proved in [6] that every claw-free graph with $\delta \geq 1$ satisfies $\gamma_{pr}(G)/\gamma_t(G) \leq 4/3$. The graph L_k of Figure 1.b shows that this bound is sharp.

Square (4,1): $\gamma/\gamma_{\times 2}$.

For every graph with $\delta \geq 1$, $\gamma(G) \leq \gamma_{\times 2}(G)$. More precisely it has been observed in [14, 10] that $\gamma(G) \leq \gamma_{\times 2}(G) - 1$. Hence $\gamma(G)/\gamma_{\times 2}(G) < 1$. The graph J_k of Figure 2 shows that the bound 1 is asymptotically sharp.

The graph J_k contains claws. It is proved in [10] that every claw-free graph with $\delta \geq 1$ satisfies $\gamma(G)/\gamma_{\times 2}(G) \leq 3/4$ (actually this inequality holds in the larger class of claw-free block graphs, that are graphs for which each block is claw-free). The claw-free graph H_k of Figure 3 shows that this bound is sharp.

Square (4,3): $\gamma_{pr}/\gamma_{\times 2}$.

Theorem 3.1. *Every graph G with $\delta \geq 1$ satisfies $\gamma_{pr}(G) \leq \max\{2\gamma_{\times 2}(G) - 4, 2\}$.*

Proof. Let S be a $\gamma_{\times 2}(G)$-set, M a maximum matching of S, $B = V(M)$ and $A = S - B$. If $A = \emptyset$, then S is a paired dominating set of G and $\gamma_{pr}(G) \leq \gamma_{\times 2}(G)$. Hence $\gamma_{pr}(G) \leq 2$ if $\gamma_{\times 2}(G) = 2$ and $\gamma_{pr}(G) \leq 2\gamma_{\times 2}(G) - 4$ if $\gamma_{\times 2}(G) \geq 4$. Assume henceforth $A \neq \emptyset$. Since the matching M is maximum in S the set A is independent and since S is a total dominating set of G each vertex of A has at least one neighbor in B. If $N(A) \subset N(B)$ then B is a paired dominating set of G of order $|S| - |A| \leq \gamma_{\times 2}(G) - 1$. As above $\gamma_{pr}(G) \leq \max\{2\gamma_{\times 2}(G) - 4, 2\}$. Assume now $N(A)$ is not contained in $N(B)$ and let $A' = N(A) - N(B) = V(G) - S - N(B)$. Since S is a 2-dominating set of G each vertex x of A' has at least two neighbors in A. Therefore for every $u \in A$, the set $A - \{u\}$ dominates A' and thus, if D is a minimum subset of A dominating A', $|D| \leq |A| - 1$. Let M' be a maximum matching between D and A'. Then $|V(M')| \leq 2|D| \leq 2|A| - 2$. If a vertex x of A' is not dominated by $V(M')$, there exists a vertex $y \in D - V(M')$ adjacent to x and $M' \cup xy$ contradicts the choice of M'. Hence $B \cup V(M')$ is a paired dominating set of G. Therefore $\gamma_{pr}(G) \leq |B| + |V(M')| \leq |B| + 2|A| - 2 = 2|S| - |B| - 2$. But since $\delta(S) \geq 1$, $|B| \geq 2$. Hence $\gamma_{pr}(G) \leq 2|S| - 4$ which completes the proof. $\square$

Corollary 3.2. *Every graph G with $\delta \geq 1$ satisfies $\gamma_{pr}/\gamma_{\times 2} < 2$.*

Proof. Obvious from Theorem 3.1 since $\gamma_{pr}(G) = \gamma_{\times 2}$ if $\gamma_{\times 2} = 2$ and $\gamma_{pr}(G) < 2\gamma_{\times 2}(G)$ otherwise. $\square$

The graph J_k shown in Figure 2 satisfies $\gamma_{pr}(J_k) = 2k - 2$ and $\gamma_{\times 2}(J_k) = k + 1$. Hence J_k is an extremal graph for Theorem 3.1 and shows that the bound 2 on $\gamma_{pr}/\gamma_{\times 2}$ is asymptotically sharp in the class of all graphs.

For claw-free graphs with $\delta \geq 1$, the bound 2 can be lowered and the authors of [4] proved that $\gamma_{pr}/\gamma_{\times 2} \leq 1$. As for the ratio $\gamma_t/\gamma_{\times 2}$, the graphs H_k of Figure 3 show that the bound 1 is sharp.

Squares (1,4),(2,4),(3,4): $\gamma_{\times 2}/\gamma$, $\gamma_{\times 2}/\gamma_t$, $\gamma_{\times 2}/\gamma_{pr}$.

In general graphs these three ratios can be arbitrarily large as can be seen on the graph $M_{k,p}$ of Figure 4.a for which $\gamma(M_{k,p}) = \gamma_t(M_{k,p}) = \gamma_{pr}(M_{k,p}) = 2k$ and $\gamma_{\times 2}(M_{k,p}) = 2pk$.

However Corollary 3.9, which will be given later, shows that $\gamma_{\times 2}(G)/\gamma(G) \leq 3$ in every claw-free graph with $\delta \geq 1$ and the following theorem shows that the other two ratios $\gamma_{\times 2}/\gamma_t$ and $\gamma_{\times 2}/\gamma_{pr}$ are also bounded in the class of claw-free graphs.

Theorem 3.3. *Every claw-free graph G with $\delta \geq 1$ satisfies $\gamma_{\times 2}(G)/\gamma_{pr}(G) \leq \gamma_{\times 2}(G)/\gamma_t(G) \leq 2$.*

Proof. Let S be a $\gamma_t(G)$-set and $u \in S$. The external S-private neighborhood epn(S, u) of u is the set of all the vertices of $V - S$ having u as their unique neighbor in S. Since G is claw-free and since u has at least one neighbor in S, epn(S, u) is either empty or induces a clique in G. Let $f(u) = \{u\}$ if epn$(S, u) = \emptyset$, $f(u) = \{u, u'\}$ with $u' \in$ epn(S, u) otherwise. Then $T = \cup_{u \in S} f(u)$ is a double dominating set of G such that $|T| \leq 2|S|$. Hence $\gamma_{\times 2}(G) \leq 2\gamma_t(G) \leq 2\gamma_{pr}(G)$. $\square$

The graph Q_k of Figure 5.a for which $\gamma(Q_k) = k$ and $\gamma_{\times 2}(Q_k) = 3k$ shows that the bound 3 in Corollary 3.9 is sharp.

The graph R_k of Figure 4.b for which $\gamma_t(R_k) = \gamma_{pr}(R_k) = 2k$ and $\gamma_{\times 2}(R_k) = 4k$ shows that the bound 2 in Theorem 3.3 is sharp for the two ratios.

Squares (1,5),(2,5),(3,5): β/γ, β/γ_t, β/γ_{pr}.

The star $K_{1,k}$ for which $\gamma(K_{1,k}) = 1$, $\gamma_t(K_{1,k}) = \gamma_{pr}(K_{1,k}) = 2$ and $\beta(K_{1,k}) = k$ shows that β/γ, β/γ_t and β/γ_{pr} can be arbitrarily large in graphs with claws.

For claw-free graphs, the bound $\beta(G)/\gamma(G) \leq 2$ was given by Sumner in [19] without proof. For the sake of completeness we give here a short proof of this property. Let G be a claw-free graph, S a $\beta(G)$-set and $D = \{x_1, \ldots, x_\gamma\}$ a $\gamma(G)$-set. Since D is dominating, $V(G) = \cup_{x \in D} N[x]$ and since G is claw-free, $|N[x] \cap S| \leq 2$ for every x in D. Hence we have

$$\beta(G) = |S| = |S \cap (\cup_{x \in D} N[x])| \leq \sum_{x \in D} |S \cap N[x]| \leq 2|D| = 2\gamma(G).$$

The graph S_k of Figure 5.b shows that the bound 2 on β/γ is sharp in claw-free graphs.

In the following theorem we show that β/γ_t and β/γ_{pr} also are bounded in claw-free graphs.

Theorem 3.4. *Every claw-free graph G with $\delta \geq 1$ satisfies*

$$\beta(G)/\gamma_{pr}(G) \leq \beta(G)/\gamma_t(G) \leq 3/2.$$

Proof. Let G be a claw-free graph with $\delta \geq 1$, S a $\gamma_t(G)$-set, I a $\beta(G)$-set and $M = \{x_1 y_1, \ldots, x_k y_k\}$ a maximum matching of S. Since G is claw-free, for every edge $x_i y_i$ the neighborhood $N(\{x_i, y_i\})$ contains at most three independent vertices. Hence $|I \cap N(\{x_i, y_i\})| \leq 3$. Moreover, since every vertex $w \in S - V(M)$ has at least one neighbor, say x_i, in $V(M)$, the set I contains at most one vertex from $N(w) - N(x_i)$. It follows that

$$|I| \leq 3|M| + (|S| - 2|M|) = |S| + |M| \leq |S| + |S|/2 = 3|S|/2. \qquad \square$$

The graph T_k of Figure 5.c shows that the bound $3/2$ on $\beta(G)/\gamma_{pr}(G)$ and $\beta(G)/\gamma_t(G)$ is sharp.

Square (4,5): $\beta/\gamma_{\times 2}$.

The graph G consisting of k triangles $a_i bc$ sharing the edges bc satisfies $\beta(G) = k$, $\gamma_{\times 2}(G) = 2$ and shows that the ratio $\beta/\gamma_{\times 2}$ is not bounded in general graphs.

This is not true for claw-free graphs as shown by the following theorem.

Theorem 3.5. *Every claw-free graph G with $\delta \geq 1$ satisfies*

$$\beta(G)/\gamma_{\times 2}(G) \leq \beta(G)/\gamma_2(G) \leq 1.$$

Proof. Let G be a claw-free graph with $\delta \geq 1$, D a $\gamma_2(G)$-set and I a $\beta(G)$-set. The set D is not strictly contained in I for otherwise the vertices of $I - D$ would not be dominated by D. If $I \subset D$ we are done. Hence we assume $I - D \neq \emptyset$ and $D - I \neq \emptyset$. Every vertex of $I - D$ has at least two neighbors in $D - I$ since D is a 2-dominating set. Every vertex of $D - I$ has at most two neighbors in $I - D$ since G is claw-free. Therefore the number $e(I - D, D - I)$ of edges of G between $I - D$ and $D - I$ satisfies $2|I - D| \leq e(I - D, D - I) \leq 2|D - I|$ which leads to $\beta(G) \leq \gamma_2(G) \leq \gamma_{\times 2}(G)$. □

The graph W_k of Figure 5.d shows that the bound 1 on $\beta/\gamma_{\times 2}$ and β/γ_2 is sharp.

Squares (5,1),(5,2),(5,3): γ/β, γ_t/β, γ_{pr}/β.

Since $\beta(G) \geq \gamma(G)$ and by the bounds in squares (1,2) and (1,3), we have $\gamma(G)/\beta(G) \leq 1$, $\gamma_t(G)/\beta(G) \leq 2$ and $\gamma_{pr}(G)/\beta(G) \leq 2$ in every graph G. The claw-free graphs Z_k of Figure 5.e, and more generally the claw-free well-covered graphs, satisfy $\beta(G)/\gamma(G) = 1$ (well-covered means that $i(G) = \beta(G)$). For the last two ratios, the graph B_k of Figure 1.a shows that the bound 2 is asymptotically sharp even in claw-free graphs.

Square (5,4): $\gamma_{\times 2}/\beta$.

Theorem 3.6. *Every graph G with $\delta \geq 1$ satisfies $\gamma_{\times 2}(G) \leq i(G) + \beta(G)$.*

Proof. Let S be an $i(G)$-set and S' a maximum independent set of $G[V(G) - S]$. Since S' is independent in G, $|S'| \leq \beta(G)$. Let $A = N(S') \cap S$ and $B = S - A$. If $B = \emptyset$ then the graph induced by $S \cup S'$ has no isolated vertex and so $S \cup S'$ double dominates G. Therefore $\gamma_{\times 2}(G) \leq |S \cup S'| \leq i(G) + \beta(G)$. Assume now that $B \neq \emptyset$. Let C be a smallest set of $V - S$ that dominates B. Such a set exists since G has no isolated vertices and $|C| \leq |B|$. Then $A \cup B \cup C \cup S'$ is a double dominating set of G. Moreover $|A| + |C| \leq |A| + |B| = i(G)$ and $B \cup S'$ is independent. Hence
$$\gamma_{\times 2}(G) \leq |A \cup B \cup C \cup S'| \leq |A \cup C| + |B \cup S'| \leq i(G) + \beta(G). \qquad \Box$$

Corollary 3.7. *Every graph G with $\delta \geq 1$ satisfies $\gamma_{\times 2}(G)/\beta(G) \leq 2$.* □

Another obvious consequence of Theorem 3.6 is the following

Corollary 3.8. *Every graph G satisfies $\gamma_2(G) \leq i(G) + \beta(G)$.* □

The graphs B_k of Figure 1.a show that Theorem 3.6 and Corollary 3.8 are sharp even for claw-free graphs. The graphs Z_k of Figure 5.e give claw-free extremal examples for Theorem 3.6 and Corollary 3.7.

Finally, since $i(G) = \gamma(G)$ and $\beta(G) \leq 2\gamma(G)$ hold for every claw-free graph [19], Theorem 3.6 also shows the following inequality, already mentioned earlier with equality for Q_k.

Corollary 3.9. *Every claw-free graph G with $\delta \geq 1$ satisfies $\gamma_{\times 2}(G)/\gamma(G) \leq 3$.* □

References

[1] R. Allen and R. Laskar, On domination and independent domination numbers of a graph. Discrete Math. 23 (2) (1978), 73–76.

[2] C. Berge, *Théorie des graphes et ses applications*. Dunod, Paris, 1958.

[3] M. Blidia, M. Chellali, T.W. Haynes and M.A. Henning, Independent and double domination in trees. Utilitas Math. (to appear).

[4] M. Chellali and T.W. Haynes, On paired and double domination in graphs. Utilitas Math. 67 (2005), 161–167.

[5] E.J. Cockayne, R.M. Dawes and S.T. Hedetniemi, Total domination in graphs. Networks 10 (1980), 211–219.

[6] R.D. Dutton and R.C. Brigham, Domination in claw-free graphs. Congr. Numer. 132 (1998), 69–75.

[7] O. Favaron, Independence and upper irredundance in claw-free graphs. Discrete Applied Math. 132 (1-3) (2001), 85–95.

[8] O. Favaron and M.A. Henning, Upper total domination in claw-free graphs. J. Graph Theory 44(2) (2003), 148–158.

[9] O. Favaron and M.A. Henning, Paired domination in claw-free cubic graphs. Graphs Combin. 20(4) (2004), 447–456.

[10] O. Favaron, M. A. Henning, J. Puech and D. Rautenbach, On domination and annihilation in graphs with claw-free blocks. Discrete Math. 231(1-3) (2001), 143–151.

[11] O. Favaron, V. Kabanov and J. Puech, The ratio of three domination parameters in some classes of claw-free graphs. JCMCC 31 (1999), 151–159.

[12] O. Favaron and D. Kratsch, Ratios of domination parameters. in *Advances in Graph Theory*, ed. V. Kulli, Vishwa International Publications, 173–182, 1991.

[13] J.F. Fink and M.S. Jacobson, On n-domination, n-dependence and forbidden subgraphs. in Y. Alavi and A.J. Schwenk, editors, Graph Theory with Applications to Algorithms and Computer Science, Wiley, 1985, 301–311.

[14] F. Harary and T.W. Haynes, Double domination in graphs. Ars Combinatoria 55 (2000), 201–213.

[15] J.H. Hattingh and M.A. Henning, The ratio of the distance irredundance and domination numbers of a graph. J. Graph Theory 18(1) (1994), 1–9.

[16] T.W. Haynes, S.T. Hedetniemi and P.J. Slater, *Fundamentals of Domination in Graphs*. Marcel Dekker, New York, 1998.

[17] T.W. Haynes and P.J. Slater, Paired-domination in graphs. Networks 32(3) (1998), 199–206.

[18] N. Sbihi, Algorithmes de recherche d'un stable de cardinalité maximum dans un graphe sans étoile. Discrete Math. 29 (1980), 53–76.

[19] D.P. Sumner, Critical concepts in domination. Discrete Math. 86 (1990), 33–46.

[20] L. Volkmann, The ratio of the irredundance and domination number of a graph. Discrete Math. 178 (1998), 221–228.

[21] V.E. Zverovich, The ratio of the irredundance number and the domination number for block-cactus graphs. J. Graph Theory 29(3), (1998) 139–149.

Mostafa Blidia and Mustapha Chellali
Département de Mathématiques
Université de Blida
B.P. 270
Blida, Algérie
e-mail: mblidia@hotmail.com
e-mail: mchellali@hotmail.com

Odile Favaron
L.R.I., URM 8623, Bât. 490
Université de Paris-Sud
F-91405-Orsay cedex, France
e-mail: of@lri.fr

Excessive Factorizations of Regular Graphs

Arrigo Bonisoli and David Cariolaro

> **Abstract.** An excessive factorization of a graph G is a minimum set $\mathcal{F}$ of 1-factors of G whose union is $E(G)$. In this paper we study excessive factorizations of regular graphs. We introduce two graph parameters related to excessive factorizations and show that their computation is NP-hard. We pose a number of questions regarding these parameters. We show that the size of an excessive factorization of a regular graph can exceed the degree of the graph by an arbitrarily large quantity. We conclude with a conjecture on the excessive factorizations of r-graphs.
>
> **Mathematics Subject Classification (2000).** Primary 05C70; Secondary 05C15.
>
> **Keywords.** 1-factor, 1-factorization, 1-factor cover, excessive factorization.

1. Introduction

In this paper all graphs are simple, finite and undirected, unless stated otherwise. Terminology and notation, not explicitly introduced here, will follow [9]. We shall occasionally say that a set S is *odd* (*even*) if its cardinality is odd (even). Accordingly, we shall say that a graph G is *odd* or *even* if $V(G)$ is odd or even, respectively. The *order* of G is $|V(G)|$. The symbol r will be always used, without further mention, to denote the *degree* of regularity of the graph under consideration. A graph G is *1-factorizable* if its edge set $E(G)$ can be partitioned into edge-disjoint 1-factors (perfect matchings). It is obvious that, if the graph G is 1-factorizable, then G is even and regular, but it is a notoriously difficult question to determine which regular graphs are 1-factorizable (we call this the 1-*Factorization Problem*). For example, the Petersen graph P is even and regular, but not 1-factorizable.

If the degree of G is sufficiently high (with respect to its order), it appears that G ought to be 1-factorizable. That is the content of the following long-standing conjecture, see [6].

The first author carried out this research within the activity of G.N.S.A.G.A. of the Italian I.N.d.A.M. with the financial support of the Italian Ministry M.I.U.R., project "Strutture geometriche, combinatoria e loro applicazioni."

Conjecture 1.1 (1-Factorization Conjecture). *Let G be a regular graph of even order $2n$ and degree $r \geq n^*$, where*

$$n^* = \begin{cases} n & \text{if } n \text{ is odd,} \\ n-1 & \text{if } n \text{ is even.} \end{cases}$$

Then G is 1-factorizable.

Notice that the bound of Conjecture 1.1 is best possible. Indeed, if n is odd, a regular graph G of order $2n$ and degree $r = n-1$ which is not 1-factorizable may be produced by taking for G two copies of K_n, and for n even a similar example of order $2n$ and degree $n-2$ can be constructed (see Fig. 1). We also notice that, to prove Conjecture 1.1, it suffices to prove that the regular graphs of order $2n$ and degree equal to n^* are 1-factorizable. Indeed every graph of larger degree can be reduced to a graph of degree exactly n^* by removing some of its 1-factors, and the 1-factorizability of the latter graph implies the 1-factorizability of the former graph. (It is easily seen, e.g., by applying Dirac's condition for Hamiltonicity[10], that every graph of order $2n$ and degree larger than n^* has a 1-factor.) This shows that the class of the n^*-regular graphs of order $2n$ is particularly important with respect to the 1-Factorization Problem and deserves special attention.

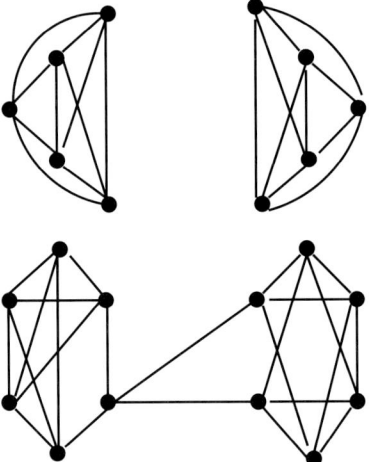

FIGURE 1. Two regular graphs of even order which show that the bound of Conjecture 1.1 is best possible.

A slightly weaker version of Conjecture 1.1 (namely the conjecture that every regular graph of order $2n$ and degree $r \geq n$ is 1-factorizable) has been proved to hold "asymptotically" (i.e., for n very large and $r \geq n + \epsilon$) by Perkovic and Reed [17]. For what concern exact results, the best result to date was obtained by Chetwynd and Hilton [7] and, independently, Niessen and Volkmann [16], who proved that even regular graphs with degree $r \geq \frac{1}{2}(\sqrt{7}-1)|V(G)| \simeq \frac{5}{6}|V(G)|$

are 1-factorizable. A partial improvement has been obtained by Cariolaro [5], who proved that, if $r \geq (0.794)|V(G)|$, then either G is 1-factorizable or G belongs to a very special (possibly empty) class of graphs. The techniques used to prove these results rely, ultimately, on the fact that, if in a graph the set of vertices of maximum degree induces a subgraph with certain properties, then the graph is Class 1 (i.e., Δ-edge colourable, where Δ is the maximum degree). It does not seem probable, at the present state of understanding, that a significant improvement upon the above results could be obtained using the same technique. Therefore, in order to solve Conjecture 1.1, new ideas need to be developed.

Partially motivated by the necessity of finding new ideas and approaches for the 1-Factorization Problem, we shall introduce a generalization of the concept of 1-factorization, called *excessive factorization*. Informally speaking an excessive factorization is a minimum "cover" of the edge set of a graph by a set of (not necessarily disjoint) 1-factors. Thus, if G is 1-factorizable, an excessive factorization is nothing else than a 1-factorization, but if G is not 1-factorizable, then an excessive factorization of G (if it exists) provides an answer to the problem: "Find a minimum set of 1-factors which cover all the edges of G"[1].

To give an example of a possible (admittingly recreational) application of the idea of excessive factorization, suppose that a tennis tournament with $2n$ participants is being organized and that there is a fixed set M of pairs of players that are supposed to play against each other during the tournament. To allow sufficient time for rest every player can play at most one match per day. However (as often happens in amateur tournaments) the director of the tournament wants to ensure that every player plays exactly one match per day, so that no player is left aside during any day of the tournament. In order to ensure that the tournament is run at the above conditions, the director is willing, if necessary, to let the same pair of players in M play against each other more than once during the tournament. It is natural to assume that the director of the tournament will be willing to run the tournament in the minimum possible number of days.

It is easy to see, by considering the graph G having the $2n$ players as vertices, with two vertices joined by an edge if and only if the corresponding pair of players is in M, that the above problem is equivalent to the problem of finding a minimum set of 1-factors which covers G, i.e., an excessive factorization of G.

Problems of this kind constitute a further motivation for the introduction and study of the concept of excessive factorization which forms the subject of this paper.

---

[1]The term "excessive factorization" is borrowed from [3], where it was used with a slightly different meaning, namely to denote a (inclusionwise) minimal family of 1-factors of K_{2n} covering all the edges of K_{2n} and consisting of precisely $2n$ 1-factors. Such a family is called an *overfull set* of 1-factors by Wallis in [21]

2. Preliminary results and definitions

Let G be a graph and let $\mathcal{F}$ be a set of 1-factors of G. We say that $\mathcal{F}$ *covers* G if
$$\bigcup\{F \mid F \in \mathcal{F}\} = E(G).$$
Although we shall be mainly interested in minimum covers, it will be convenient to have a more general notion of cover. Therefore we introduce the following definition.

Definition 2.1. A *1-factor cover* of G is a set $\mathcal{F}$ of 1-factors of G which covers G.

If $\mathcal{F}$ is a 1-factor cover, and $|\mathcal{F}| = p$, we say that $\mathcal{F}$ has *size* p. We also say that $\mathcal{F}$ is a *minimal* 1-factor cover if $\mathcal{F}$ does not contain properly any 1-factor cover. Bonisoli [3] considered minimal 1-factor covers of the complete graph of order $2n$ and consisting of precisely $2n$ 1-factors. In this paper we shall be mainly concerned with regular graphs and 1-factor covers of minimum size. Therefore we pose the following definition.

Definition 2.2. An excessive factorization of G is a 1-factor cover of G of minimum size.

An obvious question is: "Which graphs admit an excessive factorization?". This question is clearly equivalent to the question: "Which graphs admit a 1-factor cover?" and it is easy to see that such graphs are precisely those for which every edge belongs to (at least) one 1-factor of the graph. In the literature (see, e.g., [15]) connected graphs with this property are usually called *matching covered* or *1-extendable*. The restriction that G is connected is here of no particular significance and we shall use the term "1-extendable" with a slightly more general meaning, namely to denote a (not necessarily connected) graph which admits an excessive factorization.

It will be convenient to have at our disposal a notation for the size of an excessive factorization of a 1-extendable graph G. Accordingly, we pose the following definition.

Definition 2.3. Let G be a 1-extendable graph. The *excessive index* of G is defined as the size of an excessive factorization of G and denoted by $\chi'_e(G)$.

By defining $\chi'_e(G) = \infty$ for all those graphs which are not 1-extendable, we have now defined the parameter χ'_e on all graphs[2].

The question arises to compute $\chi'_e(G)$ for any graph G. It is obvious that $\chi'_e(G) \geq \Delta(G)$, where $\Delta(G)$ is the maximum degree of G. Moreover, if G is regular, then the sign of equality holds in the previous inequality if and only if G is 1-factorizable. This shows in particular that the computation of $\chi'_e(G)$ is at least

---

[2] If we assume the convention that $\min \emptyset = \infty$ then the above definition can be made in a unified way by letting
$$\chi'_e(G) = \min\{|\mathcal{F}| \mid \mathcal{F} \text{ is a 1-factor cover of } G\}.$$

as hard as the decision problem: "Is G 1-factorizable?", which is known to be NP-hard (see [14]). We are therefore led to investigate $\chi'_e(G)$ only for specific classes of graphs G. In this paper we shall only be considering (even) regular graphs.

Since the difference $\chi'_e(G) - r$ is nonnegative for any even r-regular graph G and null if and only if G is 1-factorizable, it is natural to take such difference as a measure of "how far is G from being 1-factorizable". This prompts us to introduce a new parameter.

Definition 2.4. Let G be an even regular graph with degree r. The *excessive class*[3] of G is defined as
$$\mathrm{exc}(G) = \chi'_e(G) - r.$$

If a graph G satisfies $\mathrm{exc}(G) = k$, we shall say that G is k-*excessive*. Thus the 0-excessive regular graphs are precisely those which are 1-factorizable. It is obvious from the above definition that the computation of $\mathrm{exc}(G)$ is equivalent to the computation of $\chi'_e(G)$ (given that we know the degree of G). Therefore, to compute $\mathrm{exc}(G)$ is NP-hard in general. More precisely, to decide whether $\mathrm{exc}(G) = 0$ is NP-complete for arbitrary graphs (and remains NP-complete even when restricted to cubic graphs, see Holyer [14]). The complexity of solving the decision problem $\mathrm{exc}(G) \leq k$, for $k > 0$, is, to the best of our knowledge, unknown. This is obviously equivalent to the problem of determining whether G has a 1-factor cover of size at most $r + k$, and we suspect that, at least for $k = 1$, this problem is NP-complete. We leave this as an open problem.

Problem 1: Determine the complexity of finding whether $\mathrm{exc}(G) \leq k$ for an arbitrary regular even graph G, for any constant $k > 0$.

The inequality $\chi'_e(G) \geq r$ can be slightly improved upon. Indeed, it is easy to see that, if G has a 1-factor cover $\mathcal{F}$ of size p, then G has also a p-edge colouring. (An easy way to obtain this is by labelling the edges of the 1-factors in $\mathcal{F}$ with integers $1, 2, \ldots, p$ and then by giving to each edge as colour the minimum of the labels of the 1-factors of $\mathcal{F}$ which contain it.) Therefore we have that $\chi'_e(G) \geq \chi'(G)$, where $\chi'(G)$ is the chromatic index of G.

Conjecture 1.1, when stated in terms of the excessive class, says that any regular graph G of order $2n$ has excessive class 0 if $r \geq n^*$. Having at our disposal the concept of excessive class, it is natural to formulate the following weaker form of Conjecture 1.1.

Conjecture 2.5. *Let $k > 0$ be an integer. Let G be a regular graph of order $2n$ and degree $r \geq n^*$, where n^* is as in Conjecture 1.1. Then $\mathrm{exc}(G) \leq k$.*

Unfortunately, even Conjecture 2.5 seems to be very hard (however large is the integer k), and one is tempted to go one step further and ask whether the constant k could be replaced by a function $\theta(G)$ (possibly depending on the order or other graph parameters of G). We state this as a problem.

---

[3]This definition can be generalized to the case that G is non-regular by replacing r with $\Delta(G)$, but we shall not be concerned in this paper with the investigation of non-regular graphs.

Problem 2: Find a function $\theta(G)$ such that any regular graph of order $2n$ and degree $r \geq n^*$ (where n^* is as in Conjecture 1.1) satisfies $\mathrm{exc}(G) \leq \theta(G)$. (It is obvious that what we really ask is a function $\theta(G)$ which provides the best possible upper bound for $\mathrm{exc}(G)$.)

We note here that, trivially, $\chi'_e(G) \leq |E(G)|$ for any 1-extendable graph G. Moreover, using the fact that the intersection of two distinct 1-factors in G cannot contain more than $|V(G)|/2 - 2$ edges, the above inequality can be improved without difficulty to $\chi'_e(G) \leq \max |\{|E(G)| - |V(G)|/2, 1\}|$, so that the function $\theta(G) = \max\{|E(G)| - |V(G)|/2 - \Delta(G), 0\}$ provides a first, rather crude upper bound for $\mathrm{exc}(G)$ (and this upper bound holds also for 1-extendable non-regular graphs).

3. Graphs with small excessive class

The following theorem was proved in [3] for complete even graphs, but holds, more generally, for regular even graphs.

Theorem 3.1. *Let G be a regular graph of even order. Let G have a 1-factor cover $\mathcal{F}$ of size $r+1$. Call an edge "excessive" if it belongs to more than one 1-factor in $\mathcal{F}$. Then we have the following:*

1. *each excessive edge of G belongs to exactly two 1-factors in $\mathcal{F}$;*
2. *the set of excessive edges forms a 1-factor F^* of G, which we call the "excessive 1-factor";*
3. *$\mathcal{F}$ is a minimal 1-factor cover of G if and only if $F^* \notin \mathcal{F}$.*

Proof. For any vertex u of G there are precisely r edges incident with u. Since every 1-factor is incident with u, and all the r edges incident with u are covered by at least one 1-factor in $\mathcal{F}$, it follows that there is precisely one edge incident with u which is excessive, and this edge belongs to exactly two 1-factors in $\mathcal{F}$. Since the choice of u is arbitrary, this proves Claim 1 and Claim 2. Let now F^* be the set of excessive edges, which is a 1-factor by (2). If $F^* \in \mathcal{F}$, then $\mathcal{F} \setminus \{F^*\}$ covers G, so that $\mathcal{F}$ is not a minimal 1-factor cover of G. Conversely, if $\mathcal{F}$ is a 1-factor cover of G which is not minimal and $F_1 \in \mathcal{F}$ is such that $\mathcal{F} \setminus \{F_1\}$ covers G, then $\mathcal{F} \setminus \{F_1\}$ is a 1-factorization of G, and hence F_1 coincides with the set F^* of excessive edges of $\mathcal{F}$. This proves Claim 3 and concludes the proof of the theorem. □

It would be desirable to have an extension of Theorem 3.1 to 1-factor covers of size $r+2$. In this case it is easy to see that the excessive edges do not always form a 2-factor, but it would be interesting to prove under what conditions they do.

Theorem 3.1 is useful, for instance, when we want to prove or disprove that a given graph is 1-excessive. An example is offered by the case of the Petersen graph, which is 2-excessive. To exemplify the application of Theorem 3.1, we give a detailed proof of this fact.

Proposition 3.2. *The Petersen graph is 2-excessive.*

Proof. First notice that $\text{exc}(P) > 0$, since P is not 1-factorizable. It is easy to see that there is a 1-factor cover of P of size 5, thus proving that $\chi'_e(P) \leq 5$, and hence that $\text{exc}(P) \leq 2$. To terminate the proof, it remains to be shown that $\text{exc}(P) \neq 1$. Arguing by contradiction, assume that $\text{exc}(P) = 1$. Let $\mathcal{F} = \{F_1, F_2, F_3, F_4\}$ be an excessive factorization of P. Notice that $\mathcal{F}$ is, in particular, a minimal 1-factor cover. Applying Theorem 3.1, we have that the excessive factor $F^* \notin \mathcal{F}$. Since every two distinct 1-factors of P intersect in precisely one edge, we have that

$$|F_1 \cap F^*| = |F_2 \cap F^*| = |F_3 \cap F^*| = |F_4 \cap F^*| = 1.$$

But $|F^*| = 5$, which implies that F^* contains one edge which is not in $F_1 \cup F_2 \cup F_3 \cup F_4 = E(G)$, which is a contradiction. Hence $\text{exc}(P) = 2$. □

Is there an example of a regular 1-excessive graph G? These graphs appear to be quite rare among those with small order. Cariolaro [5] gave an example of a 5-regular graph of order 18 with excessive class 1. Wallis [22] found an example of a cubic 1-excessive graph of order 18. This graph can be obtained by taking two copies of P^* (where P^* denotes the Petersen graph with one vertex deleted) and matching the three vertices of degree 2 of one copy of P^* to the three vertices of degree 2 of the other copy. However these are not the smallest regular graphs with excessive class 1. Indeed the graph P' obtained by expanding one vertex of the Petersen graph into a triangle is a regular 1-excessive graph and has order 12 (see Fig. 2). Using the known classification of the graphs of order at most 12 with respect to the chromatic index (see [2, 4, 8, 12]), it is easy to show that P' is indeed the smallest 1-excessive regular graph (in the sense that there is no other 1-excessive regular graph of order at most 12).

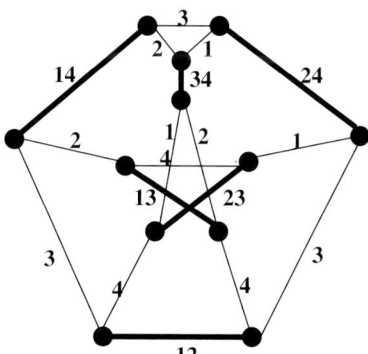

FIGURE 2. The smallest regular graph with excessive class 1. An excessive factorization consisting of four 1-factors labelled $1, 2, 3, 4$ is shown. The excessive edges are marked in bold. Notice that the excessive edges form a 1-factor.

4. Graphs of arbitrary excessive class

The examples of graphs considered thus far show, in particular, the existence of regular graphs of excessive class 0, 1 and 2. It is natural to ask if there are graphs with higher excessive class. In this section we will prove that there is no upper bound on the excessive class of regular graphs of even order (even when we exclude those with infinite excessive class, i.e., those that are not 1-extendable).

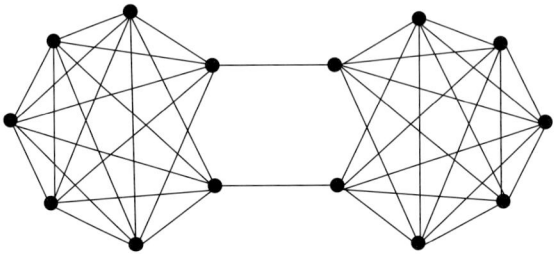

FIGURE 3. A regular graph with excessive class 4.

We will actually prove a stronger result, namely the following.

Theorem 4.1. *Let k be a positive integer. Then there is a regular graph G such that $\mathrm{exc}(G) = 2k$. Moreover G can be chosen to have order $2n$ and degree $n-1$, for $n = 2k+3$.*

Proof. Let n be an odd positive integer, $n \geq 5$, and let G_n be the graph obtained as follows: take two copies H_1, H_2 of $K_n - e$ (i.e., the complete graph of order n with one edge deleted). Let $e_1 = u_1 v_1$ be the deleted edge in H_1 and let $e_2 = u_2 v_2$ be the deleted edge in H_2. To complete the definition of G_n add the edges $u_1 u_2$ and $v_1 v_2$.(Fig. 3 shows the graph G_7.) We shall prove that $\mathrm{exc}(G_n) = n - 3$, which will prove the theorem. Notice that G_n is $(n-1)$-regular; hence our claim is equivalent to the claim that $\chi'_e(G_n) = (n-1) + (n-3) = 2n - 4$. We first prove the inequality $\chi'_e(G_n) \leq 2n - 4$ by exhibiting a 1-factor cover of G_n of size $2n - 4$. Notice that the graphs $H_1 - u_1, H_1 - v_1, H_2 - u_2, H_2 - v_2$ are complete graphs of order $n - 1$ and, since $n - 1$ is even, they are 1-factorizable. Thus in particular there exists a 1-factor cover $\mathcal{F}_1$ of $H_1 - u_1$ of size $n - 2$. Similarly there exists a 1-factor cover $\mathcal{F}_2$ of $H_2 - u_2$ of size $n - 2$. We can combine each 1-factor $F_1 \in \mathcal{F}_1$ with one 1-factor $F_2 \in \mathcal{F}_2$ and, adding the edge $u_1 u_2$, obtain a 1-factor of G_n. By doing this for all the 1-factors in $\mathcal{F}_1, \mathcal{F}_2$ we obtain a set $\mathcal{F}$ of 1-factors of G_n of size $n - 2$ which covers the edge $u_1 u_2$, all the edges of $H_1 - u_1$ and all the edges of $H_2 - u_2$. Similarly, there exists a set $\mathcal{F}'$ of 1-factors of G_n of size $n - 2$, such that $\mathcal{F}'$ covers the edge $v_1 v_2$, all the edges of $H_1 - v_1$ and all the edges of $H_2 - v_2$. It is easy to see that the set $\mathcal{F} \cup \mathcal{F}'$ has size $2n - 4$ and covers all the edges of G_n, and hence is a 1-factor cover of G_n of the required size. Thus $\chi'_e(G_n) \leq 2n - 4$.

It remains to prove that $\chi'_e(G_n) \geq 2n - 4$. Let $p = \chi'_e(G_n)$ and let $\mathcal{F}$ be a 1-factor cover of G_n of size p. It is easy to see that every 1-factor of G_n contains either the edge $u_1 u_2$ or the edge $v_1 v_2$. By symmetry we may assume that the edge $u_1 u_2$ is contained in at least $\lceil \frac{p}{2} \rceil$ of the 1-factors in $\mathcal{F}$. Let $\mathcal{F}_0$ be the set of 1-factors in $\mathcal{F}$ which contain the edge $u_1 u_2$. Let $e_1, e_2, \ldots, e_{n-2}$ be the edges incident with u_1 and distinct from the edge $u_1 u_2$. Certainly no 1-factor in $\mathcal{F}_0$ can contain any of these edges. Moreover, to cover each of these edges, we need distinct 1-factors, since these edges are mutually adjacent. Therefore there are at least $n - 2$ one-factors in $\mathcal{F} \setminus \mathcal{F}_0$. This proves that

$$p \geq \lceil \tfrac{p}{2} \rceil + n - 2, \quad \text{which gives} \quad p \geq 2(n-2),$$

as wanted. □

Notice that Theorem 4.1 reinforces the claim that the bound $r \geq n^*$ in Conjecture 1.1 is best possible, since it shows that there are regular graphs of order $2n$ and degree $n^* - 1$ which not only fail to be 1-factorizable, but, in some sense, are "very far from being 1-factorizable altogether" (and not in an obvious way as the graphs of Fig. 1 for which no excessive factorization exists).

It would be interesting to complement the construction given in the proof of Theorem 4.1 with a parallel construction that shows that all the odd positive integers can also be attained as excessive classes of regular graphs.

5. r-graphs and the perfect matching polytope

Some of the above concepts and results may be rephrased in terms of the so-called *perfect matching polytope*. The perfect matching polytope of G, denoted by $PM(G)$, is the convex hull of the incidence vectors of the 1-factors of G (viewed as 0–1 functions on the edge set of G). It is a well-known and studied concept in matching theory and combinatorial optimization (see, e.g., [15]). For example, the requirement that G has an excessive factorization is equivalent to the requirement that, if $e \in E(G)$, then $x_e > 0$ for at least one vector $\mathbf{x}$ in $PM(G)$.

By Edmonds' theorem [11], the perfect matching polytope can be described as follows. Let G be an even graph. Call an edge cut an *odd cut* if it is the set of edges joining an odd set $X \subset V(G)$ and the (odd) set $V(G) \setminus X$. Call an odd cut *trivial* if either $X = \{v\}$ or $V(G) \setminus X = \{v\}$, for a vertex $v \in V(G)$. Then $PM(G)$ is the solution to the following system of inequalities:

$$\begin{cases} \mathbf{x} \geq 0 \\ \mathbf{x}(C) = 1 & (\text{where } C \text{ is a trivial odd cut}) \\ \mathbf{x}(C) \geq 1 & (\text{where } C \text{ is a non trivial odd cut}) \end{cases}$$

(Here the notation $\mathbf{x}(C)$ is used to denote $\sum_{e \in C} x_e$.)

A class of regular graphs is naturally associated with the perfect matching polytope and it is interesting to investigate the excessive factorizations for this class. These graphs are named *r-graphs* in Seymour's paper [18]. The *r*-graphs are precisely those regular graphs for which the vector $\mathbf{x} \in \mathbb{R}^E$ equal to $1/r$ on every edge belongs to $PM(G)$. (An equivalent definition is that every odd cut of G contains at least r edges.) Thus it is immediate that, if G is an r-graph, then G is 1-extendable and hence admits an excessive factorization. (This fact could also be proved directly using Tutte's 1-factor theorem [19]). It is easy to see that every 1-factorizable graph of degree r is an r-graph. However the converse is not true since the Petersen graph is a 3-graph but is not 1-factorizable. Tutte [20] conjectured that every 3-graph with no Petersen minor is 1-factorizable, i.e., has excessive class 0. (A proof of this conjecture has been announced by Robertson, Sanders, Seymour and Thomas). If true, this implies that the Petersen graph is, in some sense, the "only" graph which causes any 3-graph to have a positive excessive class. In fact we are led to believe that there is no r-graph which has excessive class greater than 2 and we would like to pose this as a conjecture.

Conjecture 5.1. *For every r-graph G, $\mathrm{exc}(G) \leq 2$.*

Notice that Conjecture 5.1 implies Conjecture 2.5 with $k = 2$, since every regular graph of order $2n$ and degree $r \geq n^*$ is an r-graph. The case $r = 3$ of Conjecture 5.1 has an interest of its own. It claims that every cubic bridgeless graph can be covered by 5 one-factors (or less). It seems that, up to now, it is not even known if the number 5 in the previous statement can be replaced by any larger number k (see [1]).

The same statement is also a weaker statement than Fulkerson's Conjecture [13]. Indeed an equivalent formulation of Fulkerson's Conjecture is that every cubic bridgeless graph can be covered by 5 one-factors in such a way that the excessive edges (i.e., the edges which belong to more than one of the 1-factors) form a 2-factor. Accordingly it would be interesting to know when a given 1-factor cover has an excessive factor (by which term we mean the subgraph induced by the excessive edges) which forms a 2-factor, as remarked at the end of the proof of Theorem 3.1.

Conjecture 5.1 suggests a new way to classify r-graphs, introducing in particular an element of differentiation among Class 2 (i.e., non 1-factorizable) r-graphs. Specifically, Conjecture 5.1 suggests the following classification scheme for r-graphs:

1. 1-factorizable r-graphs (those with $\mathrm{exc}(G) = 0$);
2. Type I r-graphs (those with $\mathrm{exc}(G) = 1$, like the one in Fig. 2);
3. Type II r-graphs (those with $\mathrm{exc}(G) \geq 2$, like the Petersen graph).

Acknowledgements

We thank the referee for his/her careful reading of the preliminary version of this paper and for a number of valuable comments. Part of this work was done while the second author was a Ph.D. student at the University of Reading, U.K., and appears in [5, Chapter 5]. The authors are indebted to Prof. A.J.W. Hilton for useful comments and suggestions.

References

[1] D. Archdeacon, *Covering cubic graphs with perfect matchings.* Problems in Topological Graph Theory, Dan Archdeacon's Home Page, http://www.emba.uvm.edu/archdeac/

[2] L.W. Beineke and S. Fiorini, *On small graphs critical with respect to edge colourings.* Discrete Math., **16** (1976), 109–121.

[3] A. Bonisoli, *Edge covers of K_{2n} with $2n$ one-factors.* Rendiconti del Seminario Matematico di Messina, Serie II, **9** (2003), 43–51.

[4] G. Brinkmann and E. Steffen, *Chromatic index critical graphs of order* 11 *and* 12. Europ. J. Combin., **19** (1998), 889–900.

[5] D. Cariolaro, *The 1-Factorization Problem and some related Conjectures.* Ph.D. thesis, University of Reading, U.K., 2004.

[6] A.G. Chetwynd and A.J.W. Hilton, *Regular graphs of high degree are 1-factorizable.* Proc. London Math. Soc., (3) **50** (1985), 193–206.

[7] A.G. Chetwynd and A.J.W. Hilton, *1-factorizing regular graphs of high degree – an improved bound.* Discrete Math. **75** (1989), 103–112.

[8] A.G. Chetwynd and H.P. Yap, *Chromatic index critical graphs of order* 9. Discrete Math., **47** (1983), 23–33.

[9] R. Diestel, *Graph Theory.* Springer, New York, 1997.

[10] G.A. Dirac, *Some theorems on abstract graphs.* Proc. London Math. Soc., **2** (1952), 69–81.

[11] J. Edmonds, *Maximum matching and a polyhedron with $0, 1$-vertices.* J. Res. Nat. Bur. Stand. B, Math & Math. Phys. **69** B (1965), 125–130.

[12] S. Fiorini and R.J. Wilson, *Edge Colourings of Graphs.* Research Notes in Mathematics, **17**, Pitman, 1977.

[13] D.R. Fulkerson, *Blocking and anti-blocking pairs of polyhedra.* Math. Programming, **1** (1971), 168–194.

[14] I. Holyer, *The NP-completeness of edge coloring.* SIAM J. Comput., **10** (4), (1981), 718–720.

[15] L. Lovász and M.D. Plummer, *Matching Theory.* Annals of Discrete Mathematics, **29**, North-Holland, 1986.

[16] T. Niessen and L. Volkmann, *Class* 1 *conditions depending on the minimum degree and the number of vertices of maximum degree.* Journal of Graph Theory (2) **14** (1990), 225–246.

[17] L. Perkovic and B. Reed, *Edge coloring regular graphs of high degree.* Discrete Math., **165/166** (1997), 567–578.

[18] P.D. Seymour, *On multicolourings of cubic graphs and conjectures of Fulkerson and Tutte.* Proc. Lond. Math. Soc. **33** (1979), 423–460.

[19] W.T. Tutte, *The factorization of linear graphs.* J. Lond. Math. Soc. **22** (1947), 459–474.

[20] W.T. Tutte, *On the algebraic theory of graph colourings.* J. Comb. Theory, **1** (1966), 15–50.

[21] W.D. Wallis, *Overfull sets of one-factors*. Thirty-fifth Southeastern Conference on Graphs, Combinatorics and Computing, Florida Atlantic University, Boca Raton, March 3–8, 2004.

[22] W.D. Wallis, personal communication.

Arrigo Bonisoli
Dipartimento di Scienze Sociali, Cognitive e Quantitative
Università di Modena e Reggio Emilia
Via Allegri 9
I-42100 Reggio Emilia, Italy
e-mail: `bonisoli.arrigo@unimore.it`

David Cariolaro
Institute of Mathematics
Academia Sinica Nankang
Taipei 11529 Taiwan
e-mail: `davidcariolaro@hotmail.com`

Odd Pairs of Cliques

Michel Burlet, Frédéric Maffray and Nicolas Trotignon

> **Abstract.** A graph is Berge if it has no induced odd cycle on at least 5 vertices and no complement of induced odd cycle on at least 5 vertices. A graph is perfect if the chromatic number equals the maximum clique number for every induced subgraph. Chudnovsky, Robertson, Seymour and Thomas proved that every Berge graph either falls into some classical family of perfect graphs, or has a structural fault that cannot occur in a minimal imperfect graph. A corollary of this is the strong perfect graph theorem conjectured by Berge: every Berge graph is perfect. An even pair of vertices in a graph is a pair of vertices such that every induced path between them has even length. Meyniel proved that a minimal imperfect graph cannot contain an even pair. So even pairs may be considered as a structural fault. Chudnovsky et al. do not use them, and it is known that some classes of Berge graph have no even pairs.
>
> The aim of this work is to investigate an "even-pair-like" notion that could be a structural fault present in every Berge graph. An odd pair of cliques is a pair of cliques $\{K_1, K_2\}$ such that every induced path from K_1 to K_2 with no interior vertex in $K_1 \cup K_2$ has odd length. We conjecture that for every Berge graph G on at least two vertices, either one of $G, \overline{G}$ has an even pair, or one of $G, \overline{G}$ has an odd pair of cliques. We note that this conjecture is true for basic perfect graphs. By the strong perfect graph theorem, we know that a minimal imperfect graph has no odd pair of maximal cliques. In some special cases we prove this fact independently of the strong perfect graph theorem. We show that adding all edges between any 2 vertices of the cliques of an odd pair of cliques is an operation that preserves perfectness.
>
> **Keywords.** Perfect graph, graph, even pair.

1. Introduction

In this paper graphs are simple, non-oriented, with no loop and finite. Several definitions that can be found in most handbooks (for instance [11]) will not be given. A graph G is *perfect* if every induced subgraph G' of G satisfies $\chi(G') = \omega(G')$, where $\chi(G')$ is the chromatic number of G' and $\omega(G')$ is the maximum clique size in G'. Berge [2, 3] introduced perfect graphs and conjectured that the complement of a perfect graph is a perfect graph.

This conjecture was proved by Lovász:

Theorem 1.1 (Lovász, [19, 18]). *The complement of every perfect graph is a perfect graph.*

Berge also conjectured a stronger statement: *a graph is perfect if and only if it does not contain as an induced subgraph an odd hole or an odd antihole* (the Strong Perfect Graph Conjecture), where a *hole* is a chordless cycle with at least four vertices and an *antihole* is the complement of a hole. We follow the tradition of calling *Berge graph* any graph that contains no odd hole and no odd antihole. The Strong Perfect Graph Conjecture was the objet of much research (see the book [24]), until it was finally proved by Chudnovsky, Robertson, Seymour and Thomas:

Theorem 1.2 (Chudnovsky, Robertson, Seymour and Thomas [5]). *Every Berge graph is perfect.*

In fact Chudnovsky, Robertson, Seymour and Thomas [5] proved a stronger fact, conjectured by Conforti, Cornuéjols and Vušković [9]: every Berge graph either falls in a *basic class* or has a *structural fault*. Before stating this more precisely, let us say that a *basic class* of graphs is a class of graphs that are proved to be perfect by some classical coloring argument. A *structural fault* in a graph is something that cannot occur in a minimal counter-example to the perfect graph conjecture. The basic classes used by Chudnovsky et al. are the bipartite graphs, their complement, the line-graphs of bipartite graphs, their complement, and the double split-graphs. The structural faults used by Chudnovsky et al. are the *2-join* (first defined by Cornuéjols and Cunningham [10]), the *even skew partition* (a refinement of Chvátal's skew partition [7]) and the *homogeneous pair* (first defined by Chvátal and Sbihi [8]). We do not give here the precise definitions as far as we do not need them.

Despite those breakthroughs, some conjectures about Berge graphs remain open. An *even pair* in a graph G is a pair of non-adjacent vertices such that every chordless path between them has even length (number of edges). Given two vertices x, y in a graph G, the operation of *contracting* them means removing x and y and adding one vertex with edges to every vertex of $G \setminus \{x, y\}$ that is adjacent in G to at least one of x, y; we denote by G/xy the graph that results from this operation. Fonlupt and Uhry proved the following:

Theorem 1.3 (Fonlupt and Uhry [15]). *If G is a perfect graph and $\{x, y\}$ is an even pair in G, then the graph G/xy is perfect and has the same chromatic number as G.*

Meyniel also proved the following:

Theorem 1.4 (Meyniel, [21]). *Let G be a minimal imperfect graph. Then G has no even pair.*

So even pairs can be consider as a "structural fault", with respect to a proof of perfectness for some classes of graphs. This approach for proving perfectness has

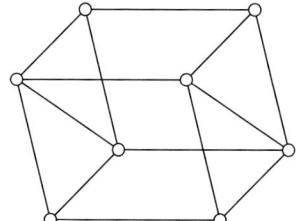 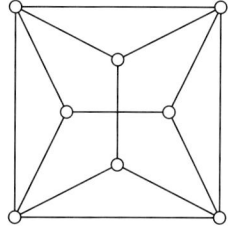

FIGURE 1. The double-diamond and $L(K_{3,3} \setminus e)$

been formalised by Meyniel [21]: a *strict quasi-parity* graph is a graph such that every induced subgraph either is a clique or has an even pair. By Theorem 1.4, every strict quasi-parity graph is perfect. Many classical families of perfect graphs, such as Meyniel graphs, weakly chordal graphs, perfectly orderable graphs, Artemis graphs, are strict quasi-parity, see [12, 20]. A *quasi-parity* graph is a graph G such that for every induced subgraph G' on at least two vertices, either G' has an even pair, or $\overline{G'}$ has an even pair. By Theorems 1.4 and 1.1, we know that quasi-parity graphs are perfect. Quasi-parity graphs graphs include every strict quasi parity graphs, and also other classes of graphs: bull-free Berge graphs [13], bull-reducible Berge graphs [14].

There are interesting open problems about quasi-parity graphs. Say that a graph is a *prism* if it consists of two vertex-disjoint triangles (cliques of size 3) with three vertex-disjoint paths between them, and with no other edges than those in the two triangles and in the three paths. (Prisms were called stretchers in [12] and 3PC(Δ, Δ)'s in [9]). A prism is said to be long if it has at least 7 vertices. The double-diamond and $L(K_{3,3} \setminus e)$ are the graphs depicted figure 1. Let us now recall a definition: a graph is *bipartisan* [6] if in G and $\overline{G}$ there is no odd hole, no long prism, no double-diamond and no $L(K_{3,3} \setminus e)$. The last 50 pages of the strong perfect theorem paper [5] are devoted to a proof of perfectness for bipartisan graphs. This part could be replaced by a proof of the following conjecture:

Conjecture 1.5 (Maffray, Thomas). *Every bipartisan graph is a quasi-parity graph.*

Why not conjecture that *every* Berge graph is a quasi-parity graph? Simply because this is false. Some counter-examples (like the smallest one: $L(K_{3,3} \setminus e)$), were known since the very beginning of the study of even-pairs. Hougardy found an infinite class of counter-examples:

Theorem 1.6 (Hougardy, [17]). *Let G be the line-graph of a 3-connected graph. Then G and $\overline{G}$ have no even pair.*

The aim of this paper is to investigate the following question: is there an "even-pair-like" notion that could be a structural fault present in every Berge graph? We know that line-graphs of bipartite graphs are likely to be without even pairs. So, they certainly form one of the first class where we have to find something.

On the other hand, a bipartite graph B with at least 3 vertices always has an even pair: consider two vertices a, b in the same side of the bipartition. What happens to this even pair $\{a, b\}$ in $L(B)$? All the edges incident to a form a clique K_a of $L(B)$, and there is a similar clique K_b. Moreover, every induced path from K_a to K_b with no interior vertices in $V(K_a) \cup V(K_b)$ has odd length. This leads us to the following definition: let K_1 and K_2 be two cliques of a graph G. We say that an induced path P is *external from K_1 to K_2* if P has one end-vertex in K_a, one end-vertex in K_b, and all the other possible vertices in $V(G) \setminus (V(K_1) \cup V(K_2))$. The pair $\{K_1, K_2\}$ is an *odd pair of cliques* if every external induced path between K_1 and K_2 has odd length. Note that if $\{K_1, K_2\}$ is an odd pair of cliques, then K_1 and K_2 are disjoint since a possible common vertex would be an external path of length 0. Similarly, we say that $\{K_1, K_2\}$ is an *even pair of cliques* when every external induced path between K_1 and K_2 has even length. We propose the following two conjectures:

Conjecture 1.7. *Let G be a Berge graph on at least two vertices. Then either:*
- *G or $\overline{G}$ has an even pair.*
- *G or $\overline{G}$ has an odd pair of cliques $\{K_1, K_2\}$ such that K_1, K_2 are maximal cliques of G.*

Conjecture 1.8. *Let G be a minimal imperfect graph. Then G has no odd pair of cliques $\{K_1, K_2\}$, such that K_1, K_2 are maximal cliques of G.*

Clearly, between two maximal cliques of an odd hole, there exists an external induced path of even length. Between two maximal cliques of an odd antihole, there exists an external induced path of length 2. But by the strong perfect graph theorem, the only minimal imperfect are the odd holes and the odd antiholes. Thus the conjecture above is true. But we would like a proof that does not use the strong perfect graph theorem.

As already mentioned, it is easy to see that Conjecture 1.7 holds for bipartite graphs, line-graphs of bipartite graphs, and their complement. Let us prove that it holds also for the last basic class: double split graphs. A *double split graph* (defined in [5]) is any graph G that can be constructed as follows. Let $m, n \geq 2$ be integers. Let $A = \{a_1, \ldots, a_m\}$, $B = \{b_1, \ldots, b_m\}$, $C = \{c_1, \ldots, c_n\}$, $D = \{d_1, \ldots, d_n\}$ be four disjoint sets. Let G have vertex set $A \cup B \cup C \cup D$ and edges in such a way that:

- a_i is adjacent to b_i for $1 \leq i \leq m$. There are no edges between $\{a_i, b_i\}$ and $\{a_{i'}, b_{i'}\}$ for $1 \leq i < i' \leq m$.
- c_j is non-adjacent to d_j for $1 \leq i \leq m$. There are all four edges between $\{c_j, d_j\}$ and $\{c_{j'}, b_{j'}\}$ for $1 \leq j < j' \leq n$.
- There are exactly two edges between $\{a_i, b_i\}$ and $\{c_j, d_j\}$ for $1 \leq i \leq m$ and $1 \leq j \leq n$ and these two edges are disjoint.

If G is a double split graph with the notation of the definition, we may assume up to a relabeling of the c_j, d_j's that a_1 sees every c_j and that b_1 sees every d_j (if this fails for some j, just swap c_j, d_j). Now it is easy to see that

FIGURE 2. The claw and the diamond

$K_a = \{a_1, c_1, \ldots, c_n\}$ and $K_b = \{b_1, d_1, \ldots, d_n\}$ are both maximal cliques of G. The only possible external induced paths of length greater than 1 from K_a to K_b, are paths from c_j to d_j for some j. But such a path must start in c_j, and then go to some a_i, and then the only option is to go to b_i, and then back to d_j. So, every external path from K_a to K_b has length 1 or 3. So, $\{K_a, K_b\}$ is an odd pair of maximal cliques. Note that there exist double split graphs that have no even pair: $L(K_{3,3} \setminus e)$ is an example and arbitrarily large examples exist. However, Conjecture 1.7 holds for every basic graph.

2. Odd pairs of cliques in line-graphs of bipartite graphs

We observed in the introduction that every line-graph of bipartite graph has an odd pair of cliques. In this section, we will see that we can say something much stronger. But we first need some information on the structure of line-graphs of bipartite graphs.

The facts stated in this paragraph need careful checking, but we do not prove them since they are well known (see [1] and [16]). Let us consider a graph G that contains no claw and no diamond. Let v be a vertex of G. Either v belongs to exactly one maximal clique of G, or v belongs to exactly two maximal cliques of G. In the second case, the intersection of the two cliques is exactly $\{v\}$. Let us build a new graph R. Every maximal clique of G is a vertex of R. Such vertices of R are called the *clique vertices* of R. Every vertex of G that belongs to a single clique of G is also a vertex of R. Such vertices of R are called *pendent vertices* of R. We add an edge between two clique vertices of R whenever the two corresponding cliques in G do intersect. We add an edge between a clique vertex u of R and a pendent vertex v of R whenever the pendent vertex v is a vertex of G that belongs to u seen as a clique of G. Note that a pendent vertex of R has always degree 1 (the converse is not true when G has a connected component that consists in a single vertex). One can check that R has no triangle and that G is isomorphic to $L(R)$. This leads us to the following well known theorem:

Theorem 2.1. *Let G be a graph. There exists a triangle-free graph R such that $G = L(R)$ if and only if G contains no claw and no diamond.*

The following theorem shows that the maximal cliques of the line-graph of a bipartite graph behave like the vertices of a bipartite graph in quite a strong sense.

Theorem 2.2. *Let G be a graph with no claw and no diamond. Then G is the line-graph of a bipartite graph if and only if the maximal cliques of G may be partitioned into two sets A and B such that for every distinct maximal cliques K_1, K_2 of G we have:*

- *If $K_1 \in A$ and $K_2 \in A$, then $\{K_1, K_2\}$ is an odd pair of cliques.*
- *If $K_1 \in B$ and $K_2 \in B$, then $\{K_1, K_2\}$ is an odd pair of cliques.*
- *If $K_1 \in A$ and $K_2 \in B$, then $\{K_1, K_2\}$ is an even pair of cliques.*

Proof. By the discussion above, we know that G is isomorphic to the line-graph of a triangle-free graph R. So, we may assume that R is built from G like in the construction described in the discussion above.

If R is bipartite, then the clique vertices of R are partitioned into two stable sets A and B. This partition is also a partition of the maximal cliques of G. So, let $K_1 \in A$ and $K_2 \in A$ be two maximal cliques of G. Note that K_1 and K_2 are also non-adjacent vertices of R. So, we know that K_1 and K_2 are disjoints cliques of G. If there exists an induced path of G, of even length, external from K_1 to K_2, then the interior vertices of this path (which has length at least 2) are the edges of the interior of a path of R of odd length, linking the vertex K_1 to the vertex K_2. This contradicts the bipartition of R. So, every external induced path in G between K_1 and K_2 is of odd length, in other words, K_1 and K_2 form an odd pair of cliques. By the same way, we prove that if $K_1 \in B$ and $K_2 \in B$, then K_1 and K_2 form an odd pair of cliques. Similarly, if $K_1 \in A$ and $K_2 \in B$, then K_1 and K_2 form an even pair of cliques.

If R is not bipartite, then R has an odd hole H of length at least 5 (because R is triangle-free). Let $v_1, v_2, \ldots v_{2k+1}$ be the vertices of H in their natural order. Every vertex of H has degree at least 2, and therefore is a clique vertex of R. So, every vertex v_i is in fact a maximal clique of G. If one manages to partition the maximal cliques of G into two sets A and B as indicated in the lemma, two consecutive cliques v_i and v_{i+1} are not disjoint. So, they cannot be both in A or both in B. So in the sequence $(v_1, \ldots, v_{2k+1}, v_1, \ldots)$ every second clique is in A and the other ones are in B. But this is impossible because there is an odd number of v_i's. $\square$

3. An operation that preserves perfectness

We know that the contraction of an even pair $\{x, y\}$ in a perfect graph G yields another perfect graph. What would be the corresponding operation in $L(G)$ for an odd pair of cliques? The edges incident to x form a clique K_x of $L(G)$, and those incident to y form a clique K_y. The contracted vertex xy in G/xy is incident to the edges that were incident to x or y in G, and so becomes in $L(G)$ a clique obtained by adding an edge between every vertex of K_x and every vertex of K_y. So let us define the following operation for any graph G and any pair $\{K_1, K_2\}$ of disjoint cliques of G: just add an edge between every vertex in K_1 and every vertex in K_2 (if they are not adjacent). The graph obtained is denoted by $G_{K_1 \equiv K_2}$. We will see that this operation preserves perfectness when applied to an odd pair of cliques. Before this, we need a technical lemma, roughly saying that in $G_{K_1 \equiv K_2}$, there is no other big clique than the clique induced by $V(K_1) \cup V(K_2)$:

Lemma 3.1. *Let $\{K_1, K_2\}$ be an odd pair of cliques in a graph G. Let K be a clique of $G_{K_1 \equiv K_2}$. There are then only two possibilities:*
- *K is a clique of G.*
- *$V(K) \subseteq V(K_1) \cup V(K_2)$.*

Proof. If K is not a clique of G, then K contains at least a vertex v_1 of K_1 and a vertex v_2 of K_2 that are not adjacent in G. Moreover, if $V(K)$ if not included in $V(K_1) \cup V(K_2)$, then K contains a vertex v that is neither in K_1, nor in K_2, and that sees v_1 and v_2. But then, $v_1 - v - v_2$ is an external induced path of G, of even length from K_1 to K_2, a contradiction. □

The proof of the next theorem looks like the proof of Fonlupt and Uhry for Theorem 1.3. For Theorem 1.3, it is needed to prove by a bichromatic exchange that some vertices may have *the same* color in some optimal coloring of a graph. The only possible obstruction to this exchange is a path of odd length between them, contradicting the definition of an even pair of vertices. In our theorem, at a certain step we will need to prove that there is an optimal coloring that gives *different* colors to some vertices. The only obstruction to this will be an induced path of even length, contradicting the definition of odd pairs of cliques.

Theorem 3.2. *Let G be a perfect graph and let $\{K_1, K_2\}$ be an odd pair of cliques of G. Then $G_{K_1 \equiv K_2}$ is a perfect graph.*

Proof. Let H' be an induced subgraph of $G_{K_1 \equiv K_2}$. Let H be the induced subgraph of G that has the same vertex-set than H'. Clearly, $V(K_1) \cap V(H)$ and $V(K_2) \cap V(H)$ form an odd pair of cliques in H and $H' = H_{(K_1 \cap H) \equiv (K_2 \cap H)}$. So, to prove the theorem, it suffices to check $\chi(G_{K_1 \equiv K_2}) = \omega(G_{K_1 \equiv K_2})$. Let us suppose that G is colored with $\omega(G)$ colors. We look for a coloring of $G_{K_1 \equiv K_2}$ with $\omega(G_{K_1 \equiv K_2})$ colors.

Let us first color the vertices that are neither in K_1 nor in K_2: we give them their color in G. If $\omega(G_{K_1 \equiv K_2}) > \omega(G)$, then by Lemma 3.1, we know that $V(K_1) \cup V(K_2)$ induces the only maximum clique of $G_{K_1 \equiv K_2}$. So, whatever the sizes of K_1, K_2, we take $\gamma = \max(0, |V(K_1) \cup V(K_2)| - \omega(G))$ new colors. We use them to color γ vertices in $V(K_1) \cup V(K_2)$. So, we are left with $|V(K_1) \cup V(K_2)| - \gamma$ vertices in $V(K_1) \cup V(K_2)$: let us give them their color in G. We may assume that there is a vertex v_1 in K_1 and a vertex v_2 in K_2 with the same color (say red) for otherwise we have an $\omega(G_{K_1 \equiv K_2})$-coloring of $G_{K_1 \equiv K_2}$ and the conclusion of the lemma holds.

So there is a color used in G (say blue) that is used neither in K_1 nor in K_2. Let C be the set of vertices of G that are red or blue. The set C induces a bipartite subgraph of G and we call C_1 the connected component of v_1 in this subgraph. If $v_2 \in C_1$, then a shortest path in C_1 from v_1 to v_2 is an induced path of G, of even length from K_1 to K_2. This path is external because there is no blue vertex in $V(K_1) \cup V(K_2)$. This contradicts the definition of an odd pair of cliques, so v_1 and v_2 are not in the same connected component C_1. So, we can exchange the colors red and blue in C_1, and give the color blue to v_1, without changing the color of v_2. We can do this again as long as there are vertices of the same color in $K_1 \cup K_2$. Finally, we obtain an $\omega(G_{K_1 \equiv K_2})$-coloring of $G_{K_1 \equiv K_2}$. □

4. Odd pairs of cliques in minimal imperfect graphs

In this section, we will see that in a minimal imperfect graph G, there is no odd pair of cliques (K_1, K_2) with $|K_1| + |K_2| = \omega(G)$. This will be proven without using the strong perfect graph theorem. We first need some results on minimal imperfect graphs.

Theorem 4.1 (Lovász, [18]). *A graph G is perfect if and only if for every induced subgraph G' we have $\alpha(G')\omega(G') \geq |V(G')|$.*

Lovász also introduced an important notion. Let $p, q \geq 1$ be two integers. A graph G is (p,q)-*partitionable* if and only if for every vertex v of G, the graph $G \setminus v$ can be partitioned into p cliques of size q and also into q stable sets of size p. The theorem to come follows from Theorem 4.1:

Theorem 4.2 (Lovász, [18]). *Let G be a minimal imperfect graph. Then G is partitionable.*

Partitionable graphs have several interesting properties (see [23] for a survey). Padberg [22] proved the following in the particular case of minimal imperfect graphs:

Theorem 4.3 (Bland, Huang, Trotter [4]). *Let G be a graph (p,q)-partitionable with $n = pq + 1$ vertices. Then:*

1. *$\alpha(G) = p$ and $\omega(G) = q$.*
2. *G has exactly n cliques of size ω.*
3. *G has exactly n stable sets of size α.*
4. *Every vertex of G belongs to exactly ω cliques of size ω.*
5. *Every vertex of G belongs to exactly α stable sets of size α.*
6. *Every clique of G of size ω is disjoint from exactly one stable set of G of size α.*
7. *Every stable set of G of size α is disjoint from exactly one clique of G of size ω.*
8. *For every vertex v of G, there is a unique coloring of $G \setminus v$ with ω colors.*

If K_1 and K_2 are two disjoint subcliques of a clique K, then they form an odd pair of cliques. In this case, we say that K_1 and K_2 form a *trivial odd pair of cliques*.

The following theorem is a particular case of Conjecture 1.8:

Theorem 4.4. *Let G be a minimal imperfect graph. Let $\{K_1, K_2\}$ be a non trivial odd pair of cliques of G. Then $|K_1| + |K_2| \neq \omega(G)$.*

Proof. Suppose $|K_1| + |K_2| = \omega(G)$. By Lemma 3.1, $\omega(G_{K_1 \equiv K_2}) = \omega(G)$. Moreover, $\alpha(G_{K_1 \equiv K_2}) \leq \alpha(G)$. And by Theorem 4.1, we have $\alpha(G)\omega(G) < |V(G)|$.

By the definition, every induced subgraph of G is perfect. So, by Theorem 3.2, every induced subgraph of $G_{K_1 \equiv K_2}$ is perfect. Note that the ω-clique $K_1 \cup K_2$ of $G_{K_1 \equiv K_2}$ is not a clique of G since $\{K_1, K_2\}$ is not a trivial odd pair of cliques. So, by counting the cliques and by the fact that G is partitionable, we know that $G_{K_1 \equiv K_2}$ is not partitionable (because of Property (2) of Theorem 4.3). All its subgraphs are perfect, so by Theorem 4.2, we know it is perfect. But we have:

$$\alpha(G_{K_1 \equiv K_2})\omega(G_{K_1 \equiv K_2}) \leq \alpha(G)\omega(G) < |V(G)| = |V(G_{K_1 \equiv K_2})|$$

This contradicts Theorem 4.1. □

By the preceding theorem, if $\{K_1, K_2\}$ is an odd pair of cliques in a minimal imperfect graph G, there are two cases:

- $|K_1| + |K_2| < \omega(G)$

  In this case, interestingly, the edges that we add when constructing $G_{K_1 \equiv K_2}$ do not create any ω-clique by Lemma 3.1. Moreover, these edges do not destroy any α-stable. Let us prove this:

  Proof. Suppose that an α-stable set of G is destroyed. This means that there exists two vertices $v_1 \in K_1$ and $v_2 \in K_2$ that are in some α-stable set S of G. By Property (7) of Theorem 4.3, there exists one ω-clique K disjoint from S. Let $v \in V(K)$. By the definition of partitionable graphs, $G \setminus v$ can be partitioned into ω stable sets of size α. At least one of these stable sets (say S') is disjoint from K, since $K \setminus v$ contains $\omega - 1$ vertices. By Property (6) of Theorem 4.3, we know that $S' = S$. So we have found in G a vertex v such that $G \setminus v$ can be optimally colored giving to v_1 and v_2 the same color, say red. But since $|K_1| + |K_2| < \omega(G)$, there exists a color (say blue) that is not used in $K_1 \cup K_2$. By a bichromatic exchange (like in the proof of Theorem 3.2), we can find a coloring of $G \setminus v$ that gives the same red color to v_1 and color blue to v_2 (if such an exchange fails, there is an external induced path of even length between K_1 and K_2, a contradiction). Finally we found two different colorings of $G \setminus v$. This contradicts Property (8) of Theorem 4.3. □

  So $G_{K_1 \equiv K_2}$ is a partitionable graph. Seemingly, this does not lead to a contradiction.

- $|K_1| + |K_2| > \omega(G)$

  In this case, by Lemma 3.1, $G_{K_1 \equiv K_2}$ has a unique maximum clique: $K_1 \cup K_2$. This graph is not partitionable, all its induced subgraphs are perfect, so it is perfect. One more time, this does not seem to lead to contradiction.

5. Odd pairs of cliques in Berge graphs

To prove Conjecture 1.7, one could try to use the approach that worked for the decomposition of Berge graphs [5]: first, consider the case when G has a "substantial" line-graph H as an induced subgraph. We know that H has an odd pair of cliques (by Theorem 2.2). Then, one could hope that this pair of cliques is likely to somehow "grow" to an odd pair of cliques of the whole graph. A *star-cutset* in a graph G is a set C of vertices such that $G \setminus C$ is disconnected and such that there exists a vertex in C that sees all the other vertices of C. Star cutsets have been introduced by Chvátal [7], who proved that they are a "structural fault" that cannot occur in minimal imperfect graph. It is known however that some non-basic Berge graphs have no star-cutset. The following lemma shows that there is something wrong in the idea of making the odd pair cliques "grow": it can work only in graphs that have a star-cutset.

Lemma 5.1. *Let $\{K_1, K_2\}$ be an odd pair of cliques of a graph G. Suppose that K_2 is a maximal clique of G. Let $K'_1 \neq K_1$ be a sub-clique of K_1. If $\{K'_1, K_2\}$ is an odd pair of cliques, then G has a star cutset.*

Proof. Let $a \in K'_1$ and $b \in K_2$ be non adjacent vertices (they exist because K_2 is maximal). Let c be any vertex of $V(K_1) \setminus V(K'_1)$. We are going to show that $\{a\} \cup N(a) \setminus \{c\}$ is a cutset of G separating c from b. To prove this, we check that every induced path P from c to b that has no interior vertex in K_1 contains a neighbor of a different of c. Indeed:

If the interior of P contains no vertex of K_2, then P has odd length because $\{K_1, K_2\}$ is an odd pair of cliques. Since $\{K'_1, K_2\}$ is an odd pair of cliques, there is a chord in the even-length path $(a, c, \ldots, b)$, and this chord is between a and a vertex of the interior of P.

If the interior of P contains a vertex of K_2, then this vertex is the neighbor of b in P: we denote it by d. We see that c–P–d has odd length because $\{K_1, K_2\}$ is an odd pair of cliques. So the path $(a, c, \ldots, d)$ has even length, and there is a chord between a and a vertex of the interior of P (this chord can be ad). □

References

[1] L.W. Beineke, *Characterisation of derived graphs*, Journal of Combinatorial Theory **9** (1970), 129–135.

[2] C. Berge, *Les problèmes de coloration en théorie des graphes*, Publ. Inst. Stat. Univ. Paris, 1960.

[3] C Berge, *Färbung von Graphen, deren sämtliche bzw. deren ungerade Kreise starr sind (Zusammenfassung)*, Wiss. Z. Martin Luther Univ. Math.-Natur. Reihe (Halle-Wittenberg), 1961.

[4] R.G. Bland, H.C. Huang, and L.E. Trotter, Jr., *Graphical properties related to minimal imperfection*, Discrete Math. **27** (1979), 11–22.

[5] M. Chudnovsky, N. Robertson, P. Seymour, and R. Thomas, *The strong perfect graph theorem*, Manuscript, 2002.

[6] M. Chudnovsky, N. Robertson, P. Seymour, and R. Thomas, *Progress on perfect graphs*, Manuscript, 2002.

[7] V. Chvátal, *Star-cutsets and perfect graphs*, J. Combin. Ser. B **39** (1985), 189–199.

[8] V. Chvátal and N. Sbihi, *Bull-free Berge graphs are perfect*, Graphs and Combinatorics **3** (1987), 127–139.

[9] M. Conforti, G. Cornuéjols, and K. Vušković, *Square-free perfect graphs*, Jour. Comb. Th. Ser. B **90** (2004), 257–307.

[10] G. Cornuéjols and W.H. Cunningham, *Composition for perfect graphs*, Disc. Math. **55** (1985), 245–254.

[11] R. Diestel, *Graph theory*, second ed., Springer, New York, 2000.

[12] H. Everett, C.M.H. de Figueiredo, C. Linhares Sales, F. Maffray, O. Porto, and B.A. Reed, *Even pairs*, in Ramírez Alfonsín and Reed [24], pp. 67–92.

[13] C.M.H. de Figueiredo, F. Maffray, and O. Porto, *On the structure of bull-free perfect graphs*, Graphs Combin. **13** (1997), 31–55.

[14] C.M.H. de Figueiredo, F. Maffray, and C.R. Villela Maciel, *Even pairs in bull-reducible graphs*, Manuscript, 2004. Res. Report 117, Laboratoire Leibniz.

[15] J. Fonlupt and J.P. Uhry, *Transformations which preserve perfectness and h-perfectness of graphs*, Ann. Disc. Math. **16** (1982), 83–85.

[16] F. Harary and C. Holzmann, *Line graphs of bipartite graphs*, Rev. Soc. Mat. Chile **1** (1974), 19–22.

[17] S. Hougardy, *Even and odd pairs in line-graphs of bipartite graphs*, European J. Combin. **16** (1995), 17–21.

[18] L. Lovász, *A characterization of perfect graphs*, J. Combin. Theory Ser. B **13** (1972), 95–98.

[19] L. Lovász, *Normal hypergraphs and the perfect graph conjecture*, Discrete Math. **2** (1972), 253–267.

[20] F. Maffray and N. Trotignon, *A class of perfectly contractile graphs*, Submited to Comb. Th. Ser. B (2003).

[21] H. Meyniel, *A new property of critical imperfect graphs and some consequences*, European J. Comb. **8** (1987), 313–316.

[22] M.W. Padberg, *Almost integral polyhedra related to certain combinatorial optimization problems*, Math. Programming **6** (1974), 180–196.

[23] M. Preissmann and A. Sebő, *Some aspects of minimal imperfect graphs*, in Ramírez Alfonsín and Reed [24], pp. 185–214.

[24] J.L. Ramírez Alfonsín and B.A. Reed (eds.), *Perfect graphs*, Series in Discrete Mathematics and Optimization, Wiley-Interscience, 2001.

Michel Burlet, Frédéric Maffray and Nicolas Trotignon
Laboratoire Leibniz
46 avenue Félix Viallet
F-38031 Grenoble cedex, France

e-mail: {michel.burlet, frederic.maffray, nicolas.trotignon}@imag.fr

Recognition of Perfect Circular-arc Graphs

Kathie Cameron, Elaine M. Eschen,
Chính T. Hoàng and R. Sritharan

> **Abstract.** We give an O($mn \log \log n + m^2$)-time algorithm to recognize perfect circular-arc graphs.
>
> **Mathematics Subject Classification (2000).** Primary 05C85; Secondary 68Q25.
>
> **Keywords.** Circular-arc graph, perfect graph.

1. Introduction and motivation

Given a family $\mathcal{F}$ of non-empty sets, the *intersection graph* $\mathcal{I}(\mathcal{F})$ of $\mathcal{F}$ has vertex-set $\mathcal{F}$ and an edge between two elements u and v of $\mathcal{F}$ exactly when $u \cap v \neq \emptyset$. The family $\mathcal{F}$ is called a model for the graph $\mathcal{I}(\mathcal{F})$. *Interval graphs* are the intersection graphs of a set of intervals on a line. *Circular-arc graphs* are the intersection graphs of a set of arcs on a circle.

Let G be a graph. A *clique* in G is a set of vertices every pair of which is joined by an edge; the size of a largest clique is denoted $\omega(G)$. A *stable set* in G is a set of vertices no two of which are joined by an edge; the size of a largest stable set is denoted $\alpha(G)$. A *colouring* of G is a partition of the vertices into stable sets; the stable sets are called the colours; the minimum number of colours required in a colouring is the *chromatic number* of G, denoted $\chi(G)$. A *clique cover* of G is a partition of the vertices into cliques; the minimum number of cliques required in a clique cover is the *clique covering number* of G, denoted $\theta(G)$.

Clearly, for any graph G, $\alpha(G) \leq \theta(G)$ and $\omega(G) \leq \chi(G)$. Berge [4] defined a graph G to be *perfect* if for every induced subgraph H of G, $\omega(H) = \chi(H)$. Lovász [13] proved that a graph is perfect if and only if its complement is perfect. Thus, a graph is perfect if and only if for every induced subgraph H of G, $\alpha(H) = \theta(H)$.

We give an algorithm to recognize perfect circular-arc graphs. Our recognition algorithm is faster than the general perfect graph recognition algorithm in [6].

This work was supported by the Natural Sciences and Engineering Research Council of Canada (NSERC), the Research Council of The University of Dayton, and Wilfrid Laurier University.

We also describe a method to optimally colour a perfect circular-arc graph, for which there is an efficient implementation using tools similar to those used by our recognition algorithm.

A *hole* is a chordless cycle with at least four vertices. An *antihole* is the complement of a hole. A hole or antihole is called odd or even depending on whether it has an odd or even number of vertices. Berge [4] conjectured and Chudnovsky, Robertson, Seymour and Thomas proved the Strong Perfect Graph Theorem:

Theorem 1.1. [7] *Graph G is perfect if and only if G does not contain an odd hole or an odd antihole as an induced subgraph.*

The Strong Perfect Graph Theorem restricted to circular-arc graphs had been proved previously by Tucker [19]. We remark that all holes and all odd antiholes are circular-arc graphs.

1.1. Notation

For vertex v in graph G, $N_G(v)$ denotes the set of neighbors of v, and $N_G[x] = N_G(x) \cup \{x\}$. The set of non-neighbors of v is $M_G(v) = V(G) - \{v\} - N_G(v)$. We use $d_G(v)$ to denote the degree of vertex v in G. We omit the subscript G when the graph is clear from the context.

For a circular-arc graph, we use A_v to denote the arc in the model that corresponds to vertex v in the graph. When we traverse the circle in the clockwise direction, the arc A_v is assumed to extend from endpoint $CC(v)$ to $CL(v)$. We can assume that in a model for a circular-arc graph, all arc endpoints are distinct. In the model for an interval graph, $l(I)$ and $r(I)$ denote, respectively, the left and right endpoints of interval I. We use I_v to denote the interval in the model that corresponds to vertex v in the graph.

We use *odd (even) chordless path* to refer to a chordless path on an odd (even) number of edges. Given the model for an interval graph, by a *chordless path between intervals I and J* we mean a chordless path between the corresponding vertices in the interval graph. We use m and n to refer to the number of edges and number of vertices, respectively, in a given graph.

2. A tool

Our recognition algorithm detects an odd hole in a circular-arc graph by testing for each vertex x, if x lies in an odd hole. This task is accomplished by searching for a pair u, v of non-adjacent vertices in $N(x)$ such that there is an odd chordless path from u to v whose interior vertices lie in $M(x)$, the set of non-neighbors of x. This sub-task is essentially a problem on an appropriately constructed interval graph. In this section we present an algorithm to solve this problem. The algorithm will be used as a subroutine by our recognition algorithm.

A similar problem that has appeared in the literature is the *parity path problem*, which asks "Given a graph G and a pair x, y of non-adjacent vertices, does there exist a chordless path of specified parity between x and y?". This problem is NP-complete for general graphs [3] and solvable in polynomial time for perfect graphs (using the recognition algorithm for perfect (Berge) graphs given in [6]). Polynomial-time algorithms have been reported for some subclasses of perfect graphs as well [17]. In addition, an O(mn)-time algorithm for circular-arc graphs [1] and an O($m+n$)-time algorithm for chordal graphs [1] have been designed for the parity path problem. The problem we consider here is slightly different. Also, our algorithm is more efficient.

We now present our algorithmic tool. While *Right* is a set of intervals, for convenience we will also refer to *Right* as a set of vertices, each of which corresponds to an interval in *Right*.

Algorithm *Scan* ($\mathcal{I}$, L, R, $\mathcal{V}$, *Right*)

Input:
  $\mathcal{I}$: a set of intervals spanning the line segment $[L, R]$
  $\mathcal{V}$: the only interval in $\mathcal{I}$ with $l(\mathcal{V}) = L$
  Right: a set of intervals each of whose right endpoint is R

Output:
  For each interval $W \in \mathcal{I}$, an odd (even) chordless path from
  $\mathcal{V}$ to W such that no interior vertex of the path belongs to *Right*,
  if such a path exists.

We first describe the information maintained by the algorithm. Each interval may be assigned one or both of the labels *odd* and *even*. An interval I is assigned the label *odd* (*even*) if and only if there exists an odd (even) chordless path from the interval $\mathcal{V}$ to the interval I such that no interior vertex of the path belongs to *Right*.

Our algorithm processes the intervals by scanning their endpoints from L to R. *ActiveEven* is a set of intervals, each of which has the following properties: i) its left endpoint has been scanned, but its right endpoint has not, ii) it is not in *Right*, and iii) it is labeled *even*. The set *ActiveOdd* is defined analogously. We say that an interval J is *active* if ($J \in$ *ActiveEven*) or ($J \in$ *ActiveOdd*).

There are two ways that an interval I can obtain the label *odd* (*even*). One is when $r(J)$ is scanned where J is an active interval with *even* (*odd*) label and $l(J) < l(I) < r(J) < r(I)$ (see Step 4a). The other is when $r(I)$ is scanned and I does not yet have an *odd* (*even*) label (see Step 4b); this is the case when interval I is contained in all active *even* (*odd*) labeled intervals with $l(J) < l(I)$. Note that in this latter case, interval I is necessarily the last interval in an odd (even) length chordless path from $\mathcal{V}$.

We now present the details of the algorithm.

Algorithm *Scan* ($\mathcal{I}$, L, R, $\mathcal{V}$, *Right*)
(0) **for** each interval I **do**
        $Label(I) \leftarrow \emptyset$
    endfor

(1) Add *even* to $Label(\mathcal{V})$
(2) Starting at L scan the line segment $[L, R]$ from L to R
(3) **when** $l(I)$ is scanned
      $LabelMeOdd(I) \leftarrow$ interval in *ActiveEven* with the smallest right endpt.
      $LabelMeEven(I) \leftarrow$ interval in *ActiveOdd* with the smallest right endpt.
endwhen
(4) **when** $r(J)$ is scanned
(4a)   **for** each interval I with $J = LabelMeOdd(I)$ or $J = LabelMeEven(I)$ **do**
        if $(J = LabelMeOdd(I))$ **then**
          Add *odd* to $Label(I)$
        endif
        if $(J = LabelMeEven(I))$ **then**
          Add *even* to $Label(I)$
        endif
    endfor
(4b)   **if** $LabelMeEven(J)$ is still active **then**
        Add *even* to $Label(J)$
    endif
    if $LabelMeOdd(J)$ is still active **then**
        Add *odd* to $Label(J)$
    endif
endwhen
endAlgorithm

We next show that the algorithm *Scan* is correct.

Lemma 2.1. *If an interval I is labeled odd (even), then there is an odd (even) chordless path from $\mathcal{V}$ to I such that no interior vertex of the path belongs to Right.*

Proof of Lemma 2.1. We will prove by induction on the order of the left endpoints of the intervals that *if an interval I is labeled odd (even), then there is an odd (even) chordless path $v_0 v_1 \ldots v_k$ from $\mathcal{V}$ to I (i.e., $I_{v_0} = \mathcal{V}$ and $I_{v_k} = I$) such that no interior vertex of the path belongs to Right and such that for even j, $LabelMeEven(I_{v_j}) = I_{v_{j-1}}$, and for odd j $LabelMeOdd(I_{v_j}) = I_{v_{j-1}}$.*

Suppose interval I is labeled *odd*; the proof when I is labeled *even* is analogous. Let Q be the interval that labeled I *odd*; that is, $Q = LabelMeOdd(I)$. Then Q was active and had label *even* when $l(I)$ was scanned, so $l(Q) < l(I) < r(Q)$. By induction, there is a chordless path $P_Q = v_0 v_1 \ldots v_{k-1}$ with $I_{v_0} = \mathcal{V}$ and $I_{v_{k-1}} = Q$ such that no interior vertex of the path belongs to *Right*, and such that $LabelMeEven(I_{v_t}) = I_{v_{t-1}}$ whenever t is even and $LabelMeOdd(I_{v_t}) = I_{v_{t-1}}$ whenever t is odd, for $t = 0, \ldots, k-1$. Since $Q = LabelMeOdd(I)$, Q was in *ActiveEven*; thus, by the definition of *ActiveEven*, $Q = I_{v_{k-1}}$ is not in *Right*. Consider the path $P_I = v_0 v_1 \ldots v_{k-1} v_k$ obtained by adding the vertex corresponding to I to P_Q (i.e., $I_{v_k} = I$). Since P_Q is chordless, any chord of P_I must have v_k as one of its endpoints; further, for $t \leq k-3$, $r(I_{v_t}) < l(Q) < l(I)$, so the only possible chord of P_I is $v_{k-2}v_k$. But, since $I_{v_{k-2}} = \textit{LabelMeEven}(Q)$ and Q had the label *even* when $l(I)$ was scanned, it follows that $r(I_{v_{k-2}}) < l(I)$. Thus, P_I is the desired path. □

Lemma 2.2. *If there is an odd (even) chordless path from the interval $\mathcal{V}$ to an interval J such that no interior vertex of the path belongs to Right, then the algorithm will label J odd (even). Moreover, if J' immediately precedes J on the path, then J' is active with label even (odd) when $l(J)$ is scanned.*

Proof of Lemma 2.2. The proof is by induction on the length of the path (which is one less than the number of vertices in the path). Assume the lemma holds for chordless paths of length at most $k-1$. We will prove the lemma holds for chordless paths of length k, where k is odd; the proof when k is even is analogous. Consider an odd chordless path $P_J = v_0 v_1 \ldots v_k$, where $\mathcal{V} = I_{v_0}$, $J = I_{v_k}$ and $k \geq 2$. By induction, $I_{v_{k-1}}$ is correctly labeled *even* and $I_{v_{k-2}}$ was active with label *odd* when $l(I_{v_{k-1}})$ was scanned. It follows that $l(I_{v_{k-2}}) < l(I_{v_{k-1}})$. Since P_J is a chordless path, I_{v_k} does not intersect $I_{v_{k-2}}$ (but intersects $I_{v_{k-1}}$, by definition). Thus, we have $l(I_{v_{k-2}}) < l(I_{v_{k-1}}) < r(I_{v_{k-2}}) < l(I_{v_k}) < r(I_{v_{k-1}})$.

So, when $r(I_{v_{k-2}})$ is scanned, either $I_{v_{k-1}}$ receives the label *even* (in Step 4a), or it already has the label. This means that when $l(I_{v_k})$ is scanned, $I_{v_{k-1}} \notin \textit{Right}$ is active with label *even* and hence is on *ActiveEven*. So, $\textit{LabelMeOdd}(I_{v_k})$ will be assigned a valid value (either $I_{v_{k-1}}$ or some other interval K with label *even* and $r(K) < r(I_{v_{k-1}})$), and eventually I_{v_k} will be labeled *odd*. □

Finally, we note that for any interval $W \in \mathcal{I}$ that is labeled *odd* (*even*), a required chordless path from $\mathcal{V}$ can easily be constructed using the information stored in *LabelMeOdd* (*LabelMeEven*).

2.1. Implementation of the algorithm *Scan*

Next, we show that the algorithm *Scan* can be implemented to run in $O(n \log \log n)$ time where n is the number of intervals in the input.

We maintain *ActiveEven* and *ActiveOdd* as priority queues of intervals where the priority of an interval I is its right endpoint $r(I)$. It is possible for an interval to belong to both the priority queues. However, neither priority queue will ever hold more than n elements.

For each interval I, we maintain the following information:

- *LabelThis(I)*: a doubly linked list containing the intervals that I is responsible for labeling, i.e., a list of those intervals J such that either *LabelMeOdd(J)* $= I$ or *LabelMeEven(J)* $= I$.
- *LabelMeEven(I)*: remembers the only interval that can label I even.
- *PtrEven(I)*: a pointer to where I is in *LabelThis(J)*, where J is the interval that is supposed to label I even.
- *LabelMeOdd(I)*: remembers the only interval that can label I odd.
- *PtrOdd(I)*: points to where I is in *LabelThis(J)*, where J is the interval that is supposed to label I odd.

When a left endpoint $l(I)$ is scanned in Step 3 of the algorithm, let J_1 = top of *ActiveEven* and J_2 = top of *ActiveOdd*. We add I to *LabelThis*(J_i), $i = 1, 2$, and set the pointers *PtrEven*(I) and *PtrOdd*(I) appropriately. We also set *LabelMeEven*(I) to J_2, and *LabelMeOdd*(I) to J_1.

When a right endpoint $r(J)$ is scanned in Step 4 of the algorithm we do the following: the intervals that J must label are readily available in *LabelThis*(J). Therefore, we implement Step 4a by simply traversing *LabelThis*(J) and labeling every interval I in the list as specified in the algorithm (i.e., test whether $J = LabelMeOdd(I)$ and whether $J = LabelMeEven(I)$). When an interval I is labeled *even* (*odd*) thus, if I does not belong to *Right*, as I has now become active, we add I to the priority queue *ActiveEven* (*ActiveOdd*) with $r(I)$ as the priority. Then, we remove J from any priority queue that it is present in.

Finally, in Step 4b of the algorithm, if the interval that is supposed to label J *even* (*odd*) finishes later than J, then J must label itself. We implement this step as follows: let $K = LabelMeEven(J)$. If $r(K) > r(J)$, then add the label *even* to J. Remove J from the list *LabelThis*(K) using *PtrEven*(J). Similarly, let $M = LabelMeOdd(J)$. If $r(M) > r(J)$, then add the label *odd* to J. Remove J from the list *LabelThis*(M) using *PtrOdd*(J).

We have the following theorem:

Theorem 2.3. *Algorithm Scan can be implemented to run in $O(n \log \log n)$ time.*

Proof of Theorem 2.3. During the entire run of the algorithm, no interval is labeled more than twice. Therefore, the number of times an interval is added to priority queues is at most two, and also the number of times an interval is added to *LabelThis* lists is at most two. Thus, the total number of operations performed on each priority queue is O(n), and the total number of times intervals are added and deleted from *LabelThis* lists is O(n). Each operation on the *LabelThis* lists takes constant time. As the priorities are integers in the range 1 through $2n$, *ActiveEven* and *ActiveOdd* can be implemented as Van Emde Boas priority queues [20], supporting each insertion and deletion in O($\log \log n$) time. Therefore, algorithm *Scan* runs in O($n \log \log n$) time. □

3. Recognition algorithm

In this section we present an O($mn \log \log n + m^2$)-time algorithm to test whether a given graph is a perfect circular-arc graph. As the recognition of circular-arc graphs and construction of a model when the input is a circular-arc graph can both now be done in O($m + n$) time [15], our task is to test whether a given circular-arc graph is perfect. Again, given Theorem 1.1, the problem is reduced to testing whether a given circular-arc graph, presented via its model, contains an odd hole or an odd antihole. We note that the current best algorithm for recognizing perfect graphs [6] runs in O(n^9) time.

Next, we present an algorithm that tests whether a given circular-arc graph contains an odd hole. The basic idea behind the algorithm is to check for each vertex x, whether there exists a pair u, v of non-adjacent vertices in $N(x)$ such that there is an odd chordless path from u to v whose interior vertices lie in $M(x)$, the set of non-neighbors of x. Note that vertices whose arcs are contained in A_x or contain A_x cannot be in an odd hole with x.

Algorithm *FindOddHole* $(\mathcal{A})$

Input:
    $\mathcal{A}$: the model for a circular-arc graph G

Output:
    An odd hole in G, or the message "no odd holes"

for each vertex x in G **do**   /* Arc A_x corresponds to vertex x */
    /* P contains arcs that overlap the counter-clockwise end of A_x only */
    $P \leftarrow \{A_w \in \mathcal{A} \mid CC(x) \in A_w \text{ and } CL(x) \notin A_w\}$
    /* Q contains arcs that overlap the clockwise end of A_x only */
    $Q \leftarrow \{A_w \in \mathcal{A} \mid CC(x) \notin A_w \text{ and } CL(x) \in A_w\}$
    /* M contains arcs that do not intersect A_x */
    $M \leftarrow \{A_w \in \mathcal{A} \mid w \in M(x)\}$

    Chop each arc $A_w \in P$ at $CC(x)$ by changing $CL(w)$ to $CC(x)$
    Chop each arc $A_u \in Q$ at $CL(x)$ by changing $CC(u)$ to $CL(x)$
    for each $A_w \in P$ **do**
        $Q' \leftarrow Q - \{A_u \in Q \mid u \in N_G(w)\}$
        $\mathcal{I} \leftarrow \{A_w\} \cup M \cup Q'$
        $L \leftarrow CC(x)$
        $R \leftarrow CL(x)$
        $\mathcal{V} \leftarrow A_w$
        $Right \leftarrow Q'$
        Scan $(\mathcal{I}, L, R, \mathcal{V}, Right)$
        if some $A_u \in Right$ is labeled *odd* **then**
            return x and the odd chordless path from $\mathcal{V}$ to A_u found
                by *Scan*
        endif
    endfor
endfor
return "no odd holes"
endAlgorithm

Next, we prove the correctness of the algorithm *FindOddHole* and establish its time complexity.

Theorem 3.1. *Algorithm FindOddHole is correct and runs in $O(mn \log \log n)$ time.*

Proof of Theorem 3.1. Suppose the given circular-arc graph has an odd hole $v_1 v_{2k+1} \ldots v_3 v_2$, $k \geq 2$, where the vertices are indexed in the order their corresponding arcs' counter-clockwise endpoints appear in a clockwise scan of the circle. Consider the algorithm when $x = v_1$; it must be that $A_{v_2} \in P$ and $A_{v_{2k+1}} \in Q$.

Now, consider the invocation of *Scan* when $\mathcal{V} = A_{v_2}$; it must be that $A_{v_{2k+1}} \in$ *Right*. Further, the set $\mathcal{I}$ of intervals computed by the algorithm will include the arcs A_{v_i}, $i = 2, \ldots, 2k+1$. Hence, an odd chordless path exists between A_{v_2} and $A_{v_{2k+1}}$ in $\mathcal{I}$ such that no interior vertex of the path belongs to *Right*. Therefore, algorithm *Scan* will label $A_{v_{2k+1}}$ *odd*. Since A_{v_2} and $A_{v_{2k+1}}$ do not intersect in the model for the graph, they do not intersect in $\mathcal{I}$ either. Thus, any odd chordless path between A_{v_2} and $A_{v_{2k+1}}$ that the algorithm *Scan* discovers must have at least three edges. Further, as no internal vertex of such a path belongs to *Right*, given the construction of $\mathcal{I}$, each such internal vertex must belong to $M(x)$. Finally, as the chordless path discovered in $\mathcal{I}$ also corresponds to a chordless path in the graph, the vertices on this path along with x induce an odd hole in the graph.

On the other hand, for some invocation of the algorithm *Scan* with $\mathcal{V} = A_u$, suppose an $A_v \in$ *Right* is labeled *odd*. Then, clearly $\{u, v\} \subseteq N(x)$ and u is not adjacent to v in the graph, so any odd chordless path between A_u and A_v in $\mathcal{I}$ must have at least three edges. Given the construction of $\mathcal{I}$, the internal vertices of the odd chordless path discovered between A_u and A_v in $\mathcal{I}$ belong to $M(x)$. Therefore, as the chordless path discovered in $\mathcal{I}$ also corresponds to a chordless path in the graph, the vertices on that path with x induce an odd hole in the graph.

For a particular vertex x, since $|P| \leq d(x)$, *Scan* is invoked $O(d(x))$ times. Therefore, the inner loop runs in $O(d(x) * n \log \log n)$ time. Since the outer loop is done once for each vertex x in the graph, the overall running time of the algorithm is $O(mn \log \log n)$. □

The $O(m + n)$ algorithm in [2] can be used to test whether there is an odd chordless path between two given vertices of an interval graph. This algorithm can be used to recognize perfect circular-arc graphs in $O(n^3(n + m))$ time in an obvious way. Our implementation of Algorithm FindOddHole gives a better time bound.

Theorem 3.2. *Perfect circular-arc graphs can be recognized in $O(m^2 + mn \log \log n)$ time.*

Proof of Theorem 3.2. Given a graph G, we can test whether G is a circular-arc graph and if so, construct a model for it in $O(m + n)$ time [15]. We can then test whether G has an odd hole in $O(mn \log \log n)$ time using algorithm *FindOddHole*. It is well known that a circular-arc graph cannot have an antihole on an even number of vertices. Therefore, any antihole present in G must have an odd number of vertices. It is shown in [16] that when a graph does not have a hole on five vertices, whether it has an antihole can be decided in $O(m^2)$ time using $O(m+n)$ space; we can then use this algorithm to determine if there are any odd antiholes present. □

4. An approach for a colouring algorithm

An approach used to optimally colour the vertices of certain perfect graphs is via the use of an even-pair in the graph. Non-adjacent vertices u and v in graph G form an *even-pair* if every chordless path between u and v has an even number of edges.

If $\{u, v\}$ is an even-pair in graph G, then G/uv is the graph obtained by adding a vertex uv such that $N(uv) = N(u) \cup N(v)$, and then deleting the vertices u and v. We say that G/uv is obtained from G by *contracting* the even-pair $\{u, v\}$.

Lemma 4.1. [10] *Let $\{u, v\}$ be an even-pair in graph G. Then, $\omega(G/uv) = \omega(G)$ and $\chi(G/uv) = \chi(G)$. Moreover, when G is perfect, G/uv is perfect.*

Suppose $\mathcal{C}$ is a class of graphs such that every member of $\mathcal{C}$ that is not a complete graph contains an even-pair. For a graph $G \in \mathcal{C}$ and even-pair $\{u, v\}$ in G, if G/uv were in $\mathcal{C}$, then one could repeat the even-pair contractions to obtain an optimal vertex colouring for G. However, it is known that there are classes of graphs, such as the class of *Meyniel* graphs, whose noncomplete members have even-pairs, but they do not have an even-pair whose contraction yields a graph in the class [9]. On the other hand, there are classes of graphs, such as *perfectly orderable* graphs and the graphs satisfying the hypothesis of Theorem 4.2, for which contraction of an appropriate even-pair preserves membership in the class [9, 14]. Next, we show that for a perfect circular-arc graph G that is not a complete graph, there is always a choice for an even-pair so that the contraction of the even-pair yields another perfect circular-arc graph. We need some definitions first.

A *stretcher* [14] is a graph that consists of two vertex-disjoint triangles with three vertex-disjoint chordless paths between them, and with no edge other than those in the two triangles and in the three paths. Two stretchers are illustrated in Figure 1. Observe that if a stretcher does not contain an odd hole, then all three chordless paths must be of the same parity.

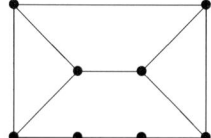
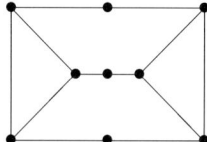

FIGURE 1. Some stretchers

Vertex x is *simplicial* if $N(x)$ is a clique. The following result is implied by a stronger theorem in [14].

Theorem 4.2. [14] *Suppose G contains no odd holes, no antiholes, and no stretcher, and vertex x is not simplicial in G. Then, there exists an even-pair $\{u, v\}$ of G such that $\{u, v\} \subseteq N(x)$.*

Since stretchers and even antiholes are not circular-arc graphs, and since odd holes and odd antiholes are not perfect, we have the following:

Corollary 4.3. *Suppose G is a perfect circular-arc graph and vertex x of G is not simplicial in G. Then, there exists an even-pair $\{u,v\}$ of G such that $\{u,v\} \subseteq N(x)$.*

Lemma 4.4. *Suppose G is a circular-arc graph, x is a vertex of G such that A_x contains no other arc in the model for G, and $\{u,v\} \subseteq N(x)$ is an even-pair of G. Then, G/uv is a circular-arc graph.*

Proof of Lemma 4.4. Let $\mathcal{A}$ be a circular-arc model for G. Let A_x be an arc that does not contain any other arc in the model. Suppose $\{u,v\} \subseteq N(x)$ is an even-pair of G.

Since u and v are non-adjacent and A_x does not contain any other arc, we can assume that A_u contains $CC(x)$ and not $CL(x)$, while A_v contains $CL(x)$ and not $CC(x)$. Consider the set of arcs $\mathcal{A}'$ obtained from $\mathcal{A}$ by replacing A_u and A_v with the arc A_{uv} that extends from $CC(u)$ to $CL(v)$ in the clockwise direction. Note that $A_{uv} = A_u \cup A_x \cup A_v$.

We claim that $\mathcal{A}'$ is a circular-arc model for G/uv.

For any pair of vertices y, z, neither of which is uv, the adjacency between them is the same in G/uv as in G and the corresponding arcs A_y and A_z are unchanged in $\mathcal{A}'$. Therefore, we need only verify that A_{uv} intersects an arc A_w ($w \neq uv$) if and only if $w \in N(u) \cup N(v)$ in G. If $w \in N(u) \cup N(v)$, then A_w intersects either A_u or A_v. In either case, A_w must then also intersect A_{uv}. Now suppose A_{uv} intersects an arc A_w. If A_w does not intersect A_u or A_v, then A_w must be contained in A_x, which contradicts the fact that A_x contains no other arc in $\mathcal{A}$. Therefore, it must be that $w \in N(u) \cup N(v)$ in G. □

The above results suggest the following recursive algorithm to optimally colour a perfect circular-arc graph. Let A_x be an arc that contains no other arc in the model for a perfect circular-arc graph G. If x is simplicial in G, we can recursively colour $G \setminus x$, after which x can be coloured easily. If x is not simplicial, by Corollary 4.3 and Lemma 4.4, we can find an appropriate even-pair $\{u,v\}$ of G in $N(x)$ and colour G/uv recursively. We can then complete the colouring of G by assigning vertices u and v the colour given to vertex uv in the colouring of G/uv.

When this paper was submitted for publication, an $O(mn^4)$-time algorithm to optimally colour a graph satisfying the hypothesis of Theorem 4.2 had been given in [14]. We could provide an implementation of the algorithm described above to optimally colour a perfect circular-arc graph, via appropriate use of algorithm *Scan*, that runs in $O(mn^2 + n^3 \log \log n)$ time. However, since then an $O(mn^2)$-time algorithm to optimally colour a graph satisfying the hypothesis of Theorem 4.2 has been presented in [12]. In view of this, we omit the details of our algorithm for optimally colouring perfect circular-arc graphs.

5. Conclusions and future work

The algorithms in this paper together with the algorithms referenced here provide a combinatorial polynomial-time algorithm for finding what the following existentially polynomial-time (EP) theorem [5] asserts to exist: For any graph G, either G contains an odd hole, or G contains an odd antihole, or G has a clique and a colouring of the same size, or G is not a circular-arc graph. Such an algorithm is what Spinrad refers to as a robust algorithm for finding a clique and a colouring of the same size in a circular-arc graph [18].

The algorithms for testing whether a given graph is a circular-arc graph construct a model in the case that the input graph is circular-arc [11, 8, 15]. However, if the graph is not circular-arc, the algorithms do not provide a polynomial-time verifiable certificate showing the graph is not circular-arc other than a trace of the execution of the algorithm. It would be very interesting to design a recognition algorithm that could provide such a certificate (a co-NP description of circular-arc graphs). It would provide a nicer robust algorithm for the above EP theorem in that there would be a nice certificate given for whichever clause of the theorem holds.

Another interesting direction for future work is to find an algorithm for determining whether a circular-arc graph has an antihole that is more efficient than the general algorithm of [16].

References

[1] S.R. Arikati, C. Pandu Rangan, and G.K. Manacher, Efficient reduction for path problems on circular-arc graphs, *BIT* **31** (1991), 182–193.

[2] S.R. Arikati and U.N. Peled, A linear algorithm for group path problem on chordal graphs, *Discrete Appl. Math.* **44** (1993), 185–190.

[3] D. Bienstock, On the complexity of testing for odd holes and induced odd paths, *Discrete Math.* **90** (1991), 85–92. See also corrigendum by B. Reed, *Discrete Math.* **102** (1992), p. 109.

[4] C. Berge, Perfect graphs, 1–21, *Six Papers on Graph Theory*, Indian Statistical Institute, 1963.

[5] K. Cameron and J. Edmonds, Existentially polytime theorems, *Polyhedral Combinatorics*, 83–100, DIMACS Ser. Discrete Math. Theoret. Comput. Sci., 1, Amer. Math. Soc., Providence, RI, 1990.

[6] M. Chudnovsky, G. Cornuéjols, X. Liu, P. Seymour, and K. Vušković, Recognizing Berge graphs, *Combinatorica* **25** (2005), 143–187.

[7] M. Chudnovsky, N. Robertson, P. Seymour, and R. Thomas, The strong perfect graph theorem, *Annals of Math.*, to appear.

[8] E.M. Eschen and J.P. Spinrad, An $O(n^2)$ algorithm for circular-arc graph recognition, *Proceedings of the Fourth Annual ACM-SIAM Symposium on Discrete Algorithms* (Austin, TX, 1993), 128–137, ACM, New York, 1993.

[9] H. Everett, C. de Figueiredo, C.L. Sales, F. Maffray, O. Porto, and B. Reed, Even pairs, in *Perfect graphs*, edited by J.L. R. Alfonsín and B. A. Reed, John Wiley & Sons, Ltd, 2001.

[10] J. Fonlupt and J.P. Uhry, Transformations which preserve perfectness and h-perfectness of graphs, *Ann. Discrete Math.* **16** (1982), 83–85.

[11] W.L. Hsu, $O(MN)$ algorithms for the recognition and isomorphism problems on circular-arc graphs, *SIAM J. Comput.* **24** (1995), 411–439.

[12] B. Lévêque, F. Maffray, B. Reed and N. Trotignon, Coloring Artemis graphs, *Manuscript*.

[13] L. Lovász, A characterization of perfect graphs, *J. Combinatorial Theory (Ser. B)* **13** (1972), 95–98.

[14] F. Maffray and N. Trotignon, A class of perfectly contractile graphs, *Manuscript*, 2002.

[15] R. McConnell, Linear-time recognition of circular-arc graphs, *Algorithmica* **37** (2003), 93–147.

[16] S.D. Nikolopoulos and L. Palios, Hole and antihole detection in graphs, *Proceedings of the Fifteenth Annual ACM-SIAM Symposium on Discrete Algorithms* (SODA), 2004, 843–852.

[17] C.R. Satyan and C. Pandu Rangan, The parity path problem on some subclasses of perfect graphs, *Discrete Appl. Math.* **68** (1996), 293–302.

[18] J.P. Spinrad, *Efficient Graph Representations*, American Mathematical Society, Providence, Rhode Island, 2003.

[19] A. Tucker, Coloring a family of circular arcs, *SIAM J. Appl. Math.* **29** (1975), 493–502.

[20] P. Van Emde Boas, Preserving order in a forest in less than logarithmic time and linear space, *Information Processing Letters* **6** (1977), 80–82.

Kathie Cameron
Department of Mathematics, Wilfrid Laurier University
Waterloo, Canada N2L 3C5
e-mail: `kcameron@wlu.ca`

Elaine M. Eschen
Lane Department of Computer Science and Electrical Engineering
West Virginia University
Morgantown, WV 26506, USA
e-mail: `eeschen@csee.wvu.edu`

Chính T. Hoàng
Department of Physics and Computer Science, Wilfrid Laurier University
Waterloo, Canada N2L 3C5
e-mail: `choang@wlu.ca`

R. Sritharan
Computer Science Department, The University of Dayton
Dayton, OH 45469, USA
e-mail: `rst@cps.udayton.edu`

On Edge-maps whose Inverse Preserves Flows or Tensions

Matt DeVos, Jaroslav Nešetřil and André Raspaud

> **Abstract.** A *cycle* of a graph G is a set $C \subseteq E(G)$ so that every vertex of the graph $(V(G), C)$ has even degree. If G, H are graphs, we define a map $\phi : E(G) \to E(H)$ to be *cycle-continuous* if the pre-image of every cycle of H is a cycle of G. A fascinating conjecture of Jaeger asserts that every bridgeless graph has a cycle-continuous mapping to the Petersen graph. Jaeger showed that if this conjecture is true, then so is the 5-cycle-double-cover conjecture and the Fulkerson conjecture.
>
> Cycle continuous maps give rise to a natural quasi-order $\succ$ on the class of finite graphs. Namely, $G \succ H$ if there exists a cycle-continuous mapping from G to H. The goal of this paper is to establish some basic structural properties of this (and other related) quasi-orders. For instance, we show that $\succ$ has antichains of arbitrarily large finite size. It appears to be an interesting question to determine if $\succ$ has an infinite antichain.

1. Introduction

Some of the most striking conjectures in structural graph theory have an algebraic flavour. These include Tutte's conjectures on flows, a variety of polynomials associated with combinatorial phaenomena, the Hedetniemi product conjecture, and Ulam's reconstruction conjecture. In all these cases not only one can *formulate* these problems involving some familiar algebraic notions and constructions but in all of these cases some of the (currently) best results were obtained after the proper algebraic context was realized, see, e.g., [9, 23, 25, 26, 27, 14, 16]. It is perhaps not surprising that many of these problems can be expressed as statements about partially ordered (or quasi-ordered) sets and classes. In some of these situations such a formulation is straightforward as the problem deals directly with the category of graphs and standard maps, such as homomorphisms. This is the case, e.g.,

---

Supported by Project LN00A056 and 1M0021620808 of the Czech Ministery of Education.
Supported by Barrande 02887WD P.A.I. Franco-Tchèque.

for the product conjecture, see, e.g., [23, 17]. However, in other situations a *different* algebraic and order-theoretic formulation is far from obvious and the right definitions were sought for a long time. Sometimes strange looking definitions are far from arbitrary as they reflect the experience gained with dealing with concrete problems (such as 4CC) and other algebraic concepts such as matroids, flows and tensions. The later notions are the subject of this paper.

In [10], Jaeger constructs a partial order on the class of graphs, and makes a fascinating conjecture concerning the atoms of this order. If true, his conjecture would immediately imply both the 5-cycle double cover conjecture, and Fulkerson's conjecture. Jaeger's interesting partial order was the starting point for this research. In the next section we shall introduce flow and tension continuous maps over rings, together with the corresponding orders. Our $\mathbb{Z}_2$-flow continuous order is quite closely related to Jaeger's order, and later in the paper, this connection is made precise. After establishing some basic properties of flow and tension continuous maps, we study the $\mathbb{Z}_2$ and $\mathbb{Z}$ flow and tension orders in detail. Two results of note are a theorem showing that the $\mathbb{Z}_2$ flow order has arbitrarily large finite antichains, and a theorem establishing the relationship between the $\mathbb{Z}$-tension order and the usual homomorphism order.

Our approach has some similarities to [13] where the authors are also interested in various maps between graphs (and mostly between edges). However despite some formal similarities our approach is very different (although it is manifested in some subtle differences): our mappings are defined by "continuous"-type condition (for example: by requiring that the preimage of every cycle is a cycle), whereas mappings in [13] are mostly "open" (for example cycle preserving). The motivation of [13] is matroid theory (and strong maps are one of the classes considered). Our motivation is flow and coloring problems (following Tutte's original approach). For these types of questions, a "continuous" approach seems to be better suited.

In this paper we concentrate on graphs (both directed and undirected). However several results one can consider in the context of matroids (and regular matroids in particular). This we postpone to another occasion.

2. Basic definitions and overview

All graphs considered in this paper are assumed to be finite unless it is explicitly stated otherwise. Graphs may have both loops and multiple edges. Frequently, we will have need to refer to both an oriented graph and the underlying undirected graph. If G is an undirected graph, then we may use $\vec{G}$ to denote an orientation of G. If $\vec{G}$ is defined to be an oriented graph, then it is understood that G is the underlying undirected graph.

Let G be a graph and let $C \subseteq E(G)$. We say that C is a *cycle* if every vertex of the graph $(V(G), C)$ has even degree. A *circuit* is a non-empty cycle which is minimal with respect to inclusion. We define the *odd-girth* $\gamma^o(G)$ of a graph G to be the size of the smallest circuit of G of odd cardinality (or ∞ if none exists). If $X \subseteq V(G)$, then we will let $\Delta(X)$ denote the set of edges with one end in X and one end in $V(G) \setminus X$. For a vertex $v \in V(G)$, we use $\Delta(v)$ to denote $\Delta(\{v\})$. Any set of edges of the form $\Delta(X)$ for some $X \subseteq V(G)$ is defined to be an *edge-cut*. A *bond* is a non-empty edge-cut which is minimal with respect to inclusion. We define $\lambda^o(G)$ to be the size of the smallest bond of G of odd cardinality (or ∞ if none exists). A single edge $e \in E(G)$ is a *cut-edge* if $\{e\}$ is an edge-cut.

If $\vec{G}$ is an oriented graph and $X \subseteq V(\vec{G})$, then we let $\Delta^+(X)$ denote the set of edges with tail in X and head in $V(\vec{G}) \setminus X$. We define $\Delta^-(X)$ to be $\Delta^+(V(\vec{G}) \setminus X)$ and as before, for a vertex $v \in V(\vec{G})$, we let $\Delta^+(v) = \Delta^+(\{v\})$ and $\Delta^-(v) = \Delta^-(\{v\})$. If $C \subseteq G$ is a circuit and $e, f \in E(\vec{G})$, then e and f are either given the same orientation relative to C or the opposite orientation relative to C. A *direction* of C is a pair (X, Y) of disjoint subsets of $E(C)$ with union $E(C)$ so that every $e \in X$ and $f \in Y$ have opposite orientation with respect to C. We call the edges in X *forward* edges and the edges in Y *backward* edges. We say that an edge $e \in E(\vec{G})$ is a *cut-edge* if the corresponding edge is a cut-edge of the underlying undirected graph G.

Let M be an Abelian group, let $\vec{G}$ be an oriented graph, and let $\phi : E(\vec{G}) \to M$ be a map. We say that ϕ is a *flow* or an *M-flow* if

$$\sum_{e \in \Delta^+(v)} \phi(e) = \sum_{e \in \Delta^-(v)} \phi(e)$$

holds for every vertex $v \in V(\vec{G})$. We say that ϕ is a *tension* or an *M-tension* if

$$\sum_{e \in A} \phi(e) = \sum_{e \in B} \phi(e)$$

holds for every circuit $C \subseteq G$ where (A, B) is the direction of C.

We now follow the framework of Jaeger in [11] by defining a restricted class of flows and tensions. Let $B \subseteq M$ and assume that $-B = B$. If $\phi : E(\vec{G}) \to M$ is a flow (tension) and $\phi(E(\vec{G})) \subseteq B$, then we say that ϕ is a *B-flow* (*B-tension*). We say that a flow (tension) ϕ is *nowhere-zero* if it is a $(M \setminus \{0\})$-flow (tension). We say that $\phi : E(G) \to \mathbb{Z}$ is a *k-flow* (*k-tension*) for a positive integer k if ϕ is a B-flow (B-tension) where $B = \{-(k-1), \ldots, -1, 0, 1, \ldots, k-1\}$. If ϕ is a B-flow (B-tension) of $\vec{G}$ and we reverse the orientation of some edge $e \in E(\vec{G})$, then by replacing $\phi(e)$ by its additive inverse, we maintain that ϕ is a B-flow (B-tension). Thus, for an unoriented graph G, we have that some orientation of G has a B-flow (B-tension) if and only if every orientation of G has a B-flow (B-tension). In this case, we say that G has a B-flow (B-tension). Similarly, we say that G has a nowhere-zero M-flow or k-flow (M-tension or k-tension) if some (and thus every) orientation of G has such a flow (tension). The following is a famous conjecture of Tutte on nowhere-zero flows.

Conjecture 2.1 (The 5-flow conjecture). *Every graph with no cut-edge has a nowhere-zero 5-flow.*

In this introduction, we will focus most of our attention on B-flows. However, we wish to mention here that the theory of B-tensions is quite rich and is closely connected with graph coloring. Indeed, it is an easy fact (see Proposition 3.4) that a graph has a B-tension if an only if it has a homomorphism to a certain Cayley graph. It follows from this that G has a nowhere-zero k-tension if and only if it is k-colorable.

Jaeger initiated the study of B-flows and B-tensions and observed that a number of important questions in graph theory may be phrased in terms of the existence of certain B-flows. Here we list three famous conjectures. For each of these problems we offer two equivalent formulations. The first is the traditional statement of the problem, the second is an equivalent statement in terms of B-flows.

Conjecture 2.2 (The five cycle double cover conjecture).
(1) *For every graph with no cut-edge, there is a list of five cycles so that every edge is contained in exactly two.*
(2) *Every graph with no cut-edge has a B-flow for the set $B \subseteq \mathbb{Z}_2^5$ consisting of those vectors with exactly two 1's.*

Conjecture 2.3 (The orientable five cycle double cover conjecture).
(1) *For every oriented graph with no cut-edge, there is a list of five 2-flows $\phi_1, \phi_2, \ldots, \phi_5$ with $\sum_{i=1}^{5} \phi_i = 0$ such that every edge is in the support of exactly two of these flows.*
(2) *Every graph with no cut-edge has a B-flow for the set $B \subseteq \mathbb{Z}^5$ consisting of those vectors with exactly three 0's, one 1, and one -1.*

Conjecture 2.4 (Fulkerson).
(1) *For every cubic graph with no cut-edge, there is a list of 6 perfect matchings so that every edge is contained in exactly two.*
(2) *Every graph with no cut-edge has a B-flow for the set $B \subseteq \mathbb{Z}_2^6$ consisting of those vectors with exactly four 1's.*

For the history of these conjectures see, e.g., [2, 24, 11, 8] and also the original papers [5, 22, 3]. In addition to defining B-flows, Jaeger defined a type of mapping between graphs which is closely related to one we give here. We will discuss the relationship between our and Jaeger's definitions in Section 9 of this paper. Next we give the central definition for this paper.

Definition. Let $\vec{G}$ and $\vec{H}$ be oriented graphs, let M be an Abelian group, and let $f : E(\vec{G}) \to E(\vec{H})$. We say that f is M-*flow-continuous* (M-*tension-continuous*) if $\phi \circ f$ is a M-flow (M-tension) of $\vec{G}$ for every M-flow (M-tension) ϕ of $\vec{H}$ (see Figure 1).

The name flow-continuous (tension-continuous) is used here since in such a map every flow (tension) of $\vec{H}$ lifts to a flow (tension) of $\vec{G}$. Note that if f is a flow-continuous (tension-continuous) map from $\vec{G}$ to $\vec{H}$ and we reverse the direction of

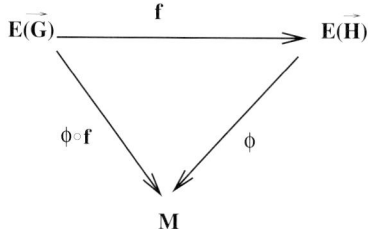

FIGURE 1. M-flow(tension)-continuous

an arc $e \in E(\vec{H})$, then by reversing the arcs $f^{-1}(\{e\})$ in $\vec{G}$, we maintain that f is flow-continuous (tension-continuous). Also note that if f is M-flow-continuous (M-tension-continuous), then f is also M^n-flow-continuous (M^n-tension-continuous) for every positive integer n. Here we mention the key property of flows and continuous maps which is motivation for our study.

Proposition 2.5. *If there is a M-flow-continuous (M-tension-continuous) map from $\vec{G}$ to $\vec{H}$ and H has a B-flow (B-tension) for some $B \subseteq M^n$, then G also has a B-flow (B-tension).*

Proof. We prove the proposition in the flow-continuous case. The tension-continuous case follows by the same argument. If $f : E(\vec{G}) \to E(\vec{H})$ is M-flow-continuous, then it is also M^n-flow-continuous. Thus, if $\phi : E(\vec{H}) \to B$ is a B-flow of H, then $\phi \circ f$ is a B-flow of G. □

The above proposition is especially interesting because it suggests a different approach to showing the existence of a B-flow. To prove that G has a B-flow, it suffices to show that some orientation of G has an M-flow-continuous map to an orientation of a graph H which is known to have a B-flow. Based on this property, we now define for every Abelian group M the relations $\succ_M^f$ and $\succ_M^t$ as follows. For any two unoriented graphs G, H, we write $G \succ_M^f H$ ($G \succ_M^t H$) if there exists an M-flow-continuous (M-tension-continuous) map between some orientation of G and some orientation of H. We write $G \not\succ_M^f H$ or $G \not\succ_M^t H$ if no such map exists. A relation is a *quasi-order* if it is reflexive and transitive.

Proposition 2.6. *The relations $\succ_M^f$ and $\succ_M^t$ are quasi-orders on the class of finite graphs.*

Proof. We give the proof only for $\succ_M^f$, a similar argument works for $\succ_M^t$. For any graph G, and any orientation $\vec{G}$ of G, the identity map from $E(\vec{G})$ to $E(\vec{G})$ is M-flow-continuous, so $G \succ_M^f G$. To prove that $\succ_M^f$ is transitive, let F, G, H be graphs with $F \succ_M^f G \succ_M^f H$. Then there exists a flow-continuous map f from an orientation $\vec{F}$ of F to an orientation $\vec{G}$ of G and a flow-continuous map f' from an orientation $\check{G}$ of G to an orientation $\vec{H}$ of H. By possibly reversing the direction of some arcs in $\vec{G}$ and reversing the corresponding arcs in $\vec{F}$ (as described above), we may assume that $\vec{G} = \check{G}$. Now, for any M-flow ϕ of $\vec{H}$, the map $\phi \circ f'$ is a flow of $\vec{G} = \check{G}$ and the map $\phi \circ f' \circ f$ is a K-flow of F. Thus, the map $f' \circ f$ is a M-flow continuous map from $\vec{F}$ to $\vec{H}$ and we have that $F \succ_M^f H$ as required. □

We review now some simple characteristics of quasi-orders which we will investigate. Let $\succ$ be a quasi-order on S. Two elements $x, y \in S$ are *comparable* if either $x \succ y$ or $y \succ x$. We say that x *dominates* y if $x \succ y$ and we say that x and y are *equivalent* if $x \succ y$ and $y \succ x$. An element $x \in S$ is *maximal* (*minimal*) if x is equivalent to every element y for which $y \succ x$ ($x \succ y$). A set $Y \subseteq S$ is an *antichain* if no two elements in Y are comparable. A set $X \subseteq S$ is a *chain* if every two elements in X are comparable but not equivalent. An *increasing* (*decreasing*) *chain* is a sequence $\{x_n\}_{n=1}^\infty$ such that $x_j \succ x_i$ if and only if $j \geq i$ ($j \leq i$). An increasing chain $\{x_n\}_{n=1}^\infty$ is said to be a *scaling* chain if every $y \in S$ which is not maximal is dominated by x_i for some $i \geq 1$. An element $y \in S$ is an *atom* if y is not minimal and every element which is dominated by y but is not equivalent to y, is minimal. Finally, we say that a function $f : S \to \mathbb{Z}$ is *monotone* if either $x \succ y$ implies $f(x) \geq f(y)$ or $x \succ y$ implies $f(x) \leq f(y)$.

A quasi-order $\succ$ is said to be *well founded* if it does not contain an infinite (strictly) descending chain $X_1 \succ X_2 \succ \cdots$. We say that it is a *well quasi-order* (shortly WQO) if it is well founded and it does not contain an infinite antichain. The following proposition shows that these two concepts are equivalent for the orders $\succ_M^f$ and $\succ_M^t$.

Proposition 2.7. *For any Abelian group M, the orders $\succ_M^f$ and $\succ_M^t$ are WQO if and only if they are well founded.*

Proof. We need only prove that well founded implies WQO as the other direction is clear. Assume the contrary and let $G_1, G_2, \ldots$ be an infinite antichain. Then letting $+$ denote the disjoint union, we have that $G_1 \succ_M^f G_1 + G_2 \succ_M^f G_1 + G_2 + G_3 \cdots$ is a strictly decreasing chain, contradicting our assumption. □

Next we state an important open problem concerning the flow orders.

Problem 2.8. *Is $\succ_M^f$ WQO for any group M?*

This problem is related to the global conjectures by means of the following:

Theorem 2.9. *If $\succ_M^f$ is WQO then $\succ_M^f$ has a single atom.*

We shall see (Section 3) that a graph G is an atom of $\succ_M^f$ if G contains no bridge and there is no bridgeless graph H satisfying $G \succ_M^f H$ and $H \not\succ_M^f G$.

Proof. Assume that $\succ_M^f$ is WQO. By the well-foundedness property for every bridgeless graph G there exists an atom H with $G \succ_M^f H$. Let $\mathcal{A}$ be the set of all (mutually non-equivalent) atoms. Assume that $|\mathcal{A}| > 1$ and let H, H' be atoms satisfying $H \not\succ_M^f H' \not\succ_M^f H$. Then consider the graph $H'' = H + H'$ (disjoint union of H and H'; 1-sum could be used too). Then $H \succ_M^f H''$ and $H' \succ_M^f H''$ (by inclusion), while obviously $H'' \not\succ_M^f H$, $H'' \not\succ_M^f H'$. Thus H (and H') fails to be atoms, a contradiction. □

Theorem 2.9 explains why all of the global conjectures have only a single atom, and is another indication of the importance of these atoms in the study of the flow orders.

The main purpose of this paper is to investigate the structure of the quasi-orders $\succ_M^f$ and $\succ_M^t$. We will establish some basic properties and connections between these orders and raise some new open problems. Here we mention a fascinating conjecture equivalent to a conjecture of Jaeger concerning the order $\succ_{\mathbb{Z}_2}^f$ which we view as powerful motivation for the study of the flow-continuous quasi-orders. We use P_{10} to denote the Petersen graph (see Figure 2).

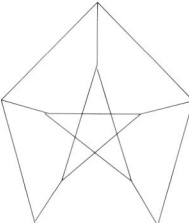

FIGURE 2. P_{10}

Conjecture 2.10. *If G has no cut-edge, then $G \succ_{\mathbb{Z}_2}^f P_{10}$.*

If this conjecture is true, then so is the five cycle double cover conjecture and Conjecture 2.4 of Fulkerson. This implication follows immediately from Proposition 2.5 and the second formulations of these conjectures given in the introduction and the fact that these conjectures hold for the Petersen graph.

This paper is organized as follows. In the next section we establish some general properties of flow and tension continuous maps. In section 4, we compare the various quasi-orders and prove that homomorphisms are precisely the cut-tension continuous maps. Following this are four short sections in which each of the orders $\succ_{\mathbb{Z}_2}^f$, $\succ_{\mathbb{Z}}^f$, $\succ_{\mathbb{Z}_2}^t$, and $\succ_{\mathbb{Z}}^t$ is given a quick investigation. Finally, a section is devoted to Jaeger's order $\prec_J$ and its comparison to our approach (which we believe is more streamlined).

3. Flow/tension-continuous maps over rings

Before we study the orders generated by some particular groups, we wish to mention some general properties satisfied by all flow or tension-continuous maps over rings. This will shorten and unify some of our statements below. Our approach is the standard one.

Throughout this section, we will assume that K is a ring. For every oriented graph $\vec{G}$, we regard $K^{E(\vec{G})}$ as a module over K. It follows from the definitions that the set of K-tensions and K-flows are both submodules of $K^{E(\vec{G})}$. If $\phi, \psi \in K^{E(\vec{G})}$, we say that ϕ and ψ are *orthogonal*, written $\phi \perp \psi$, if $\sum_{e \in E(\vec{G})} \phi(e)\psi(e) = 0$. It is an elementary fact that every K-flow is orthogonal to every K-tension. Furthermore, a map $\psi : E(\vec{G}) \to K$ is a K-flow (K-tension) if and only if it is orthogonal to every K-tension (K-flow).

Next we state a key equivalence. If $f : X \to Y$ and $\psi : X \to K$, then we let $\psi_f : Y \to K$ be given by the rule $\psi_f(y) = \sum_{x \in f^{-1}(\{y\})} \psi(x)$.

Theorem 3.1. *Let $\vec{G}$ and $\vec{H}$ be oriented graphs and let $f : E(\vec{G}) \to E(\vec{H})$. Then f is K-flow-continuous (K-tension-continuous) if and only if ψ_f is a K-tension (K-flow) of $\vec{H}$ for every K-tension (K-flow) ψ of $\vec{G}$.*

Proof. We prove the theorem only in the flow-continuous case. The tension continuous case follows by a similar argument. Let $f : E(\vec{G}) \to E(\vec{H})$ be a map, let ψ be a K-tension of G and let ϕ be a K-flow of H. Then we have the following equations :

$$\sum_{e \in E(H)} \phi(e)\psi_f(e) = \sum_{e \in E(H)} \phi(e) \sum_{s \in f^{-1}(\{e\})} \psi(s)$$
$$= \sum_{e \in E(H)} \sum_{s \in f^{-1}(\{e\})} \phi(f(s))\psi(s) = \sum_{s \in E(G)} (\phi \circ f)(s)\psi(s).$$

If we assume that f is K-flow-continuous, then $(\phi \circ f)$ is a flow on G, so the last line in the above equation evaluates to zero. In this case, we have that $\psi_f \perp \phi$. Since ϕ was an arbitrary flow, it follows that ψ_f is orthogonal to every flow, so ψ_f is a tension as desired.

If we assume that ψ_f is a tension of H, then the first line in the above equation evaluates to zero. In this case, we have that $\psi \perp (\phi \circ f)$. Since ψ was an arbitrary tension, it follows that $\phi \circ f$ is orthogonal to every tension, so $\phi \circ f$ is a flow as desired. □

Next we prove that for every graph H, there is a subspace $B \subseteq K^n$ with $-B = B$, such that $G \succ_K^f H$ ($G \succ_K^t H$) if and only if G has a B-flow (B-tension). This useful fact was first discovered by Jaeger (in the case where $K = \mathbb{Z}_2$). Let A be a matrix with entries in K and columns indexed by $E(G)$, and let $a_1, a_2, \ldots, a_n$ denote the rows of A. We say that A *represents the cycle-space (cocycle-space) of* G if every a_i is a K-flow (K-tension) and for every K-flow (K-tension) ψ of G, there exist $x_1, x_2, \ldots, x_n \in K$ such that $\psi = \sum_{i=1}^n x_i a_i$.

Theorem 3.2. *Let $\vec{H}$ be an oriented graph, let K be a ring, let A be an $n \times m$ matrix over K which represents the cycle-space (cocycle-space) of $\vec{H}$, and let $B = \{x \in K^n \mid x \text{ or } -x \text{ is a column of } A\}$. Then for every graph G, we have that $G \succ_K^f H$ ($G \succ_K^t H$) if and only if G has a B-flow (B-tension).*

Proof. Again, we prove the statement only in the case when A represents the cycle-space of H. A similar argument proves the statement when A represents the cocycle-space of H. Let $a_1, a_2, \ldots, a_n$ denote the row vectors of A. We think of a_i as a map from $E(H)$ to K and we let $\phi : E(H) \to K^n$ be the map given by the rule $\phi(e) = (a_1(e), a_2(e), \ldots, a_n(e))$. Let $\vec{G}$ be an orientation of G, and let $f : E(\vec{G}) \to E(\vec{H})$ be a map. Next we establish the following claim.

Claim. f is K-flow-continuous if and only if $\phi \circ f$ is a flow.

Proof. If f is K-flow-continuous, then $a_i \circ f$ is a flow for $1 \le i \le n$, so $\phi \circ f$ is also a flow. On the other hand, if $\phi \circ f$ is a flow and $\psi : E(\vec{H}) \to K$ is any K-flow of H, then we may choose $x_i \in K$ for $1 \le i \le n$ so that $\psi = \sum_{i=1}^n x_i a_i$. Since $\phi \circ f$ is a flow, $a_i \circ f$ is a flow for $1 \le i \le n$ and we find that $\psi \circ f = \sum_{i=1}^n x_i a_i \circ f$ is also a flow. Since ψ was an arbitrary flow of $\vec{H}$, it follows that f is K-flow-continuous. This completes the proof of the claim.

Let B_0 denote the set of columns of the matrix A. It follows from the above claim that $G \succ_K^f H$ if and only if there exists an orientation $\vec{G}$ of G and a map $\psi : E(\vec{G}) \to B_0$ so that ψ is a flow. By reversing edges, the latter condition is equivalent to the statement that G has a B-flow. This completes the proof. □

Let M be an Abelian group and let $p : V(\vec{G}) \to M$ be a map (p for potential). We define the *coboundary* of p to be the map $\delta p : E(\vec{G}) \to M$ given by the rule $\delta p(e) = p(v) - p(u)$ if e is directed from u to v. It is easy to see that δp is always a tension. The following well-known lemma shows that every tension arises in this manner from a potential.

Lemma 3.3. *For every tension $\phi : E(\vec{G}) \to M$, there exists a map $p : V(G) \to M$ so that $\delta p = \phi$. Further, if G is connected and $\delta p = \phi = \delta p'$, then there is a fixed $x \in M$ so that $p(v) - p'(v) = x$ for every $e \in E(\vec{G})$.*

Proof. Define the *height* of a walk W to be the sum of ϕ on the forward edges of W minus the sum of ϕ on the backward edges of W. Since ϕ is a tension, the height of every closed walk is zero. Now, choose a vertex u and define the map $p : V(\vec{G}) \to M$ by the rule $p(v) =$ the height of a walk from u to v. It follows from the fact that every closed walk has height zero that p is well defined. Furthermore, by construction $\delta p = \phi$. To prove the second statement in Lemma 3.3, let $p' : V(\vec{G}) \to K$ satisfy $\delta p' = \phi$. Now for any edge e directed from u to v, we have that $p'(v) - p'(u) = \phi(e) = p(v) - p(u)$. Thus, $p(v) - p'(v) = p(u) - p'(u)$ and Lemma 3.3 follows. □

If G, H are undirected graphs. A *homomorphism* from G to H is a map $f : V(G) \to V(H)$ with the property that $f(u) \sim f(v)$ whenever $u \sim v$. It is easy to see that there is a homomorphism from G to K_n if and only if G is n-colorable. Thus, we may view homomorphisms as a generalization of graph coloring.

For any Abelian group M and any subset $B \subseteq M$ with $B = -B$, we let $Cayley(M, B)$ denote the simple (undirected, but not necessarily loopless) graph with vertex set M in which two vertices $x, y \in M$ are adjacent if and only if $x - y \in B$. Note that $Cayley(M, B)$ is an infinite graph if M is infinite. The following proposition is a well-known equivalence which we sketch a proof of for completeness.

Proposition 3.4. *Let M be an Abelian group and let $B \subseteq M$ with $-B = B$. Then a graph G has a B-tension if and only if there is a homomorphism from G to $Cayley(M, B)$.*

Proof. Let $\vec{G}$ be an orientation of G. If there is a homomorphism p from G to $Cayley(M, B)$, then the map $\psi = \delta p$ is a B-tension. If $\psi : E(\vec{G}) \to M$ is a B-tension, then by the above lemma, we may choose a map $p : V(\vec{G}) \to M$ so that $\delta p = \psi$. Since $V(\vec{G}) = V(G)$, the map p is a homomorphism from G to $Cayley(M, B)$ as desired. □

Based on the above proposition and Theorem 3.2, we now have the following corollary.

Corollary 3.5. *Let $\vec{H}$ be an oriented graph, let K be a ring, let A be an $n \times m$ matrix over K which represents the cocycle-space of $\vec{H}$, and let $B = \{x \in K^n \mid x \text{ or } -x \text{ is a column of } A\}$. Then $G \succ_K^t H$ if and only if $G \succ_{\mathrm{hom}} Cayley(K^n, B)$.*

Proof. By Theorem 3.2, $G \succ_K^t H$ if and only if G has a B-tension. By Proposition 3.4 this is equivalent to the existence of a graph homomorphism from G to $Cayley(K^n, B)$. □

4. Comparing the quasi-orders

In this section we compare the quasi-orders induced by flow-continuous (tension-continuous) maps over different groups. At the end of this section, we introduce a quasi-order based on graph homomorphisms and we compare this to the tension-continuous orders. We begin with a definition of circuit-flows and cut-tensions followed by an easy (folkloristic) proposition which we prove for the sake of completeness.

Let $\vec{G}$ be an oriented graph and let M be an Abelian group. For every $X \subseteq V(\vec{G})$ and every $z \in M$, define the map $\gamma_X^z : E(\vec{G}) \to M$ by the rule

$$\gamma_X^z(e) = \begin{cases} z & \text{if } e \in \Delta^+(X) \\ -z & \text{if } e \in \Delta^-(X) \\ 0 & \text{otherwise.} \end{cases}$$

For any map $\phi : E(\vec{G}) \to M$, we say that ϕ is a *cut-tension* if there exist $X \subseteq V(G)$ and $z \in M$ so that $\phi = \gamma_X^z$. If such an X, z exist with the added property that $\Delta(X)$ is a bond, then we say that ϕ is a *bond-tension*. The following observation follows from the definitions.

Observation 4.1. *If $\vec{G}$ is connected and ϕ is a cut-tension of $\vec{G}$, then every $p : V(\vec{G}) \to M$ which satisfies $\delta p = \phi$ must have $|p(V(\vec{G}))| \leq 2$.*

If $C \subseteq G$ is a circuit and (A, B) is a direction of C, and then $z \in M$, the map $\phi_C^z : E(\vec{G}) \to M$ given by the rule

$$\psi_C^z(e) = \begin{cases} z & \text{if } e \in A \\ -z & \text{if } e \in B \\ 0 & \text{otherwise} \end{cases}$$

is a flow. We define any flow of this form to be a *circuit-flow*. Circuit-flows are dual to bond-tensions. Since we will not need the flow analogue of a cut-tension, we will not define it.

Proposition 4.2. *For every flow (tension) ϕ of $\vec{G}$, there exist circuit-flows (bond-tensions) $\phi_1, \phi_2, \ldots, \phi_n$ such that $\phi = \sum_{i=1}^n \phi_i$.*

Proof. We prove the proposition in the case that ϕ is a flow. The case when ϕ is a tension follows by a similar argument. We proceed by induction on $Supp(\phi)$. The proposition is trivially true if $Supp(\phi) = \emptyset$, so we may assume that this is not so and choose an edge $e \in E(\vec{G})$ with $\phi(e) = x \neq 0$. It follows immediately from the definitions that there is no edge-cut C containing e with the property that $\phi(f) = 0$ for every $f \in C \setminus \{e\}$. Thus, we may choose a circuit D with $e \in D$ such that $D \subseteq Supp(\phi)$. Let $\phi_1 : E(\vec{G}) \to M$ be a circuit-flow with $Supp(\phi_1) = D$ and with $\phi_1(e) = x$. By induction, we may choose a list of circuit flows $\phi_2, \phi_3, \ldots, \phi_n$ with $\sum_{i=2}^n \phi_i = \phi - \phi_1$. By construction, $\phi_1, \phi_2, \ldots, \phi_n$ is a list of circuit flows with the required properties. □

Using this we can characterize the minimal and maximal elements in the flow and tension continuous orders over every group. In particular, the following shows that the minimal elements in the flow (tension) continuous order are independent of the group.

Theorem 4.3. *A graph is minimal in $\succ_M^f$ ($\succ_M^t$) if and only if it contains a cut-edge (loop). A graph G is maximal in $\succ_M^f$ ($\succ_M^t$) if and only if there is an orientation $\vec{G}$ of G such that every constant map from $E(\vec{G})$ to M is a flow (tension).*

Proof. We prove the proposition only for the flow order $\succ_M^f$. The same argument works for $\succ_M^t$ if we replace every occurrence of "flow" with "tension" and interchange the use of the words "loop" and "cut-edge".

Let $\vec{H}$ be an oriented graph with a cut-edge s and let $\vec{G}$ be any oriented graph. We claim that the map $f : E(\vec{G}) \to E(\vec{H})$ given by the rule $f(e) = s$ for every $e \in E(\vec{G})$ is M-flow-continuous. To see this, let ϕ be a flow of $\vec{H}$. Then $\phi(s) = 0$, so $\phi \circ f$ is identically zero and we have that it is a flow.

To see that these are the only minimal graphs, let $\vec{G}$ be an oriented graph without a cut-edge and let $\vec{H}$ be an oriented graph with a single edge s which is a cut-edge. By the above argument $G \succ_M^f H$. We claim that $H \not\succ_M^f G$. To see this, let $f : E(\vec{H}) \to E(\vec{G})$ be a map and let $e = f(s)$. Since e is not a cut-edge of $\vec{G}$, there is a circuit containing e, so we may choose a circuit-flow $\phi : E(\vec{G}) \to M$ with $e \in Supp(\phi)$. Now $\phi \circ f$ is not a flow of $\vec{H}$.

Let G be a graph with an orientation $\vec{G}$ such that every constant map from $\vec{G}$ to M is a flow. Let $\vec{H}$ be an oriented graph, let $s \in E(\vec{H})$ be an edge and let $f : E(\vec{G}) \to E(\vec{H})$ be the map given by the rule $f(e) = s$ for every $e \in E(\vec{G})$. Then for every flow $\phi : E(\vec{H}) \to M$, the map $\phi \circ f$ is constant, so by assumption it is a flow.

To see that these are the only maximal graphs, let H be a graph with no orientation satisfying the property above and let $\vec{H}$ be an orientation of H. Let $\vec{G}$ be an orientation of a graph with a single edge s which is a loop. By the above argument $G \succ^f_M H$. We claim that $H \not\succ^f_M G$. To see this, let $f : E(\vec{H}) \to E(\vec{G})$ be a map and choose $x \in M$ such that the function on $E(\vec{H})$ which is constantly x is not a flow. Then the map $\phi : E(\vec{G}) \to M$ given by the rule $\phi(s) = x$ is a flow, but $\phi \circ f$ is not. The claim follows. $\square$

We can also prove that the order $\succ^f_\mathbb{Z}$ ($\succ^t_\mathbb{Z}$) is the most restrictive among the flow (tension) continuous orders.

Theorem 4.4. *If $G \succ^f_\mathbb{Z} H$ ($G \succ^t_\mathbb{Z} H$), then $G \succ^f_M H$ ($G \succ^t_M H$) for every Abelian group M.*

Proof. We prove the proposition only for the case $G \succ^t_\mathbb{Z} H$. The flow-continuous case follows by a similar argument. Let $f : E(\vec{G}) \to E(\vec{H})$ be a $\mathbb{Z}$-tension continuous map from an orientation of G to an orientation of H and let $\phi : E(\vec{H}) \to M$ be a tension. By Proposition 4.2, we may choose bond-tensions $\phi^{z_1}_{X_1}, \phi^{z_2}_{X_2}, \ldots, \phi^{z_n}_{X_n}$ of $\vec{H}$ such that $\phi = \sum_{i=1}^n \phi^{z_i}_{X_i}$. Since $\phi \circ f = \sum_{i=1}^n \phi^{z_i}_{X_i} \circ f$, it suffices to show that $\phi^{z_i}_{X_i} \circ f$ is a tension for $1 \le i \le n$. Let $i \in \{1, 2, \ldots, n\}$ and consider the bond-tension $\gamma^1_{X_i} : E(\vec{G}) \to \mathbb{Z}$. By assumption, $\gamma^1_{X_i} \circ f$ is a $\mathbb{Z}$-tension of $\vec{H}$, but it follows immediately from this that $\phi^{z_i}_{X_i} \circ f$ is an M-tension of $\vec{H}$. Since $1 \le i \le n$ was arbitrary, we have that $\phi \circ f$ is M-tension-continuous as required. $\square$

Next we turn our attention to the quasi-order related to graph homomorphisms: we define the relation $\succ_{\text{hom}}$ by the rule $G \succ_{\text{hom}} H$ if there exists a homomorphism from G to H. Since the identity map is a homomorphism and the composition of two homomorphisms is a homomorphism, $\succ_{\text{hom}}$ is a quasi-order.

We also define homomorphisms between oriented graphs as mappings preserving direction of arcs. Each homomorphism $f : \vec{G} \to \vec{H}$ induces a mapping $f^\sharp : E(\vec{G}) \to E(\vec{H})$ defined by $f^\sharp(x, y) = (f(x), f(y))$. This induced mapping will be called the *chromatic mapping* $E(\vec{G}) \to E(\vec{H})$ induced by f. It is easy to see that for every homomorphism $p : V(G) \to V(H)$ and every orientation $\vec{H}$ of H, there exists an orientation $\vec{G}$ of G and a map $f : E(\vec{G}) \to E(\vec{H})$ which is chromatic and is induced by p. So in particular, $G \succ_{\text{hom}} H$ if and only if there is a chromatic map from some orientation of G to some orientation of H. The following proposition gives a key property of chromatic maps.

Proposition 4.5. *Let M be an Abelian group and let $f : E(\vec{G}) \to E(\vec{H})$ be a chromatic map. Then we have the following.*

(i) *$\phi \circ f$ is a cut-tension of $\vec{G}$ for every cut-tension $\phi : E(\vec{H}) \to M$ of $\vec{H}$.*

(ii) *f is M-tension continuous for every Abelian group M.*

Proof. To prove (i), let $p : V(G) \to V(H)$ be a homomorphism so that f is induced by p, and choose $z \in M$ and $X \subseteq V(\vec{G})$ so that $\phi = \gamma_X^z$. It follows from the definitions that $\phi \circ f = \gamma_{p^{-1}(X)}^z$, so $\phi \circ f$ is an M-cut-tension as required.

To prove (ii), let $\phi : E(\vec{H}) \to M$ be a tension. By Proposition 4.2 we may choose cut-tensions $\phi_1, \phi_2, \ldots, \phi_n$ such that $\sum_{i=1}^n \phi_i = \phi$. By (i), $\phi_i \circ f$ is a cut-tension, so in particular it is a tension. Thus, $\phi \circ f = \sum_{i=1}^n \phi_i \circ f$ is a tension of $\vec{G}$. Since ϕ was arbitrary, it follows that f is M-tension continuous as required. □

The following proposition proves perhaps a surprising converse of Proposition 4.5, thus giving an equivalent formulation of graph homomorphisms in terms of "cut-tension-continuous maps".

Theorem 4.6. *Let $\vec{G}$ and $\vec{H}$ be connected oriented graphs and let $f : E(\vec{G}) \to E(\vec{H})$. Then f is chromatic if and only if $\phi \circ f$ is a cut-tension of $\vec{G}$ for every cut-tension $\phi : E(\vec{H}) \to \mathbb{Z}$ of $\vec{H}$.*

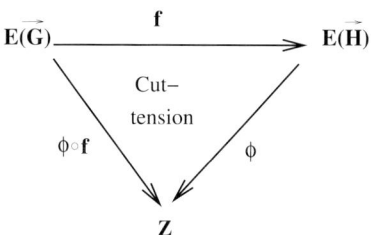

FIGURE 3. Cut-tension

Proof. The "only if" part of the proof is an immediate consequence of Proposition 4.5. To prove the "if" part, we will assume that $\phi \circ f$ is a cut-tension of $\vec{G}$ for every cut-tension $\phi : E(\vec{H}) \to \mathbb{Z}$ of $\vec{H}$. For every vertex $v \in V(\vec{H})$ we have that $\gamma_{\{v\}}^1 : E(\vec{H}) \to \mathbb{Z}$ is a cut-tension, so by our assumption, we may assume that there exists $Y_v \subseteq V(\vec{G})$ so that $\gamma_{\{v\}}^1 \circ f = \gamma_{Y_v}^1$.

Claim 1. *If $u, v \in V(\vec{H})$ and $u \neq v$, then either $Y_u \cap Y_v = \emptyset$ or $Y_u \cup Y_v = V(\vec{G})$.*

Proof of Claim 1. For every $A \subseteq V(\vec{G})$, let $\chi_A : V(\vec{G}) \to \{0, 1\}$ be the characteristic map given by the rule $\chi_A(v) = 1$ if $v \in A$ and $\chi_A(v) = 0$ otherwise. Clearly

$\delta\chi_A = -\gamma_A^1 = \gamma_{V(G)\setminus A}^1 = \gamma_{\bar{A}}^1$ (by $\bar{A}$ we denoted the complement of A). Now, $\gamma_{\{u,v\}}^1 = \gamma_u^1 + \gamma_v^1$, so we have that

$$\begin{aligned}\gamma_{\{u,v\}}^1 \circ f &= \gamma_{\{u\}}^1 \circ f + \gamma_{\{v\}}^1 \circ f \\ &= \gamma_{Y_u}^1 + \gamma_{Y_v}^1 \\ &= \delta\chi_{\bar{Y}_u} + \delta\chi_{\bar{Y}_v} \\ &= \delta(\chi_{\bar{Y}_u} + \chi_{\bar{Y}_v}).\end{aligned}$$

Since $\gamma_{\{u,v\}}^1 \circ f$ is a cut-tension, and $\vec{G}$ is connected, we have by Observation 4.1 that $p = \chi_{\bar{Y}_u} + \chi_{\bar{Y}_v}$ takes on at most two distinct values. If p does take on two distinct values, then these values must differ by exactly one since $\delta p = \gamma_{\{u,v\}}^1 \circ f$. By definition every $w \in \bar{Y}_u \cap \bar{Y}_v$ must satisfy $p(w) = 2$ and every $w \in V(\vec{G}) \setminus (\bar{Y}_u \cup \bar{Y}_u)$ must satisfy $p(w) = 0$, so it follows that at least one of these sets must be empty as required.

Fix distinct vertices $u, v \in V(\vec{G})$. By possibly switching the orientations of every edge in $\vec{G}$ and then replacing Y_w with $V(\vec{G}) \setminus Y_w$ for every $w \in V(G)$, we may assume that $Y_u \cap Y_v = \emptyset$. The next claim shows that now, Y_w and $Y_{w'}$ are disjoint whenever $w \neq w'$.

Claim 2. $Y_w \cap Y_{w'} = \emptyset$ if $w \neq w'$.

Proof of Claim 2. If there is a vertex w so that $Y_w \cap Y_u \neq \emptyset$, then by Claim 1, $Y_w \cup Y_u = V(\vec{G})$, so in particular $Y_v \subseteq Y_w$. But then by Claim 1 we must have either $Y_v = \emptyset$ or $Y_w = V(\vec{G})$ and either possibility contradicts our assumption. Thus, we find that Y_w is disjoint from Y_u for every $w \in V(\vec{G}) \setminus \{u\}$. If there exist vertices $w, w' \in V(\vec{G}) \setminus \{u\}$ so that $Y_w \cap Y'_w \neq \emptyset$, then $Y_w \cup Y_{w'} = V(\vec{G})$, by Claim 1 so either $Y_w \cap Y_u \neq \emptyset$ or $Y_{w'} \cap Y_u \neq \emptyset$. This contradiction implies that $Y_w \cap Y_{w'} = \emptyset$ whenever $w \neq w'$ as required.

The following observation follows immediately from our construction.

Observation. If $e \in E(\vec{H})$ is directed from u to v, then $f^{-1}(\{e\}) \subseteq \Delta^+(Y_u) \cap \Delta^-(Y_v)$.

Since H does not have any isolated vertices, Claim 2 and the above observation imply that $\{Y_w \mid w \in V(\vec{H})\}$ is a partition of $V(\vec{G})$. Now, let $p : V(\vec{G}) \to V(\vec{H})$ be given by the rule $p(v) = u$ if $v \in Y_u$. It follows from the above observation that f is chromatic with respect to p. This completes the proof. □

Let us rephrase Theorem 4.6 in terms of homomorphisms:

Corollary 4.7. *Given two connected oriented graphs $\vec{G}$ and $\vec{H}$, a mapping $f : V(\vec{G}) \to V(\vec{H})$ is a homomorphism $\vec{G} \to \vec{H}$ if and only if $\phi \circ f^\sharp$ is a cut-tension of $\vec{G}$ whenever ϕ is a cut-tension of $\vec{H}$.*

This is indicated by Figure 4.

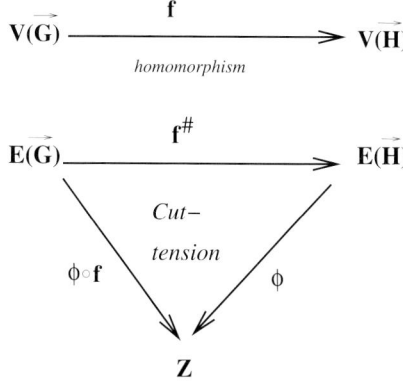

FIGURE 4. Homomorphism and cut-tension

The cut-tension continuous mappings form a quasi-order on their own which may be denoted by $\succ_{ct}$. The above Theorem 4.6 then means that $\succ_{ct} = \succ_{\text{hom}}$. It follows that (despite its similarity to $\succ_{\mathbb{Z}}^t$) the cut-tension order is very rich (and indeed countable universal) quasi-order, see [17]. However note that $\succ_{ct}$ is a proper subset of $\succ_{\mathbb{Z}}^t$. Examples are abundant. For example for any oriented bipartite graph $\vec{G}$ we have $\vec{G} \succ_{\mathbb{Z}}^t \vec{K}_2$ (here K_2 denotes the complete graph on two vertices) however the oriented bipartite graphs $\vec{G}$ satisfying $\vec{G} \not\succ_{\text{hom}} \vec{K}_2$ induce a universal poset on their own (for example $\vec{P}_3 \not\succ_{\text{hom}} \vec{K}_2$ where P_3 denotes the two edge path).

5. The order $\succ_{\mathbb{Z}_2}^f$

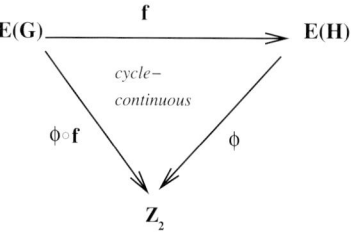

FIGURE 5. Cycle-continuous

We start our investigation of particular orders by the order which has perhaps the most intuitive appeal.

In the ring $\mathbb{Z}_2$, addition and subtraction are the same operation. As such, the orientation of the graph does not play a role, and we define a map $\phi : E(G) \to \mathbb{Z}_2$ to be a *flow* of the unoriented graph G if $\sum_{e \in \Delta(v)} \phi(e) = 0$ for every $v \in V(G)$.

Note that this definition is consistent with our earlier definitions since a map $\phi : E(G) \to \mathbb{Z}_2$ is a flow if and only if it is a flow of some (and thus every) orientation of G. A set $X \subseteq E(G)$ is a cycle if and only if it is the support of a $\mathbb{Z}_2$-flow. Therefore, a map $f : E(G) \to E(H)$ is $\mathbb{Z}_2$-flow-continuous if and only if $f^{-1}(C)$ is a cycle of G for every cycle $C \subseteq E(H)$. Based on this link, we call such a map *cycle-continuous*. The following is a corollary of Theorem 4.3.

Corollary 5.1. *A graph G is maximal in $\succ_{\mathbb{Z}_2}^{f}$ if and only if every vertex of G has even degree. A graph G is minimal in $\succ_{\mathbb{Z}_2}^{f}$ if and only if it has a cut-edge.*

Based on the above corollary, we can now restate conjecture 2.10 as follows.

Conjecture 5.2 (Jaeger). *P_{10} is the only atom in the cycle-continuous order.*

In the first paragraph of this section, we observed that a set of edges is a cycle if and only if it is the support of a $\mathbb{Z}_2$-flow. Similarly, a set of edges $D \subseteq E(G)$ is an edge-cut if and only if it is the support of a $\mathbb{Z}_2$-tension. With the help of this observation, Theorem 3.1 now gives us a monotone invariant of $\succ_{\mathbb{Z}_2}^{f}$.

Proposition 5.3. *If $G \succ_{\mathbb{Z}_2}^{f} H$ then $\lambda^o(G) \geq \lambda^o(H)$.*

Proof. If $\lambda^o(G) = \infty$ then there is nothing to prove, so we may assume that $\lambda^o(G)$ is finite and choose an edge-cut $C \subseteq E(G)$ of size $\lambda^o(G)$. Let $\psi : E(G) \to \mathbb{Z}_2$ be given by the rule $\psi(e) = 1$ if $e \in C$ and $\psi(e) = 0$ otherwise. Now, ψ is a tension of G, so by Theorem 3.1, ψ_f is a tension of H. Let D be the support of ψ_f. Now, $|D|$ is odd since $|C|$ was odd, and D is an edge-cut of H. Thus, we have that $\lambda^o(H) \leq \lambda^o(G)$ as desired. □

For every positive integer h, we let K_2^h denote the graph on two vertices consisting of h edges in parallel. For any graph G, we say that a set of edges $J \subseteq E(G)$ is a *postman join* if $E(G) \setminus J$ is a cycle. The following proposition gives a characterization of when $G \succ_{\mathbb{Z}_2}^{f} K_2^{2a+1}$ and when $K_2^{2a+1} \succ_{\mathbb{Z}_2}^{f} G$.

Proposition 5.4.
1. $K_2^{2a+1} \succ_{\mathbb{Z}_2}^{f} G$ if and only if $\lambda^o(G) \leq 2a + 1$.
2. $G \succ_{\mathbb{Z}_2}^{f} K_2^{2a+1}$ if and only if $E(G)$ may be written as a disjoint union of $2a+1$ postman joins.

Proof. Since $\lambda^o(K_2^{2a+1}) = 2a + 1$, Proposition 5.3 gives us the "only if" direction of (1). To prove the "if" direction, let G be a graph with $\lambda^o(G) \leq 2a+1$ and choose an odd edge-cut C of G of size $\leq 2a+1$. Next choose a map $f : E(K_2^{2a+1}) \to E(G)$ with the property that $|f^{-1}(\{e\})|$ is odd for every $e \in C$ and $f^{-1}(E(G) \setminus C) = \emptyset$. It follows easily that this map is cycle-continuous.

To see the "only if" direction of (2) let $f : E(G) \to E(K_2^{2a+1})$ be cycle-continuous, and note that $f^{-1}(\{e\})$ is a postman join for every $e \in E(K_2^{2a+1})$ since $f^{-1}(E(K_2^{2a+1}) \setminus \{e\})$ is a cycle. To see the "if" direction, let $J_1, J_2, \ldots, J_{2a+1}$ be a list of disjoint postman joins with union $E(G)$, let $E(K_2^{2a+1}) = \{e_1, e_2, \ldots, e_{2a+1}\}$, and consider the map $f : E(G) \to E(K_2^{2a+1})$ given by the rule $f(s) = e_i$ if $s \in J_i$. It is easily verified that this map is cycle-continuous. This completes the proof. $\square$

Based on this proposition, we have the following scaling chain.

Proposition 5.5. *The graphs $K_2^1, K_2^3, K_2^5, \ldots$ form a scaling chain in $\succ_{\mathbb{Z}_2}^f$.*

Proof. It follows immediately from (1) of the previous proposition that $K_2^1, K_2^3, K_2^5, \ldots$ is an increasing chain. If G is any non maximal graph in $\succ_{\mathbb{Z}_2}^f$, then $\lambda^o(G) < \infty$ and we have by (1) of the previous proposition that there exists a positive integer $2a + 1$ such that $K_2^{2a+1} \succ_{\mathbb{Z}_2}^f G$. $\square$

Proposition 5.4 above shows a connection between the cycle-continuous order and the problem of partitioning the edge set of a graph into postman joins. We now state a special case of a conjecture of Rizzi [20] concerning postman joins. In the language of the cycle-continuous order, his conjecture asserts that the graph K_2^{2a+1} is comparable with every other graph.

Conjecture 5.6 (Rizzi). *If $K_2^{2a+1} \not\succ_{\mathbb{Z}_2}^f G$, then $G \succ_{\mathbb{Z}_2}^f K_2^{2a+1}$.*

Since every graph dominates K_2^1, this conjecture obviously holds for $a = 0$. A graph can be partitioned into 3 postman joins if and only if it has a nowhere-zero 4-flow. With this, the above conjecture for $a = 1$ follows from Jaeger's 4-flow theorem. A recent result of DeVos and Seymour asserts that Rizzi's conjecture holds with an added factor of two.

Theorem 5.7 (DeVos, Seymour). *If $K_2^{4a-1} \not\succ_{\mathbb{Z}_2}^f G$ then $G \succ_{\mathbb{Z}_2}^f K_2^{2a+1}$.*

If Jaeger's conjecture 2.10 is correct, then every graph is comparable with P_{10}. If Rizzi's conjecture 5.6 is correct, then every graph is comparable with K_2^{2a+1} for every nonnegative integer a. In light of these conjectures it may not be surprising that it is tricky to construct antichains in this order. In particular, we cannot solve the following problem.

Problem 5.8. *Does $\succ_{\mathbb{Z}_2}^f$ contain an infinite antichain?*

We can prove that the order $\succ_{\mathbb{Z}_2}^f$ does contain finite antichains of arbitrary size. We have two closely related families of graphs which demonstrate this fact. One of these families comes from a clever construction of Xuding Zhu and will be given below. The second family will be described here, but we will postpone the proof of its validity to the full paper [4].

It is easy to see that an r-regular graph G satisfies $G \succ_{\mathbb{Z}_2}^f K_2^r$ if and only if it is r-edge colorable. Let G be an r-regular graph with $\lambda^o(G) = r < \infty$ (so in particular, r is odd). If every edge-cut of G of size r is of the form $\Delta(x)$ for some $x \in V(G)$ and $G \not\succ_{\mathbb{Z}_2}^f K_2^r$, then we say that G is an r-snark. If $G \setminus e \succ_{\mathbb{Z}_2}^f K_2^r$ for every edge $e \in E(G)$, then we say that G is *critical*. Both of our families of antichains are based on the following proposition.

Proposition 5.9. *If G, H are non-isomorphic critical r-snarks with $|E(G)| = |E(H)|$, then G and H are incomparable in the order $\succ_{\mathbb{Z}_2}^{f}$.*

Proof. Suppose the proposition is false and let $f : E(G) \to E(H)$ be a cycle-continuous map. First we establish the following claim.

Claim. *The map f is a bijection.*

Proof of the Claim. Since G and H have the same number of edges by assumption, it will suffice to prove that f is onto. Suppose (for a contradiction) that f is not onto and choose an edge $e \in E(H)$ which is not in the image of $E(G)$. It follows that f is a cycle-continuous map from G to $H \setminus e$. But then we have that $G \succ_{\mathbb{Z}_2}^{f} H \setminus e \succ_{\mathbb{Z}_2}^{f} K_2^r$ which contradicts the assumption that G is an r-snark.

Continuing the proof, let $v \in V(G)$ and consider the edge-cut $\Delta_G(v)$. The image of $\Delta_G(v)$ is another edge-cut of odd size, so by assumption, it must be equal to $\Delta_H(v')$ for some vertex $v' \in V(H)$. If u and v are distinct vertices of G and $f(\Delta_G(u)) = \Delta_H(u')$ and $f(\Delta_G(v)) = \Delta_H(v')$ for $u', v' \in V(H)$, then it follows from the fact that f is a bijection that $u' \neq v'$. Thus the map which sends every $v \in V(G)$ to the corresponding $v' \in V(H)$ is an isomorphism between G and H and we have a contradiction. □

Let M be a perfect matching of the Petersen graph P_{10}. Define the graph $P(a, b)$ to be the graph obtained from P_{10} by adding $a - 1$ parallel edges to every edge not in M and adding $b - 1$ parallel edges to every edge in M. Our first family of antichains is as follows.

Theorem 5.10 (Zhu). *For every nonnegative integer k, the set $\{P(2j+1, 6k-4j+1) \mid 0 \leq j \leq k\}$ is an antichain in $\succ_{\mathbb{Z}_2}^{f}$ of size $k+1$.*

Proof. Let $\mathcal{F}_k = \{P(2j+1, 6k-4j+1) \mid 0 \leq j \leq k\}$ and let $G \in \mathcal{F}_k$. It follows from our construction that G is $(6k+3)$-regular, $\lambda^o(G) = 6k+3$, and every odd edge-cut of G of size $6k+3$ is of the form $\Delta(x)$ for some $x \in V(G)$. Now, for an odd integer r, and an r-regular graph G satisfies $G \succ_{\mathbb{Z}_2}^{f} K_2^r$ if and only if G is r-edge-colorable. It is well known that none of the graphs in $\mathcal{F}_k$ are $(6k+3)$-edge-colorable. Thus, we find that every graph in $\mathcal{F}_k$ is a $(6k+3)$-snark with the same number of edges. Now, it follows from a theorem of Rizzi [20] that for every $e \in E(G)$ the edges of the graph $G \setminus e$ may be partitioned into $(6k+3)$-disjoint postman joins, so $G \setminus e \succ_{\mathbb{Z}_2}^{f} K_2^{(6k+3)}$. It follows from this and Proposition 5.9 that $\mathcal{F}_k$ is an antichain as desired. □

The above construction gives us antichains of arbitrary size, but requires graphs with λ^o large. Next we describe a construction for antichains of arbitrary size with λ^o bounded (and thus bounded in $\succ_{\mathbb{Z}_2}^{f}$ by K_2^3).

Let G, H be cubic graphs, let $st \in E(G)$ and let x_1, x_2, t and x_3, x_4, s be the neighbors of s and t respectively. Let $y_1 y_2, y_3 y_4 \in E(H)$ be nonadjacent edges. Let F be a graph obtained from the disjoint union of $G \setminus \{s, t\}$ and $H \setminus \{y_1 y_2, y_3 y_4\}$

by adding new edges with ends x_i, y_i for $1 \leq i \leq 4$, F is again a cubic graph. We say that F is a *dot product* of G with H. In [4], the following theorem is proved.

Theorem 5.11. *If G is a critical 3-snark, then every dot product of G with P_{10} is a critical 3-snark.*

Since the Petersen graph P_{10} is a critical 3-snark, any dot product of P_{10} with itself is critical. There are two nonisomorphic graphs on 18 vertices known as Blanusa's snarks [1] which are obtained as dot products of P_{10} with itself. By Proposition 5.9, these two graphs are incomparable in the order $\succ^f_{\mathbb{Z}_2}$. It is straightforward to iterate this operation to create large families of nonisomorphic 3-snarks on the same number of edges, which by Proposition 5.9 are antichains. Equivalently, there are arbitrarily large antichains of graphs under K^3_2 (in $\succ^f_{\mathbb{Z}_2}$).

6. The order $\succ^t_{\mathbb{Z}_2}$

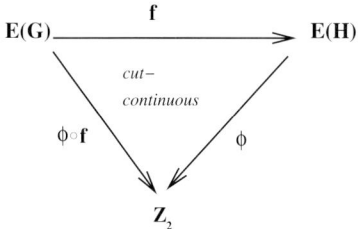

FIGURE 6. Cut-continuous

As was the case in the order $\succ^f_{\mathbb{Z}_2}$, the orientation of the edges does not play any role in the order $\succ^t_{\mathbb{Z}_2}$, and thus $\succ^t_{\mathbb{Z}_2}$ relates to undirected graphs. As in the previous section, we define a map $\phi : E(G) \to \mathbb{Z}_2$ to be a *tension* of the undirected graph G if $\sum_{e \in E(C)} \phi(e) = 0$ for every circuit C of G. As observed earlier, a set of edges is an edge-cut if and only if it is the support of a $\mathbb{Z}_2$-tension. Thus, a map $f : E(G) \to E(H)$ is $\mathbb{Z}_2$-tension-continuous if and only if $f^{-1}(C)$ is an edge-cut of G for every edge-cut C of H. Based on this property, we call such a map *cut-continuous*. We begin by stating a corollary of Theorem 4.3 which gives us the maximal and minimal elements in this order.

Corollary 6.1. *The maximal elements in $\succ^t_{\mathbb{Z}_2}$ are the bipartite graphs. The minimal elements are the graphs which contain loops.*

As was the case with the cycle-continuous order, Theorem 3.1 implies that odd girth is a monotone invariant:

Proposition 6.2. *If $G \succ^t_{\mathbb{Z}_2} H$ then $\gamma^o(G) \geq \gamma^o(H)$.*

Proof. If $\gamma^o(G) = \infty$ then there is nothing to prove, so we may assume that $\gamma^o(G)$ is finite and choose a cycle $C \subseteq E(G)$ of size $\gamma^o(G)$. Let $\psi : E(G) \to \mathbb{Z}_2$ be given by the rule $\psi(e) = 1$ if $e \in C$ and $\psi(e) = 0$ otherwise. Let f be a $\mathbb{Z}_2$-tension-continuous map from $E(G)$ to $E(H)$. Now, ψ is a flow of G, so by Theorem 3.1, ψ_f is a flow of H. Let D be the support of ψ_f. Now, $|D|$ is odd since $|C|$ was odd, and D is a cycle of H. Thus, we have that $\gamma^o(H) \leq \gamma^o(G)$ as desired. $\square$

We let C_n denote the circuit of length n for every $n \geq 1$. Let Q_n denote the graph of the n-cube for $n \geq 1$. The vertex set of Q_n is $\{0,1\}^n$ and two vertices are adjacent if and only if they differ in a single coordinate. Let Q_{2n}^+ denote the graph obtained from the $2n$-cube by adding all edges between vertices which differ in every coordinate. Now, we have the following proposition which characterizes when a graph dominates and is dominated by C_{2a+1}.

Proposition 6.3.
1. $C_{2a+1} \succ_{\mathbb{Z}_2}^t G$ if and only if $\gamma^o(G) \leq 2a + 1$
2. $G \succ_{\mathbb{Z}_2}^t C_{2a+1}$ if and only if $G \succ_{\text{hom}} Q_{2a}^+$.

Proof. Since $\gamma^o(C_{2a+1}) = 2a+1$, Proposition 6.2 gives us the "only if" direction of (1). To prove the "if" direction, let G be a graph with $\gamma^o(G) \leq 2a+1$ and choose an odd cycle C of G of size $\leq 2a+1$. Next choose a map $f : E(C_{2a+1}) \to E(G)$ with the property that $|f^{-1}(\{e\})|$ is odd for every $e \in C$ and $f^{-1}(E(G) \setminus C) = \emptyset$. It follows easily that this map is tension-continuous.

For (2), let $B \subseteq \mathbb{Z}_2^{2a}$ be the subset consisting of all vectors with exactly one 1 together with the vector $(1,1,1,\ldots,1)$. Now, $Q_{2a}^+ \cong Cayley(\mathbb{Z}_2^{2a}, B)$, so by Corollary 3.5, it suffices to see that the following $2a \times (2a+1)$ matrix represents the $\mathbb{Z}_2$-cocycle-space of C_{2a+1}.

$$\begin{bmatrix} 1 & & & & 1 \\ & 1 & 0 & & 1 \\ & & \ddots & & \vdots \\ 0 & & & 1 & 1 \end{bmatrix}$$

$\square$

As was the case with the graphs K_2^{2a+1} in the cycle-continuous order, the odd circuits C_{2a+1} form a scaling chain in the cut-continuous order.

Proposition 6.4. *The odd circuits $C_3, C_5, C_7, \ldots$ form a scaling chain in the cut-continuous order $\succ_{\mathbb{Z}_2}^t$.*

Proof. It follows immediately from (1) of the previous proposition that $C_3, C_5, \ldots$ is an increasing chain. If G is any non maximal graph in $\succ_{\mathbb{Z}_2}^t$, then $\gamma^o(G) < \infty$ and we have by (1) of the previous proposition that there exists a positive integer $2a+1$ such that $C_{2a+1} \succ_{\mathbb{Z}_2}^t G$. $\square$

For any graph G, we let G^n denote the graph with vertex set $V(G)$ in which two vertices are adjacent if and only if they are distance n in G. The following proposition gives a condition similar to that in part 2 of Proposition 6.3 for graphs dominating a complete graph. Our setting provides a short proof :

Proposition 6.5 (Linial, Meshulam, Tarsi [13]). $G \succ^t_{\mathbb{Z}_2} K_n$ if and only if $G \succ_{\hom} Q_n^2$.

Proof. Let $B \subseteq \mathbb{Z}_2^n$ be the set of all vectors with exactly two 1's. Then $Q_n^2 \cong Cayley(\mathbb{Z}_2^n, B)$. Let A be the incidence matrix of K_n considered as a matrix over $\mathbb{Z}_2$. Then A represents the $\mathbb{Z}_2$-cocycle-space of K_n and
$$\{x \in \mathbb{Z}_2^n \mid x \text{ is a column of } A\} = B.$$
Thus, the proposition now follows from Corollary 3.5. □

The above proposition 6.5 demonstrates that $G \succ^t_{\mathbb{Z}_2} K_3$ if and only if $G \succ_{\hom} K_4$. We have a similar relation for the complete graphs with 2^n vertices.

Proposition 6.6. $G \succ^t_{\mathbb{Z}_2} K_{2^n}$ if and only if $G \succ_{\hom} K_{2^n}$.

Proof. Let f be a cut-continuous map from G to K_{2^n} and choose edge-cuts $D_1, D_2, \ldots, D_n$ of K_{2^n} so that $\bigcup_{i=1}^n D_i = E(K_{2^n})$. Now $f^{-1}(D_i)$ for $1 \leq i \leq n$ is a list of n edge-cuts of G containing every edge. It follows that $\chi(G) \leq 2^n$, so $G \succ_{\hom} K_{2^n}$. If $G \succ_{\hom} K_{2^n}$, then by Proposition 4.5 we have that $G \succ^t_{\mathbb{Z}_2} K_{2^n}$. □

Thus, we find that K_3 and K_4 are equivalent and that the sequence K_4, K_8, K_{16}, ... is a strict descending chain in $\mathbb{Z}_2^t$. For every graph G we define the log-*chromatic number* $\chi_{\log}(G) = \lceil \log_2 \chi(G) \rceil = \min\{n : G \succ_{\hom} K_{2^n}\}$. Our next proposition shows that $\chi_{\log}$ is a monotone invariant with respect to the cut continuous order.

Proposition 6.7. If $G \succ^t_{\mathbb{Z}_2} H$, then $\chi_{\log}(G) \leq \chi_{\log}(H)$.

Proof. Let $\chi_{\log}(H) = n$. Then we have that $G \succ^t_{\mathbb{Z}_2} H \succ^t_{\mathbb{Z}_2} K_{2^n}$ so by Proposition 6.6 $G \succ_{\hom} K_{2^n}$ so $\chi_{\log}(G) \leq \chi_{\log}(H)$ as desired. □

With this last proposition, we are ready to construct an infinite antichain in the cut-continuous order.

Proposition 6.8. *The order $\succ^t_{\mathbb{Z}_2}$ contains an infinite antichain.*

Proof. Let G_0 be an arbitrary graph. To create G_{i+1} given $G_0, G_1, \ldots, G_i$, choose G_{i+1} to be a graph with $\gamma^o(G_{i+1}) > \gamma^o(G_i)$ and with $\chi_{\log}(G_{i+1}) > \chi_{\log}(G_i)$ (such a graph always exists due to the existence of graphs with arbitrarily high girth and chromatic number). It now follows from Proposition 6.2 that there is no cut-continuous map from G_j to G_{i+1} for every $1 \leq j \leq i$ and from proposition 6.7 that there is no cut-continuous map from G_{i+1} to G_j for every $1 \leq j \leq i$. □

We have two monotone invariants in the cut-continuous order, namely γ^o and $\chi_{\log}$. The above construction uses these two invariants to build an infinite antichain. It would be interesting to know if graphs of high chromatic number are essential for this construction. In particular, we offer the following problem.

Problem 6.9. *Does there exist an infinite antichain of graphs in the cut-continuous order which have a bounded chromatic number?*

7. The order $\succ_{\mathbb{Z}}^{f}$

The order $\succ_{\mathbb{Z}}^{f}$ is similar to the flow order $\succ_{\mathbb{Z}_2}^{f}$ except that the orientations of edges begin to play a strong role.

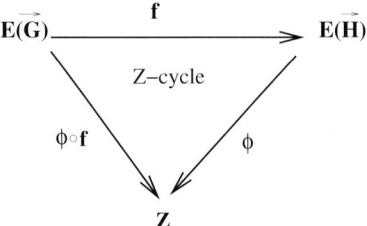

FIGURE 7. Z-cycle

A corollary of Theorem 4.3 gives us the maximal and minimal elements in this order.

Corollary 7.1. *The maximal elements of $\succ_{\mathbb{Z}}^{f}$ are the graphs in which every vertex has even degree. The minimal elements are the graphs with cut-edges.*

We say that an oriented graph $\vec{G}$ is a (mod p) orientation (of G) for a positive integer p if $\deg^+(v) - \deg^-(v) \equiv 0$ (modulo p) for every $v \in V(G)$. We say that an undirected graph G has a *mod p orientation* if there is an orientation of G which is a mod p orientation. Note that if G has a mod p orientation for an even integer p, then every vertex of G has even degree. It is worth noting that a graph has a mod $2k+1$ orientation if and only if it has circular flow number at most $2 + \frac{1}{k}$ (see [6] for more details). With this definition, we are ready to characterize when $K_2^{2a+1} \succ_{\mathbb{Z}}^{f} G$ and when $G \succ_{\mathbb{Z}}^{f} K_2^{2a+1}$.

Proposition 7.2.
1. $K_2^{2a+1} \succ_{\mathbb{Z}}^{f} G$ if and only if $\lambda^o(G) \leq 2a+1$.
2. $G \succ_{\mathbb{Z}}^{f} K_2^{2a+1}$ if and only if G has a mod $(2a+1)$ orientation.

Proof. If $K_2^{2a+1} \succ_{\mathbb{Z}}^{f} G$, then $K_2^{2a+1} \succ_{\mathbb{Z}_2}^{f} G$, so by Proposition 5.3, $\lambda^o(G) \leq 2a+1$. If $\lambda^o(G) \leq 2a+1$, then choose an odd edge-cut $\Delta(X)$ of G of size $2b+1 \leq 2a+1$. Let $\vec{G}$ be an orientation of G such that $\Delta^-(X) = \emptyset$. Let u, v be the vertices of K_2^{2a+1}, let $e_1, e_2, \ldots, e_{2a+1}$ be the edges of K_2^{2a+1} and let $\vec{K_2}^{2a+1}$ be an orientation with edges $e_1, e_2, \ldots, e_{a+b+1}$ directed from u to v and with edges $e_{a+b+2}, \ldots, e_{2a+1}$ directed from v to u. Choose a map $f : E(\vec{K_2}^{2a+1}) \to E(\vec{G})$ such that f maps $\{e_1, e_2, \ldots, e_{2b+1}\}$ injectively onto the set $\Delta^+(X)$ and such that $f(\{e_{2b+2}, e_{2b+3}, \ldots, e_{2a+1}\}) = \{e\}$ for some $e \in E(\vec{G})$. For every flow $\phi : E(\vec{G}) \to \mathbb{Z}$ the map $\phi \circ f$ is a flow of $\vec{K_2}^{2a+1}$, so we have that $K_2^{2a+1} \succ_{\mathbb{Z}}^{f} G$ as desired.

If $G \succ_{\mathbb{Z}}^{f} K_2^{2a+1}$, then let $\vec{H}$ be an orientation of K_2^{2a+1} such that every edge has the same head and choose an orientation $\vec{G}$ of G and a $\mathbb{Z}$-flow-continuous map $f : E(\vec{G}) \to E(\vec{H})$. We claim that $\vec{G}$ is a mod $2a+1$ orientation of G. Let $\{e_1, e_2, \ldots, e_{2a+1}\} = E(\vec{H})$ and let $X_i = f^{-1}(\{e_i\})$ for $1 \leq i \leq 2a+1$. Now for every $2 \leq i \leq 2a+1$, the map $\phi_i : E(\vec{H}) \to \mathbb{Z}$ given by the rule

$$\phi_i(e) = \begin{cases} 1 & \text{if } e = e_1 \\ -1 & \text{if } e = e_i \\ 0 & \text{otherwise} \end{cases}$$

is a flow. Thus $\phi_i \circ f$ is a 2-flow of $\vec{G}$ and for every $v \in V(G)$ we have that $\sum_{e \in \Delta^+(v)} \phi_i(e) = \sum_{e \in \Delta^-(v)} \phi_i(e)$. It follows from this that

$$|X_1 \cap \Delta^+(v)| - |X_1 \cap \Delta^-(v)| = |X_i \cap \Delta^+(v)| - |X_i \cap \Delta^-(v)|$$

holds for every $v \in V(G)$. Since the above equation holds for every $1 \leq i \leq 2a+1$ we have that $|\Delta^+(v)| - |\Delta^-(v)| = (2a+1)(|X_i \cap \Delta^+(v)| - |X_i \cap \Delta^-(v)|)$. Thus $\vec{G}$ is a mod $2a+1$ orientation of G as desired.

Let $\vec{G}$ be a mod $2a+1$ orientation of G. Suppose that there is a vertex $v \in V(G)$ with $\Delta^+(v) \neq \emptyset$ and $\Delta^-(v) \neq \emptyset$. Choose an edge $e \in \Delta^-(v)$ with tail u and an edge $e' \in \Delta^+(v)$ with head w and form a new oriented graph $\vec{G_1}$ by deleting the edges e, e' and adding a new edge directed from u to w. Now $\vec{G_1}$ is still a mod $2a+1$ orientation. Further, any M-flow continuous map from G_1 to H naturally extends to a M-flow continuous map from G to H. Thus, by repeating this operation, we may assume that either $\Delta^+(v)$ or $\Delta^-(v)$ is empty for every $v \in V(G)$. Suppose that $v \in V(G)$ with $|\Delta^+(v)| > 2a+1$. Then we may form a new oriented graph $\vec{G_2}$ by replacing v by two new vertices v_1, v_2 so that every edge incident with v now attaches to one of v_1, v_2 and such that $|\Delta_{G_2}^+(v_1)|$ and $|\Delta_{G_2}^+(v_2)|$ are both positive multiples of $(2a+1)$. As before, $\vec{G_2}$ is a mod $2a+1$ orientation. Further, any M-continuous map from G_2 to H can easily be extended to a M-continuous map from G to H. Thus, by repeating this operation, we may assume that $|\Delta^+(v)| = 2a+1$ and $\Delta^-(v) = \emptyset$ or $|\Delta^-(v)| = 2a+1$ and $\Delta^+(v) = \emptyset$ for every $v \in V(G)$. Let $X = \{v \in V(G) \mid \Delta^-(v) = \emptyset\}$ and let $Y = V(G) \setminus X$. Then G is a $(2a+1)$-regular bipartite graph with bipartition (X, Y). By König's theorem, there exists a partition of $E(G)$ into perfect matchings $\{Z_1, Z_2, \ldots, Z_{2a+1}\}$. Let $\vec{H}$ be an orientation of K_2^{2a+1} so that every edge has the same head and let $\{e_1, e_2, \ldots, e_{2a+1}\}$ be arcs of $\vec{H}$. Define the map $f : E(\vec{G}) \to E(\vec{H})$ by the rule $f(e) = e_i$ if $e \in Z_i$. It follows easily that f is $\mathbb{Z}$-flow continuous. This completes the proof. $\square$

Based on this proposition, we find an infinite chain of graphs of the form K_2^{2a+1} as before.

Proposition 7.3. *The graphs $K_2^1, K_2^3, K_2^5, \ldots$ form an scaling chain.*

Proof. This follows immediately from part 1 of the preceding proposition. $\square$

Jaeger [12] has conjectured that every $4p$-edge-connected graph has a mod $(2p+1)$ orientation. He proved that if this conjecture is true, then both Tutte's 3-Flow and 5-Flow conjectures are also true. The following is a slight extension of this conjecture made by Zhang [28]. Evidence for this stronger conjecture is provided by some work of Zhang (see [29]) and Zhu (see [30]).

Conjecture 7.4. *If $K_2^{4p-1} \not\succ_{\mathbb{Z}}^{f} G$, then $G \succ_{\mathbb{Z}}^{f} K_2^{2p+1}$.*

A graph G has a nowhere-zero 3-flow if and only if it has a mod 3 orientation which exists if and only if $G \succ_{\mathbb{Z}}^{f} K_2^3$. The following proposition shows that G has a nowhere-zero 4-flow if and only if $G \succ_{\mathbb{Z}}^{f} K_4$.

Proposition 7.5. *A graph G has a nowhere-zero 4-flow if and only if $G \succ_{\mathbb{Z}}^{f} K_4$.*

Proof. Tutte proved that G has a nowhere-zero 4-flow if and only if it has a B-flow where $B \subseteq \mathbb{Z}^4$ is the set of all vectors with two 0's, one 1, and one -1. Thus, the proposition follows from Theorem 3.2 and the observation that the matrix

$$\begin{bmatrix} 1 & 1 & 1 & 0 & 0 & 0 \\ -1 & 0 & 0 & 1 & 1 & 0 \\ 0 & -1 & 0 & -1 & 0 & 1 \\ 0 & 0 & -1 & 0 & -1 & -1 \end{bmatrix}$$

represents the $\mathbb{Z}$-cycle-space of K_4. □

Based on the above propositions, we have that the Petersen graph does not dominate K_2^3 or K_4 and that K_2^3 does dominate the Petersen. It is not difficult to verify the the Petersen graph and K_4 are $\succ_{\mathbb{Z}}^{f}$-incomparable. Viewing this one could suggest that $K_4 + P_{10}$ is an atom of the order $\succ_{\mathbb{Z}}^{f}$. However this is not so as the following graph Q is strictly below both K_4 and P_{10}:

Let Q be the graph obtained from P_{10} by splitting a vertex of degree 3 to form three vertices of degree 1 and then placing a triangle on these three vertices.

Claim 7.6. *Q does not dominate P_{10} in the order $\succ_{\mathbb{Z}}^{f}$.*

Proof. First observe the following:

Observation 7.7. *If $G \succ_{K}^{f} H$ in the K-flow order for any Abelian group K, and e is an edge of G, then $G/e \succ_{K}^{f} H$ in the K-flow order (here G/e denotes the graph obtained from G by contracting e).*

The proof of this observation is left to the reader. By an earlier result (Theorem 3.2) a graph F will dominate H in the order $\succ_{\mathbb{Z}}^{f}$ if and only if F has a B-flow for a certain carefully constructed set B. If G dominates H, then G has a B-flow, but then this gives a B-flow of G/e, so G/e dominates H as well.

This observation is tricky to apply. We have to be careful to keep the multiple edges around after contracting – it doesn't work if we delete them.

If the graph Q above dominated P_{10} then anything obtained from Q by pure contraction would also dominate P_{10}. But it is possible to contract edges in Q

to form a graph which is isomorphic to K_4 with some loops added on a vertex. Since K_4 does not dominate P_{10} we have by the above observation that Q does not dominate P_{10}. This completes the proof of the claim. □

The ordering $\succ_{\mathbb{Z}}^{f}$ does suggest some other questions. For instance, there is a 3 elements chain consisting of $K_2^3 \succ_{\mathbb{Z}}^{f} V_8 \succ_{\mathbb{Z}}^{f} K_4$ (one can prove that K_4 and V_8, see figure 8, are not equivalent in the order $\succ_{\mathbb{Z}}^{f}$). Jaeger's 4-flow theorem asserts that every 4-edge-connected graph dominates K_4, while a conjecture of Jaeger (actually a weak version of Tutte's 3-flow conjecture) asserts that for some k every k-edge-connected graph dominates K_2^3. The following problem is a natural weakening of that conjecture.

Problem 7.8. *Does there exist a fixed integer k so that every k-edge-connected graph dominates V_8?*

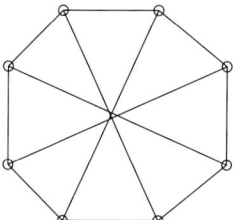

FIGURE 8. V_8

8. The order $\succ_{\mathbb{Z}}^{t}$

The order $\succ_{\mathbb{Z}}^{t}$ is similar to the order $\succ_{\mathbb{Z}_2}^{t}$ except that the orientations of edges begin to play a strong role and it is more related to the homomorphism order $\succ_{\text{hom}}$.

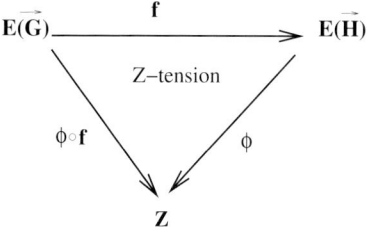

FIGURE 9. Z-tension

Applying Theorem 4.3, we establish the maximal and minimal elements.

Corollary 8.1. *The maximal graphs in the order $\succ_\mathbb{Z}^t$ are the bipartite graphs. The minimal graphs are the graphs which contain loop edges.*

As before, odd-girth γ^o is a monotone invariant.

Proposition 8.2. *If $G \succ_\mathbb{Z}^t H$ then $\gamma^o(G) \geq \gamma^o(H)$*

Proof. If $G \succ_\mathbb{Z}^t H$, then $G \succ_{\mathbb{Z}_2}^t H$, so by Proposition 6.2 $\gamma^o(G) \geq \gamma^o(H)$ as required. □

Proposition 6.6 showed that $G \succ_{\mathbb{Z}_2}^t K_{2^n}$ if and only if $G \succ_{\text{hom}} K_{2^n}$. The following proposition gives a similar equivalence for the order $\succ_\mathbb{Z}^t$ but for a much richer class of graphs.

Theorem 8.3. *For every positive integer n, $G \succ_\mathbb{Z}^t Cayley(\mathbb{Z}_n, B)$ if and only if $G \succ_{\text{hom}} Cayley(\mathbb{Z}_n, B)$.*

Proof. The "if" direction is an immediate consequence of Proposition 4.5. To prove the "only if" direction, let G be a graph with $G \succ_\mathbb{Z}^t Cayley(\mathbb{Z}_n, B)$. Then $G \succ_{\mathbb{Z}_n}^t Cayley(\mathbb{Z}_n, B)$. Now, $Cayley(\mathbb{Z}_n, B)$ has a B-tension by Proposition 3.4, so G has a B-tension. But then by Proposition 3.4 we have that $G \succ_{\text{hom}} Cayley(\mathbb{Z}_n, B)$. □

Based on this theorem, we have the following corollary.

Corollary 8.4.
1. $G \succ_\mathbb{Z}^t K_n$ if and only if $G \succ_{\text{hom}} K_n$.
2. $G \succ_\mathbb{Z}^t C_n$ if and only if $G \succ_{\text{hom}} C_n$.

Proof. 1. follows from $K_n \cong Cayley(\mathbb{Z}_n, \mathbb{Z}_n \setminus \{0\})$.
2. follows from $C_n \cong Cayley(\mathbb{Z}_n, \{-1, 1\})$. □

Based on this proposition, we have the following infinite chains.

Proposition 8.5. *In the order $\succ_\mathbb{Z}^t$, the graphs $C_3, C_5, C_7, \ldots$ form a scaling chain and $K_3, K_4, K_5, \ldots$ form a decreasing chain.*

Proof. This follows immediately from the above corollary. □

9. Jaeger's order

As we stated in the introduction one of our main motivations for this paper was provided by Jaeger's work. However rather than following his actual definitions we tried to follow his ideas and it is in this final section where we carefully compare our approach (which we believe is a more streamlined one) to Jaeger's original definitions.

In [11] Jaeger defined the following relation: Let $G_1 = (E_1, V_1)$ and $G_2 = (V_2, E_2)$ be two graphs. We say that $G_2 \succ_J G_1$ if and only if there exits a subdivision $G_1' = (V_1', E_1')$ of G_1 and a bijective mapping f from E_2 to E_1' such that, for each $\mathbb{Z}_2$-flow ϕ of G'_1, $\phi \circ f$ is a $\mathbb{Z}_2$-flow of G_2. We write $G_1 \simeq_J G_2$ if $G_2 \succ_J G_1$ and $G_1 \succ_J G_2$. It is easy to see that $\succ_J$ is a quasi-order. We now have the following proposition relating $\succ_J$ with $\succ_{\mathbb{Z}_2}^f$.

Proposition 9.1. *If G_1, G_2 are graphs then $G_1 \succ_J G_2$ if and only if there exists a cycle-continuous map $f : E(G_1) \to E(G_2)$ which is onto. In particular, $G_1 \succ_J G_2$ implies that $G_1 \succ_{\mathbb{Z}_2}^f G_2$.*

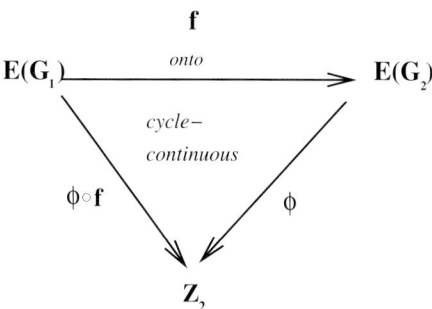

FIGURE 10. $G_1 \succ_J G_2$

Proof. This is easy to see. The "if" direction is clear. For the "only if" direction observe that it suffices to subdivide every $e \in E(G_2)$ by $|f^{-1}(e)| - 1$ vertices. □

It follows from this proposition that the order $\succ_J$ is a subset of $\succ_{\mathbb{Z}_2}^f$. The following proposition is an easy consequence of this fact.

Proposition 9.2. *Let $B \subseteq \mathbb{Z}_2^k$ for some positive integer k. If $G \succ_J H$ and H has a B-flow, then G also has a B-flow.*

Proof. This follows immediately from Propositions 9.1 and 2.5. □

If G is a graph we will denote by $\mu(G)$ the dimension of its cycle space and we denote by $\tau(G)$ the maximum number of edge-disjoint circuits of G. The following proposition proved by Jaeger now follows from the definitions:

Proposition 9.3. *If $G_2 \succ_J G_1$ then the following holds:*
  i) $|E(G_1)| \le |E(G_2)|$;
  ii) $\mu(G_1) \le \mu(G_2)$;
  iii) $\tau(G_1) \le \tau(G_2)$

The monotone parameters appearing in the above proposition make the construction of arbitrarily large antichains for $\succ_J$ quite simple. For instance, to form an antichain of size k, choose graphs $G_1, G_2, \ldots, G_k$ so that $\lambda^o(G_1) > \lambda^o(G_2) > \cdots > \lambda^o(G_k)$ and then subdivide edges of these graphs so that $|E(G_1)| < |E(G_2)| < \cdots < |E(G_k)|$. Despite these simple constructions we do not know if there exists an infinite antichain in Jaeger's order (which is equivalent to $\succ_J$ being WQO). Next we establish a scaling chain in $\succ_J$.

Proposition 9.4. *The graphs* $K_2^1, K_2^3, K_2^5, \ldots$ *form a scaling chain in* $\succ_J$.

Proof. To show that $K_2^1, K_2^3, K_2^5, \ldots$ is a chain we will show that $K_2^{2a+1} \succ_J K_2^{2a-1}$ for every $a \geq 1$. To see this, replace one edge of K_2^{2a-1} by a path of length three to form the graph $(K_2^{2a-1})'$. Now, let $f : E(K_2^{2a+1}) \to E((K_2^{2a-1})')$ be a bijection. It follows easily that f satisfies Jaeger's condition.

Let G be a graph with a vertex v of degree $2k+1$ and let $m = |E(G \setminus v)|$. We claim that $K_2^{2m+2k+1} \succ_J G$. To see this, subdivide every edge $e \in E(G)$ not incident with v to form the graph G' and let ϕ be a bijection from $E(K_2^{2m+2k+1})$ to $E(G')$. By construction, every cycle of G' has an even number of edges. Thus ϕ demonstrates that $K_2^{2m+2k+1} \succ_J G$ and we conclude that $K_2^1, K_2^3, K_2^5, \ldots$ is a scaling chain. $\square$

If A is a matrix over $\mathbb{Z}_2$ which represents the cycle-space of P_{10} and B is the set of columns in A, then we say that a B-flow of a graph G is a *Petersen-flow* of G. In Jaeger's original article, he conjectured that every bridgeless graph G must satisfy $G \succ_J P_{10}$, $G \succ_J K_2^3$, or $G \succ_J K_1^1$. This is equivalent to Conjecture 2.10, and Jaeger showed that it is also equivalent to the conjecture that every bridgeless graph has a Petersen-flow. These equivalent conjectures are collected in the following theorem.

Theorem 9.5. *For every graph G, the following statements are equivalent.*
 (i) $G \succ_J P_{10}$ *or* $G \succ_J K_2^3$ *or* $G \succ_J K_1^1$.
 (ii) $G \succ_{\mathbb{Z}_2}^f P_{10}$
 (iii) G *has Petersen-flow.*

Proof. The equivalence between (ii) and (iii) is an immediate consequence of Theorem 3.2. Since $K_1^1 \succ_{\mathbb{Z}_2}^f K_2^3 \succ_{\mathbb{Z}_2}^f P_{10}$, it follows that (i) implies (ii). To see that (ii) implies (i), let $f : E(G) \to E(P_{10})$ be cycle-continuous. If f is onto, then $G \geq_J P_{10}$ and we are finished. Otherwise, there is an edge $e \in E(P_{10})$ not in the image of f and we find that $G \succ_{\mathbb{Z}_2}^f P_{10} \setminus e \succ_{\mathbb{Z}_2}^f K_2^3$. If g is a cycle-continuous map from G to K_2^3, then either g is onto and $G \geq_J K_2^3$ or there is an edge $e' \in E(K_2^3)$ not in the image of g and we find that $G \succ_{\mathbb{Z}_2}^f K_2^3 \setminus e' \succ_{\mathbb{Z}_2}^f K_1^1$. In this case we must have $G \succ_J K_1^1$ so we are done. $\square$

If G is a cubic graph, then a *Petersen edge-coloring* of G is a coloring of the edges of G using edges of P_{10} so that any three adjacent edges of G map to three adjacent edges of P_{10}.

Proposition 9.6. *If G is a cubic graph, then the following statements are equivalent.*
 (i) $G \succ_{\mathbb{Z}_2}^f P_{10}$
 (ii) G *has a Petersen edge-coloring*

Proof. It follows immediately that (ii) implies (i). To see the reverse direction, note that by Proposition 5.3, in every cycle-continuous mapping from G to P_{10}, the image of every vertex star must be a vertex star. □

In Jaeger's original article, he showed that Conjecture 2.10 could be reduced to cubic graphs, thus establishing another form of his conjecture :

Conjecture 9.7 (Jaeger). *Every bridgeless cubic graph has a Petersen edge-coloring.*

References

[1] D.A. Holton, J. Sheehan, *The Petersen Graph*, Australian Mathematical Society Series, 7. Cambridge University Press, 1993

[2] A.J. Bondy, Basic Graph Theory: Paths and Circuits. *Handbook of Combinatorics*, edited by R. Graham, M. Grötschel and L. Lovász. (1995), 3–110

[3] U. Celmins, On cubic graphs that do not have an edge 3-coloring. Ph. D. Thesis. University of Waterloo (1984)

[4] M. DeVos, J. Nešetřil, A. Raspaud, On flow and tension-continuous maps. KAM-DIMATIA Series 2002-567.

[5] D.R. Fulkerson, Blocking and anti-blocking pairs of polyhedra. Math. Prog. 1 (1971), 168–194.

[6] L.A. Goddyn, M. Tarsi, C-Q. Zhang, On (k,d)-colorings and fractional nowhere zero flows, J. Graph Theory 28 (1998), 155–161.

[7] P. Hell and J. Nešetřil, *Graphs and Homomorphisms*, Oxford University Press, 2004.

[8] F. Jaeger, A survey of the cycle double cover conjecture. Cycles in Graphs, Ann. Discrete Mathematics 27, North-Holland, Amsterdam, 1985, pp. 1–12.

[9] F. Jaeger, Flows and generalized coloring theorems in graphs, J. Comb. Theory, Ser. B 26, 205–216 (1979).

[10] F. Jaeger, On graphic-minimal spaces, Ann. Discrete Math. 8, 123–126 (1980).

[11] F. Jaeger, Nowhere zero-flow problems. *Selected topics in Graph Theory 3* Academic Press, London 1988, 71–95.

[12] F. Jaeger, On circular flows in graphs in *Finite and Infinite Sets*, volume 37 of Colloquia Mathematica Societatis Janos Bolyai, edited by A. Hajnal, L. Lovasz, and V.T. Sos. North-Holland (1981) 391-402.

[13] N. Linial, R. Meshulam, M. Tarsi, Matroidal bijections between graphs J. Comb. Theory, Ser. B 45, No.1, 31–44 (1988).

[14] L. Lovász, Operations with structures, Acta Math. Acad. Sci. Hung. 18 (1967), 321–329.

[15] M. Mihail, P. Winkler, On the number of Eulerian orientations of a graph, Proc. of the 3rd ACM-SIAM Symp. on Discrete Algorithms (1992), pp. 138–145.

[16] V. Müller, The edge reconstruction hypothesis is true for graphs with more than $n \log n$ edges, J. Comb. Th. B, 22 (1977), 281–183.

[17] J. Nešetřil, Aspects of Structural Combinatorics, Taiwanese J. Math. 3, 4 (1999), 381–424.

[18] J. Nešetřil, C. Tardif, Duality Theorems for Finite Structures (Characterizing Gaps and Good Characterizations), J. Comb. Th. B 80 (2000), 80–97.
[19] J. Nešetřil, X. Zhu, Paths homomorphisms, Proc. Cambridge Phi. Soc. 120 (1996), 207–220.)
[20] R. Rizzi, On packing T-joins, manuscript.
[21] G. C. Rota, On the foundations of combinatorial theory. I. Z. Wahrscheinlichkeitstheorie und Verw. Gebiete 2 1964 340–368 (1964).
[22] M. Preissmann, Sur les colorations des arêtes des graphes cubiques. Thèse de Doctorat de 3ème cycle. Grenoble (1981)
[23] P. Seymour, Nowhere-zero 6-flows, J. Comb. Theory, Ser. B 30, 130–135 (1981).
[24] P.D. Seymour, Nowhere-zero flows. *Handbook of Combinatorics*, edited by R. Graham, M. Grötschel and L. Lovász. (1995), 289–299
[25] W.T. Tutte, A contribution to the theory of chromatic polynomials, Can. J. Math. 6, 80–91 (1954).
[26] W.T. Tutte, A class of Abelian groups, Can. J. Math. 8, 13–28 (1956).
[27] D. Welsh, *Complexity: knots, colourings and counting.* London Mathematical Society Lecture Note Series, 186. Cambridge University Press, Cambridge, 1993.
[28] C.Q. Zhang, Circular flows of nearly Eulerian graphs and vertex-splitting. J. Graph Theory 40 (2002), no. 3, 147–161.
[29] C.Q. Zhang, *Integer flows and cycle covers of graphs*, Pure and Applied Mathematics, Marcel Dekker. 205. New York, NY: Marcel Dekker.
[30] X. Zhu, Circular Chromatic Number of Planar Graphs with large odd Girth, Electronic J. Comb. 2001, #25.

Matt DeVos
Applied Math Department
Princeton University
Princeton, NJ 08544, USA
e-mail: `matdevos@math.princeton.edu`

Jaroslav Nešetřil
Department of Applied Mathematics (KAM)
Institut for Theoretical Computer Sciences (ITI)
Charles University, 11800 Prague, Czech Republic
e-mail: `nesetril@kam.ms.mff.cuni.cz`

André Raspaud
LaBRI, Université Bordeaux I
F-33405 Talence Cedex, France
e-mail: `raspaud@labri.fr`

On the Extremal Number of Edges in 2-Factor Hamiltonian Graphs

Ralph J. Faudree, Ronald J. Gould and Michael S. Jacobson

Abstract. In this paper we consider the question of determining the maximum number of edges in a Hamiltonian graph of order n that contains no 2-factor with more than one cycle, that is, 2-factor Hamiltonian graphs. We obtain exact results for both bipartite graphs, and general graphs, and construct extremal graphs in each case.

Mathematics Subject Classification (2000). Primary 05C45; Secondary 05C38.

Keywords. 2-factor, Hamiltonian, size.

1. Introduction

In this paper, we determine the maximum number of edges in a Hamiltonian graph of order n containing no 2-factor with more than one component. The question of the structure of Hamiltonian graphs with no 2-factors with more than one component has been receiving attention lately, for example see [2], [3], [4] and [5]. A Hamiltonian cycle is interpreted as a 2-factor with one component. In [4], the question of the minimum degree in a Hamiltonian graph sufficient to ensure the existence of a 2-factor with two cycles is considered. A 4-regular Hamiltonian graph with no other 2-factor with less than $n/5$ cycles is shown. However, the exact minimum degree condition remains an open question. Hendry [6] provided sharp results for the maximum number of edges in a graph with a unique 2-factor.

We say a graph is *2-factor isomorphic* if it contains a 2-factor X, but contains no 2-factor that is not isomorphic to X. If X is a Hamiltonian cycle, then of course, there are no 2-factors with more than one cycle. In this instance we will refer to such graphs as 2-factor Hamiltonian graphs.

The following is a special case of a result in [1].

Theorem 1.1. *If G is a Hamiltonian graph with $\delta(G) \geq 8$, then G is not 2-factor Hamiltonian.*

While in [3] the following was shown.

Theorem 1.2. *Let G be a 2-factor Hamiltonian k-regular graph. Then $k \leq 3$.*

We consider the nonregular case for 2-factor Hamiltonian graphs and determine the maximum number of edges in such graphs. In addition, we present examples of the extremal graphs and show that when $n \equiv 2 \bmod 4$ and bipartite, the extremal graph is unique. The extremal graphs are shown not to be unique in all other cases studied here.

Let G be a graph. We denote the minimum degree of G by $\delta(G)$. For a vertex x of G, we denote by $N(x)$ and $\deg x$ the neighborhood of x and the degree of x in G, respectively. Given a vertex x on a cycle C with an orientation, $\vec{C}$, then the successor of x on C will be denoted by x^+ and the predecessor by x^-.

For convenience we establish the following notation. Let C be a cycle with a given orientation and $v \in V(C)$. A *t-chord associated with v* will be an edge $e = v^{+(t-1)/2}v^{-(t-1)/2}$ such that e forms a t-cycle containing v and the cycle uses only the edge e and edges of the cycle C. Note, this is only defined for odd t. Similarly, a *t-chord associated with an edge* $f = xy \in E(C)$ is an edge $e = x^{-(t-2)/2}y^{+(t-2)/2}$ such that e forms a t-cycle containing f and the cycle uses only the edge e and edges of the cycle C.

2. Extremal graph constructions

In this section we present several different constructions of graphs which will be shown to be 2-factor Hamiltonian and attain the maximum size in Sections 3 (the bipartite case) and 4 (the nonbipartite case).

Let B_n be bipartite of even order n with partite sets $\{u_1, u_2, \ldots, u_{2m}\}$ and $\{v_1, v_2, \ldots, v_{2m}\}$ if $n \equiv 0 \bmod 4$ and $\{u_1, u_2, \ldots, u_{2m+1}\}$ and $\{v_1, v_2, \ldots, v_{2m+1}\}$ if $n \equiv 2 \bmod 4$. Define the adjacencies in B_n as follows:

$N(u_1) = \{v_1, v_2\}$,
$N(u_2) = \{v_1, v_2, v_3\}, \quad N(u_3) = \{v_1, v_2, v_4\}$,
$N(u_4) = \{v_1, v_2, v_3, v_4, v_5\}, \quad N(u_5) = \{v_1, v_2, v_3, v_4, v_6\}, \ldots$
$N(u_{2j}) = \{v_1, v_2, \ldots, v_{2j}, v_{2j+1}\}, \quad N(u_{2j+1}) = \{v_1, v_2, \ldots, v_{2j}, v_{2j+2}\}, \ldots,$
$N(u_{2m-2}) = \{v_1, v_2, \ldots, v_{2m-2}, v_{2m-1}\}, \quad N(u_{2m-1}) = \{v_1, v_2, \ldots, v_{2m-2}, v_{2m}\}$

and

$N(u_{2m}) = \{v_1, v_2, \ldots, v_{2m}\}$ (if $n \equiv 0 \bmod 4$)

while

$N(u_{2m}) = N(u_{2m+1}) = \{v_1, v_2, \ldots, v_{2m+1}\}$ (if $n \equiv 2 \bmod 4$).

Extremal graphs for the nonbipartite case when $n \equiv 0 \bmod 4$ can be obtained by inserting all possible edges into either one of the partite sets of a copy of B_n of the appropriate order. We designate these two graphs as $S_{n,u}$ ($S_{n,v}$) when the set $\{u_1, \ldots, u_{2m}\}$ ($\{v_1, \ldots, v_{2m}\}$) is complete. When $n \equiv 2 \bmod 4$ the graphs $S_{n,u}$ and $S_{n,v}$ are isomorphic.

Next we consider the case for odd n. When $n \equiv 1 \mod 4$ we form the graph O_n as follows: take a copy of $S_{n-1,v}$ along with a new vertex x and we join x to each vertex of the complete set $\{v_1, \ldots, v_{2m}\}$ and we join x to u_{2m}. When $n \equiv 3 \mod 4$ we form O_n by taking a copy of $S_{n-1,v}$ along with a new vertex x where x is joined to each of $v_1, \ldots, v_{2m}$ and u_{2m+1}.

3. Bipartite 2-factor Hamiltonian graphs

We now turn to the question of establishing the upper bounds on the size of a bipartite 2-factor Hamiltonian graph.

Theorem 3.1. *If G is a bipartite 2-factor Hamiltonian graph of order $n \equiv 0 \mod 4$, then*

$$|E(G)| \leq n^2/8 + n/2$$

and the bound is sharp.

Proof. Assume G is a bipartite 2-factor Hamiltonian graph of order $n = 2k$ (we use a more general condition to establish a setting useful in the subsequent theorem as well) with partite sets $\{u_1, u_2, \ldots, u_k\}$ and $\{v_1, v_2, \ldots, v_k\}$. Let $C^* : v_1, u_k, v_2, u_{k-1}, \ldots, v_k, u_1, v_1$ be a Hamiltonian cycle in G.

Define parallel classes of pairs of vertices as follows:

$$P_1 = \{u_1v_1, u_2v_2, \ldots, u_kv_k\}$$

and let P_i be the parallel class containing the pair u_1v_i, obtained from the rotation of P_1. Note that every edge of the complete graph on these vertices is in exactly one parallel class.

As G is 2-factor Hamiltonian, it follows that at most $\lceil \frac{(n/2-2)}{2} + 2 \rceil = \lceil \frac{n}{4} + 1 \rceil$ of the pairs in any parallel class can then be edges of G, for otherwise a 2-factor with two cycles formed using consecutive parallel edges would clearly result. It now follows that $|E(G)| \leq (n/4 + 1)n/2 = n^2/8 + n/2$, when $n \equiv 0 \mod 4$.

To see that this is optimal consider B_n, $n \equiv 0 \mod 4$, as defined in the previous section. The cycle $C : u_1, v_2, u_3, v_4, \ldots, u_{2m-1}, v_{2m}, u_{2m}, v_{2m-1}, u_{2m-2}, \ldots, v_3, u_2, v_1, u_1$ shows that this graph is Hamiltonian. To see that there is no nonisomorphic 2-factor observe that the edges u_1v_1 and u_1v_2 must be in any 2-factor. If the edges u_2v_3 and u_3v_4 are not in the 2-factor, then one of u_2 or u_3, say u_2, would be adjacent to v_1 and v_2 in the 2-factor. However, this would imply that u_3 would have degree one in the remaining graph and a 2-factor could not be formed. Now a similar argument applies to u_4 and u_5 forcing the edges u_4v_5 and u_5v_6 to be used. Subsequently, $u_{2t}v_{2t+1}$ and $u_{2t+1}v_{2t+2}$ would also be used. Therefore, any 2-factor must contain the path $v_{2m-1}, u_{2m-2}, v_{2m-3}, \ldots, u_2, v_1, u_1, v_2, u_3, \ldots, v_{2m}$. Hence, it follows that the only possible 2-factor is a Hamiltonian cycle. Furthermore, this graph has $n^2/8 + n/2$ edges, demonstrating the extremal number. □

Theorem 3.2. *If G is a bipartite 2-factor Hamiltonian graph of order $n \equiv 2 \bmod 4$, then*
$$|E(G)| \leq n^2/8 + n/2 + 1/2.$$
Further, the graph B_n is the unique extremal graph in this case.

Proof. Consider B_n, $n \equiv 2 \bmod 4$, as defined in the previous section. The cycle $C^* : u_1, v_2, u_3, v_4, \ldots, u_{2m-1}, v_{2m}, u_{2m+1}, v_{2m+1}, u_{2m}, v_{2m-1}, u_{2m-2}, \ldots, v_3, u_2, v_1, u_1$ shows that this graph is Hamiltonian.

To see that there is no nonisomorphic 2-factor, observe that the edges u_1v_1 and u_1v_2 must be in any 2-factor. If the edges u_2v_3 and u_3v_4 are not in the 2-factor, then one of u_2 or u_3, say u_2, would be adjacent to v_1 and v_2 in the 2-factor. However, this would imply that u_3 would have degree one in the remaining graph and a 2-factor could not be formed. Now a similar argument applies to u_4 and u_5 forcing the edges u_4v_5 and u_5v_6 to be used. Subsequently, $u_{2t}v_{2t+1}$ and $u_{2t+1}v_{2t+2}$ would also be used. Therefore, any 2-factor must contain the path

$$u_{2m}, v_{2m-1}, u_{2m-2}, v_{2m-3}, \ldots, u_2, v_1, u_1, v_2, u_3, \ldots, v_{2m}.$$

Hence, it follows that the only possible 2-factor is a Hamiltonian cycle. Furthermore, this graph has $n^2/8 + n/2 + 1/2$ edges, demonstrating the extremal number of edges is at least this number.

Now let G be a bipartite 2-factor Hamiltonian graph of order $n = 4m + 2$ containing the extremal number of edges. Let C be a Hamiltonian cycle in G with the given ordering $v_1, u_k, v_2, \ldots, v_k, u_1, v_1$, and note, to avoid a 2-factor with two cycles, each parallel class as defined in the previous theorem admits at most $m + 2$ edges into the graph G.

For each edge e of the Hamiltonian cycle, let F_e be the family of edges of the parallel class containing e which are included in the graph G. In this case, we call the edge e *strong* if $|F_e| = m + 2$, and *weak* otherwise. When e is weak, $|F_e| \leq m + 1$. Also note that when e is strong, F_e contains the 4-chord associated with e and alternate edges of the parallel class must also be in F_e, in particular, the 8-chord associated with e must also be in F_e. Observe that if e' is the antipodal edge of e on C, e is strong if and only if e' is strong. Note by the edge count, at least $2m + 2$ of the edges must be strong.

There cannot be three consecutive strong edges or a 2-factor with two cycles results (see Figure 1).

If consecutive edges e_1 and e_2 are strong, then the families

$$F_{e_3}, F_{e_5}, F_{e_7}, \ldots, F_{e_{2m+1}}, F_{e_{2m+4}}, F_{e_{2m+6}}, \ldots, F_{e_{4m+2}}$$

cannot contain the associated 4-chords. See Figure 2 for the e_5 and e_7 cases. All other cases work similarly, using one chord from the strong parallel class of e_1, one chord from the strong parallel class of e_2 and the 4-chord in question from e_{2j+1} for an appropriate j, to produce a 2-factor with two cycles. Thus, each of these edges must be weak. This implies there are at most $2m + 2$ strong edges but as we discussed above, we can conclude that there are precisely $2m + 2$ strong edges.

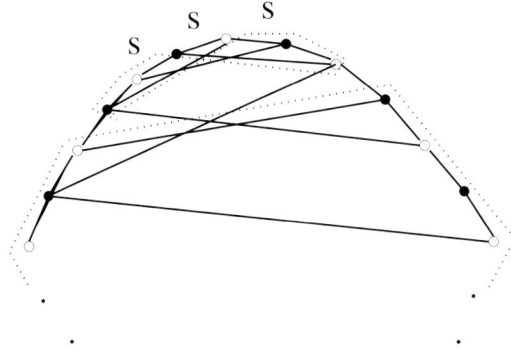

FIGURE 1. For $n \geq 10$, 3 consecutive strong edges produce a contradiction.

Each of the families associated with the remaining weak edges must have precisely $m + 1$ edges and no 4-chords. Thus, G has size at most $n^2/8 + n/4 + 1/2$.

The extremal case arises when strong and weak edges alternate, with the exception of two consecutive strong edges (and their corresponding two consecutive strong antipodal edges). Furthermore, the weak classes are completely determined as alternating edges in the family. Thus, this graph is unique and hence must be isomorphic to B_n. □

See Figure 3 for the $n = 14$ case. This figure shows the graph for this case, displays a Hamiltonian cycle C, as well as the strong and weak edges. Consequently, the graph of Figure 3 and B_n restricted to the case when $n = 14$, are seen to be isomorphic.

We conclude this section with a summary of the results.

Theorem 3.3. *If G is a bipartite 2-factor Hamiltonian graph of order n, then*
$$|E(G)| \leq \begin{cases} n^2/8 + n/2 & \text{if } n \equiv 0 \bmod 4, \\ n^2/8 + n/2 + 1/2 & \text{if } n \equiv 2 \bmod 4, \end{cases}$$
and the bounds are sharp in each case.

4. The general case

We now consider the question of establishing the upper bound on the size of a 2-factor Hamiltonian graph of order n.

Theorem 4.1. *If G is a 2-factor Hamiltonian graph of order n, then*
$$|E(G)| \leq \lceil n^2/4 + n/4 \rceil$$
and the bound is sharp for all $n \geq 6$.

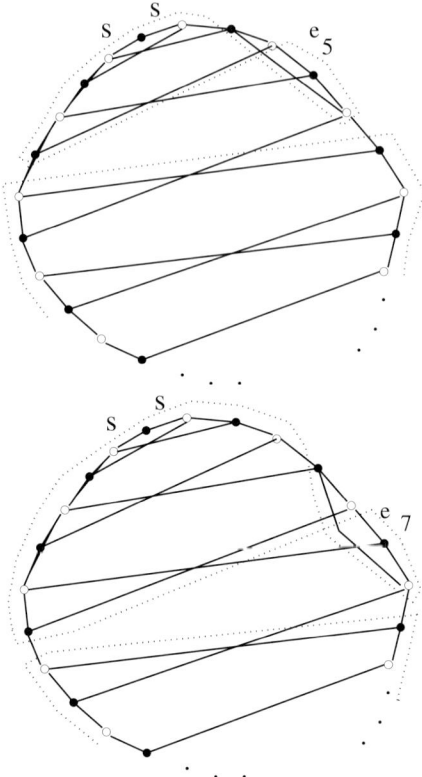

FIGURE 2. A 4-chord in F_{e_5} and F_{e_7} cases.

Proof. First suppose that n is even. Let $V(G) = \{v_1, v_2, \ldots, v_n\}$ and let $C : v_1, v_2, \ldots, v_n, v_1$ be a Hamiltonian cycle in G. We next partition $V(G)$ into two sets, $V_1 = \{2, 4, \ldots, n-2, n\}$ and $V_2 = \{1, 3, 5, \ldots, n-1\}$.

Besides the parallel classes for edges as in the bipartite case, we now must also consider parallel classes of edges within the sets V_1 and V_2. That is, parallel classes defined relative to a vertex rather than an edge.

Hence, we recognize two types of parallel classes. Define the classes relative to edges of C (hence relative to the bipartite structure of V_1 and V_2) just as we did in the bipartite case. From Theorem 3.3 we have a bound on the maximum number of these edges that may be included in G.

Next, define the classes relative to a vertex of C as: $P_n = \{(1, n-1), (2, n-2), (3, n-3), \ldots\}$ and let P_i be obtained by a translation of P_n to contain the pair $(i-1, i+1)$. These edges partition the edges whose ends both are within the set V_1 or both within the set V_2. Let Q_i be the edges in G that are in P_i. Note that in this case a parallel class Q_i has at most $\lfloor n/4 \rfloor$ edges.

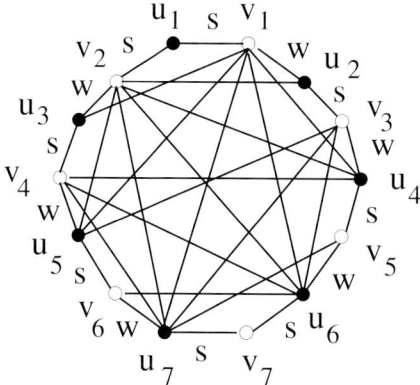

FIGURE 3. The graph when $n = 14$.

First suppose that $n \equiv 2 \mod 4$. Now the bipartite parallel classes may contribute a total of at most $n^2/8 + n/2 + 1/2$ edges. The $n/2$ distinct classes Q_i may contribute at most $(n-2)/4$ edges each. Hence, we may have at most:

$$|E(G)| \leq n^2/8 + n/2 + 1/2 + ((n-2)/4)(n/2) \qquad (4.1)$$
$$= n^2/8 + n/2 + 1/2 + n^2/8 - n/4 \qquad (4.2)$$
$$= n^2/4 + n/4 + 1/2. \qquad (4.3)$$

Note that the split graphs S_n (and similarly O_n) are 2-factor Hamiltonian because the corresponding B_n are 2-factor Hamiltonian. Since the split graph S_n, $n \equiv 2 \mod 4$ achieves this size, this bound is sharp.

If $n = 4m$, then the $n/2$ classes Q_i cannot all contain $n/4$ edges, as a 2-factor with two cycles is then easily produced. We say a parallel class Q_i is *full* (F) if it contains m edges. Clearly, in this case, Q_i cannot contain more than m edges, for otherwise a 2-factor with two cycles results. Note that if Q_i is full, then the antipodal class $Q_{i+n/2}$ is also full. A class Q_i is called *near-full* (N) if it contains exactly $m-1$ edges and is called *partially full* (P) if it contains at most $m-2$ edges. Observe that for any i, families Q_i, Q_{i+1} and Q_{i+2} cannot all be full (see Figure 4). In fact, if Q_{i+1} is full, then we know that at least one of Q_i and Q_{i+2} must not contain the associated 3-chord.

Suppose we have two consecutive full classes, without loss of generality say Q_n and Q_1. Now consider the class Q_2. Since Q_2 does not contain the 3-chord associated with vertex 2, we know Q_2 is not full. We now show that Q_2 is a partially full class. Assume to the contrary that it contains more than $m-2$ edges. By the remarks above, $Q_{n/2+2}$ cannot contain its associated 3-chord. Hence, for Q_2 to be near-full, it must contain precisely the $5, 9, 13, \ldots$-chords associated with vertex 2. Thus, we see that $C_1 : 1, 2, 3, , n-3, n-2, 4, n, n-1, 1$ and $C_2 : 5, 6, 7, n-7, n-6, 10, 11, n-11, n-10, \ldots, n/2-2, n/2-1, n-(n/2-1), n-(n/2-$

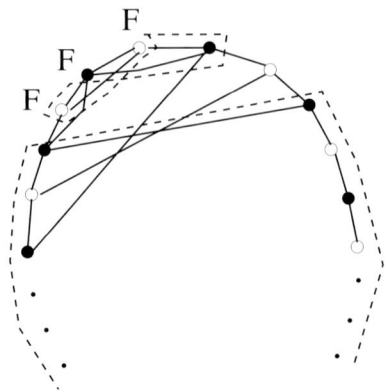

FIGURE 4. Three consecutive full families

2), $n/2, n-(n/2-4), n-(n/2-3), n/2-3, n/2-4, \ldots, n-8, n$ 9, 9, 8, $n-4, n-5, 5$ forms a 2-factor with two cycles, a contradiction. Hence, Q_2 must be a partially full class.

Therefore, the vertices can be partitioned into intervals around C containing one, two or three consecutive vertices of C into patterns of the form FN, FFP, P or N. Each of the intervals must average $m - 1/2$ edges in order to achieve the size of S_n.

Thus, we see that the edge average for each of these patterns is: $m - 2/3$ for FFP, $m - 1/2$ for FN, $m - 1$ for N, and finally at most $m - 2$ for P.

Hence, the upper bound on the size can only be obtained by having the pattern FN repeated around the cycle. Thus, we must have $n/4$ full classes and $n/4$ near-full classes with $(n/4 - 1)$ edges each, implying

$$\begin{align}
|E(G)| &\leq n^2/8 + n/2 + (n/4)(n/4) + (n/4)(n/4 - 1) \tag{4.4} \\
&= n^2/8 + n/2 + n^2/16 + n^2/16 - n/4 \tag{4.5} \\
&= n^2/4 + n/4. \tag{4.6}
\end{align}$$

Again the split graph on $n = 4m$ vertices with the bipartite structure from the previous section achieves this bound.

Now suppose that $n \equiv 3 \mod 4$. Let C be a Hamiltonian cycle. In this case the parallel classes defined relative to a vertex may contain at most $(n+1)/4$ edges. Since there are n such classes, we see that $|E(G)| \leq n(n+1)/4 = n^2/4 + n/4$ as desired. The graph O_n shows that the bound is sharp in this case.

Finally, suppose that $n \equiv 1 \mod 4$, say $n = 4m + 1$. First, we call a vertex strong provided its parallel class contains exactly $(n+3)/4$ edges. It is weak otherwise. Note that this includes the antipodal edge on C.

As before, there cannot be three consecutive strong vertices, or a 2-factor with two cycles is immediate. Next we show that there are at most $2m + 1 = (n+1)/2$

strong vertices. Otherwise, if there were more than $(n+1)/2$ strong vertices, then either there are three consecutive strong vertices, or there are at least two places around the cycle where there are two consecutive strong vertices. Therefore, there exists a strong vertex, separated from two consecutive strong vertices by an even number of vertices. For convenience let the two consecutive vertices be labeled v_n and v_1 and suppose there is another strong at vertex v_{2t}. Observe that the chord $c = v_{2t+(n-3)/2} v_{2t-(n-3)/2}$ is an edge of G since v_{2t} is strong. Note that there is a chord c_1 from the parallel class of vertex v_1 between the vertices $v_{(n-5)/2-2t+1}$ and $v_{2t+(n+5)/2}$. Further, there is a chord c_2 from the parallel class of v_n between the vertices $v_{2t+(n-1)/2}$ and $v_{(n-5)/2-2t+2}$. We now form a 2-factor with two cycles as follows. For one cycle we take the chord c_1 and the path on the Hamiltonian cycle between the ends of c_1 and containing v_1. The second cycle is formed by taking the chord c_2, following the Hamiltonian cycle back to $v_{2t+(n+3)/2}$, then taking the chord c, and now following the Hamiltonian cycle back to $v_{(n-5)/2-2t+2}$.

Therefore, if there are two consecutive vertices, say v_n and v_1, then by our previous observation, vertices $v_2, v_4, \ldots, v_{n-1}$ must all be weak. Hence, there can be at most $(n+1)/2$ strong vertices. (If there are not two consecutive strong vertices, then we can have at most $(n-1)/2$ strong vertices.) Consequently,

$$|E(G)| \leq \left(\frac{n+1}{2}\right)\left(\frac{n+3}{4}\right) + \left(\frac{n-1}{2}\right)\left(\frac{n-1}{4}\right).$$

Hence, $|E(G)| \leq \lceil n^2/4 + n/4 \rceil$ as desired. The graph O_n for this case shows that this bound is sharp. □

5. Remarks

We conclude with a few remarks. First, it is clear from our proofs that any Hamiltonian graph with more than the extremal number of edges, must contain a 2-factor with exactly two cycles.

Further, note that B_n for $n \equiv 2 \mod 4$ was the only unique extremal graph. This is easily seen since in all other cases there were parallel classes which allowed some flexibility in exactly where the edges were placed. This flexibility allows the existence of nonisomorphic extremal graphs in these cases. For example, in the graph B_{4m}, the edge $u_{2m-2}v_{2m-2}$ can be removed and then the edge $u_{2m-1}v_{2m-1}$ can be inserted, forming B'. The graph B' can be further altered by removing $u_{2m-4}v_{2m-4}$ and inserting $u_{2m-3}v_{2m-3}$. Each of these graphs can easily be seen to be 2-factor Hamiltonian. Further, we can continue this edge exchange process until the edge u_2v_2 is removed and u_3v_3 inserted. At this point a graph isomorphic to B_{4m} has been constructed, with the role of the partite sets interchanged.

Finally, consider S_n. Here we note that the edge $v_{n/2-1}v_{n/2}$ can be removed and the edge $u_{n/2-1}u_{n/2}$ can be inserted. These two graphs are clearly nonisomorphic and it is again easy to see the new graph is 2-factor Hamiltonian.

References

[1] M. Abreu, R.E.L. Aldred, M. Funk, B. Jackson, D. Labbate, J. Sheehan, Graphs and digraphs with all 2-factors isomorphic. J. Combin. Theory Ser. B, 92(2004), no. 2, 395–404.

[2] R.E.L. Aldred, M. Funk, B. Jackson, D. Labbate, and J. Sheehan, Regular bipartite graphs with all 2-factors isomorphic. J. Combin. Theory Ser. B, 92(2004), no. 1, 151–161.

[3] M. Funk, B. Jackson, D. Labbate, and J. Sheehan, 2-Factor Hamiltonian graphs. J. Combin. Theory. Ser. B, 87(2003), no. 1, 138–144.

[4] R.J. Faudree, R.J. Gould, M.S. Jacobson, L. Lesniak and A. Saito, A note on 2-factors with two components in Hamiltonian graphs. Discrete Math. 300(2995) no. 1-3, 218–224.

[5] R.J. Gould, Advances on the Hamiltonian Problem – A Survey, Graphs and Combinatorics 19(2003), 7–52.

[6] G R.T. Hendry, Maximum graphs with a unique k-factor. J. Combin. Theory Ser. B, 37(1984), no. 1, 53–63.

Ralph J. Faudree
Office of the Provost
University of Memphis
Memphis, TN 38152, USA
e-mail: `rfaudree@memphis.edu`

Ronald J. Gould
Department of Mathematics and Computer Science
Emory University
Atlanta, GA 30322, USA
e-mail: `rg@mathcs.emory.edu`

Michael S. Jacobson
Department of Mathematics
University of Colorado at Denver
Denver, CO 80217, USA
e-mail: `msj@math.cudenver.edu`

Generalized Colourings (Matrix Partitions) of Cographs

Tomás Feder, Pavol Hell and Winfried Hochstättler

Abstract. Ordinary colourings of cographs are well understood; we focus on more general colourings, known as matrix partitions. We show that all matrix partition problems for cographs admit polynomial time algorithms and forbidden induced subgraph characterizations, even for the list version of the problems. Cographs are the largest natural class of graphs that have been shown to have this property. We bound the size of a biggest minimal M-obstruction cograph G, both in the presence of lists, and (with better bounds) without lists. Finally, we improve these bounds when either the matrix M, or the cograph G, is restricted.

1. Introduction

Cographs are a well-understood class of graphs [4, 5, 14, 18]. A recursive definition is as follows. The one-vertex graph K_1 is a cograph; if G' and G'' are cographs, then so are the *disjoint union* $G' \cup G''$ and their *join* $G' + G''$ (obtained from $G' \cup G''$ by adding all edges joining vertices of G' to vertices of G''). It follows that the complement of a cograph is a cograph, and in fact the join of G' and G'' is the complement of the disjoint union of $\overline{G'}$ and $\overline{G''}$. It is not hard to show that G is a cograph if and only if it contains no induced path with four vertices [18]. Cographs can be recognized in linear time [5], and they can be represented, in the same time, by their *cotree* [5], which embodies the sequence of binary operations $\cup, +$, from the recursive definition, used in their construction. Many combinatorial optimization problems can be efficiently solved on the class of cographs, using the cotree representation [4, 5, 14]. This includes computing the chromatic number, and, more specifically, deciding if a cograph G is k-colourable. This suggests looking at more general colouring problems for the class of cographs. In fact, such investigations have already begun in [6, 19].

In [3, 7, 10, 11], a framework was developed, which encompasses many generalizations of colourings. Let M be a symmetric m by m matrix over $0, 1, *$.

An *M-partition* of a graph G is a partition of the vertex set $V(G)$ into m parts $V_1, V_2, \ldots, V_m$ such that V_i is a clique (respectively independent set) whenever $M(i,i) = 1$ (respectively $M(i,i) = 0$), and there are all possible edges (respectively no edges) between parts V_i and V_j whenever $M(i,j) = 1$ (respectively $M(i,j) = 0$). Thus the diagonal entries prescribe when the parts are cliques or independent sets, and the off-diagonal entries prescribe when the parts are completely adjacent or nonadjacent (with $*$ meaning no restriction). A graph G that does not admit an M-partition is called an *M-obstruction*, and is also said to *obstruct M*. A minimal *M-obstruction* is a graph G which is an M-obstruction, but such that every proper induced subgraph of G admits an M-partition. If $\mathcal{M}$ is a set of matrices, we say that G is a minimal $\mathcal{M}$-obstruction if it is an M-obstruction for all $M \in \mathcal{M}$, but every proper induced subgraph of G admits an M-partition for some $M \in \mathcal{M}$.

Given a graph G, we sometimes associate lists with its vertices: a list $L(v)$ of a vertex v is a subset of $\{1, 2, \ldots, m\}$, and it prescribes the parts to which v can be placed. In other words, a *list M-partition* of G (with respect to the lists $L(v), v \in V(G)$) is an M-partition of G in which each vertex v belongs to a part V_i with $i \in L(v)$. Note that the trivial case when all lists are $L(v) = \{1, 2, \ldots, m\}$ corresponds to the situation when no lists are given. M-obstructions and minimal M-obstructions (as well as $\mathcal{M}$-obstructions and minimal $\mathcal{M}$-obstructions) for graphs G with lists L are defined in the obvious way.

In the *(list) M-partition problem*, we have a fixed matrix M, and are asked to decide whether or not a given graph G (with lists) does or does not admit a (list) M-partition (with respect to the given lists).

We shall mostly focus on matrices M which have no diagonal $*$'s. If M has a diagonal $*$, then every graph G admits an M-partition; however, if lists are involved we will allow diagonal $*$'s. A matrix without diagonal $*$'s may be written in a block form, by first listing the rows and columns with diagonal 0's, then those with diagonal 1's. The matrix falls into four blocks, a k by k diagonal matrix A with a zero diagonal, an ℓ by ℓ diagonal matrix B with a diagonal of 1's, and a k by ℓ off-diagonal matrix C and its transpose. We shall say that M is a *constant matrix*, if the off-diagonal entries of A are all the same, say equal to a, the off-diagonal entries of B are all the same, say b, and all entries of C are the same, say c. In this case, we also say that M is an (a, b, c)-*block matrix*. Note that we may assume that $a \neq 0$ and $b \neq 1$, or else we can decrease k or ℓ.

Let M be a fixed matrix; if we prove that all cographs that are minimal M-obstructions have at most K vertices, then we can characterize M-partitionability of cographs by a finite set of forbidden induced subgraphs.

The *complement* $\overline{M}$ of a matrix M has all 0's changed to 1's and vice versa. It is clear that G admits an M-partition if and only if $\overline{G}$ admits an $\overline{M}$-partition, and that this also applies in the obvious way to M-partitions with lists, and to $\mathcal{M}$-partitions.

If the matrix M is a $(*, *, *)$-block matrix, then an M-partition of G is precisely a partition of the vertices of G into k independent sets and ℓ cliques. Such partitions have been introduced in [1] (see also [10, 11, 17]), and further studied in

[15, 16] for the class of chordal graphs (see also [12, 13]) and in [9] for the class of perfect graphs. More recently, they have been studied (without lists) for the class of cographs in [6, 19].

Suppose M is an m by m matrix; we shall refer to the integers $1, 2, \ldots, m$ as *parts*, since they index the set of parts in any M-partition of a graph. Given two sets of parts, $P, Q \subseteq \{1, 2, \ldots, m\}$, we define $M_{P,Q}$ to be the submatrix of M obtained by taking the rows in P and the columns in Q. We also let M_P denote $M_{P,P}$.

2. List partition problems

We first prove that for every matrix M the list M-partition problem for cographs can be solved in polynomial time, and characterized by finitely many forbidden induced subgraphs (with lists). By contrast, it is shown in [12, 13] that there exist matrices M for which the M-partition problem restricted to *chordal* graphs is NP-complete, even without lists.

Many of our arguments use the following observation. A disconnected graph $G = G_1 \cup G_2$ has an $\mathcal{M}$-partition if and only if G_1 has an M_P-partition and G_2 has an M_Q-partition, for some matrix $M \in \mathcal{M}$ and sets P, Q of parts such that $M_{P,Q}$ contains no 1. Of course the argument applies also with lists, if we view G_1, G_2 as inheriting the corresponding lists. We shall state this in the contrapositive form as follows.

Lemma 2.1. *Let $\mathcal{M}$ be fixed, and let $G = G_1 \cup G_2$ be a disconnected graph, with lists.*

Then G is an $\mathcal{M}$-obstruction if and only if for any matrix $M \in \mathcal{M}$ and any two sets P, Q of parts from M such that $M_{P,Q}$ does not contain a 1, the graph G_1 (with the corresponding lists) is an M_P-obstruction, or the graph G_2 (with the corresponding lists) is an M_Q-obstruction. $\square$

Suppose $\mathcal{M}$ is fixed, and $G = G_1 \cup G_2$ is disconnected.

Let $\mathcal{M}_1$ be a set of matrices M_P, where $M \in \mathcal{M}$ and P is a set of parts in M, such that G_1 is an M_P-obstruction, and let $\mathcal{M}_2$ be a set of matrices M_Q, where $M \in \mathcal{M}$ and Q is a set of parts in M, such that G_2 is an M_Q-obstruction. If, for any $M \in \mathcal{M}$, and any sets of parts P, Q of M such that $M_{P,Q}$ does not contain a 1, we have $M_P \in \mathcal{M}_1$ or $M_Q \in \mathcal{M}_2$, then the lemma ensures that for any subgraphs G_1' of G_1 and G_2' of G_2 which are $\mathcal{M}_1$-obstruction and $\mathcal{M}_2$-obstruction respectively, the subgraph $G' = G_1' \cup G_2'$ of G is also an $\mathcal{M}$-obstruction. Thus the minimality of G also implies the minimality of G_1, G_2. Such sets $\mathcal{M}_1, \mathcal{M}_2$ can be always chosen – for instance as the sets of *all* matrices M_P such that G_1 is an M_P-obstruction, respectively all matrices M_Q such that G_2 is an M_Q-obstruction.

Corollary 2.2. *Let $\mathcal{M}$ be fixed, and let $G = G_1 \cup G_2$ be a disconnected graph, with lists. Let $\mathcal{M}_1$ and $\mathcal{M}_2$ be chosen as described above.*

Then G is an $\mathcal{M}$-obstruction if and only if G_1 is an $\mathcal{M}_1$-obstruction and G_2 is an $\mathcal{M}_2$-obstruction.

Moreover, if G is a minimal $\mathcal{M}$-obstruction, then G_1 is a minimal $\mathcal{M}_1$-obstruction, and G_2 is a minimal $\mathcal{M}_2$-obstruction. □

Let $f(m)$ be the smallest integer such that for every m by m matrix M and every minimal M-obstruction cograph G with lists, G has at most $f(m)$ vertices. (In other words, $f(m)$ is the largest *size*, i.e., number of vertices, of a minimal M-obstruction cograph, over all m by m matrices M.)

Theorem 2.3. *For every integer m, we have*
$$f(m) \leq a^m m!$$
where $a = \frac{1}{ln(3/2)}$.

Proof. We apply Corollary 2.2 with $\mathcal{M}$ consisting of the single matrix M. Clearly a minimal $\mathcal{M}$-obstruction has size at most equal to the sum of the sizes of minimal M'-obstructions for all $M' \in \mathcal{M}$; thus we have
$$f(m) \leq 2 \sum_{i<m} \binom{m}{i} f(i).$$

By induction, letting $a = 1/\ln(3/2)$, we have
$$f(m) \leq 2m! a^m \sum_{0<j\leq m} 1/(j! a^j) \leq 2m! a^m (e^{1/a} - 1) = a^m m!. \qquad \square$$

Lemma 2.1 also yields an efficient algorithm to solve the list M-partition problem in the class of cographs. We consider the cotree of G, associating with each node t of the cotree (corresponding to a cograph G_t involved in the construction of G) a family of matrices $\mathcal{M}_t$. The family $\mathcal{M}_t$ consists of all matrices M_X, for $X \subseteq \{1, 2, \ldots, m\}$, such that G_t obstructs M_X. If t is a node of the cotree with children t', t'' corresponding to $G_t = G_{t'} \cup G_{t''}$, we know that G_t obstructs M_X if and only if for any $P \subseteq X, Q \subseteq X$ with $M_{P,Q}$ not containing 1, the graph $G_{t'}$ obstructs M_P or the graph $G_{t''}$ obstructs M_Q. Thus from the families $\mathcal{M}_{t'}, \mathcal{M}_{t''}$ we can compute the family $\mathcal{M}_t$. If $G_t = G_{t'} + G_{t''}$, we use complementation, as discussed earlier. Since the leaves of the cotree are single vertex cographs, each leaf t has $\mathcal{M}_t = \emptyset$. Then the given cograph G, is at the root r of the cotree, $G = G_r$, and we conclude G has a list M-partition if and only if $M \notin \mathcal{M}_r$.

Each set $\mathcal{M}_t$ has at most 2^m members, since there are at most 2^m subsets of $\{1, 2, \ldots, m\}$. Thus we obtain the following bound. (Note that $2^{O(m)}$ accounts also for the time to check if M is one of the matrices in $\mathcal{M}_r$.)

Corollary 2.4. *Every list M-partition problem for cographs can be solved in time $2^{O(m)} n$, linear in n.* □

We could, of course, proceed similarly, to solve the cograph list $\mathcal{M}$-partition problem for a *family* $\mathcal{M}$ of matrices.

We note that in [6] there are efficient algorithms solving related partition problems for cographs, for special matrices M, but not necessarily of fixed size.

We now derive a lower bound on $f(m)$. The special m by m matrix M_m has m diagonal zeros, and all off-diagonal entries $*$. Thus a list M_m-partition of G is precisely a list m-colouring of G. It turns out that there are very large cograph minimal M_m-obstructions. Since we are dealing with list colourings, we shall use the corresponding terminology.

Theorem 2.5. *For every positive integer m, there exists a minimal M_m-obstruction cograph G, with lists, of size $(e - 1 - \epsilon(m))m!$, where $1 \geq \epsilon(m) = o(1)$.*

Proof. We shall construct a cograph G, with lists from the set $\{1, \ldots, m\}$ of colours, that does not have a list colouring, but each of its proper induced subgraphs does. The construction will be done recursively. For each subset of colours, $K \subseteq \{1, 2, \ldots, m\}$, we shall construct a graph $G(K)$, with lists from $\{1, \ldots, m\}$, such that

- $G(K)$ is list colourable with colours from a set $S \subseteq \{1, \ldots, m\}$ if and only if $|S| \geq |K|$ and $S \neq K$, and,
- for each $v \in V(G)$, the subgraph $G(K) \setminus v$ is list colourable with colours from the set K.

Then $G = G(\{1, 2, \ldots, m\})$ will be a minimal M_m-obstruction, as desired.

The recursion starts with sets K consisting of a single element i. The graph $G(\{i\})$ is a single vertex with list $\{1, \ldots, m\} \setminus i$. This graph clearly satisfies the above conditions. The graph $G(K)$ with $K \subseteq \{1, \ldots, m\}$ and $|K| \geq 2$ is recursively defined as the disjoint union of all graphs $G(K \setminus j)$ for $j \in K$, together with an additional vertex v_K, with list $\{1, \ldots, m\}$, that is adjacent to all other vertices. Note that each $G(K)$ is a cograph, by induction.

Let S be a set of colours such that $G(K)$ has a list colouring with colours from S, and let j_0 denote the colour of v_K in such a colouring. Then each graph $G(K \setminus j)$ has a list colouring using the colours from $S \setminus j_0$, and hence, by induction, $|S \setminus j_0| \geq |K \setminus j|$, and $S \neq K$. On the other hand, if we remove v_K, all components $G(K \setminus j)$ are colourable with colours from K by induction, and if we remove any other vertex $v \in G(K \setminus j_0)$, then, again by induction, we can colour $G(K \setminus j_0) \setminus v$ and all $G(K \setminus j)$, for $j \neq j_0$, with colours from $K \setminus j_0$, and colour v_K by j_0.

Thus $G = G(\{1, \ldots, m\})$ is a minimal M_m-obstruction (with lists). Let $g(k)$ denote the number of vertices of a graph $G(K)$ with $|K| = k$. Then $g(1) = 1$ and $g(k) = 1 + kg(k-1)$, and hence

$$g(m) = \sum_{i=0}^{m-1} \frac{m!}{(m-i)!} = m! \sum_{1 \leq i \leq m} 1/i! = m!(e - 1 - \epsilon(m)),$$

where $1 \geq \epsilon(m) = \sum_{i=m+1}^{\infty} \frac{1}{i!} = o(1)$. □

Corollary 2.6. *For every integer m, we have*

$$(e - 1 - o(1))m! \leq f(m) \leq a^m m!$$

for $a = 1/\ln(3/2)$. □

3. Partition problems without lists

For the remainder of the paper, we shall focus on the $\mathcal{M}$-partition problem without lists. This implies that we now think of M in the block form, having k diagonal 0's and ℓ diagonal 1's, with $m = k + \ell$. Specifically, the parts $1, 2, \ldots, k$ will be independent sets, and the parts $k + 1, k + 2, \ldots, k + \ell = m$ will be cliques.

Given that we have no lists, we can improve the general bounds on the size of cograph minimal M-obstructions G. This is what we shall do in the present section. In the following two sections we shall obtain even better bounds when either the matrices M, or the cographs G, are restricted.

Lemma 3.1. *Let $\mathcal{M}$ be a collection of matrices, each of size at most m.*

If G is a minimal $\mathcal{M}$-obstruction cograph with maximum clique size r, then G has at most $g(m,r) \leq 2\binom{m+r}{r} + \binom{m+r-1}{r-1} - \binom{m+r-2}{r-2} - m - 1$ vertices.

The same conclusion applies if G has maximum independent set size r.

Proof. Suppose G has maximum clique size r. Since G is a cograph, its vertices can be partitioned into three graphs G_0, G_1, G_2 with no edges between G_0 and G_1, G_2, and with all edges between G_1 and G_2, where G_1 and G_2 are non-empty. We may assume that $G' = G_1 + G_2$ contains a clique of r vertices; in particular, there exists an integer $1 \leq t \leq r - 1$ such that the maximum clique size in G_1 is $r - t$ and in G_2 is at most t. We now consider how many vertices are needed to ensure that G does not admit an M-partition for any matrix $M \in \mathcal{M}$. Note that no matrix $M \in \mathcal{M}$ can contain the submatrix M_r (defined above Theorem 2.5), since G is perfect, and hence r-colourable.

Let $g(m,r)$ denote the maximum number of vertices in a minimal $\mathcal{M}$-obstruction cograph G with maximum clique size r. We derive a recurrence on $g(m,r)$ by estimating separately G_0, G_1, and G_2. If G_0 is not empty, then G' has an M-partition for some $M \in \mathcal{M}$, and since M does not contain M_r, each clique of size r in G' is placed in some set P of $t \leq r$ parts such that M_P contains a 1. This ensures that G_0 cannot use at least one part of M. Thus G_0 can be described as a minimal $\mathcal{M}'$-obstruction where all matrices in $\mathcal{M}'$ have size at most $m - 1$, i.e., G_0 has at most $g(m-1, r)$ vertices. On the other hand, G_1 and G_2 have at most $g(m, r-t)$ respectively $g(m, t)$ vertices, as noted above. We obtain the recurrence

$$g(m,r) \leq g(m-1, r) + g(m, r-t) + g(m, t),$$

$$g(0, r) = 1, g(m, 1) \leq m + 1.$$

In order to bound $g(m,r)$ we consider the quantity

$$h(m,r) = 2\binom{m+r}{r} + \binom{m+r-1}{r-1} - \binom{m+r-2}{r-2}.$$

Using the well-known identity $\binom{n}{k} - \binom{n-1}{k-1} = \binom{n-1}{k}$ we find that

$$h(m,r) - h(m, r-1) = h(m-1, r),$$

and thus
$$(h(m,r) - h(m, r-1)) - (h(m, r-1) - h(m, r-2)) = h(m-1, r) - h(m-1, r-1)$$
$$= 2\binom{m+r-2}{r} + \binom{m+r-3}{r-1} - \binom{m+r-4}{r-2} \geq 0.$$

(Note that $\binom{m+r-2}{r} \geq \binom{m+r-4}{r-2}$.) Therefore $h(m, r-(t+1)) + h(m, t+1) \leq h(m, r-t) + h(m, t)$ for $t+1 \leq r-t$, and so $h(m, r-t) + h(m, t) \leq h(m, r-1) + h(m, 1)$. Using the recursion for $g(m, r)$ we conclude inductively that $g(m, r) \leq h(m, r) - m - 1$, namely

$$\begin{aligned} g(m, r) &\leq h(m-1, r) - m - 2 + h(m, r-t) - m - 1 + h(m, t) - m - 1 \\ &\leq h(m-1, r) + h(m, r-1) + h(m, 1) - 3m - 4 \\ &= h(m, r) + 2(m+1) + 1 - 3m - 4 = h(m, r) - m - 1. \end{aligned}$$

The case of maximum independent set size r follows by complementation. $\square$

Theorem 3.2. *Any minimal M-obstruction cograph G has at most $O(8^m/\sqrt{m})$ vertices.*

Proof. We shall consider a cotree for G, and associate with each node t of the cotree a set $\mathcal{M}_t$ of submatrices of M, obstructed by the graph G_t corresponding to the node t, and such that G_t obstructs $\mathcal{M}_t$ if and only if the two graphs $G_{t'}, G_{t''}$, corresponding to the two children t', t'' of t in the cotree, obstruct $\mathcal{M}_{t'}$ and $\mathcal{M}_{t''}$ respectively. This is analogous to the algorithm inherent in Corollary 2.2. The root t_0 of our cotree will have $\mathcal{M}_{t_0}$ consisting of the one (given) matrix M, and the corresponding (given) graph $G_{t_0} = G$. The total number of vertices of G is precisely the number of leaves in the cotree. If G_t has maximum clique size at most $\tilde{m}$, and if all matrices in $\mathcal{M}_t$ have size at most $\tilde{m}$, then the entire branch of the cotree rooted at t contains at most $2\binom{2\tilde{m}}{\tilde{m}} + \binom{2\tilde{m}-1}{\tilde{m}-1} - \binom{2\tilde{m}-2}{\tilde{m}-2} - \tilde{m} - 1$ leaves, by the above lemma. If $G_t = G_{t'} \cup G_{t''}$, and if both $G_{t'}$ and $G_{t''}$ contain a clique of size greater than k (the number of diagonal 0's in M), then we can choose $\mathcal{M}_{t'}$ and $\mathcal{M}_{t''}$ to consist of matrices with maximum size smaller than the maximum size of a matrix in $\mathcal{M}_t$. Indeed, any M'-partition of $G_{t'}$ (or of $G_{t''}$), with $M' \in \mathcal{M}_t$, uses a part j of M which is a clique ($j > k$), and which therefore cannot be used by $G_{t''}$ (respectively $G_{t'}$); thus it suffices to certify the non-partitionability of $G_{t'}$ and $G_{t''}$ for matrices of strictly smaller size. Similarly, if $G_t = G_{t'} + G_{t''}$, and both $G_{t'}$ and $G_{t''}$ contain an independent set of size greater than ℓ, it suffices to certify their non-partitionability for matrices of size strictly smaller than the maximum size of a matrix in $\mathcal{M}_t$.

We let $g(\tilde{m})$ denote the maximum size of a minimal $\mathcal{M}$-obstruction cograph G, over all sets $\mathcal{M}$ consisting of matrices of size at most $\tilde{m}$. Suppose $G = G' \cup G''$ (the case $G = G' + G''$ is similar), and the maximum clique sizes in G', G'' are c', c'' respectively, with $c' \geq c''$. We have observed above that if $c' \geq c'' > k$, then G has at most $2g(\tilde{m}-1)$ vertices. If $c'' \leq c' \leq \tilde{m}$, then both G' and G'' have size at most $2\binom{2\tilde{m}}{\tilde{m}} + \binom{2\tilde{m}-1}{\tilde{m}-1} - \binom{2\tilde{m}-2}{\tilde{m}-2} - \tilde{m} - 1$ by the lemma, whence G has size at most

$2[2\binom{2\tilde{m}}{\tilde{m}} + \binom{2\tilde{m}-1}{\tilde{m}-1} - \binom{2\tilde{m}-2}{\tilde{m}-2} - \tilde{m} - 1]$. If $c' > \tilde{m}$ and $c'' \leq k$ we continue exploring the cotree, obtaining a sequence of graphs $G'_0, G'_1, \ldots, G'_s$, where $G'_i = G'_{i+1} \cup G''_{i+1}$ or $G'_i = G'_{i+1} + G''_{i+1}$. We always assume that if $G'_i = G'_{i+1} \cup G''_{i+1}$, then G'_{i+1} has a clique of size greater than $\tilde{m}$ and G''_{i+1} has maximum clique size at most k, and if $G'_i = G'_{i+1} + G''_{i+1}$, then G'_{i+1} has an independent set of size greater than $\tilde{m}$ and G''_{i+1} has maximum independent set size at most ℓ. We now argue that the sequence cannot be too long, namely, that $s \leq 2^{\tilde{m}}$. Indeed, we may assume that the sets $\mathcal{M}_t$ are *reduced*, in the sense that no $M_P, M_Q \in \mathcal{M}_t$ have $P \subseteq Q$ (as any graph obstructing M_P also obstructs M_Q). If we let N_i denote the set of all maximal sets P of parts (out of the m parts of M) such that $M_P \in \mathcal{M}_t$ corresponding to G'_i, then we see that $N_{i+1} \neq N_i$, otherwise G''_{i+1} is not needed. Thus one maximal set is dropped in each step from N_i to N_{i+1}. This implies that $s \leq 2^{\tilde{m}}$, and we obtain the general recurrence

$$g(\tilde{m}) \leq 2^{\tilde{m}+1}[2\binom{2\tilde{m}}{\tilde{m}} + \binom{2\tilde{m}-1}{\tilde{m}-1} - \binom{2\tilde{m}-2}{\tilde{m}-2} - \tilde{m} - 2] + 2g(\tilde{m}-1)$$

$$\leq O(2^{3\tilde{m}}/\sqrt{\tilde{m}}) + 2g(\tilde{m}-1),$$

which solves to $g(m) \leq O(8^m/\sqrt{m})$. □

We now define $F(m)$ to be the size (number of vertices) of a largest minimal M-obstruction cograph G without lists, for any m by m matrix M. From the above theorem we have an upper bound on $F(m)$; the following lower bound will follow from Theorem 5.2.

Corollary 3.3. *We have*

$$m^2/4 \leq F(m) \leq O(8^m/\sqrt{m}).$$

□

4. Constant matrices

In this section we prove that for each *constant* matrix M with k diagonal 0's and ℓ diagonal 1's, all cograph minimal M-obstructions have size at most $(k+1)(\ell+1)$. These M-partitions for constant matrices M (i.e., for (a,b,c)-block matrices M) have been investigated in the classes of perfect and chordal graphs in [9, 12, 13], and, in the case of $(*,*,*)$-block matrices (corresponding precisely to partitions into k independent sets and ℓ cliques), in [6, 15, 16, 19]. Recall that we do not consider lists in this section.

We illustrate the technique in the special case of $(*,*,*)$-block matrices, proving the following result; special cases of this result have been proved, by a different technique, in [6], cf. also [19].

Theorem 4.1. *Let M be a $(*,*,*)$-block matrix. Then each minimal M-obstruction cograph is $(k+1)$-colourable, and partitionable into $\ell+1$ cliques.*

Proof. When $\ell = 0$, each minimal M-obstruction is a minimal cograph G that is not k-colourable. Since cographs are perfect, $G = K_{k+1}$, which is both $(k+1)$-colourable, and partitionable to $(0+1)$ cliques. The case $k = 0$ follows by complementation, and we can proceed by induction on $k + \ell$. Let the cograph G be a minimal M-obstruction; we may assume that G is disconnected, $G = G_1 \cup G_2$ (or we can consider $\overline{G}$ instead). We shall now use Corollary 2.2, with the set $\mathcal{M}$ consisting of the single matrix M, and with all lists equal to $\{1, 2, \ldots, m\}$ (i.e., without lists); we shall be taking into account the special form of M to choose particular families $\mathcal{M}_1, \mathcal{M}_2$.

Specifically, let j be the smallest integer such that G_1 has a partition into k independent sets and j cliques. (Note that $0 \leq j \leq \ell$, by the minimality of G.) Since G_1 has a partition into k independent sets and j cliques, G_2 does not have a partition into k independent sets and $\ell - j$ cliques (otherwise G is not an M-obstruction). Let M_1 be the $(*, *, *)$-block matrix with k diagonal 0's and $j - 1$ diagonal 1's, and let M_2 be the $(*, *, *)$-block matrix with k diagonal 0's and $\ell - j$ diagonal 1's. We now let $\mathcal{M}_1$ consist of M_1 and all its submatrices, and let $\mathcal{M}_2$ consist of M_2 and all its submatrices. It is easy to check that these classes $\mathcal{M}_1, \mathcal{M}_2$ satisfy the conditions stated below Lemma 2.1. Indeed, if P, Q are such that $M_P \notin \mathcal{M}_1, M_Q \notin \mathcal{M}_2$, then M_P has at least j diagonal 1's (parts that are cliques), and M_Q has at least $\ell - j + 1$ diagonal 1's (parts that are cliques). This means that some part $i, i > k$, (part that is a clique) lies in both P and Q, whence $M_{P,Q}$ contains a 1.

We conclude, by Corollary 2.2, that G_1 is a minimal $\mathcal{M}_1$-obstruction, and G_2 is a minimal $\mathcal{M}_2$-obstruction, and hence a minimal M_1-obstruction and a minimal M_2-obstruction respectively (because of the special form of $\mathcal{M}_1, \mathcal{M}_2$).

Now, by the induction hypothesis, G_1 is $(k+1)$-colourable and partitionable into j cliques, while G_2 is $(k+1)$-colourable and partitionable into $\ell - j + 1$ cliques. It follows that G is both $(k+1)$-colourable and partitionable into $\ell + 1$ cliques. □

Note that a clique can meet an independent set in at most one vertex. Thus we have an upper bound on the size of a minimal M-obstruction. In fact, we can conclude that a minimal M-obstruction cograph G can be described as follows. The vertices of G are $v_{i,j}, i = 0, 1, \ldots, k, j = 0, 1, \ldots, \ell$, with any two $v_{i,j}, v_{i',j}$ adjacent, and no two $v_{i,j}, v_{i,j'}$ adjacent. (There are additional constraints on when arbitrary $v_{i,j}, v_{i',j'}$ are adjacent, arising from the fact that G is a cograph. This aspect is examined in [6, 19].)

Corollary 4.2. *Let M be a $(*, *, *)$-block matrix. Then each cograph minimal M-obstruction has exactly $(k+1)(\ell+1)$ vertices.* □

We shall prove the general result in a form better able to support induction. Instead of obstructions to one single (a, b, c)-block matrix M with k diagonal 0's and ℓ diagonal 1's, we shall consider collections $\mathcal{M}$ consisting of (a, b, c)-block matrices $M_0, M_1, M_2, \ldots, M_r$, each having k_i diagonal 0's and ℓ_i diagonal 1's. We shall further assume that the collection $\mathcal{M}$ is *staircase-like*, meaning that $k_i \leq k_j$

and $\ell_i \geq \ell_j$ for all $i < j$. If we have strict inequality everywhere, we call the collection *strictly staircase-like*. Clearly every collection of (a,b,c)-block matrices $\mathcal{N}$ contains a staircase-like subcollection $\mathcal{M}_1$ as well as a strictly staircase-like subcollection $\mathcal{M}_2$, such that a graph G is an $\mathcal{N}$-obstruction if and only if it is an $\mathcal{M}_1$-obstruction if and only if it is an $\mathcal{M}_2$-obstruction.

For notational convenience we shall allow matrices with $k_i = -1$ or $\ell_i = -1$. In this case we view each graph G as obstructing such a matrix. In particular, we shall set $k_{-1} = \ell_{r+1} = -1$.

Theorem 4.3. *Let a, b, c be fixed. Let $\mathcal{M} = \{M_i\}_{i=0}^r$ be a staircase-like collection of (a,b,c)-block matrices.*

Then the maximum size of a minimal $\mathcal{M}$-obstruction cograph is at most

$$f(\mathcal{M}) = \sum_{i=0}^{r}(k_i - k_{i-1})(\ell_i + 1) = \sum_{i=0}^{r}(\ell_i - \ell_{i+1})(k_i + 1).$$

Proof. Since the values of a, b, c are fixed, the matrices M_i are fully described by their parameters k_i, ℓ_i. To simplify the discussion, we shall write each M_i in the more descriptive form $M[k_i, \ell_i]$, and also write the bounding function $f(\mathcal{M})$ in the more descriptive form $f(\{(k_i, \ell_i)\}_{i=0}^r)$.

Let G be a minimal $\mathcal{M}$-obstruction. We may again suppose that G is disconnected, say $G = G_1 \cup G_2$, and shall derive an upper bound on G from upper bounds on G_1, G_2, using Corollary 2.2. Recall that we may assume that $a \neq 0$ and $b \neq 1$. We shall distinguish two main cases – when $c \neq 1$ and when $c = 1$.

Case 1: $c \neq 1$.
We first consider the subcase when $a = *$. Thus $a = *, b \neq 1, c \neq 1$, and the matrices in $\mathcal{M}$ have no 1's, except those on the main diagonal. As in the proof of Theorem 4.1, the graph G obstructs $M[k_i, \ell_i]$ if and only if there exists some $0 \leq j_i \leq \ell_i + 1$ such that G_1 obstructs $M[k_i, j_i - 1]$ and G_2 obstructs $M[k_i, \ell_i - j_i]$ (and, moreover, if G is a minimal $M[k_i, \ell_i]$-obstruction, then G_1 is a minimal $M[k_i, j_i - 1]$-obstruction, and G_2 a minimal $M[k_i, \ell_i - j_i]$-obstruction). As $M[k_i, d]$ is a submatrix of $M[k_{i+1}, d]$ we can choose j_i so that $j_i \geq j_{i+1}$ and $\ell_i - j_i \geq \ell_{i+1} - j_{i+1}$. Using induction, and setting $j_{r+1} = 0$, we compute

$$\begin{aligned}
f(\{(k_i,\ell_i)\}_{i=0}^r) &= f(\{(k_i, j_i - 1)\}_{i=0}^r) + f(\{(k_i, \ell_i - j_i)\}_{i=0}^r) \\
&= \sum_{i=0}^{r}((j_i - j_{i+1})(k_i + 1) + (\ell_i - j_i - \ell_{i+1} + j_{i+1})(k_i + 1)) \\
&= \sum_{i=0}^{r}(\ell_i - \ell_{i+1})(k_i + 1).
\end{aligned}$$

Now we consider the other subcase, when $a = 1$. Here $a = 1, b \neq 1, c \neq 1$, and there are off-diagonal ones between any parts j, j' that are independent sets ($j, j' \leq k$). Thus any two vertices that are placed in different independent sets must be adjacent. We can derive the following conditions from Corollary 2.2, or by the arguments given below.

The graph $G = G_1 \cup G_2$ has an $M[k_i, \ell_i]$-partition if and only if it has a partition where all parts i that are independent sets ($i \leq k$) are in one of G_1, G_2, or a partition in which there is only one part that is an independent set, and that set intersects both G_1 and G_2 (for this we must have $k_i \geq 1$). Equivalently, G obstructs $M[k_i, \ell_i]$ if and only if the following three conditions hold:

1. there exists a u_i with $0 \leq u_i \leq \ell_i + 1$ such that G_1 obstructs $M[0, u_i - 1]$ and G_2 obstructs $M[k_i, \ell_i - u_i]$,
2. symmetrically, there exists a v_i with $0 \leq v_i \leq \ell_i + 1$ such that G_2 obstructs $M[0, v_i - 1]$ and G_1 obstructs $M[k_i, \ell_i - v_i]$, and
3. if $k_i \geq 1$, there exists a w_i with $0 \leq w_i \leq \ell_i + 1$ such that G_1 obstructs $M[1, w_i - 1]$ and G_2 obstructs $M[1, \ell_i - w_i]$.

Note that we always can choose u_i and v_i such that

$$u_i + v_i \geq \ell_i + 1 \quad \text{for all } 0 \leq i \leq r. \tag{1}$$

If x denotes the largest value such that G_1 obstructs $M[0, x - 1]$ we may actually assume that $u_i = \min\{x, \ell_i + 1\}$ and $v_i = \min\{y, \ell_i + 1\}$, where y denotes the largest value such that G_2 obstructs $M[0, y - 1]$. In particular, this implies $u_i \geq u_{i+1}$ and $v_i \geq v_{i+1}$ for $0 \leq 1 \leq r - 1$. Similarly, if i_0 is the smallest index such that $k_{i_0} \geq 1$ we may assume that $w_i = \min\{w_{i_0}, \ell_{i_0} + 1\}$.

Thus, in order to meet both conditions, it suffices that G_1 obstructs $M[0, u_0 - 1]$, $M[1, w_{i_0} - 1]$ and $M[k_i, \ell_i - v_i]$ for $i \geq 0$, and G_2 obstructs $M[0, v_0 - 1]$, $M[1, \ell_{i_0} - w_{i_0}]$ and $M[k_i, \ell_i - u_i]$. We may assume that the parameters k_i are strictly increasing, for if $k_i = k_{i+1}$ then, as $\ell_i > \ell_{i+1}$ any graph that obstructs $M[k_i, \ell_i]$ also must obstruct $M[k_{i+1}, \ell_{i+1}]$ and, furthermore $(k_{i+1} - k_i)(\ell_{i+1} + 1) = 0$. By Corollary 4.2, we may assume that $r \geq 1$ or $k_0 \geq 1$.

If $k_0 = 0$ and $k_1 = 1$, we may assume that $x = u_0$ and $y = v_0$ have been chosen such that $u_0 + v_0 = \ell_0 + 1$ (x and y not necessarily maximal). Also we may assume that $w_1 = \ell_1 - v_1$ as well as $\ell_1 - w_1 = \ell_1 - u_1$. Thus, also $u_0 - 1 = \ell_0 - v_0$, $w_1 = \ell_1 - v_1$ and $v_0 - 1 = \ell_0 - u_0$ and using induction we compute the size of G as the sum of the sizes of G_1 and G_2, at most

$$\sum_{i=0}^{r}(k_i - k_{i-1})(\ell_i - v_i + 1) + \sum_{i=0}^{r}(k_i - k_{i-1})(\ell_i - u_i + 1)$$
$$= \sum_{i=0}^{r}(k_i - k_{i-1})\left((\ell_i + 1) - (u_i + v_i - \ell_i - 1)\right)$$
$$\leq \sum_{i=0}^{r}(k_i - k_{i-1})(\ell_i + 1) = f(\mathcal{M}).$$

In order to complete this case it suffices to additionally consider the first three summands in the induction step. Assume first, that $k_0 \geq 2$. Then we have the first three summands in $f(G_1)$ are $u_0 + w_0 + (k_0 - 1)(\ell_0 + 1 - v_0)$ and for G_2 we have $v_0 + \ell_0 + 1 - w_1 + (k_0 - 1)(\ell_0 + 1 - u_0)$. Adding up these numbers yields

$$(k_0 + 1)(\ell_0 + 1) - (k_0 - 2)(u_0 + v_0 - \ell_0 - 1) \leq (k_0 + 1)(\ell_0 + 1).$$

If $k_0 = 1$ similar to the first case we may assume $w_0 = u_0 = \ell_0 + 1 - v_0$ and we compute

$$u_0 + w_0 + (k_1 - k_0)(2\ell_1 + 2 - v_1 - u_1) + v_0 + (\ell_0 + 1 - w_0)$$
$$= (k_1 - k_0)(\ell_1 + 1) + 2(\ell_0 + 1) - (k_1 - k_0)(u_1 + v_1 - \ell_1 - 1)$$
$$\leq (k_0 + 1)(\ell_0 + 1) + (k_1 - k_0)(\ell_1 + 1).$$

Finally, if $k_0 = 0$ and $k_1 \geq 2$ again we may assume $x + y = u_0 + v_0 = \ell_0$ and compute

$$u_0 + w_1 + (k_1 - 1)(2\ell_1 + 2 - v_1 - u_1) + v_0 + \ell_1 + 1 - w_1$$
$$= (k_0 + 1)(\ell_0 + 1) + (k_1 - k_0)(\ell_1 + 1) - (k_1 - 1)(u_1 + v_1 - \ell_1 - 1)$$
$$\leq (k_0 + 1)(\ell_0 + 1) + (k_1 - k_0)(\ell_1 + 1).$$

Thus, in any case G has at most $f(\mathcal{M})$ vertices.

Case 2: $c = 1$.

In this case $a \neq 0, b \neq 1, c = 1$, and a disconnected graph $G = G_1 \cup G_2$ has an $M[k, \ell]$-partition if and only if it has an $M[0, \ell]$-partition, or an $M[k, 0]$-partition. It follows from facts proved in [8], and is easy to see directly, that the only minimal $M[k, 0]$-obstruction is K_{k+1}, except in the case when $a = 1$ and $k \geq 2$, when the disjoint union of K_1 and K_2 is the only other minimal $M[k, 0]$-obstruction. Complements of these graphs are all the minimal $M[0, \ell]$-obstructions, the complement of $K_1 \cup K_2$, i.e., the path P_3 with three vertices, only if $\ell \geq 2$ and $b = 0$.

Suppose now G is an $\mathcal{M}$-obstruction. Let k, ℓ be largest integers such that G obstructs $M[k, 0], M[0, \ell]$; note that $k_r \leq k, \ell_0 \leq \ell$. We claim that G contains a disconnected induced subgraph H which obstructs $M[k_r, 0]$ and $M[0, \ell_0]$ and has size

$$k_r + \ell_0 + 1 = f(\{(0, \ell_0), (k_r, 0)\}) \leq f(\{(k_i, l_i)\}_{i=0}^{r}).$$

We may assume that both k_r and ℓ_0 are positive, as in case $k_r = 0$ or $\ell_0 = 0$ the claim holds trivially using the minimal $M[k_r, 0]$-obstructions and the minimal $M[0, \ell_0]$-obstructions.

If G contains K_{k_r+1} and $\overline{K_{\ell_0+1}}$ then, since in a cograph any maximum clique meets any maximum independent set (see, for instance Theorem 11.3.3 in [2]), the union of any two such sets can serve as H (with $k_r + \ell_0 + 1$ vertices).

Next, we consider the case that G contains K_{k_r+1} and P_3 and $\ell_0 \geq 2$. If these obstructions are in different components, then we let $H = K_{k_r+1} \cup P_3$, of size $k_r + 4 \leq (\ell_0 + 1) + k_r$, unless $\ell_0 = 2$. In the latter case we remove the midpoint v of P_3. Then $H \setminus v$ has the right size and contains $\overline{K_3}$. If K_{k_r+1} and P_3 are in the same component, then this component is not a clique. Hence, by connectivity, it contains a clique K of size $k_0 + 1 \geq 2$ and a vertex w which is adjacent to some vertex of K and non-adjacent to another. Now, $K + w$ contains a P_3 and has $k_r + 2 < (\ell_0 + 1) + k_r$ vertices.

If G contains $\overline{K_{\ell_0+1}}$ and $K_1 + K_2$ and $k_r \geq 2$, then we correspondingly find an independent set I with $\ell_0 + 1$ vertices, and a vertex w adjacent to some vertex in I and non-adjacent to another. Hence $I + w$ also contains $K_1 + K_2$.

Finally, if G contains P_3 as well as $K_1 + K_2$, then the P_3 plus a vertex from a different component yields an obstruction of size $4 < (\ell_0 + 1) + k_r$. □

Corollary 4.4. *If M is a constant matrix and G a minimal M-obstruction cograph, then G has at most $(k+1)(\ell+1)$ vertices.* □

If $c \neq 1$ and $b = *$, or if $c \neq 0$ and $a = *$, the bound from Theorem 4.3 is tight. We give a minimal obstruction of size $f(\mathcal{M})$ for the first case, the second follows by taking complements. Let G consist of the disjoint union of $\ell_r + 1$ cliques of size $k_r + 1$ and $\ell_i - \ell_{i+1}$ cliques of size $k_i + 1$ for $0 \leq i \leq r - 1$. We show that G cannot be partitioned into k_i independent sets and ℓ_i cliques. Assume it had such a partition. There are $\ell_i + 1$ cliques of size at least $k_i + 1$. At least one vertex of each of these cliques has to be mapped to a clique, a contradiction. In order to show that G is minimal let v be a vertex in a $k_i + 1$ clique. Then we have ℓ_i cliques of size $> k_i$ and the other cliques can be partitioned into k_i independent sets.

Theorem 4.3 also implies efficient algorithms for the M-partition problem, where M is an (a, b, c)-block matrix. Thus suppose that a, b, c are fixed; given a cograph G, we can find the strictly staircase-like collection dominating all the matrices M_i to which G is an obstruction, in time $O((k+\ell)n)$. Given a staircase-like collection of matrices $\mathcal{M}$, such that G contains an $\mathcal{M}$-obstruction, we can find an induced subgraph H of G, such that H has size at most $f(\mathcal{M})$ and H also contains an $\mathcal{M}$-obstruction, in time $O((k+\ell)n)$. (We always assume the cograph G is given by its cotree; note that the cotree can be found in linear time [5].) The algorithms find all minimal pairs (k, ℓ) such that a corresponding partition exists (along the boundary of the staircase) for each node in the cotree, testing each one in constant time as indicated by the cases in the proof, given the corresponding staircases for the two children in the cotree. Since the length of the boundary of the staircase is $O(k + \ell)$, and there are n nodes in the cotree, the time $O((k + \ell)n)$ follows.

We remark that the upper bound $(k+1)(\ell+1)$ does not hold in general even for the class of trees. For instance, in the case $k = 1$, $b = 0$, $c = *$, there is a tree with $(\ell/3)^2$ vertices that is a minimal M-obstruction [12, 13]. The more general bound $f(\mathcal{M})$ does not hold for trees even in the case $a = b = c = *$: take the stair-like collection $\mathcal{M}$ of two matrices M_0, M_1 with $k_0 = 0, k_1 = 1, \ell_0 = 7, \ell_1 = 4$ – we have $f(\mathcal{M}) = 13$, but there is a minimal $\mathcal{M}$-obstruction with 14 vertices which is a tree, namely an edge $e = uv$ plus four attached paths of length 3, two attached at u and two attached at v. However, it is shown in [15] that the upper bound $(k+1)(\ell+1)$ does apply to collections consisting of one matrix, in the case of chordal graphs.

5. Unions of cliques

In this section we study minimal obstructions that are unions of cliques. Unions of cliques are an interesting subclass of cographs – while cographs are precisely those graphs not containing an induced path on four vertices, unions of cliques are precisely those graphs not containing an induced path on three vertices.

Recall that we are no longer considering lists. We start the simplest case of a non-constant matrix.

Proposition 5.1. *Let M be an $m \times m$ matrix which has only 0's on the main diagonal, one off-diagonal 1 and $*$'s elsewhere. Then M has just two minimal obstructions that are cographs, namely K_{m+1} and $K_m \cup K_{m-1}$.*

Proof. An M-partition of a graph G is an m-colouring of G, in which two special colour classes are completely adjacent (each vertex of one is adjacent to each vertex of the other). Clearly both K_{m+1} and $K_m \cup K_{m-1}$ are minimal M-obstructions. Suppose G is an M-obstruction cograph not containing K_{m+1}. Then its maximum clique size must be m, as otherwise G, as a cograph, and hence a perfect graph, would be $m-1$ colourable, and so would admit an M-partition. Let A be a clique of size m in G.

Suppose $e = uv$ is any edge of A. The graph $G - u - v$ must have a clique B_e of size $m - 1$, or else $G - u - v$ would be $m - 2$- colourable, and u and v could be placed as the only vertices in the two classes that are completely adjacent, yielding an M-partition of G.

Suppose G is a minimal M-obstruction cograph. We now claim that the cliques B_e can be chosen so that no clique B_e can contain a vertex w adjacent to exactly one vertex of the edge $e = uv$, say w adjacent to v. (In other words, each vertex $w \in B_e$ is adjacent to either both or to neither of u, v.) Otherwise, let G_e be a smallest induced subgraph of G containing A and B_e without an M-colouring placing u and v as the only vertices in the special classes that are completely adjacent: indeed, considering the cotree of G_e we find that the $\cup$-node where the directed paths from u resp. w to the root meet must be a descendent of the $+$-node, where both meet the path from v to the root. Let U, W be the graphs defined by the children of that $\cup$-node such that $u \in U$ and $w \in W$ and $v \in S$ the graph defined by the child of the $+$node. The minimality of G_e implies that $G_e \setminus W$ can be placed, and the maximality of the clique A in graph G implies that the largest clique in W is no larger than the clique $U \cap A$. Given the placement for $G_e \setminus W$, we may then place W in the parts where the clique $U \cap A$ is placed, since these parts are joined by $*$, and W can be colored with $|U \cap A|$ colors, thus placing all of G_e, a contradiction.

We may choose e in A joining two sets S and S' closest to the root of the cotree of G. If A and B_e are in different components of G then $A \cup B_e = K_m \cup K_{m-1}$ is an obstruction, while if A and B_e are in the same component of G then each vertex w in B_e is adjacent to at least one endpoint of e, and thus to both, giving the obstruction $B_e \cup \{u, v\} = K_{m+1}$. □

For general matrices M (with k diagonal 0's and ℓ diagonal 1's) we derive the following bounds on possible M-obstructions that are unions of cliques. Recall that we view M as a block matrix with a diagonal matrix A (having zero diagonal) and B (having a diagonal of 1s), and an off-diagonal matrix C and its transpose.

We shall consider the function
$$f(k,\ell) = \begin{cases} (k+1)(\ell+1) & \text{if } k \leq \ell+2 \\ k(\ell+2) - 1 & \text{if } \ell+2 \leq k \leq 2\ell+4 \\ \lfloor (k+2\ell+4)^2/8 - 1 \rfloor & \text{if } k \geq 2\ell+4. \end{cases}$$

We note that $f(k,\ell) = \max((k+1)(\ell+1), \Theta(k^2))$, i.e., there exists a function $h(k) = \Theta(k^2)$ such that $f(k,\ell) = \max((k+1)(\ell+1), h(k))$.

Theorem 5.2. *For each k and ℓ there exists a matrix M with k 0s and ℓ 1s on the diagonal, which admits a minimal M-obstruction G with $f(k,\ell)$ vertices that is a union of cliques.*

Proof. The case $k \leq \ell+2$ follows from Corollary 4.2, thus assume $k \geq \ell+2$. Let $2 \leq 2r \leq k$ and start M with r blocks $\begin{pmatrix} 0 & 1 \\ 1 & 0 \end{pmatrix}$ on the diagonal. This is followed by a constant matrix of size $k - 2r$ with 0's on the diagonal and 1's off-diagonal. All other off-diagonal entries are $*$'s. Let G be a disjoint union of $\ell + r$ cliques of size $k - 2r + 2$ and one of size $k - 2r + 1$. This is an obstruction since removing ℓ of the cliques, we are left with r cliques of size $k - 2r + 2$ that we can partition at best into an edge and a clique of size $k - 2r$. As the cliques are pairwise non-adjacent each of these edges has to use a different block, leaving one vertex of the clique of size $2k - 2r + 1$ pending. This already shows how to partition $G \setminus v$ if v is in the smaller clique. If v belongs to a clique of size $k - 2r + 2$ then we can map one vertex of each of the cliques of size $k - r + 1$ to the same element of one of the small blocks and, otherwise, proceed as above. Now, choosing $r = \max\{1, \lceil (k - 2(l+1))/4 \rceil\}$ yields the desired bounds. $\square$

Theorem 5.3. *Let f be defined as above. If M is any matrix with k 0's and ℓ 1's on the diagonal, then each minimal M-obstruction that is a disjoint union of cliques has at most $f(k,\ell) + (k+1)\ell$ vertices.*

If the block C of M contains no 1, then each minimal M-obstruction that is a disjoint union of cliques has at most $f(k,\ell)$ vertices.

Proof. Suppose a minimal M-obstruction G is the disjoint union of cliques $K_1, \ldots, K_t$, with $|K_i| \geq |K_j|$ for $i \leq j$. Removing $v \in K_t$ we have a partition assigning each K_i to a set of parts S_i for $i < t$. The submatrix M_i corresponding to S_i must have at least one 1, as otherwise each part of S_i would be an independent set, i.e., it would contain at most one vertex from K_i; since $|K_t| \leq |K_i|$, we could additionally assign K_t to parts of S_i and not have an obstruction. If M_i has a 1 on the diagonal, we may assume that $S_i = \{s_i\}$ consists of this entry alone. The sets S_i, thus, are partitioned into a collection T of S_i's all having $S_i = \{s_i\}$ and $M_i = (1)$ and a collection R of S_j's where M_j has 0's on the diagonal and some 1 off-diagonal

$M(r_j, c_j) = 1$. Note, that part S_j must have at least one vertex, that is in class r_j as well as one that is in c_j. We may further assume that for $S_i \in T$ and $S_j \in R$ we always have $|K_i| \geq |K_j|$, since s_i can absorb any clique. Hence, the clique $K_t \setminus v$ is assigned to parts in R and no pair of cliques from $K_1, \ldots, K_{t-1}, K_t \setminus v$ may share the parts s_i, r_j, c_j. Since G is minimal the K_i in T are of size at most $k+1$, as this already enforces them to use a 1 on the diagonal of M. Let $r = |R|$. A set $S_j \in R$ must not use r_k, c_k for $k \neq j$, since K_j is non-adjacent to K_k. Hence, such an S_j has size at most $k - 2r + 2$ and K_t has size at most $k - 2r + 1$, for $K_t - v$ avoids all pairs c_j, r_j (note, that $j < t$).

If part C of M has no 1's, then also the $S_i = \{s_i\} \in T$ correspond to cliques of size at most $k - 2r + 2$. For if, say K_i has size $k_i > k - r + 2$ and $v \in K_i$, then $G \setminus v$ has an M-partition where we may assume that $K_i \setminus v$ is one of the ℓ cliques, contradicting G being an M-obstruction. Therefore, in this case, $|V(G)| \leq (\ell + r + 1)(k - 2r + 2) - 1$ which is maximized at $f(k, \ell)$. If $r = 0$ we have ℓ cliques in T of size $k+1$ and $|K_t| \leq k+1$ adding up to $(\ell+1)(k+1) = f(k, \ell)$. This proves the upper bound for this special case.

Continuing with the general case, the $S_i = \{s_i\} \in T$ correspond to cliques of size at most $k+1$, giving at most $(k+1)\ell$ additional vertices, so $|V(G)| \leq f(k, \ell) + (k+1)\ell$. □

Theorem 5.4. *There exists a matrix M with the k by k block A with 0 diagonal having no off-diagonal 0s, the ℓ by ℓ block B with 1 diagonal having all off-diagonal entries $*$, and the k by ℓ block C having all entries $*$, such that M admits a minimal M-obstruction with $f'(k, \ell) = \Theta(k\ell + k^{1.5})$ vertices, that is a disjoint union of cliques.*

To be more precise, letting

$$r = \max(1, \lceil -1/2 - \ell/3 + \sqrt{(1/2 + \ell/3)^2 + 2(k+1)/3} \rceil),$$

so that $r = \Theta(1)$ if $k \leq \ell$, $r = \Theta(k/\ell)$ if $\ell \leq k \leq \ell^2$, and $r = \Theta(\sqrt{k})$ if $k \geq \ell^2$, and $t = k + 1 - (r^2 + r)/2$, so that $t = \Theta(k)$, we have

$$f'(k, \ell) = (t+r)\ell + tr + (r^2 + r)/2.$$

Proof. We may interpret any matrix $A = (k)_{ij}$ with no off-diagonal zeros as a kind of adjacency matrix $A(H)$ of a simple graph H on the k vertices $\{1, 2, \ldots, k\}$: two vertices i, j are adjacent if $k_{ij} = 1$ and nonadjacent if $k_{ij} = *$. Vice versa, with any simple graph H we can associate this way a unique matrix $A(H)$ of the described type.

Let t, r be positive integers, and H be the disjoint union of t isolated vertices and $r - 1$ cliques of sizes $2, 3, \ldots, r$ respectively. The corresponding matrix $A = A(H)$ is a $k \times k$-matrix where $k = t - 1 + (r^2 + r)/2$.

Now let G be the graph that is the disjoint union of r cliques of sizes $t + r, t + r - 1, \ldots, t + 1$ respectively, and an additional ℓ cliques of size $t + r$. Thus $|V(G)| = q = (t+r)\ell + tr + (r^2 + r)/2$. First, we show, by induction on r, that G is an obstruction for the matrix M with the A part as described above. If G

had an M-partition, then each of the ℓ parts corresponding to a 1 diagonal can be used for a clique of G, and we may put in such parts the largest cliques possible, that is, the ℓ additional cliques of size $t + r$. The remaining r cliques must go to A, so we reduce the problem to A-partition after removing the ℓ additional cliques of size $t + r$ from G. The clique K_G of size $t + 1$ in G had to use at least one vertex of a non-trivial clique K_H of H. Since G is the disjoint union of cliques, the other cliques of G may use one and only one vertex of H if and only if K_G uses only one vertex of K_H. Let $\tilde{G}$ arise from G by deleting K_G and one vertex of each of the other non-trivial cliques of G. Then G has an M-partition only if $\tilde{G}$ has an $M(H \setminus K_H)$ partition, which is not the case by inductive assumption (G does not have an $M(\tilde{H})$ partition for any graph $\tilde{H}$ consisting of t isolated vertices and $r - 1$ non-trivial cliques; the base case $r = 1$ has G consisting of a clique of size $t + 1$ but $\tilde{H}$ has no non-trivial cliques).

We still have to show that the obstruction G is minimal. Assume v is a vertex in the clique of size $t + r - i$, then $G \setminus v$ has $r - i$ cliques of size at most $t + r - i - 1$. These can be mapped to into the t isolated vertices of H and to one vertex of each of the $r - i - 1$ cliques of size at most $r - i$ of H. From each of the remaining cliques K_j of size $t + r - j$, $0 \le j \le i - 1$ of G we map t vertices each to the isolated vertices of H and the remaining $r - j$ vertices of K_j to the clique of size $r - j$ in H.

It remains to choose r to maximize

$$q = q_r = (k + 1 - r^2/2 + r/2)\ell + (k + 3/2 - r^2/2)r.$$

We note that

$$q_{r+1} - q_r = -3/2(r^2 + 2(1/2 + \ell/3)r - 2(k+1)/3),$$

so the maximum occurs at

$$r = \max(1, \lceil -1/2 - \ell/3 + \sqrt{(1/2 + \ell/3)^2 + 2(k+1)/3} \rceil). \qquad \square$$

Theorem 5.5. *Suppose the block submatrix A with 0 diagonal has no 0 off diagonal. Let f' be defined as above, satisfying $f'(k, \ell) = \Theta(k\ell + k^{1.5})$, and let*

$$g'(k, \ell) = f'(k, \ell) + (k + 1)\ell.$$

If M has k 0's and ℓ 1's on the diagonal, then any minimal M-obstruction that is a disjoint union of cliques has at most $g'(k, \ell)$ vertices.

If in addition the block C contains no 1, then any obstruction that is a disjoint union of cliques has at most $f'(k, \ell)$ vertices.

Proof. We proceed as in the proof of Theorem 5.3, and assume a minimal M-obstruction G is the disjoint union of cliques $K_1, \ldots, K_t$ with $|K_i| \ge |K_j|$ for $i \le j$. Removing $v \in K_t$ we have a partition assigning each K_i to parts from S_i for $i < t$. The sets S_i are partitioned into a collection T of S_i's having $S_i = \{s_i\}$ and $M_i = (1)$ and a collection R of S_j's where M_j has 0's on the diagonal and 1,* off-diagonal. Let U_j be the set of indices that are used exclusively by $S_j \in R$ and D be the set of indices that are used by at least two $S_i \in R$. We may order the sets $U_1, \ldots, U_{r-1}$ nonincreasingly. Then $|U_i| \ge r + 1 - i$, since otherwise we may U be a set of size

$r-i$ consisting of one element from each of $U_i,\ldots,U_{r-1}$, and assign the cliques $K_i,\ldots,K_{r-1}$ and K_t to $U \cup D$, contrary to the fact that G is an obstruction. If we let t be the number of parts in A that do not correspond to $r+1-i$ chosen elements out of U_i, then $k \geq t+2+3+\cdots+r = t-1+(r^2+r)/2$ so $t \leq k+1-(r^2+r)/2$.

If part C of M has no 1's, then also the $S_i = \{s_i\} \in T$ correspond to cliques of size at most $t+r$. For if, say K_i has size $k_i > t+r$ and $v \in K_i$, then $G \setminus v$ has an M-partition where we may assume that $K_i \setminus v$ is one of the ℓ cliques, contradicting G being an obstruction. Therefore $|V(G)| \leq (t+r)\ell + tr + (r^2+r)/2$ which is maximized at $f'(k,\ell)$.

Continuing with the general case, we estimate the largest clique by $k+1$, so $|V(G)| \leq (k+1)\ell + tr + (r^2+r)/2 \leq (k+1)\ell + f'(k,\ell) = g'(k,\ell)$. □

We are indebted to an anonymous referee for a careful reading of the manuscript.

References

[1] A. Brandstädt, Partitions of graphs into one or two stable sets and cliques, *Discrete Math.* 152 (1996) 47–54.

[2] A. Brandstädt, V.B. Le, and J.P. Spinrad, **Graph Classes: A Survey**, SIAM Monographs on Discrete Math. and Applications 1999.

[3] K. Cameron, E.M. Eschen, C.T. Hoang, and R. Sritharan, The list partition problem for graphs, *SODA 2004*.

[4] D.G. Corneil, H. Lerchs, and L. Stewart Burlingham, Complement reducible graphs, *Discrete Applied Math.* 3 (1981) 163–174.

[5] D.G. Corneil, Y. Perl, and L.K. Stewart, A linear recognition algorithm for cographs, *SIAM. J. Computing* 14 (1985) 926–934.

[6] M. Demange, T. Ekim, and D. de Werra, Partitioning cographs into cliques and stable sets, *Discrete Optimization* 2 (2005) 145–153.

[7] C.M.H. de Figueiredo, S. Klein, Y. Kohayakawa, and B.A. Reed, Finding skew partitions efficiently, *Journal of Algorithms* 37 (2000) 505–521.

[8] T. Feder and P. Hell, On realizations of point determining graphs, and obstructions to full homomorphisms, manuscript 2004.

[9] T. Feder and P. Hell, Matrix partitions of perfect graphs, *Claude Berge Memorial Volume* (2004).

[10] T. Feder, P. Hell, S. Klein, and R. Motwani, Complexity of list partitions, *Proc. 31st Ann. ACM Symp. on Theory of Computing* (1999) 464–472.

[11] T. Feder, P. Hell, S. Klein, and R. Motwani, List partitions, *SIAM J. on Discrete Math.* 16 (2003) 449–478.

[12] T. Feder, P. Hell, S. Klein, L. Tito Nogueira, and F. Protti, List matrix partitions of chordal graphs, LATIN 2004, *Lecture Notes in Computer Science* 2976 (2004) 100–108.

[13] T. Feder, P. Hell, S. Klein, L. Tito Nogueira, and F. Protti, List matrix partitions of chordal graphs. *Theoretical Computer Science* 349 (2005) 52–66.

[14] M.C. Golumbic, **Algorithmic Graph Theory and Perfect Graphs**, Academic Press, New York, 1980.

[15] P. Hell, S. Klein, L.T. Nogueira, F. Protti, Partitioning chordal graphs into independent sets and cliques, *Discrete Applied Math.* 141 (2004) 185–194.

[16] P. Hell, S. Klein, L.T. Nogueira, F. Protti, Packing r-cliques in weighted chordal graphs, *Annals of Operations Research* 138, No. 1 (2005) 179–187.

[17] P. Hell and J. Nešetřil, **Graphs and Homomorphisms**, Oxford University Press, 2004.

[18] S. Seinsche, On a property of the class of n-colourable graphs, *J. Combin. Theory B* 16 (1974) 191–193.

[19] R. de Souza Francisco, S. Klein, and L. Tito Nogueira, Characterizing (k,ℓ)-partitionable cographs, *Electronic Notes in Discrete Mathematics* 22 (2005), 277–280.

Tomás Feder
268 Waverley St.
Palo Alto, CA 94301, USA
e-mail: `tomas@theory.stanford.edu`

Pavol Hell
School of Computing Science
Simon Fraser University
Burnaby, B.C., Canada V5A 1S6
e-mail: `pavol@cs.sfu.ca`

Winfried Hochstättler
FernUniversität in Hagen
D-58084 Hagen, Germany
e-mail: `Winfried.Hochstaettler@fernuni-hagen.de`

A Note on $[k,l]$-sparse Graphs

Zsolt Fekete and László Szegő

> **Abstract.** In this note we provide a Henneberg-type constructive characterization theorem of $[k,l]$-sparse graphs, that is, the graphs for which the number of induced edges in any subset X of nodes is at most $k|X|-l$. We consider the case $0 \le l \le k$.
>
> **Mathematics Subject Classification (2000).** 05C05; 05C40.
>
> **Keywords.** Sparse graph, constructive characterization.

1. Introduction

In this paper we consider undirected graphs and we allow parallel edges and loops. Let $G = (V,E)$ be a graph. If $u,v \in V$ and $e \in E$, then $e = uv$ denotes that edge e has endnodes u and v (there may be other edges parallel to e).

For a subset $X \subseteq V$, $i_G(X)$ denotes the number of induced edges in X, i.e., $i_G(X) := |\{e \in E : e = uv \text{ with } u,v \in X\}|$. If $v \in V$, then $i_G(v) := i_G(\{v\})$ is the number of loops on v. If $X,Y \subseteq V$, then $d_G(X,Y) := |\{e \in E : e = uv \text{ with } u \in X-Y, v \in Y-X\}|$. $d_G(X) := d_G(X, V-X)$. For a node $v \in V$ $\deg_G(v)$ will denote the degree of v, that is, $\deg_G(v) := d_G(\{v\}, V-\{v\}) + 2i_G(v)$ (note that a loop contributes 2 to the degree). We note that with this convention $\sum_{v \in V} \deg(v) = 2|E|$ holds.

Here we give the definition of the graphs for which we provide certain characterizations in the paper. Let l, k be integers and $k \ge 1$. If $l \le k$, the we say that a graph $G = (V,E)$ is $[k,l]$-sparse in $\emptyset \ne Z \subseteq V$ if $i(X) \le k|X|-l$ holds for every $\emptyset \ne X \subseteq Z$. If $k+1 \le l \le 2k-1$ holds, then we say that a graph $G = (V,E)$ is $[k,l]$-sparse in $\emptyset \ne Z \subseteq V$ if G is loopless and $i(X) \le k|X|-l$ holds for every $X \subseteq Z, |X| \ge 2$.

Research is supported by OTKA grants T 037547 and TS 049788, by European MCRTN Adonet, Contract Grant No. 504438 and by the Egerváry Research Group of the Hungarian Academy of Sciences.

We say that a graph $G = (V, E)$ is a $[k, l]$-graph if $|E| = k|V| - l$ and G is $[k, l]$-sparse in V. Remark that if $l < k$, then there can be (at most $k - l$) loops incident to any node in a $[k, l]$-graph.

Nash-Williams [6] proved the following theorem concerning coverings by trees.

Theorem 1.1 (Nash-Williams). *A graph $G = (V, E)$ is the union of k edge-disjoint forests if and only if G is $[k, k]$-sparse in V.*

A consequence of this theorem is that a graph is a $[k, k]$-graph if and only if its edgeset is a disjoint union of k spanning trees.

Frank [1] observed that a combination of a theorem of Mader and a theorem of Tutte gives rise to the following characterization. The main operation will be *pinching* a set F of edges into a new node z which means that we split every edge $e \in F$ with a node e_z and identify these with one node z. (If F is the emptyset then pinching F into a node z will simply mean adding a new isolated vertex z to the graph.)

Theorem 1.2 (Frank). *An undirected graph $G = (V, E)$ is a $[k, k]$-graph if and only if G can be built from a single node by the following two operations:*

1. *add a new node z and k new non-loop edges ending at z,*
2. *pinch j $(1 \leq j \leq k - 1)$ existing edges with a new node z, and add $k - j$ new edges connecting z with existing nodes.*

We mention that there is known an analogous constructive characterization theorem for $[k, k+1]$-graphs (see Theorem 3.2).

If $0 \leq l \leq k$, then the class of $[k, l]$-graphs can be characterized by packings of trees and pseudotrees. A *pseudotree* is a set of edges which is connected and contains exactly one cycle. Let F be an edgeset on vertex set V. We say that F is a *spanning pseudoforest* if every component of (V, F) is a pseudotree. Now we show how they are related to $[k, l]$-graphs, where $0 \leq l \leq k$. Whiteley [12] proved the following characterization.

Theorem 1.3 (Whiteley). *If $G = (V, E)$ is a graph and $0 \leq l \leq k$, then the following are equivalent.*

1. *G is a $[k, l]$-graph,*
2. *E is the disjoint union of l spanning trees and $(k - l)$ spanning pseudoforests.*

If we are given a graph $G = (V, E)$, then the system of the edgesets $F \subseteq E$ for which (V, F) is $[k, l]$-sparse in V is the collection of the independent sets of a matroid on ground set E (see, e.g., [13]).

The notion of sparsity is strongly related to rigidity theory. Laman's theorem [4] says that the minimally rigid graphs in the plane are exactly the $[2, 3]$-graphs. The rigidity of bar-and-body structures in arbitrary dimension can be characterized by $[k, k]$-graphs [10], and $[k, k+1]$-graphs has also connections to rigidity theory [8, 9]. Whitely [12] conjectures that if $0 \leq l \leq k$ then, the $[k, l]$-graphs are the minimally rigid graphs on certain surfaces.

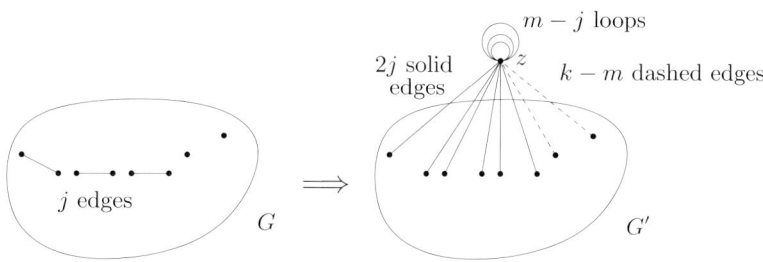

FIGURE 1. G' is obtained from G by operation $K(k,m,j)$.

Henneberg [3] proved a constructive characterization for minimally rigid graphs in the plane. With Laman's theorem this result gives a constructive characterization of $[2,3]$-graphs. In the above mentioned examples the constructive characterization theorems serve useful tools in proving properties of the graphs in question. (See Section 3 about $[k,l]$-graphs for $l > k$.)

We will prove a Henneberg-type construction of $[k,l]$-graphs for $0 \leq l \leq k$. We will use the following operations.

Definition 1.4. Let $0 \leq j \leq m \leq k$. $K(k,m,j)$ will denote the following operation. Choose j edges of G, pinch into a new node z. Put $m - j$ loops on z and link it with other nodes by $k - m$ new edges. (After this operation the resulting graph has k more edges than the original graph and the new node z has degree $(k+m)$. See Figure 1.)

The graph on one node with l loops will be denoted by P_l.

Lemma 1.5. *If graph G is obtained by operations $K(k,m,j)$ for which $j \leq m \leq k$, $m - j \leq k - l$ starting from P_{k-l}, then it is a $[k,l]$-graph.*

Proof. This can be seen directly from the definition. (Or using Theorem 1.3: one can easily construct the bases in G' if the bases in G are given.) □

We will prove the following theorem.

Theorem 1.6. *Let $G = (V,E)$ be a graph and $1 \leq l \leq k$. Then G is a $[k,l]$-graph if and only if G can be obtained starting from P_{k-l} with operations $K(k,m,j)$ where $j \leq m \leq k-1, m-j \leq k-l$.*

Let $G = (V,E)$ be a graph. Then G is a $[k,0]$-graph if and only if G can be constructed from P_k with operations $K(k,m,j)$ where $j \leq m \leq k, m - j \leq k$.

We remark that loopless $[k,l]$-graphs cannot be obtained by operations above via a sequence of loopless $[k,l]$-graphs. (See Figure 2 for an example.)

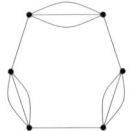

FIGURE 2. A loopless $[2,0]$-graph, which cannot be obtained by a sequence of loopless $[2,0]$-graphs.

2. Proof of Theorem 1.6

The if part of Theorem 1.6 is Lemma 1.5. To prove the other direction we need the following lemma.

Lemma 2.1. *Let $G = (V, E)$ be a $[k,l]$-graph.*
1. *If $l > 0$ then $\exists v \in V$ such that $k \leq \deg(v) \leq 2k - 1$.*
2. *If $l = 0$ then $\exists v \in V$ such that $k \leq \deg(v) \leq 2k$.*
3. *If $l < 0$ and $|V| \geq 2l + 1$ then $\exists v \in V$ such that $k \leq \deg(v) \leq 2k$.*

By *splitting off* a pair of edges $e = uv, f = uw \in E$ we mean the operation of replacing e and f by a new edge connecting v and w. Denote $G^{ef} = (V, E - e - f + vw)$ the obtained graph. We say that the new edge vw is a *split edge*.

The following will give the "only if" part of Theorem 1.6.

Theorem 2.2. *Let $0 \leq l \leq k$. Let $G = (V + s, E)$ be a $[k,l]$-graph and let m, j be integers such that $\deg_G(s) = k+m, i_G(s) = m-j$ where $j \leq m \leq k$, $m-j \leq k-l$. Then we can split off j pairs of edges so that after deleting s the remaining graph is a $[k,l]$-graph.*

We will use the following simple lemma. Let $b(X)$ denote the following $b(X) = b_G(X) := k|X| - l - i_G(X)$. We remark that a graph G is $[k,l]$-sparse in V if and only if $b_G(Z) \geq 0$ for all $\emptyset \neq Z \subseteq V$.

If $G = (V + s, E)$ is $[k,l]$-sparse in V and $e = sv, f = sw \in E$, then splitting off e and f is called *admissible* if G^{ef} is $[k,l]$-sparse in V.

Lemma 2.3. *Let $G = (V, E)$ be a graph and $X, Y \subseteq V$. Then*
1. $i(X) + i(Y) + d(X,Y) = i(X \cap Y) + i(X \cup Y)$.
2. $b(X) + b(Y) = b(X \cap Y) + b(X \cup Y) + d(X,Y)$.
3. *Let $l \leq k$. If G is a $[k,l]$-graph, then $b(X) = b(Y) = 0, X \cap Y \neq \emptyset$ implies $b(X \cup Y) = b(X \cap Y) = 0$.*
4. *Let $l \leq 2k - 1$. If G is a $[k,l]$-graph, then $b(X) = b(Y) = 0, |X \cap Y| \geq 2$ implies $b(X \cup Y) = b(X \cap Y) = 0$.*
5. *If $G = (V + s, E)$ is $[k,l]$-sparse in V and $e = sv, f = sw$ are edges incident to s ($v, w \in V$), then the pair e, f is admissible if and only if $\nexists X \subseteq V$ such that $v, w \in X$ and $b(X) = 0$.*

We omit the proof of the lemma. Now we give the proof of Theorem 2.2.

Proof. First we remark if $j = 0$, then it is clear that $G - s$ is a $[k, l]$-graph (since k edges were deleted). So we can assume that $j \geq 1$. Now assume on the contrary that we cannot split off j pairs of edges so that the resulting graph is $[k,l]$-sparse in V. Split off as many pairs as possible. We split off say $p < j$ pairs of edges and denote the resulting graph by G'. Let $e_1 = sv_1, \ldots, e_\alpha = sv_\alpha$ be the non-loop edges incident to s in G' where $\alpha = k + m - 2(m-j) - 2p = k - m + 2j - 2p \geq 2$. By Lemma 2.3 we know that for every v_ν, v_μ ($1 \leq \nu < \mu \leq \alpha$) there exists an $X_{\nu\mu} \subseteq V$ such that $v_\nu, v_\mu \in X_{\nu\mu}$ and $b_{G'}(X_{\nu\mu}) = 0$. Using the second statement of Lemma 2.3 we get that there exists an $X \subseteq V$ such that $v_\nu \in X$ for every ν and $b_{G'}(X) = 0$. Let $X_{G'}$ be a maximal set having these properties.

Now consider every G' which can be obtained by splitting off p pairs of edges at s in G. For each G' we have a set $X_{G'}$. Choose $G_1 := G'$ so that $|X_{G'}|$ is maximal. Let $X := X_{G_1}$.

Claim 2.4. *There is a split edge $e = vw$ in G_1 such that $v, w \notin X$.*

Proof. Assume on the contrary that for every split edge $e = vw$, $v \in X$ or $w \in X$. Let $\beta := |\{e : e = vw \text{ is a split edge and } v, w \in X\}|$. $b_{G_1}(X) = 0$ implies $b_G(X) = \beta$. $b_G(X+s) = b_G(X) + k - \gamma_G(s) - d_G(s, X) = b_G(X) + k - (m-j) - (k-m+2j-(p-\beta)) = \beta + k - m + j - (k - m + 2j - p + \beta) = \beta + k - m + j - k + m - 2j + p - \beta = p - j < 0$. A contradiction. $\square$

Let $e = vw$ be an edge given by the claim. Let $G_2 := G_1 - e + sv + sw$. We state that sv, sv_1 is an admissible splitting off in G_2. Because if $v, v_1 \in Y \subseteq V$ and $b_{G_2}(Y) = 0$, then $b_{G_1}(Y) \leq b_{G_2}(Y) = 0$ so $b_{G_1}(Y) = 0$. But $X \cap Y \neq \emptyset$ (since $v_1 \in X \cap Y$) hence $b_{G_1}(X \cup Y) = 0$ by Lemma 2.3, which contradicts the maximality of $|X_{G_1}|$.

Let $G_3 := G_2 - sv - sv_1 + vv_1$. We state that sw, sv_2 is an admissible splitting off in G_3. Assume on the contrary that $w, v_2 \in Z \subseteq V$ and $b_{G_3}(Z) = 0$.

If $v \notin Z$ or $v_1 \notin Z$, then $b_{G_1}(Z) \leq b_{G_2}(Z) = b_{G_3}(Z) = 0$ so $b_{G_1}(Z) = 0$. But $X \cap Z \neq \emptyset$ (since $v_2 \in X \cap Z$) hence $b_{G_1}(X \cup Z) = 0$, which contradicts the maximality of $|X_{G_1}|$.

If $v, v_1 \in Z$, then $b_{G_1}(Z) = b_{G_2}(Z) - 1 = b_{G_3}(Z) = 0$ so $b_{G_1}(Z) = 0$. But $X \cap Z \neq \emptyset$ (since $v_2 \in X \cap Z$) hence $b_{G_1}(X \cup Z) = 0$ but this contradicts the maximality of $|X_{G_1}|$.

We proved that sw, sv_2 is an admissible splitting off in G_3. This contradicts the maximality of p. $\square$

We mention that our proof is algorithmic. The only nontrivial claim we need is that we can algorithmically decide if a graph is $[k,l]$-sparse in V, and give back a set $i(X) > k|X| - l$ if not. This can be tested by the following simple claim.

Claim 2.5. *Let $G = (V, E)$ be a graph and $k, l \geq 1$ be integers. $i(X) \leq k|X| - l$ holds for every $X \subseteq V$ if and only if for every $e = uv \in E$: $i_{G_e}(X) \leq k|X|$ holds for every $X \subseteq V$, where the graph G_e is obtained by adding l uv-edges to G.*

The condition $i(X) \leq k|X|$ can be checked by the orientation theorem of Hakimi: a graph has an orientation such that every indegree is at most k if and only if $i(X) \leq k|X|$ for every set X of vertices.

Proof of Theorem 1.6. Lemma 1.5 shows the "if" part. To prove the other direction we observe that the only $[k,l]$-graph with one node is P_{k-l}. Let G be an arbitrary $[k,l]$-graph with at least two nodes. By Lemma 2.1 there exists a node s of degree at most $2k-1$ if $l > 0$ or a node of degree at most $2k$ if $l = 0$.

Let $m := \deg(s) - k$. It is clear that $0 \leq m \leq k$. By the definition of the degree, $i(s) = \deg(s) - (|E| - i(V-s))$ hold. We have that $|E| - i(V-s) \geq k|V| - l - (k|V| - k - l) = k$ because G is a $[k,l]$-graph. This implies that $i(s) \leq \deg(s) - k = m$. Let $j := m - i(s)$. Now $0 \leq j \leq m$ and $m - j = i(s) \leq k - l$ holds because G is a $[k,l]$-graph. Thus these m and j satisfy the conditions of Theorem 2.2.

Theorem 2.2 claims that G is obtained from a graph G' by an operation $K(k,m,j)$. By induction we know that G' can be constructed from P_{k-l}, this implies that G can be constructed from P_{k-l} too. □

By the remark after the proof of Theorem 2.2 we can determine an inductive construction of a $[k,l]$-graph ($0 \leq l \leq k$) algorithmically in polynomial time.

3. Partial results for other k and l values

In this section k, l will be integers and $k \geq 1$, but l can be negative. First we remark that Theorem 2.2 remains true without assumption $l \geq 0$ (the proof is the same). Thus for $l < 0$ the following version of Theorem 1.6 follows (using 3. of Claim 2.1).

Theorem 3.1. *Let $G = (V, E)$ be a graph and $l < 0 < k$. Then G is a $[k,l]$-graph if and only if G can be obtained from a $[k,l]$-graph on at most $2|l|$ vertices by operations $K(k,m,j)$ where $0 \leq j \leq m \leq k, m - j \leq k - l$.*

An undirected graph is called *k-tree-connected* if it contains k edge-disjoint spanning trees. Remark that a graph is minimally k-tree-connected if and only if it is a $[k,k]$-graph. Two variants of the notion of k-tree-connectivity were considered by Frank and Szegő in [2]. One of them is the following: a loopless graph G (with at least 2 nodes) is called *nearly k-tree-connected* if G is not k-tree-connected but adding any new edge to G results in a k-tree-connected graph. It is easy to see that a graph is nearly k-tree-connected if and only if it is a $[k, k+1]$-graph.

Let K_2^t denote the graph on two nodes with t parallel edges. Based on the work of Henneberg [3] and Laman [4], Tay and Whiteley gave a proof of the following theorem in the special case of $k = 2$ in [11].

Theorem 3.2 (Frank and Szegő). *An undirected graph $G = (V, E)$ is nearly k-tree-connected if and only if G can be built from K_2^{k-1} by applying the following operations:*

1. *add a new node z and k new edges ending at z so that no k parallel edges can arise,*
2. *choose a subset F of j existing edges $(1 \leq j \leq k-1)$, pinch the elements of F with a new node z, and add $k-j$ new edges connecting z with other nodes so that there are no k parallel edges in the resulting graph.*

In [8] Tay proved for inductive reasons that a node of degree at most $2k-1$ either can be "split off", or "reduced" to obtain a smaller nearly k-tree-connected graph. Theorem 3.2 says that there always exists a node which can be "split off".

The following theorem follows easily from the definition of $[k,l]$-graphs.

Proposition 3.3. *Let $k+1 \leq l \leq \frac{3k}{2}$. If an undirected graph $G = (V, E)$ can be built up from K_2^{2k-l} by applying the following operations, then it is a $[k,l]$-graph.*

(P1) *add a new node z and k new edges ending at z so that no $k-l+1$ parallel edges can arise.*

(P2) *Choose a subset F of j existing edges $(1 \leq j \leq k-1)$, pinch the elements of F with a new node z, and add $k-j$ new edges connecting z with other nodes so that there are no $k-l+1$ parallel edges in the resulting graph.*

Inspiring by Theorem 3.2 we would conjecture that the reverse of the proposition above is also true for all k and l satisfying $k+1 \leq l \leq \frac{3k}{2}$. But as it was shown in [7], this is not true if $k + \frac{k+2}{3} \leq l$, still we think the following holds.

Conjecture 3.4. Let $k+1 \leq l < k + \frac{k+2}{3}$. An undirected graph $G = (V, E)$ is a $[k,l]$-graph if and only if G can be built from K_2^{2k-l} by applying the operations (P1) and (P2).

At last after a lemma we give a weaker form of Theorem 2.2 for $l \leq \frac{3k}{2}$.

Lemma 3.5. *Assume $l \leq \frac{3k}{2}$ and $G = (V, E)$ is a $[k,l]$-graph. Let $X, Y, Z \subseteq V$. If $b(X) = b(Y) = b(Z) = 0$ and $|X \cap Y| = |X \cap Z| = |Y \cap Z| = 1$, $|X \cap Y \cap Z| = 0$ then $b(X \cup Y \cup Z) = 0$ and $l = \frac{3k}{2}$.*

Proof. $0 \leq b(X \cup Y \cup Z) = k|X \cup Y \cup Z| - l - i(X \cup Y \cup Z) \leq k(|X| + |Y| + |Z| - 3) - l - i(X) - i(Y) - i(Z) = k|X| - l - i(X) + k|Y| - l - i(Y) + k|Y| - l - i(Y) - 3k + 2l = b(X) + b(Y) + b(Z) - 3k + 2l = 2l - 3k \leq 0$. □

Theorem 3.6. *Assume $l \leq \frac{3k}{2}$. Let $G = (V + s, E)$ be a $[k,l]$-graph and let m, j be integers such that $\deg_G(s) = k + m$, $i_G(s) = m - j$ where $j \leq m \leq k, m - j \leq l$. Then there exist a j-element edgeset F on the neighbors of s such that $(G - s) + F$ is a $[k,l]$-graph.*

Proof. Let $N \subseteq V$ denote the neighbors of s. For arbitrary X containing N: $i_G(s) = m - j$, $\deg_G(s) = k + m$ and $i_G(X + s) = k|X + s| - l$ implies that $b_G(X) \geq j$. ($i_G(X) = i_G(X + s) - i_G(s) - d_G(s, X) \leq k(|X| + 1) - l - (m - j) - (\deg_G(s) - 2(m - j)) = k|X| + k - l - m + j - (k + m - 2m + 2j) = k|X| - l - j$).

We prove the following claim by induction on ν.

Claim 3.7. For every $0 \leq \nu \leq j$ there exist an F_ν ν-element edgeset on N such that $(G-s)+F'$ is a $[k,l]$-sparse in V.

Proof. If $\nu = 0$, then it is trivial. Suppose that there is a $(\nu - 1)$-element edgeset $F_{\nu-1}$, such that $i_{G+F_{\nu-1}}(X) \leq k|X| - l$ for all $\emptyset \neq X \subseteq V$. Now we prove that we add can one more edge.

Suppose on the contrary that for every $uv \in E$, $u, v \in N$ there exists an X_{uv} such that $u, v \in X_{uv}$: $\gamma_{G+F_{\nu-1}}(X_{uv}) = k|X_{uv}| - l$, i.e., $b_{G+F_{\nu-1}}(X_{uv}) = 0$. We claim that there exist a set X, such that $N \subseteq X \subseteq V$ and $b_{G+F_{\nu-1}}(X) = 0$. If $|N| = 1$, then $X := X_{uu}$ (where $N = \{u\}$) is appropriate. If $|N| \geq 2$, then let $u, w \in N, u \neq w$ and let $X \subseteq V$ be a maximal set satisfying $X_{uw} \subseteq X$ and $b_{G+F_{\nu-1}}(X) = 0$. We claim that $N \subseteq X$. Suppose that $v \in N - X$. If $|X_{vu} \cap X| \geq 2$ or $|X_{vw} \cap X| \geq 2$, then X cannot be maximal by Lemma 2.3. If $|X_{vu} \cap X_{vw}| \geq 2$, then $b_{G+F_{\nu-1}}(X_{vu} \cup X_{vw}) = 0$ and $|(X_{vu} \cup X_{vw}) \cap X| = |\{u,w\}| = 2$ implies $b_{G+F_{\nu-1}}(X_{vu} \cup X_{vw} \cup X) = 0$, this contradicts the maximality of X.

But then we have $|X_{vu} \cap X| = |X_{vw} \cap X| = |X_{vu} \cap X_{vw}| = 1$ and by Lemma 3.5 $b(X_{vu} \cup X_{vw} \cup X) = 0$ contradicting the maximality of X.

Now we have $0 = b_{G+F_{\nu-1}}(X) = b_G(X) - (\nu - 1) \geq b_G(X) - (j - 1)$ contradicting the remark at the beginning of the proof, which said $b_G(X) \geq j$. □

□

Acknowledgment

The authors are grateful to an anonymous referee for her/his useful comments and suggestions.

References

[1] A. Frank, *Connectivity and network flows*, in: R. Graham, M. Grötschel and L. Lovász, eds., Handbook of Combinatorics (Elsevier Science B.V., 1995), 111–177.

[2] A. Frank and L. Szegő, *Constructive Characterizations on Packing and Covering by Trees*, Discrete Applied Mathematics **131** No. 2 (2003), 347–371.

[3] L. Henneberg, Die graphische Statik der starren Systeme, Leipzig, 1911.

[4] G. Laman, *On graphs and rigidity of plane skeletal structures*, Journal of Engineering Math. **4** (1970), 331–340.

[5] C.St.J.A. Nash-Williams, *Edge-disjoint spanning-trees of finite graphs*, J. London Math. Soc. **36** (1961), 445–450.

[6] C.St.J.A. Nash-Williams, *Decomposition of finite graphs into forests*, J. London Math. Soc. **39** (1964), 12.

[7] L. Szegő, *On constructive characterizations of (k,l)-sparse graphs*, EGRES technical report, TR-2003-10, www.cs.elte.hu/egres/tr/egres-03-10.ps

[8] Tiong-Seng Tay, *Linking $(n-2)$-dimensional panels in n-space I: $(k-1,k)$-graphs and $(k-1,k)$-frames*, Graphs and Combinatorics **7** (1991), 289–304.

[9] Tiong-Seng Tay, *Linking $(n-2)$-dimensional panels in n-space II: $(n-2,2)$-frameworks and body and hinge structures*, Graphs and Combinatorics **5** (1989), 245–273.

[10] Tiong-Seng Tay, *Henneberg's method for bar and body frameworks*, Structural Topology **17** (1991), 53–58.

[11] T.-S. Tay and W. Whiteley, *Generating isostatic frameworks*, Structural Topology **11** (1985), 21–69.

[12] Walter Whiteley, *The union of matroids and the rigidity of frameworks*, SIAM J. Disc. Math. **1** No. 2 (1988), 237–255.

[13] Walter Whiteley, *Some matroids from discrete applied geometry*, Matroid theory (Seattle, WA, 1995), volume 197 of Contemp. Math., pages 171–311. Amer. Math. Soc., Providence, RI, 1996.

Zsolt Fekete
Department of Operations Research
Eötvös University
Pázmány Péter sétány 1/C
H-1117, Budapest, Hungary

and

Communication Networks Laboratory
Pázmány sétány 1/A
H-1117 Budapest, Hungary
e-mail: `fezso@cs.elte.hu`

László Szegő
Department of Operations Research
Eötvös University
Pázmány Péter sétány 1/C
H-1117 Budapest, Hungary

and

Department of Combinatorics and Optimization
University of Waterloo
200 University Avenue West
Waterloo, Ontario, N2L 3G1, Canada
e-mail: `szego@cs.elte.hu`

Even Pairs in Bull-reducible Graphs

Celina M.H. de Figueiredo, Frédéric Maffray and
Claudia Regina Villela Maciel

Abstract. A bull is a graph with five vertices a, b, c, d, e and five edges ab, bc, cd, be, ce. A graph G is bull-reducible if no vertex of G lies in two bulls. An even pair is a pair of vertices such that every chordless path joining them has even length. We prove that for every bull-reducible Berge graph G with at least two vertices, either G or its complementary graph $\overline{G}$ has an even pair.

1. Introduction

A graph is *perfect* if for every induced subgraph H of G the chromatic number of H is equal to its clique number. Perfect graphs were defined by Claude Berge [1]. The study of perfect graphs led to several interesting and difficult problems. The first one is their characterization. Berge conjectured that a graph is perfect if and only if it contains no odd hole and no odd antihole, where a hole is a chordless cycle of length at least 4, and an antihole is the complementary graph of a hole. It has become customary to call *Berge graph* any graph that contains no odd hole and no antihole, and to call the above conjecture the "Strong Perfect Graph Conjecture". This conjecture was proved by Chudnovsky, Robertson, Seymour, and Thomas [4] in 2002. A second problem is the existence of a polynomial-time algorithm to color optimally the vertices of a perfect graph. This problem was solved in 1984 by Grötschel, Lovász and Schrijver [10] with an algorithm based on the ellipsoid method for linear programming. A third problem is the existence of a polynomial-time algorithm to decide if a graph is Berge. This was solved by Chudnovsky, Cornuéjols, Liu, Seymour and Vušković [3] in 2002. There remains a number of interesting open problems in the context of perfect graphs. Some of them are related to the concept of even pair.

Even pairs: An *even pair* [18] in a graph G is a pair of vertices such that every chordless path between them has even length. A graph G is called a *quasi-parity*

This research was partially supported by CNPq, CAPES (Brazil)/COFECUB (France), project number 359/01.

graph [18] if for every induced subgraph H of G on at least two vertices, either H has an even pair or $\overline{H}$ has an even pair. A graph G is called a *strict quasi-parity graph* [18] if every induced subgraph of G on at least two vertices has an even pair. Clearly, strict quasi-parity graphs are quasi-parity graphs. Meyniel [18] proved that every quasi-parity graph is perfect. The concept of even pair turned out to be very useful for proving that certain classes of Berge graphs are perfect and for designing optimization algorithms on special classes of perfect graphs. See [8] for a survey on this matter. Some questions of particular interest are the characterization of quasi-parity graphs and of strict-quasi-parity graphs. Hougardy [14, 15] (see also [8]) made two conjectures: (1) there is a family F of line-graphs of bipartite graphs such that a graph is a strict quasi-parity graph if and only if it does not contain an odd hole, an antihole, or a graph in F; (2) there is a family F' of line-graphs of bipartite graphs such that a graph is a quasi-parity graph if and only if it does not contain an odd hole, an odd antihole, or a graph in F'. These two conjectures are still unsolved.

Bull-free graphs: A *bull* is a graph with five vertices r, y, x, z, s and five edges ry, yx, yz, xz, zs; see Figure 1. We will frequently use the notation $r - yxz - s$ for such a graph. Chvátal and Sbihi [6] proved in 1987 that every bull-free Berge graph is perfect. Subsequently Reed and Sbihi [20] gave a polynomial-time algorithm for recognizing bull-free Berge graphs. De Figueiredo, Maffray and Porto [9] proved that every bull-free Berge graph is a quasi-parity graph, and that every bull-free Berge graph with no antihole is a strict quasi-parity graph. Hayward [11] proved that every bull-free graph with no antihole if perfectly orderable (see [5, 13] for this definition), as conjectured by Chvátal. These results also settled Hougardy's above two conjectures for bull-free graphs.

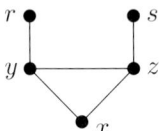

FIGURE 1. The bull $r - yxz - s$.

Bull-reducible graphs: A graph G is called *bull-reducible* if every vertex of G lies in at most one bull of G. Clearly, bull-free graphs are bull-reducible. Everett, de Figueiredo, Klein and Reed [7] proved that every bull-reducible Berge graph is perfect. Although this result now follows directly from the Strong Perfect Graph Theorem [4], the proof given in [7] is much simpler and leads moreover to a polynomial-time recognition algorithm for bull-reducible Berge graphs whose complexity is lower than that given for all Berge graphs in [3]. Here we will prove:

Theorem 1. *Let G be a bull-reducible Berge graph with at least two vertices. Then either G or $\overline{G}$ has an even pair.*

We note that this theorem settles Hougardy's above two conjectures in the case of bull-reducible graphs. The proof of this theorem is given in Section 3, while Section 2 presents some technical lemmas. We tend to follow the standard terminology of graph theory [2], but we will use the verb "sees" instead of "is adjacent to" and "misses" instead of "is not adjacent to".

2. Some technical lemmas

As in [20], call *wheel* a graph made of an even hole of length at least 6 plus a vertex that sees all vertices of this hole. Say that a proper subset H of vertices of a graph G is *homogeneous* if every vertex of $V(G) \setminus H$ either sees all vertices of H or misses all vertices of H and $2 \leq |H| \leq |V(G)| - 1$. We recall two lemmas from [7].

Lemma 2 ([7])**.** *Let G be a bull-reducible odd hole-free graph, and let C be a shortest even hole of length at least 6 in G, with its vertices colored alternately red and blue. Let v be any vertex in $V(G) \setminus V(C)$. Then v satisfies exactly one of the following conditions:*

- $N(v) \cap V(C) = \emptyset$;
- $N(v) \cap V(C) = V(C)$, *so C and v form a wheel*;
- $N(v) \cap V(C)$ *consists in either all red vertices and no blue vertex or all blue vertices and no red vertex;*
- $N(v) \cap V(C)$ *consists in either one, or two consecutive or three consecutive vertices of C;*
- $N(v) \cap V(C)$ *consists in two vertices at distance 2 along C;*
- C *has length 6 and $N(v) \cap V(C)$ consists in four vertices such that exactly three of them are consecutive.* □

Lemma 3 (Wheel Lemma [7])**.** *Let G be a bull-reducible odd hole-free graph. If G contains a wheel, then G contains a homogeneous set.* □

Now we give a few more lemmas that will be useful in the proof of the main result.

Lemma 4. *Let G be a bull-reducible odd hole-free graph. Let $P = u_0\text{-}\cdots\text{-}u_r$ be a chordless path of G of odd length $r \geq 5$, and let c be a vertex of $V(G) \setminus V(P)$ that sees u_0 and u_r. Then up to symmetry we have either:*

1. $N(c) \cap V(P) = V(P)$;
2. $N(c) \cap V(P) = \{u_0, u_1, u_r\}$ *or* $\{u_0, u_1, u_3, u_r\}$, *and in this case there is a bull* $u_r - cu_0u_1 - u_2$;
3. $r = 5$ *and* $N(c) \cap V(P) = \{u_0, u_1, u_2, u_3, u_5\}$, *and in this case there is a bull* $u_0 - cu_2u_3 - u_4$.

Proof. Since G contains no odd hole, c has two consecutive neighbors along P. If outcome 1 of the Lemma does not hold, then up to symmetry there exists an integer $i \in \{0, \ldots, r\}$ such that c sees u_i, u_{i+1} and misses u_{i+2}. Clearly $i \leq r - 3$.

Suppose i is odd. So $i \leq r - 4$. We find a first bull $u_r - cu_iu_{i+1} - u_{i+2}$. Then $i = 1$, for otherwise we find a second bull $u_0 - cu_iu_{i+1} - u_{i+2}$ containing c. Then c misses every u_j with $5 \leq j \leq r - 1$, for otherwise we find a second bull $u_j - cu_1u_2 - u_3$ containing c. Then c sees u_4 for otherwise $\{c, u_2, u_3, \ldots, u_r\}$ induces an odd hole. Then $r < 7$ for otherwise $\{c, u_4, u_5, \ldots, u_r\}$ induces an odd hole. So $r = 5$. But then we find a second bull $u_0 - cu_5u_4 - u_3$ containing c. Thus i is even.

Suppose $i = 0$. Then we find a first bull $u_r - cu_0u_1 - u_2$; and then c misses every c_j with $4 \leq j \leq r - 1$, for otherwise we find a second bull $u_j - cu_0u_1 - u_2$ containing c. So we obtain outcome 2.

Suppose i is even and $i \geq 2$. Then we find a first bull $u_0 - cu_iu_{i+1} - u_{i+2}$. Then $i = r - 3$, for otherwise we find a second bull $u_r - cu_iu_{i+1} - u_{i+2}$ containing c. Then c sees u_{i-1}, for otherwise we find a second bull $u_{i-1} - u_icu_{i+1} - u_{i+2}$ containing c. If $r = 5$ we have outcome 3. So suppose $r \geq 7$, so $i \geq 4$. Then c sees u_{i-2}, for otherwise we find a second bull $u_{i-2} - u_{i-1}u_ic - u_r$ containing c. But then we find a second bull $u_{i-2} - cu_iu_{i+1} - u_{i+2}$ containing c. This completes the proof of the lemma. □

A P_4 is a chordless path on four vertices. We call *double broom* the graph made of a P_4 (called the central P_4 of the double broom), plus two non-adjacent vertices a, b that see all vertices of the P_4, plus a vertex a' that sees only a and a vertex b' that sees only b. See Figure 2.

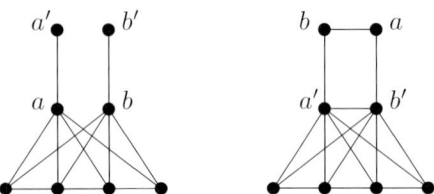

FIGURE 2. A double broom and its complement

Lemma 5. *Let G be a bull-reducible Berge graph. Let P be a chordless odd path of G of length at least 5, and let a, b, a', b' be four vertices of G such that aa' and bb' are edges, ab' and ba' are not edges, a, b see the two endpoints of P, and a', b' miss the two endpoints of P. Then G or $\overline{G}$ contains a double broom.*

Proof. Note that a, b do not lie on P. On the other hand, a', b' may be interior vertices of P. Put $P = u_0\text{-}u_1\text{-}\cdots\text{-}u_r$, with odd $r \geq 5$. Note that each of ab and $a'b'$ may be an edge or not. More precisely, if ab is an edge then $a'b'$ is an edge, for otherwise we find two intersecting bulls $a'-au_0b-b'$ and $a'-au_rb-b'$. Conversely, if $a'b'$ is an edge then ab is an edge, for otherwise $\{a', a, u_0, b, b'\}$ induces an odd hole.

We can apply Lemma 4 to P and each of a, b. If we have outcome 2 for one of a, b, say for a, then (regardless of symmetry) there is a bull containing a, u_0, u_r;

and then we do not have outcome 2 or 3 for b, for otherwise there would be a second bull containing one of u_0, u_r. If we have outcome 3 for both a, b, then we find two bulls containing u_2, u_3. Therefore we must have outcome 1 for at least one of a, b, say for b, that is, b sees all vertices of P. It follows that a' does not lie on P. We claim that a' misses every vertex of P. For suppose the contrary. Then, up to symmetry, a' sees u_i and misses u_{i-1} with $1 \le i \le (r-1)/2$, and we find a bull $a' - u_i u_{i-1} b - u_r$. Then a' sees u_{r-1}, for otherwise we find a second bull $a' - u_i u_{i-1} b - u_{r-1}$ containing b. But then we find a second bull $a' - u_{r-1} u_r b - u_0$ containing b, a contradiction. So the claim holds.

If we have outcome 2 for a, then $u_r - au_0 u_1 - u_2$ and $a' - au_0 u_1 - u_2$ are two intersecting bulls, a contradiction. If we have outcome 3 for a, then $u_0 - au_2 u_3 - u_4$ and $a' - au_2 u_3 - u_4$ are two intersecting bulls. So a sees all vertices of P, which restores the symmetry between a and b, and thus b' does not lie on P and misses every vertex of P. Now, if both $ab, a'b'$ are non-edges, then $\{u_0, u_1, u_2, u_3, a, b, a', b'\}$ induces a double broom in G, while if both are edges, the same subset induces a double broom in $\overline{G}$. This completes the proof of the lemma. $\square$

Lemma 6. *Let G be a bull-reducible C_5-free graph that contains a double broom. Then G has a homogeneous set that contains the central P_4 of the double broom.*

Proof. Pick any double broom of G, and label its vertices $w_1, w_2, w_3, w_4, a, b, a', b'$ so that its edges are $w_1 w_2, w_2 w_3, w_3 w_4, aw_1, aw_2, aw_3, aw_4, bw_1, bw_2, bw_3, bw_4, aa', bb'$. Vertices w_1, w_2, w_3, w_4 form the central P_4 of the double broom and we write $W = \{w_1, w_2, w_3, w_4\}$. We partition the vertices of $V(G) \setminus W$ as follows:
- Let T be the set of vertices of $V(G) \setminus W$ that see all of w_1, w_2, w_3, w_4.
- Let P be the set of vertices of $V(G) \setminus W$ that see at least one but not all of w_1, w_2, w_3, w_4.
- Let F be the set of vertices of $V(G) \setminus W$ that see none of w_1, w_2, w_3, w_4.

Clearly the four sets W, T, P, F are pairwise disjoint and their union is $V(G)$. Note that $a, b \in T$ and $a', b' \in F$. We define some subsets of T as follows:
$$\begin{aligned} A &= \{t \in T \mid ta' \in E, tb' \notin E\}; \\ B &= \{t \in T \mid ta' \notin E, tb' \in E\}; \\ C &= \{t \in T \mid ta' \in E, tb' \in E\}. \end{aligned}$$

Note that A, B, C are pairwise disjoint and that $a \in A, b \in B$.

Claim 6.1. *There is no edge between A and B.*

Proof. For suppose there is an edge uv with $u \in A$, $v \in B$. Then $a' - uw_i v - b'$ is a bull, for every $i = 1, \ldots, 4$, so a' belongs to four bulls, a contradiction. $\square$

Claim 6.2. *If $p \in P$, then:*
1. *There exist adjacent vertices $w_g, w_h \in W$ such that p sees w_g and misses w_h;*
2. *There exist nonadjacent vertices $w_r, w_s \in W$ such that p sees w_r and misses w_s.*

Proof. This follows directly from the definition of P and the fact that W induces a connected subgraph in G and in $\overline{G}$. □

Claim 6.3. *Every vertex of P sees all of $A \cup B \cup C$ and none of a', b'.*

Proof. Consider any $p \in P$ and $u \in A$. We first prove that p sees u. Suppose on the contrary that p misses u.

Case 1: There is a subpath w-w'-w'' of W such that p sees both w, w' and misses w''. If p misses a', then $p - w'w''u - a'$ is a bull, while if p sees a', then $a' - pww' - w''$ is a bull. In either case, p must miss b' for otherwise $b' - pww' - w''$ is a second bull containing p, a contradiction. Then p sees b or else $p - w'w''b - b'$ is a second bull containing p. But then $u - w'pb - b'$ is a second bull containing p. So p sees u.

Case 2: p sees exactly one of w_1, w_2 and misses w_4. Then $\{p, w_1, w_2, u, w_4\}$ induces a bull. This implies $pb \in E$, for otherwise $\{p, w_1, w_2, b, w_4\}$ induces a second bull containing p, a contradiction. Then p sees a', for otherwise $\{p, w_1, w_2, u, a'\}$ induces a second bull containing p. But then $\{a', p, w_g, b, w_4\}$, where $g \in \{1, 2\}$ is such that p sees w_g, induces a second bull containing p, a contradiction. So p sees u. The case where p sees exactly one of w_3, w_4 and misses w_1 is symmetric.

It is easy to see that if we are not in one of the above two cases, and up to symmetry, then p sees w_1, w_4 and misses w_2, w_3; but then $\{p, w_1, w_2, w_3, w_4\}$ induces a C_5, a contradiction. Thus we have proved that p sees u, and so p sees every vertex of $A \cup B$.

Now we prove that p misses both a' and b'. By Claim 6.2 there are two nonadjacent vertices $w_r, w_s \in W$ such that p sees w_r and misses w_s. Suppose that p sees a'. Then $a' - pw_rb - w_s$ is a bull. Now if p sees b', then $b' - pw_ra - w_s$ is a second bull containing p; while if p misses b', then $a' - pw_rb - b'$ is a second bull containing p, in either case a contradiction. So p misses a' and by symmetry it misses b'.

Finally, we prove that p sees every vertex $c \in C$. Recall that c sees both a', b'. By Claim 6.2, there are two adjacent vertices $w_g, w_h \in W$ such that p sees w_g and misses w_h. Then p sees c for otherwise we find two bulls $p - w_gw_hc - a'$ and $p - w_gw_hc - b'$ that contain p, a contradiction. Thus Claim 6.3 holds. □

Now we define subsets X, Z of F and a subset Y of $T \setminus (A \cup B \cup C)$ as follows:
- $x \in X$ if $x \in F$ and there exists in G a path p-x_1-$\cdots$-x_i, with $p \in P$, $i \geq 1$, $x_1, x_2, \ldots, x_i \in F$ and $x = x_i$. Any such path will be called a *forcing sequence* for x.
- $y \in Y$ if $y \in T \setminus (A \cup B \cup C)$ and there exists in $\overline{G}$ a path x-y_1-$\cdots$-y_j, with $x \in P \cup X$, $j \geq 1$, $y_1, y_2, \ldots, y_j \in T \setminus (A \cup B \cup C)$, and $y = y_j$. Note that if x is not in P there exists a forcing sequence p-x_1-$\cdots$-x_i for $x = x_i$. In this case the sequence p-x_1-$\cdots$-x_i-y_1-$\cdots$-y_j will be called a forcing sequence for y. In case $x \in P$ the sequence x-y_1-$\cdots$-y_j will be called a forcing sequence for y. In either case a forcing sequence for y can be denoted by x_0-$\cdots$-x_i-y_1-$\cdots$-y_j with $i \geq 0$ and $j \geq 1$.

- $z \in Z$ if $z \in F \setminus X$ and there exists in G a path y-z_1-$\ldots$-z_k, with $y \in Y$, $k \geq 1$, $z_1, z_2, \ldots, z_k \in F \setminus X$, and $z = z_k$. Note that there exists a forcing sequence x_0-x_1-$\cdots$-x_i-y_1-$\cdots$-y_j for $y = y_j$, with $i \geq 0$ and $j \geq 1$. The sequence x_0-x_1-$\cdots$-x_i-y_1-$\cdots$-y_j-z_1-$\cdots$-z_k will be called a forcing sequence for z.

Naturally we can consider for each $v \in X \cup Y \cup Z$ a shortest forcing sequence. Such sequences have notable properties which we express in the following claims.

Claim 6.4.
1. If $x \in X$ and p-x_1-$\cdots$-x_i is a shortest forcing sequence for $x = x_i$ then it is a chordless path of G.
2. If $y \in Y$ and x_0-x_1-$\cdots$-x_i-y_1-$\cdots$-y_j is a shortest forcing sequence for $y = y_j$, with the above notation, then x_0-x_1-$\cdots$-x_i is a chordless path of G, x_i-y_1-$\cdots$-y_j is a chordless path of $\overline{G}$, and, if $i \geq 1$, each of $x_0, x_1, \ldots, x_{i-1}$ sees each of $y_1, \ldots, y_j$.
3. If $z \in Z$ and p-x_1-$\cdots$-x_i-y_1-$\cdots$-y_j-z_1-$\cdots$-z_k is a shortest forcing sequence for $z = z_k$, with the above notation, then p-x_1-$\cdots$-x_i is a chordless path of G, x_i-y_1-$\cdots$-y_j is a chordless path of $\overline{G}$, y_j-z_1-$\cdots$-z_k is a chordless path of G, each of $p, x_1, \ldots, x_{i-1}$ sees each of $y_1, \ldots, y_j$, and each of $p, x_1, \ldots, x_i, y_1, \ldots, y_{j-1}$ misses each of $z_1, \ldots, z_k$.

Proof. The claim follows routinely from the definition of X, Y, Z and from the definition of a shortest forcing sequence. Details are omitted. □

Claim 6.5. *If $y \in Y$, a shortest forcing sequence for y contains at most two vertices of X.*

Proof. For suppose on the contrary that there exists a shortest forcing sequence $S = p$-x_1-$\cdots$-x_i-y_1-$\cdots$-y_j with $j \geq 1$ and $i \geq 3$. Then S satisfies the properties stated in Claim 6.4, part 2. Then for each $h = 1, \ldots, 4$ we find a bull $w_h - y_1 x_{i-2} x_{i-1} - x_i$ that contains y_1, so y_1 lies in four bulls, a contradiction. □

Claim 6.6. *If $z \in Z$, a shortest forcing sequence for z contains no vertex of X.*

Proof. For suppose on the contrary that $S = p$-x_1-$\cdots$-x_i-y_1-$\cdots$-y_j-z_1-$\cdots$-z_k is a shortest forcing sequence for $z = z_k$ with $i \geq 1$. Recall that S satisfies the properties stated in Claim 6.4, part 3. By Claim 6.2, there are nonadjacent vertices $w_r, w_s \in W$ such that p sees w_r and misses w_s. By the preceding claim we have $i \leq 2$. Suppose $i = 1$. Then $w_s - y_1 w_r p - x_1$ is a bull. If $j = 1$, then $z_1 - y_1 w_r p - x_1$ is a second bull containing p; if $j = 2$, then $z_1 - y_2 x_1 p - y_1$ is a second bull containing p; if $j \geq 3$, then $z_1 - y_j y_{j-2} p - y_{j-1}$ is a second bull containing p; in either case we have a contradiction. So $i = 2$. Then $w_s - y_1 p x_1 - x_2$ is a bull. If $j = 1$, then $z_1 - y_1 p x_1 - x_2$ is a second bull containing p; if $j = 2$, then $z_1 - y_2 x_2 x_1 - y_1$ is a second bull containing x_1; if $j \geq 3$, then $z_1 - y_j y_{j-2} p - y_{j-1}$ is a second bull containing p; in either case we have a contradiction. Thus the claim holds. □

Claim 6.7. *If $z \in Z$, a shortest forcing sequence for z contains at most two vertices of Y.*

Proof. For let $S = p\text{-}y_1\text{-}\cdots\text{-}y_j\text{-}z_1\text{-}\cdots\text{-}z_k$ be a shortest forcing sequence for $z = z_k$. The sequence S satisfies the properties stated in Claim 6.4, part 3, and it contains no vertex of X by Claim 6.6. Suppose that $j \geq 3$. Then $z_1 - y_j y_{j-2} w_h - y_{j-1}$ is a bull that contains z_1 for each $h = 1, \ldots, 4$, a contradiction. So $j \leq 2$, and the claim holds. □

Let H be the set of vertices that form the connected component of $G \setminus (T \setminus Y)$ that contains W.

Claim 6.8. $H = W \cup P \cup X \cup Y \cup Z$.

Proof. Put $H' = W \cup P \cup X \cup Y \cup Z$. First we prove that $H' \subseteq H$. Clearly, $W \subseteq H$. We also have $P \cup X \cup Y \subset H$ since every vertex of $P \cup X \cup Y$ is linked to W by a path in $G \setminus (T \setminus Y)$. Consequently $Z \subset H$, since every vertex of Z is linked to Y by a path in $G \setminus (T \setminus Y)$. So we have $H' \subseteq H$.

Conversely, let $h \in H$. Recall that $V(G)$ is partitioned into the four sets W, P, T, F. If $h \in W \cup P$ then $h \in H'$. If $h \in T$, then, by the definition of H, we have $h \in Y$. If $h \in F$, then, by the definition of H, there exists a path in $G \setminus (T \setminus Y)$ from h to W. Along this path, let v be the first vertex, starting from h, that is not in F. Then v must be in $P \cup W \cup Y$. If $v \in P \cup W$, then $h \in X$. If $v \in Y$, then $h \in Z$. So we have $H \subseteq H'$, and the claim holds. □

Claim 6.9. *Every vertex of H sees all of $T \setminus (A \cup B \cup C \cup Y)$.*

Proof. Consider any $t \in T \setminus (A \cup B \cup C \cup Y)$. So t sees all of W by the definition of T. In addition, t sees all of $P \cup X \cup Y$, for otherwise t would be in Y. Now suppose that t misses a vertex z of Z. There exists a shortest forcing sequence S for z, and by Claims 6.6 and 6.7 we have $S = p\text{-}y_1\text{-}\cdots\text{-}y_j\text{-}z_1\text{-}\cdots\text{-}z_k$ with $z = z_k$ and with $j \in \{1, 2\}$. We may also choose z such that k is as small as possible, so t sees all vertices of $S \setminus z_k$. Let w_g, w_h be two adjacent vertices of W such that p sees w_g and misses w_h. Suppose $j = 1$. Then we find a first bull $p - w_g w_h y_1 - z_1$. If $k = 1$, then $p - t w_h y_1 - z_1$ is a second bull containing p; if $k = 2$, then $p - t y_1 z_1 - z_2$ is a second bull containing p; if $k \geq 3$, then $p - t z_{k-2} z_{k-1} - z_k$ is a second bull containing p; in either case there is a contradiction. So $j = 2$. Then we find a first bull $y_1 - w_g p y_2 - z_1$. If $k = 1$, then $y_1 - t p y_2 - z_1$ is a second bull containing y_1; if $k = 2$, then $y_1 - t y_2 z_1 - z_2$ is a second bull containing y_1; If $k \geq 3$, then $y_1 - t z_{k-2} z_{k-1} - z_k$ is a second bull containing y_1; in either case there is a contradiction. Thus the claim holds. □

Claim 6.10. *Every vertex of X sees all of $A \cup B \cup C$ and none of a', b'.*

Proof. Consider any $x \in X$. By the definition of X, there exists a shortest forcing sequence $S = p\text{-}x_1\text{-}\cdots\text{-}x_i$ for $x = x_i$, with $i \geq 1$, $p \in P$, and $x_1, \ldots, x_{i-1} \in X$. Then S satisfies the properties stated in Claim 6.4, part 1, i.e., S is a chordless path. Let w_r, w_s be nonadjacent vertices of W such that p sees w_r and misses w_s. We argue by induction on i.

Assume $i = 1$. Let $u \in A \cup C$, and suppose that x misses u. We find a first bull $w_s - u w_r p - x$. Then x sees b, for otherwise $w_s - b w_r p - x$ is a second bull containing x. Then x misses a', for otherwise $a' - xpb - w_s$ is a second bull containing x. But then $a' - uw_rp - x$ is a second bull containing x. Hence x sees every vertex of $A \cup C$. Analogously, x sees every vertex of B. Suppose that x sees a'. So we find a first bull $a' - xpb - w_s$. Then x misses b', for otherwise we find a second bull $b' - xpa - w_s$ containing x. But then $b' - bpx - a'$ is a second bull containing x. Hence x misses a', and analogously, x misses b'.

Now assume $i \geq 2$. So vertices x_{i-1} and x_{i-2} are defined, with $x_{i-1} \in X$ and $x_{i-2} \in P \cup X$. Let $u \in A \cup C$, and suppose that x_i misses u. By the induction hypothesis u sees x_{i-1} and x_{i-2}, and we obtain a first bull $x_i - x_{i-1}x_{i-2}u - w_s$. Then x_i sees b, for otherwise $x_i - x_{i-1}x_{i-2}b - w_s$ is a second bull containing x_i. Then x_i misses a', for otherwise $a' - x_ix_{i-1}b - w_s$ is a second bull containing x_i. But then $x_i - x_{i-1}x_{i-2}u - a'$ is a second bull containing x_i. Hence, x_i sees every $u \in A \cup C$. Analogously, x_i sees every vertex of B. Suppose that x_i sees a'. By the induction hypothesis, x_{i-1} sees b. Hence $a' - x_ix_{i-1}b - w_r$ and $a' - x_ix_{i-1}b - w_s$ are two intersecting bulls, a contradiction. Hence, x_i misses a', and analogously, x_i misses b'. Thus the claim holds. □

Claim 6.11. *Every vertex of Y sees all of $A \cup B \cup C$ and none of a', b'.*

Proof. Consider any $y \in Y$. By the definition of Y, there exists a shortest forcing sequence $S = x_0$-$\cdots$-x_i-y_1-y_2-$\cdots$-y_j for $y = y_j$, with $j \geq 1$, and by Claim 6.5 we have $i \leq 2$. Since $Y \subseteq T \setminus (A \cup B \cup C)$, y misses a' and b'. Consider any $u \in A \cup C$. Pick a vertex w as follows: If $i = 0$ then $x_i \in P$ and x_i sees a vertex $w \in W$. If $i > 0$ then we take $w = x_{i-1}$. By Claims 6.3 and 6.10, x_i and w see both u, b and miss both a', b'. Also w sees all of $y_1, \ldots, y_j$ by the definition of Y. We prove by induction on j that y sees u. Suppose the contrary.

Assume $j = 1$. So we find a first bull $a' - ux_iw - y$. If u sees b', we find a second bull $b' - ux_iw - y$, a contradiction. So u misses b', so $u \in A$, so u misses b by Claim 6.1. Then y sees b, for otherwise we find a second bull $b' - bx_iw - y$ containing y. But then we find a second bull $b' - byw - u$ containing y. Hence, y sees every $u \in A \cup C$. Analogously, y sees every vertex of B.

Assume $j \geq 2$. By the induction hypothesis, y_{j-1} sees u and b. Then we find a first bull $a' - uy_{j-1}w - y$. If u sees b', we find a second bull $b' - uy_{j-1}w - y$, a contradiction. So u misses b', so $u \in A$, so u misses b by Claim 6.1. Then y sees b, for otherwise we find a second bull $b' - by_{j-1}w - y$ containing y. But then we find a second bull $b' - byw - u$ containing y. Hence, y sees every $u \in A \cup C$. Analogously, y sees every vertex of B. Thus the claim holds. □

Claim 6.12. *Every vertex of Z sees all of $A \cup B \cup C$ and none of a', b'.*

Proof. Consider any $z \in Z$. By Claims 6.6 and 6.7, there exists a shortest forcing sequence $S = p$-y_1-$\cdots$-y_j-z_1-$\cdots$-z_k for $z = z_k$ with $1 \leq j \leq 2$; and S satisfies the properties given in Claim 6.4, part 3. Consider any $u \in A \cup B \cup C$. So u sees all of W and, by the preceding claims, u sees all of $p, y_1, \ldots, y_j$. As usual there exist adjacent vertices $w_g, w_h \in W$ such that p sees w_g and misses w_h. We prove that z sees u and misses a', b' by induction on k.

Assume $k = 1$. If $j = 1$, we find a bull $z - y_1 w_h w_g - p$. Then z sees u for otherwise we find a second bull $z - y_1 w_h u - p$ containing z. So z sees all of $A \cup B \cup C$. Then z misses a', for otherwise we find a second bull $a' - zy_1 b - p$ containing z. Likewise z misses b'. If $j = 2$, we find a bull $z - y_2 p w_g - y_1$. Then z sees u, for otherwise we find a second bull $z - y_2 p u - y_1$ containing z. So z sees all of $A \cup B \cup C$. Then z misses a', for otherwise we find a second bull $a' - zy_2 b - y_1$ containing z. Likewise z misses b'. So the claim holds when $k = 1$.

Assume $k \geq 2$. By the induction hypothesis, u sees all of $z_1, \ldots, z_{k-1}$. If $j = 1$, we find a bull $p - w_g w_h y_1 - z_1$. Then z sees u, for otherwise we find a second bull $p - uz' z_{k-1} - z$ containing z, where $z' = z_{k-2}$ if $k \geq 3$ and $z' = y_1$ if $k = 2$. If $j = 2$, we find a bull $z_1 - y_2 p w_g - y_1$. Then z sees u, for otherwise we find a second bull $z - z_{k-1} z' u - y_1$ containing y_1, where $z' = z_{k-2}$ if $k \geq 3$ and $z' = y_2$ if $k = 2$. So z sees all of $A \cup B \cup C$. In either case ($j = 1$ or 2), z misses a', for otherwise we find a second bull $a' - zz_{k-1}b - p$ containing p. Likewise z misses b'. Thus the claim holds. □

Claim 6.13. *H is a homogeneous set.*

Proof. Since H is a component of $G \setminus (T \setminus Y)$, it suffices to prove the property that every vertex $v \in H$ sees every vertex $t \in T \setminus Y$. Claim 6.9 establishes this property when $t \in T \setminus (A \cup B \cup C \cup Y)$. Suppose $t \in A \cup B \cup C$. Then when $v \in W$ the property follows from the definition of A, B, C; and when $v \in P, X, Y, Z$ the property follows respectively from Claims 6.3, 6.10, 6.11 and 6.12. Thus the claim holds. □

This completes the proof of Lemma 6. □

3. Even pairs

Recall that a graph is *weakly triangulated* if G and $\overline{G}$ contain no hole of length at least 5. In the case of weakly triangulated the desired result is already known as it was proved by Hayward, Hoàng and Maffray [12] in a stronger form. Say that two non-adjacent vertices form a *2-pair* if every chordless path joining them has length 2.

Theorem 7 ([12])**.** *Let G be a weakly triangulated graph that is not a clique. Then G has a 2-pair.*

Now we are ready to prove our main result, which we state again:

Theorem 8. *Let G be a bull-reducible Berge graph with at least two vertices. Then either G or $\overline{G}$ has an even pair.*

Proof. We prove Theorem 8 by induction on the number of vertices of the graph G. First, suppose that G and $\overline{G}$ contain no hole of length at least 5. Then G is weakly triangulated. In that case the result follows from Theorem 7. So suppose that G is not weakly triangulated. Suppose that G has a homogeneous set. By the induction hypothesis, the subgraph H induced by this set has two vertices a, b that form an even pair in H or in $\overline{H}$. Since every vertex of $G \setminus H$ either sees both a, b or misses both a, b, it follows that a, b also form an even pair in G or in $\overline{G}$.

Now suppose that G has no homogeneous set and that one of $G, \overline{G}$ contains a hole of length at least 5. By Lemma 3, G and $\overline{G}$ contain no wheel. By Lemma 6, G and $\overline{G}$ contain no double broom. Let l be the number of vertices of a shortest hole of length at least 5 in G or $\overline{G}$. By symmetry, we may assume that G contains a hole of length l. Note that $l \geq 6$ and l is even since G is Berge. So $V(G)$ contains l pairwise disjoint and non-empty subsets $V_1, \ldots, V_l$ such that, for each $i = 1, \ldots, l$ (with subscript arithmetic modulo l), every vertex of V_i sees every vertex of $V_{i-1} \cup V_{i+1}$ and misses every vertex of $V_{i+2} \cup V_{i+3} \cup \cdots \cup V_{i-3} \cup V_{i-2}$. We write $V^* = V_1 \cup V_2 \cup \cdots \cup V_l$. We can choose these sets so that V^* is maximal. Given these subsets, we define some further subsets:

- Let A_1 be the set of vertices of $V(G) \setminus V^*$ that see all of $V_2 \cup V_4 \cup \cdots \cup V_l$ and miss all of $V_1 \cup V_3 \cup \cdots \cup V_{l-1}$;
- Let A_2 be the set of vertices of $V(G) \setminus V^*$ that see all of $V_1 \cup V_3 \cup \cdots \cup V_{l-1}$ and miss all of $V_2 \cup V_4 \cup \cdots \cup V_l$;
- For each $i = 1, \ldots, l$, let X_i be the set of vertices of $V(G) \setminus (V^* \cup A_1 \cup A_2)$ that see all of $V_{i-1} \cup V_{i+1}$ and miss all of $V_{i-2} \cup V_{i+2}$;
- Let $Z = V(G) \setminus (V^* \cup A_1 \cup A_2 \cup X_1 \cup \cdots \cup X_l)$.

Clearly, the sets $V_1, \ldots, V_l, A_1, A_2, X_1, \ldots, X_l, Z$ are pairwise disjoint and their union is $V(G)$. Let us now establish some useful properties of these sets. In the following claims, for each $i = 1, \ldots, l$, we let v_i be an arbitrary vertex of V_i.

Claim 8.1. *For $i = 1, \ldots, l$, if $X_i \neq \emptyset$ then $l = 6$ and every vertex of X_i has a neighbor in V_{i+3}. Moreover, if a vertex of X_i sees all of V_{i+3} then it has a neighbor in V_i.*

Proof. For simpler notation put $i = 3$. Let x be any vertex of X_3. So x sees all of $V_2 \cup V_4$ and misses all of $V_1 \cup V_5$. Then x must have a neighbor in $V_6 \cup \cdots \cup V_l$, for otherwise we could add x to V_3, which would contradict the maximality of V^*. Let h be the smallest index such that x has a neighbor y in V_h with $6 \leq h \leq l$. If $h \geq 7$, the set $\{x, v_4, \ldots, v_{h-1}, y\}$ induces a hole of length $h - 2$, with $5 \leq h - 2 \leq l - 2$, which contradicts G being Berge (if h is odd) or the definition of l (if h is even). So $h = 6$. Suppose $l \geq 8$. Then we can apply Lemma 2 to the hole induced by $\{v_1, v_2, v_3, v_4, v_5, y, \ldots, v_l\}$ and to x, which implies that x sees every v_j with even $j \neq 6$ and misses every v_j with odd j. Then applying Lemma 2 to the hole induced by $\{v_1, \ldots, v_l\}$ implies that x also sees every $v_6 \in V_6$. But then we have $x \in A_1$, which contradicts the definition of X_3. Thus the first part of the claim holds.

To prove the second part, let x be a vertex of X_3 that sees all of V_6. Thus $l = 6$. So x sees all of $V_2 \cup V_4 \cup V_6$ and misses all of $V_1 \cup V_5$. By Lemma 2, if x has no neighbor in V_3 then x must be in A_1, which contradicts the definition of X_3. So x has a neighbor in V_3. Thus the claim holds. $\square$

Claim 8.2. *For $i = 1, \ldots, l$, there is no P_4 in $V_i \cup X_i$.*

Proof. For if there is a P_4 in $V_i \cup X_i$, then its four vertices together with v_{i-1}, $v_{i-2}, v_{i+1}, v_{i+2}$ induce a double broom, a contradiction. □

Claim 8.3. *For $i = 1, \ldots, l$, if i is odd there is no edge between $V_i \cup X_i$ and A_1; and if i is even there is no edge between $V_i \cup X_i$ and A_2.*

Proof. Up to symmetry and for simpler notation we may take $i = 3$ and suppose that there exists an edge da with $d \in V_3 \cup X_3$ and $a \in A_1$. The definition of A_1 implies $d \in X_3$ and so, by Claim 8.1, we have $l = 6$ and d has a neighbor $u_6 \in V_6$. If d has a neighbor $u_3 \in V_3$ then we find two bulls $u_3 - dau_6 - v_5$ and $u_3 - dau_6 - v_1$ containing d, a contradiction. So d has no neighbor in V_3, and so, by Claim 8.1, d has a non-neighbor $w_6 \in V_6$. Then we find two bulls $v_3 - v_4 da - w_6$ and $v_3 - v_2 da - w_6$ containing d, a contradiction. Thus the claim holds. □

Claim 8.4. *For $i = 1, \ldots, l$, there is no edge between $V_i \cup X_i$ and $V_{i+2} \cup X_{i+2}$.*

Proof. Put $i = 3$, and suppose that there is an edge xy with $x \in V_3 \cup X_3$ and $y \in V_5 \cup X_5$. Since x has a neighbor in $V_5 \cup X_5$ we have $x \notin V_3$, so $x \in X_3$; and then, by Claim 8.1, we have $l = 6$ and x has a neighbor $u_6 \in V_6$. Likewise, y is in X_5 and has a neighbor $u_2 \in V_2$. If x has a non-neighbor $w_6 \in V_6$ and y has a non-neighbor $w_2 \in V_2$ then $\{x, y, w_6, v_1, w_2\}$ induces a C_5, a contradiction. So we may assume, up to symmetry, that x sees all of V_6. Then, by Claim 8.1, x has a neighbor $w_3 \in V_3$. So we find a first bull $w_3 - xyu_6 - v_1$. If y has a neighbor $w_5 \in V_5$, then we find a second bull $w_5 - yxu_2 - v_1$ containing x, a contradiction. So y has no neighbor in V_5, and, by Claim 8.1, y has a non-neighbor $w_2 \in V_2$. But then we find a second bull $v_1 - w_2 w_3 x - y$, a contradiction. Thus the claim holds. □

Claim 8.5. *For $i = 1, \ldots, l$, let x be a vertex that has a neighbor and a non-neighbor in $V_i \cup X_i$. If x has a neighbor in V_{i-1}, then it misses all of V_{i+2}. Likewise, if it has a neighbor in V_{i+1}, then it misses all of V_{i-2}.*

Proof. Put $i = 3$ and let a, b respectively be a neighbor and a non-neighbor of x in $V_3 \cup X_3$. Recall that a, b see all of $V_2 \cup V_4$ and miss all of $V_1 \cup V_5$. Suppose up to symmetry that x has neighbors $u_2 \in V_2$ and $u_5 \in V_5$. Then x sees every $v_4 \in V_4$, for otherwise $\{x, u_2, b, v_4, u_5\}$ induces an odd hole. Then Lemma 2, applied to x and the hole induced by $\{v_1, u_2, v_3, v_4, u_5, v_6, \ldots, v_l\}$ for every $v_3 \in V_3$, $v_6 \in V_6$, $v_1 \in V_1$, and the fact that G contains no wheel, implies that $l = 6$ and that x sees every vertex of $V_6 \cup V_4$ and none of $V_1 \cup V_3$. So $x \in A_1 \cup X_5$; and since x has a neighbor, we have $x \in X_5$; but then the edge xa contradicts Claim 8.4. Thus the claim holds. □

Claim 8.6. *For $i = 1, \ldots, l$, there is no chordless odd path of G of length at least 5 whose two endpoints are in $V_i \cup X_i$.*

Proof. For suppose that there is such a path P. Then its two endpoints see both v_{i-1}, v_{i+1} and miss both v_{i-2}, v_{i+2}, and so we can apply Lemma 5 in G to P and vertices $v_{i-1}, v_{i+1}, v_{i-2}, v_{i+2}$, which implies that G or $\overline{G}$ contains a double broom, a contradiction. □

Claim 8.7. *For $i = 1, \ldots, l$, there is no chordless odd path in $\overline{G}$ of length at least 5 whose two endpoints are in $V_i \cup X_i$.*

Proof. For suppose that there is such a path Q in $\overline{G}$. Then, in $\overline{G}$, its two endpoints see both v_{i-2}, v_{i+2} and miss both v_{i-1}, v_{i+1}, and so we can apply Lemma 5 in $\overline{G}$ to Q and vertices $v_{i-1}, v_{i+1}, v_{i-2}, v_{i+2}$, which implies that G or $\overline{G}$ contains a double broom, a contradiction. □ □

Claim 8.8. *For $i = 1, \ldots, l$, suppose that there exists a chordless path x-a-b-y in G with $a, b \in V_i \cup X_i$. Then one of x, y is in $V_i \cup X_i$.*

Proof. Put $i = 3$, and suppose that x sees v_2, v_4. By Claim 8.5, x misses all of $V_1 \cup V_5$. If x has a non-neighbor $w_2 \in V_2$, we find two intersecting bulls $v_1 - w_2ba - x$ and $w_2 - axv_4 - v_5$. So x sees all of V_2; likewise x sees all of V_4. So $x \in V_3 \cup X_3 \cup A_2$; actually, since x sees a and by Claim 8.3, we have $x \in V_3 \cup X_3$. So the claim holds in this case. It holds similarly if y sees v_2, v_4.

Suppose now that x does not see both v_2, v_4, and the same for y. At least one of x, y must see at least one of v_2, v_4, for otherwise we find two intersecting bulls $x - av_2b - y$ and $x - av_4b - y$. So assume x sees v_2 and misses v_4. By Claim 8.5, x misses v_5, and so we find a bull $x - abv_4 - v_5$. Then y sees v_4, for otherwise we find a second bull $x - av_4b - y$ containing a. Then y misses v_1 by Claim 8.5 and v_2 by the preceding paragraph. But then we find a second bull $y - bav_2 - v_1$ containing a. Thus the claim holds. □

Claim 8.9. *For $i = 1, \ldots, l$, suppose that there exists a chordless path a-u-v-b in G with $a, b \in V_i \cup X_i$. Then one of u, v is in $V_i \cup X_i$.*

Proof. Put $i = 3$. So a, b see all of $V_2 \cup V_4$ and miss all of $V_1 \cup V_5$.

First consider the case where one u, v, say u, has a neighbor in each of V_2, V_4. Let $u_2 \in V_2, u_4 \in V_4$ be neighbors of u. By Claim 8.5, u misses all of $V_1 \cup V_5$. Suppose that u has a non-neighbor $w_2 \in V_2$. Then we find a first bull $w_2 - auu_4 - v_5$. Vertex v sees w_2, for otherwise $\{w_2, a, u, v, b\}$ induces an odd hole. Then, by Claim 8.5, v misses all of V_5. Vertex v sees v_1, for otherwise we find a second bull $v_1 - u_2au - v$ containing a. Then, by Claim 8.5, v misses all of V_4. But then we find a second bull $v_5 - u_4au - v$ containing a. So u sees all of V_2, and similarly u sees all of V_4. So u is in $V_3 \cup X_3 \cup A_1$; and the definition of V_3, X_3 and Claim 8.3 imply $u \in V_3 \cup X_3$. So in this case the claim holds.

In the remaining case, we may assume that u misses all of V_4, and so v sees all of V_4 (for otherwise $\{w_4, a, u, v, b\}$ induces an odd hole for any $w_4 \in V_4 \setminus N(v)$), and so v misses all of V_2, and so u sees all of V_2. By Claim 8.5, u misses all of V_5, and v misses all of V_1. If u misses any $w_1 \in V_1$, we find two intersecting bulls $w_1 - v_2ua - v_4$ and $w_1 - v_2au - v$, a contradiction. So u sees all of V_1. Likewise, v sees all of V_5. By Lemma 2 applied to u and to the hole induced by $\{v_1, \ldots, v_l\}$, and since u sees v_1, v_2 and misses v_4, v_5, we have $N(u) \cap \{v_6, \ldots, v_l\} \subseteq \{v_l\}$.

Likewise we have $N(v) \cap \{v_6, \ldots, v_l\} \subseteq \{v_6\}$. Suppose $l \geq 8$. If u misses v_l and v misses v_6 then $\{v_1, u, v, v_5, v_6, \ldots, v_l\}$ induces a hole of odd length $l - 1$. If u sees v_l and v sees v_6 then $\{u, v, v_6, \ldots, v_l\}$ induces a hole of odd length $l - 3$. If u sees v_l and v misses v_6, then $\{u, v, v_5, v_6, \ldots, v_l\}$ induces an even hole of length $l - 2$, a contradiction to the definition of l. A similar contradiction occurs if u misses v_l and v sees v_6. So we must have $l = 6$. Then every v_6 sees one of u, v, for otherwise $\{v_1, u, v, v_5, v_6\}$ induces an odd hole. Up to symmetry let us assume that v has a neighbor $u_6 \in V_6$. Then v misses every $v_3 \in V_3$, for otherwise $\{v, v_3, v_2, v_1, u_6\}$ induces an odd hole. Suppose that v also has a non-neighbor $w_6 \in V_6$. Then, u sees w_6, for otherwise $\{w_6, v_1, u, v, v_5\}$ induces an odd hole; and u misses every $v_3 \in V_3$, for otherwise $\{u, v_3, v_4, v_5, w_6\}$ induces an odd hole; but then $\{v_2, u, v, v_4, v_3\}$ induces an odd hole, a contradiction. Thus v sees all of V_6. Now the fact that v sees all of $V_4 \cup V_5 \cup V_6$ and misses all of $V_1 \cup V_3$ implies that v is in $V_5 \cup X_5$; but then the edge vb contradicts Claim 8.4. Thus the claim holds. □

Claim 8.10. *If for some $i = 1, \ldots, l$, the set $V_i \cup X_i$ is not a clique then it contains an even pair of G or an even pair of $\overline{G}$.*

Proof. Put $i = 3$. For any two vertices $a, b \in V_3 \cup X_3$, put $N_{in}(a, b) = N(a) \cap N(b) \cap (V_3 \cup X_3)$. Choose a pair $\{a, b\}$ of non-adjacent vertices of $V_3 \cup X_3$ that maximizes the size of $N_{in}(a, b)$ (such a pair exists since $V_3 \cup X_3$ is not a clique). If the claim does not hold, $\{a, b\}$ is not an even pair of G, so there exists a chordless odd path of G with endpoints a, b. By Claim 8.6 this path has length 3, so we can write it as a-u-v-b. By Claim 8.9, we may assume up to symmetry that $u \in V_3 \cup X_3$. Consider any $d \in N_{in}(a, b)$. Then d sees u, for otherwise u-a-d-b is a P_4 in $V_3 \cup X_3$, which contradicts Claim 8.2. So we have $N_{in}(a, b) \subseteq N_{in}(u, b)$, and the choice of $\{a, b\}$ implies $N_{in}(a, b) = N_{in}(u, b)$. We claim that $\{a, u\}$ is an even pair of $\overline{G}$. For suppose that there exists a chordless odd path Q in $\overline{G}$ with endpoints a, u. By Claim 8.7, Q has length 3. So we can write $Q = a$-x-y-u in $\overline{G}$, which means that in G we have a chordless path y-a-u-x. By Claim 8.8, one of x, y is in $V_3 \cup X_3$. By symmetry we may assume that $x \in V_3 \cup X_3$. Then x misses b, for otherwise we have $x \in N_{in}(u, b) \setminus N_{in}(a, b)$. Then x sees every $d \in N_{in}(a, b)$, for otherwise x-u-d-b is a P_4 in $V_3 \cup X_3$, which contradicts Claim 8.2. But then we have $N_{in}(a, x) \supseteq N_{in}(a, b) \cup \{u\}$, which contradicts the choice of $\{a, b\}$. Thus the claim holds. □

Claim 8.11. *If for some $i = 1, \ldots, l$, the set $V_i \cup X_i$ induces a clique of size at least 2 then any two vertices of $V_i \cup X_i$ form an even pair of $\overline{G}$.*

Proof. For suppose that there is a chordless odd path Q in $\overline{G}$ with endpoints a, b in $V_i \cup X_i$. By Claim 8.7, Q has length 3, so we can write $Q = a$-x-y-b in $\overline{G}$, and so we have a chordless path y-a-b-x in G. By Claim 8.8, one of x, y is in $V_i \cup X_i$; but this contradicts the fact that $V_i \cup X_i$ is a clique. Thus the claim holds. □

Claim 8.12. *Suppose that for every $i = 1, \ldots, l$, the set $V_i \cup X_i$ has size 1. Then $\{v_i, v_{i+2}\}$ is an even pair of G for every i.*

Proof. For suppose on the contrary and up to symmetry that $\{v_1, v_3\}$ is not an even pair; so there is a chordless odd path $P = x_0\text{-}x_1\text{-}\cdots\text{-}x_r$ with $v_1 = x_0$, $v_3 = x_r$ and $r \geq 3$. Since $V(P) \cup \{v_2\}$ cannot induce an odd hole (when $r = 3$), and by Lemma 4 (when $r \geq 5$), and up to symmetry, we may assume that v_2 sees x_1. If x_1 sees v_l, then x_1 misses v_{l-1} by Lemma 2, and we have $x_1 \in V_1 \cup X_1$, a contradiction. So x_1 misses v_l, and we find a bull $v_l - v_1 x_1 v_2 - v_3$. Then v_2 misses x_{r-1}, for otherwise by symmetry we find a second bull $v_4 - v_3 x_{r-1} v_2 - v_1$. If $r = 3$, then v_l sees x_2, for otherwise we find a second bull $v_l - v_1 v_2 x_1 - x_2$ containing v_2; but then $\{v_l, v_1, v_2, v_3, x_2\}$ induces an odd hole. So $r \geq 5$. Since v_2 misses x_{r-1}, we have outcome 2 or 3 of Lemma 4, and in either case Lemma 4 states that there is a second bull containing v_2, a contradiction. Thus the claim holds. □

Claims 8.10, 8.11 and 8.12 complete the proof of the theorem. □

4. Comments

For any integer $k \geq 0$, let $\mathcal{B}_k$ be the class of graphs in which every vertex belongs to at most k bulls. So $\mathcal{B}_0$ is the class of bull-free graphs, and $\mathcal{B}_1$ is the class of bull-reducible graphs. One can consider the following statements:

Statement A_k: For every Berge graph G in $\mathcal{B}_k$ with at least two vertices, either G or $\overline{G}$ has an even pair.

Statement A'_k: For every Berge graph G in $\mathcal{B}_k$ that contains no antihole, either G is a clique or G has an even pair.

Statement A''_k: For every Berge graph G in $\mathcal{B}_k$ that contains no antihole, G is perfectly orderable.

Statements A_0 and A'_0 are theorems proved in [9]. Statement A''_0 is a theorem proved in [11]. Statement A_1 is the main result in this article. Statements A'_1 and A'_2 are theorems, as they can be obtained easily as corollaries of the main result in [17]. On the other hand, consider the graph H_{12} with 12 vertices $v_1, \ldots, v_{12}$ such that $v_1\text{-}v_2\text{-}\cdots\text{-}v_8\text{-}v_1$ is a hole, vertex v_9 is adjacent to v_1, v_2, v_{11}, vertex v_{10} is adjacent to v_3, v_4, v_{12}, vertex v_{11} is adjacent to v_5, v_6, v_9, and vertex v_{12} is adjacent to v_7, v_8, v_{10}. Then it is easy to see that H_{12} is a Berge graph (it is actually the line-graph of a bipartite graph), it contains no antihole, it is in $\mathcal{B}_5$, and H_{12} and its complement have no even pair. So H_{12} is a counterexample to statements A_k, A'_k for any $k \geq 5$. Moreover, the graph "E" in [13, p. 142, Fig. 7.1] is a counterexample to A''_3. We do not have a proof or a counterexample for any of the remaining statements $A_2, A_3, A_4, A'_3, A'_4$ and A''_1, A''_2.

References

[1] C. Berge. Les problèmes de coloration en théorie des graphes. *Publ. Inst. Stat. Univ. Paris* 9 (1960) 123–160.

[2] C. Berge. *Graphs*. North-Holland, Amsterdam/New York, 1985.

[3] M. Chudnovsky, G. Cornuéjols, X. Liu, P. Seymour, K. Vušković. Recognizing Berge graphs. *Combinatorica* 25 (2005) 143–187.

[4] M. Chudnovsky, N. Robertson, P. Seymour, R. Thomas. The strong perfect graph theorem. Manuscript, 2002. To appear in *Annals of Mathematics*.

[5] V. Chvátal. Perfectly ordered graphs. In *Topics on Perfect Graphs (Ann. Discrete Math. 21)*, C. Berge and V. Chvátal eds., North-Holland, Amsterdam, 1984, 63–65.

[6] V. Chvátal, N. Sbihi. Bull-free Berge graphs are perfect. *Graphs and Combinatorics* 3 (1987) 127–139.

[7] H. Everett, C.M.H. de Figueiredo, S. Klein, B. Reed. The perfection and recognition of bull-reducible Berge graphs. *RAIRO Theoretical Informatics and Applications* 39 (2005) 145–160.

[8] H. Everett, C.M.H. de Figueiredo, C. Linhares Sales, F. Maffray, O. Porto, B.A. Reed. Even pairs. In [19], 67–92.

[9] C.M.H. de Figueiredo, F. Maffray, O. Porto. On the structure of bull-free perfect graphs. *Graphs and Combinatorics* 13 (1997) 31–55.

[10] M. Grötschel, L. Lovász, A. Schrijver. Polynomial algorithms for perfect graphs. In *Topics on Perfect Graphs (Ann. Discrete Math. 21)*, C. Berge and V. Chvátal eds., North-Holland, Amsterdam, 1984, 325–356.

[11] R. Hayward. Bull-free weakly chordal perfectly orderable graphs. *Graphs and Combinatorics* 17 (2000) 479–500.

[12] R. Hayward, C.T. Hoàng, F. Maffray. Optimizing weakly triangulated graphs. *Graphs and Combinatorics* 5 (1989), 339–349. See erratum in vol. 6 (1990) 33–35.

[13] C.T. Hoàng. Perfectly orderable graphs: a survey. In [19], 139–166.

[14] S. Hougardy, *Perfekte Graphen*, PhD thesis, Institut für Ökonometrie und Operations Research, Rheinische Friedrich Wilhelms Universität, Bonn, Germany, 1991.

[15] S. Hougardy, Even and odd pairs in line-graphs of bipartite graphs, *European J. Combin.* 16 (1995) 17–21.

[16] L. Lovász. Normal hypergraphs and the weak perfect graph conjecture. *Discrete Math.* 2 (1972) 253–267.

[17] F. Maffray, N. Trotignon. A class of perfectly contractile graphs. To appear in *J. Comb. Th. B* (2006) 1–19.

[18] H. Meyniel. A new property of critical imperfect graphs and some consequences. *Eur. J. Comb.* 8 (1987) 313–316.

[19] J.L. Ramírez-Alfonsín, B.A. Reed. *Perfect Graphs*. Wiley Interscience, 2001.

[20] B. Reed, N. Sbihi. Recognizing bull-free perfect graphs. *Graphs and Combinatorics* 11 (1995) 171–178.

Celina M.H. de Figueiredo
Departamento de Ciência da Computação
Instituto de Matemática
Universidade Federal do Rio de Janeiro
Caixa Postal 68530
21945-970 Rio de Janeiro, RJ, Brazil
e-mail: `celina@cos.ufrj.br`

Frédéric Maffray
CNRS, Laboratoire Leibniz
46 avenue Félix Viallet
F-38041 Grenoble Cedex, France
e-mail: `frederic.maffray@imag.fr`

Claudia Regina Villela Maciel
Instituto de Matemática
Universidade Federal Fluminense
Rua Mário Santos Braga
s/n Praça do Valonguinho
CEP 24.020-140, Niterói, RJ, Brazil
e-mail: `crvillela@globo.com`

Kernels in Orientations of Pretransitive Orientable Graphs

Hortensia Galeana-Sánchez and Rocío Rojas-Monroy

Abstract. Let D be a digraph, $V(D)$ and $A(D)$ will denote the sets of vertices and arcs of D, respectively. A kernel N of D is an independent set of vertices such that for every $w \in V(D) - N$ there exists an arc from w to N. A digraph D is called *right-pretransitive* (resp. left-pretransitive) when $(u,v) \in A(D)$ and $(v,w) \in A(D)$ implies $(u,w) \in A(D)$ or $(w,v) \in A(D)$ (resp. $(u,v) \in A(D)$ and $(v,w) \in A(D)$ implies $(u,w) \in A(D)$ or $(v,u) \in A(D)$). These concepts were introduced by P. Duchet in 1980. Let G be a graph, an orientation of G is a digraph obtained from G by directing each edge of G in at least one of the two possible directions; an orientation D of G is: a right (resp. left)-pretrantive orientation of G if D is a right (resp. left)-pretransitive digraph; and D is a Meyniel-orientation or M-orientation of G, if every directed cycle of length 3 of D has at least two symmetrical arcs. In this paper the following result is proved: Let G be a simple (possible infinite) graph, and D an M-orientation of G. If there exists a right (resp. left)-pretransitive orientation T of G such that T has no infinite outward path nor infinite inward path either, and $\text{Sym}(T) = \text{Sym}(D)$, then D has a kernel. Previous results are generalized.

Mathematics Subject Classification (2000). 05C20.

Keywords. Kernel, kernel-perfect digraph, right-pretransitive digraph, left-pretransitive digraph.

1. Introduction

For general concepts we refer the reader to [2]. In the paper we write digraph to mean 1-digraph in the sense of Berge [2]. In this paper D will denote a possibly infinite digraph. Often we shall write $u_1 u_2$ instead of (u_1, u_2). An arc $u_1 u_2 \in A(D)$ is called asymmetrical (resp. symmetrical) if $u_2 u_1 \notin A(D)$ (resp. $u_2 u_1 \in A(D)$). The asymmetrical part of D (resp. symmetrical part of D), which is denoted by $\text{Asym}(D)$ (resp. $\text{Sym}(D)$), is the spanning subdigraph of D whose arcs are the asymmetrical (resp. symmetrical) arcs of D. We recall that a subdigraph D_1 of D

is a spanning subdigraph if $V(D_1) = V(D)$. If S is a nonempty subset of $V(D)$ then the subdigraph $D[S]$ induced by S is the digraph with vertex set S and whose arcs are those arcs of D which join vertices of S. A directed path is a finite sequence $(x_1, x_2, \ldots, x_n)$ of distinct vertices of D such that $(x_i, x_{i+1}) \in A(D)$ for each $i \in \{1, \ldots, n-1\}$. When D is infinite, we say that a sequence $(x_i)_{i \in \mathbb{N}}$ is an infinite outward (resp. inward) path if for each $i \in \mathbb{N}$ we have $(x_i, x_{i+1}) \in A(D)$ (resp. $(x_{i+1}, x_i) \in A(D)$) and for every $\{i, j\} \subseteq \mathbb{N}$ we have $x_i \neq x_j$. Let S_1 and S_2 be subsets of $V(D)$, a finite directed path $(x_1, \ldots, x_n)$ will be called an $S_1 S_2$-directed path whenever $x_1 \in S_1$ and $x_n \in S_2$ in particular when the directed path is an arc.

Definition 1.1. A set $I \subseteq V(D)$ is independent if $A(D[I]) = \emptyset$. A kernel N of D is an independent set of vertices such that for each $z \in V(D) - N$ there exists a zN-arc in D. A digraph D is called kernel-prefect when every induced subdigraph of D has a kernel.

The concept of kernel was introduced by Von Neumann and Morgenstern [14] in the context of Game Theory. The problem of the existence of a kernel in a given digraph has been studied by several authors in particular by Von Neumann and Morgenstern [14], Richardson [16, 17], Berge [2], Berge and Duchet [4], Duchet and Meyniel [11], Duchet [8, 10], Galeana-Sánchez and Neumann-Lara [12]. It is well known that a finite transitive digraph is kernel-perfect and a finite symmetrical digraph is kernel-perfect (see for example [2]). (We recall that a digraph D is transitive whenever $(u,v) \in A(D)$ and $(v,w) \in A(D)$ implies $(u,w) \in A(D)$.)

Definition 1.2 (Duchet [8]). A digraph D is called right (resp. left)-pretransitive digraph if every nonempty subset B of $V(D)$ possesses a vertex $t(B) = b$ such that $(x, b) \in A(D)$ and $(b, y) \in A(D)$ implies $(x, y) \in A(D)$ or $(y, b) \in A(D)$ (resp. $(x, b) \in A(D)$ and $(b, y) \in A(D)$ implies $(x, y) \in A(D)$ or $(b, x) \in A(D)$), for any two vertices $x, y \in V(D)$.

Clearly taking $B = \{b\}$ for each $b \in V(D)$ (taking all the possible singletons of $V(D)$) in Definition 1.2, we obtain that Definition 1.2 is equivalent to those given in the abstract, which for technical reasons will be used in this paper.

Theorem 1.1 (Duchet [8]). *A finite right-pretransitive (resp. left-pretransitive) digraph is kernel-perfect.*

The result proved in this paper generalize Theorem 1.1 (as any right (resp. left)-pretransitive digraph D satisfies that every directed cycle of length 3 in D has at least two symmetrical arcs), and the following result of Champetier [6].

Theorem 1.2 (Champetier [6]). *Every M-orientation of a comparability finite graph G is kernel-perfect.*

We recall that a graph G is a comparability graph whenever there exists an asymmetrical orientation D of G which is a transitive digraph. In [1] C. Berge defined the perfect graphs as follows: A graph G is perfect whenever for any induced subgraph H of G the chromatic number of H, $\chi(H)$, is equal to its clique number, $w(H)$. In [3] C. Berge proved that a comparability graph is a perfect graph. More about perfect graphs can be found in [3, 9] and [15]. In 1960 C. Berge stated the famous Strong Perfect Graph Conjecture which asserts that a graph G is perfect iff G contains nor C_{2n+1} neither $\overline{C}_{2n+1}$, for $n \geq 2$, as an induced subgraph. This Conjecture is now proved.

Theorem 1.3 (Chudnovsky et al. [7]**).** *A graph G is perfect iff G contains neither C_{2n+1} nor $\overline{C}_{2n+1}$, for $n \geq 2$, as a induced subgraph.*

A graph G is said to be solvable if every orientation of G is kernel-perfect provided that all its cliques have a kernel

Theorem 1.4 (C. Berge and P. Duchet [4]**).** *Any induced subgraph of a solvable graph is also solvable.*

Theorem 1.5. *The if part.* **(C. Berge and P. Duchet** [4]**)** *C_{2n+1} (resp. $\overline{C}_{2n+1}$) is not solvable for any $n \geq 2$.*

In [4] C. Berge and P. Duchet conjectured the following alternative characterization of perfect graphs: A graph G is perfect if and only if G is solvable. This conjecture was proved in its only if part by Boros and Gurvich in [5], and the part if is a direct consequence of Theorems 1.3, 1.4 and 1.5; so we have the following result:

Theorem 1.6. *A graph G is perfect if and only if is solvable.*

Clearly Theorem 1.2 is a particular case of Theorem 1.6. (As comparability graphs are perfect and in every M-orientation every clique has a kernel). We say that a graph G is M-solvable whenever every M-orientation of G is kernel-perfect. In this paper we prove that right (resp. left)-pretransitive orientable graphs satisfy that every M-orientation whose symmetrical part coincides with the symmetrical part of a right (resp. left)-pretransitive orientation is kernel-perfect and we can construct an infinite class of non perfect graphs with this property. This result and the Berge-Duchet Theorem (Theorem 1.5) lead us to propose the following problem: Characterize M-solvable graphs.

2. Kernels, M-orientations and pretransitive orientations

The main result of this section is Theorem 2.1. To prove this result we use a method related to the one of Sands et al. [18]. The following two Lemmas will be useful in the proof of Theorem 2.1.

Lemma 2.1. [13] *Let D be a right-pretransitive or left-pretransitive digraph. If $(x_1, x_2, \ldots, x_n)$ is a sequence of vertices such that $(x_i, x_{i+1}) \in A(D)$ and $(x_{i+1}, x_i) \notin A(D)$, then the sequence is a directed path and for each $i \in \{1, \ldots, n-1\}$, $(x_i, x_j) \in A(D)$ and $(x_j, x_i) \notin A(D)$, for every $j \in \{i+1, \ldots, n\}$.*

Lemma 2.2. [13] *Let D be a right-pretransitive digraph or left-pretransitive digraph. If D has no infinite outward path, and $\emptyset \neq U \subseteq V(D)$, then there exists $x \in U$ such that $(x, y) \in A(D)$ with $y \in U$ implies $(y, x) \in A(D)$.*

Theorem 2.1. *Let G be a possibly infinite graph. And D an M-orientation of G. If there exists a right (resp. left)-pretransitive orientation T of G, such that: T has no infinite outward path and no infinite inward path, and $\mathrm{Sym}(D) = \mathrm{Sym}(T)$, then D has a kernel.*

Proof. Given a digraph D, we denote by D^{-1} the inverse of D, which is the digraph obtained from D by reversing the direction of each one of its arcs. Let T be a right-pretransitive or a left-pretransitive orientation of G, which has no infinite outward path and no infinite inward path, and $\mathrm{Sym}(T) = \mathrm{Sym}(D)$. Since T is a right (resp. left)-pretransitive digraph iff T^{-1} is a left (resp. right)-pretransitive digraph, we may assume that T is a left-pretransitive digraph and so, T^{-1} is a right-pretransitive digraph. Notice that T as well as T^{-1} has no infinite outward path; and also observe that $\mathrm{Sym}(T) = \mathrm{Sym}(T^{-1})$ and hence $\mathrm{Sym}(T^{-1}) = \mathrm{Sym}(D)$. For distinct vertices x, y of D, $x \to y$ will mean that the arc $(x, y) \in A(D)$.

Suppose $x \to y$; we will write $x \stackrel{red}{\to} y$ when $(x, y) \in A(T)$ and $x \stackrel{blue}{\to} y$ whenever $(x, y) \in A(T^{-1})$. For $S \subseteq V(D)$, $x \to S$ will mean that there exists an arc in D from x toward a vertex in S. The negation of $x \to y$ (resp. $x \stackrel{red}{\to} y$, $x \stackrel{blue}{\to} y$, and $x \to S$) will be denoted $x \not\to y$ (resp. $x \stackrel{red}{\not\to} y$, $x \stackrel{blue}{\not\to} y$, and $x \not\to S$). Also we write $x \stackrel{T}{\to} y$ to mean $(x, y) \in A(T)$. Note that $x \stackrel{red}{\to} y$ and $x \stackrel{blue}{\to} y$ refer to arcs of $A(D)$ while $x \stackrel{T}{\to} y$ and $x \stackrel{T^{-1}}{\to} y$ do not, for $S \subseteq V(D)$ the notations $x \stackrel{T}{\to} S$, $x \stackrel{T}{\not\to} S$, $x \stackrel{T}{\not\to} y$ are analogous to the previous ones; and we can use a similar notation for T^{-1}. Notice that if $x \to y$ then $x \stackrel{red}{\to} y$ or $x \stackrel{blue}{\to} y$; also notice that if $(x, y) \in A(\mathrm{Sym}(D))$ then $x \stackrel{red}{\to} y$, $x \stackrel{blue}{\to} y$, $y \stackrel{red}{\to} x$ and $y \stackrel{blue}{\to} x$; finally if $x \stackrel{T}{\to} y$ (resp. $x \stackrel{T^{-1}}{\to} y$) and $x \stackrel{red}{\not\to} y$ (resp. $x \stackrel{red}{\not\to} y$) then $y \stackrel{blue}{\to} x$ (resp. $y \stackrel{blue}{\to} x$). Let $\mathcal{U}$ be the family of independent sets of vertices S of G such that $S \stackrel{red}{\to} x$ implies $x \to S$. We define the following binary relation in $\mathcal{U}$, $\leq$: $S \leq R$ if and only if for each $s \in S$ there exists $r \in R$ such that either $s = r$ or $s \stackrel{T^{-1}}{\to} r$ and $r \stackrel{T^{-1}}{\not\to} s$. Observe that if S and R are independent sets belonging to $\mathcal{U}$ with $S \subseteq R$, then $S \leq R$.

(1) The family $\mathcal{U}$ is partially ordered by $\leq$.

$\leq$ is reflexive.

This follows from the fact $S \subseteq S$.

$\leq$ is transitive.

Let S, Q and R sets belonging to $\mathcal{U}$ such that $S \leq Q$ and $Q \leq R$, and let $s \in S$. Since $S \leq Q$ there exists $q \in Q$ such that

$$\text{either} \quad s = q \quad \text{or} \quad (s \stackrel{T^{-1}}{\to} q \text{ and } q \stackrel{T^{-1}}{\not\to} s); \tag{I}$$

and $Q \leq R$ implies that there exists $r \in R$ such that

$$\text{either} \quad q = r \quad \text{or} \quad (q \stackrel{T^{-1}}{\to} r \text{ and } r \stackrel{T^{-1}}{\not\to} q). \tag{II}$$

If $s = q$ or $q = r$, then it follows from (II) or (I) respectively that either $s = r$ or $(s \stackrel{T^{-1}}{\to} r$ and $r \stackrel{T^{-1}}{\not\to} s)$, with $r \in R$; otherwise we have, $(s \stackrel{T^{-1}}{\to} q$ and $q \stackrel{T^{-1}}{\not\to} s)$ and $(q \stackrel{T^{-1}}{\to} r$ and $r \stackrel{T^{-1}}{\not\to} q)$, and since T^{-1} is a right-pretransitive digraph it follows from Lemma 2.1 on the succession (s, q, r) that $s \stackrel{T^{-1}}{\to} r$ and $r \stackrel{T^{-1}}{\not\to} s$. Thus $R \leq S$.

$\leq$ is antisymmetrical.

Let S and R sets belonging to $\mathcal{U}$ such that $S \leq R$ and $R \leq S$; we will prove that $S = R$. Let $s \in S$; since $S \leq R$ there exists $r \in R$ such that either, $s = r$ or $(s \stackrel{T^{-1}}{\to} r$ and $r \stackrel{T^{-1}}{\not\to} s)$. Suppose $s \neq r$; the fact $R \leq S$ implies that there exists $s' \in S$ such that either, $r = s'$ or $(r \stackrel{T^{-1}}{\to} s'$ and $s' \stackrel{T^{-1}}{\not\to} r)$. When $r = s'$ we obtain $s \stackrel{T^{-1}}{\to} s'$ contradicting that S is an independent set; so $r \neq s'$ and $(r \stackrel{T^{-1}}{\to} s'$ and $s' \stackrel{T^{-1}}{\not\to} r)$. Now applying Lemma 2.1 on the sequence (s, r, s'), we have $s \stackrel{T^{-1}}{\to} s'$ contradicting that S is an independent set. We conclude $r = s$ and consequently $s \in R$ and $S \subseteq R$. Analogously it can be proved $R \subseteq S$.

(2) $(\mathcal{U}, \leq)$ has maximal elements.

(3) $\mathcal{U} \neq \emptyset$. Since T is a left-pretransitive digraph, which has no infinite outward path it follows from Lemma 2.2 (taking $D = T$ and $U = V(T)$) there exists a vertex $y \in V(D)$ such that $y \stackrel{T}{\to} x$ implies $x \stackrel{T}{\to} y$, and in this case (x, y) is a symmetrical arc of T; since $\text{Sym}(T) \subseteq \text{Sym}(D)$ we have that $(x, y) \in \text{Sym}(D)$. Thus $y \stackrel{red}{\to} x$ implies $x \to y$ and then $\{y\} \in \mathcal{U}$.

(4) Every chain in $(\mathcal{U}, \leq)$ is upper bounded.

Let $\mathcal{C}$ be a chain in $(\mathcal{U}, \leq)$ and define $S^\infty = \{s \in \bigcup_{S \in \mathcal{C}} S \mid$ there exists $S \in \mathcal{C}$ such that $s \in R$ whenever $R \in \mathcal{C}$ and $R \geq S\}$ (S^∞ consists of all vertices of D that belong to every member of $\mathcal{C}$ from some point on). We will prove that S^∞ is an upper bound of $\mathcal{C}$.

(5) $S^\infty \neq \emptyset$, and for each $S \in \mathcal{C}$, $S^\infty \geq S$. Let $S \in \mathcal{C}$ and $t_0 \in S$, we will prove that there exists $t \in S^\infty$ such that either $t_0 = t$ or $(t_0 \stackrel{T^{-1}}{\to} t$ and $t \stackrel{T^{-1}}{\not\to} t_0)$. If $t_0 \in S^\infty$ we are done, so assume $t_0 \notin S^\infty$. We proceed by contradiction; suppose that if $t \in V(D)$ with $(t_0 \stackrel{T^{-1}}{\to} t$ and $t \stackrel{T^{-1}}{\not\to} t_0)$, then $t \notin S^\infty$. Take $R_0 = S$; since $t_0 \notin S^\infty$ there exists $R_1 \in \mathcal{C}$, $R_1 \geq R_0$ such that $t_0 \notin R_1$. Hence there exists $t_1 \in R_1$ such that $t_0 \stackrel{T^{-1}}{\to} t_1$ and $t_1 \stackrel{T^{-1}}{\not\to} t_0$; and our assumption implies $t_1 \notin S^\infty$. The fact $t_1 \notin S^\infty$ implies $t_1 \notin R_2$ for some $R_2 \in \mathcal{C}$, $R_2 \geq R_1$, and there exists $t_2 \in R_2$ such

that $t_1 \xrightarrow{T^{-1}} t_2$ and $t_2 \not\xrightarrow{T^{-1}} t_1$, since T^{-1} is a right-pretransitive digraph, it follows Lemma 2.1 on the sequence $\tau_2 = (t_0, t_1, t_2)$ that $t_0 \xrightarrow{T^{-1}} t_2$ and $t_2 \not\xrightarrow{T^{-1}} t_0$, and our assumption implies $t_2 \notin S^\infty$. We may continue that way and we obtain, for each $n \in \mathbb{N}$, $R_n \in \mathcal{C}$, $t_n \in R_n$, $(t_0 \xrightarrow{T^{-1}} t_n$ and $t_n \not\xrightarrow{T^{-1}} t_0)$ and $t_n \notin S^\infty$, hence there exists $R_{n+1} \in \mathcal{C}$ such that $R_{n+1} \geq R_n$ and $t_n \notin R_{n+1}$, so there exists $t_{n+1} \in R_{n+1}$ with $t_n \xrightarrow{T^{-1}} t_{n+1}$ and $t_{n+1} \not\xrightarrow{T^{-1}} t_n$. Since T^{-1} is a right-pretransitive digraph, and $(t_n \xrightarrow{T^{-1}} t_{n+1}$ and $t_{n+1} \not\xrightarrow{T^{-1}} t_n)$ for each $n \in \mathbb{N}$; it follows from Lemma 2.1 (on the sequence $(t_0, t_1, \ldots, t_n)$) that $\tau_{n+1} = (t_0, t_1, \ldots, t_{n+1})$ is a directed path in T^{-1} and $(t_0 \xrightarrow{T^{-1}} t_{n+1}$ and $t_{n+1} \not\xrightarrow{T^{-1}} t_0)$. And our assumption implies $t_{n+1} \notin S^\infty$. Now consider the sequence $\tau = (t_n)_{n \in \mathbb{N}}$, for each $n \in \mathbb{N}$. We have $t_n \xrightarrow{T^{-1}} t_{n+1}$, and for $n < m$, $\{t_n, t_m\} \subseteq V(\tau_m)$; and since τ_m is a directed path in T^{-1} we have $t_n \neq t_m$. Hence τ is an infinite outward path contained in T^{-1}, a contradiction. We conclude that there exists $t \in S^\infty$ such that $(t_0 \xrightarrow{T^{-1}} t$ and $t \not\xrightarrow{T^{-1}} t_0)$. Thus $S^\infty \geq S$ and $S^\infty \neq \emptyset$.

(6) S^∞ is an independent set. Let $s_1, s_2 \in S^\infty$ and suppose without loss of generality that $S_1, S_2 \in \mathcal{C}$ are such that: $s_1 \in S$ whenever $S \in \mathcal{C}$ and $S \geq S_1$; $s_2 \in S_2$ and $S_1 \leq S_2$. Then $s_1 \in S_2$ and since S_2 is independent there is no arc between s_1 and s_2 in D.

(7) $S^\infty \in \mathcal{U}$.

Suppose that $S^\infty \xrightarrow{red} y$, we will prove $y \to S^\infty$. We proceed by contradiction, assume that $y \not\to S^\infty$. Let $s \in S^\infty$ be such that $s \xrightarrow{red} y$, and let $S_1 \in \mathcal{C}$ be such that $s \in W$ for every $W \in \mathcal{C}$, $W \geq S_1$. First we prove the following assertion:

(8) If $y \xrightarrow{blue} s'$ for some $s' \in R$, $R \in \mathcal{C}$, $R \geq S_1$, then there exists $t \in S^\infty$ such that:

(i) $t \xrightarrow{red} s'$, and,

(ii) If $R' \in \mathcal{C}$ with $R' \geq R$ and $t \in R'$ then for some $s'' \in R'$ we have $s' \xrightarrow{blue} s''$, $s'' \not\xrightarrow{T^{-1}} s'$ and $y \xrightarrow{blue} s''$.

Since we are assuming that $y \not\to S^\infty$ then $s' \notin S^\infty$, from (5) we have $S^\infty \geq R$, thus there exists $t \in S^\infty$ such that $s' \xrightarrow{T^{-1}} t$ and $t \not\xrightarrow{T^{-1}} s'$; we will prove that t satisfies (i) and (ii).

Proof of (i). Since $y \xrightarrow{blue} s'$ we have $y \xrightarrow{T^{-1}} s'$. Now, the facts that T^{-1} is a right-pretransitive digraph and $t \not\xrightarrow{T^{-1}} s'$ imply that $y \xrightarrow{T^{-1}} t$. Since $y \not\to S^\infty$ then $y \not\xrightarrow{blue} t$ and $t \xrightarrow{red} y$. □

(9) $s' \not\to t$: If $s' \to t$ then (y, s', t, y) is a directed triangle contained in D; now observe that (s', t) is not a symmetrical arc of T^{-1} and then it is not a symmetrical arc of D; and $y \overset{blue}{\not\to} t$ which implies $(t, y) \notin A(\mathrm{Sym}(D))$; so this directed triangle has at most one symmetrical arc; a contradiction. Thus $s' \not\to t$, $s' \overset{blue}{\not\to} t$ and consequently $t \overset{red}{\to} s'$ (recall $s' \overset{T^{-1}}{\to} t$, so s' and t are adjacent in D).

Proof of (ii). Let $R' \in \mathcal{C}$ be such that $R' \geq R$ and $t \in R'$; since $t \overset{red}{\to} s'$, then $R' \overset{red}{\to} s'$ and thus $s' \to R'$, that is $s' \to s''$ for some $s'' \in R'$ (notice that $s'' \neq t$ because we have proved in the proof of (i) that $s' \not\to t$, see (9)). □

(10) $s' \overset{red}{\not\to} s''$: If $s' \overset{red}{\to} s''$ then $s' \overset{T}{\to} s''$; and since $t \overset{T}{\to} s'$ we have that $t \overset{T}{\to} s''$ or $s' \overset{T}{\to} t$ (as T is a left-pretransitive digraph). Now $t \overset{T}{\not\to} s''$ because $\{t, s''\} \subseteq R'$ and R' is an independent set; so: $s' \overset{T}{\to} t$ and $(s', t) \in \mathrm{Sym}(T) \subseteq \mathrm{Sym}(D)$ (recall $t \overset{red}{\to} s'$), a contradiction. (As we have proved $s' \not\to t$ (9)).

(11) $s' \overset{blue}{\to} s''$: Since $s' \to s''$, we have that $s' \overset{red}{\to} s''$ or $s' \overset{blue}{\to} s''$, and the assertion follows from (10).

(12) $s'' \overset{T^{-1}}{\not\to} s'$: Assume by contradiction that $s'' \overset{T^{-1}}{\to} s'$; it follows from the previous assertion (11) that $(s', s'') \in \mathrm{Sym}(T^{-1}) = \mathrm{Sym}(T) \subseteq \mathrm{Sym}(D)$ which implies that $s' \overset{red}{\to} s''$ a contradiction to the previous assertion (10).

(13) $y \overset{T^{-1}}{\to} s''$: Since $y \overset{blue}{\to} s'$ and $s' \overset{blue}{\to} s''$ we have $y \overset{T^{-1}}{\to} s'$ and $s' \overset{T^{-1}}{\to} s''$. Now the assertion follows from (12) and the fact that T^{-1} is a right-pretransitive digraph.

(14) $y \overset{blue}{\to} s''$. From (13) we have that y and s'' are adjacent in D; and clearly we have two possibilities: $y \to s''$ or $s'' \to y$. If $s'' \to y$ then (y, s', s'', y) is a directed triangle contained in D and it follows from the hypothesis that it has at least two symmetrical arcs. Recall $s'' \overset{T^{-1}}{\not\to} s'$ (12), so, $(s', s'') \notin \mathrm{Sym}(T^{-1}) \subseteq \mathrm{Sym}(D)$; and we conclude that $\{(y, s'), (s'', y)\} \subseteq \mathrm{Sym}(D)$. Thus in any case we obtain $y \to s''$; and from (13) we finally conclude that $y \overset{blue}{\to} s''$. Assertion (8) follows from (10), (12) and (14). Now, since $s \overset{red}{\to} y$ and $s \in S_1$, then $S_1 \overset{red}{\to} y$ which implies $y \to S_1$ (as $S_1 \in \mathcal{C} \subseteq \mathcal{U}$). Let $s_1 \in S_1$ such that $y \to s_1$. We have the following property:

(15) $y \overset{blue}{\to} s_1$: $y \to s_1$ implies $y \overset{red}{\to} s_1$ or $y \overset{blue}{\to} s_1$. Assume by contradiction that $y \overset{red}{\to} s_1$, since $s \overset{red}{\to} y$ we have $y \overset{T}{\to} s_1$ and $s \overset{T}{\to} y$; consequently $s \overset{T}{\to} s_1$ or $y \overset{T}{\to} s$ (as T is a left-pretransitive digraph). The fact $s \overset{T}{\to} s_1$ is impossible because $\{s, s_1\} \subseteq S_1$ which is an independent set; thus $y \overset{T}{\to} s$ and $(y, s) \in \mathrm{Sym}(T) \subseteq \mathrm{Sym}(D)$. It follows that $y \to s$ with $s \in S^\infty$; a contradiction to our assumption. Now, from (8) taking $s' = s_1$ and $R = S_1$ we have that there exists $t_1 \in S^\infty$ such that:

(16) $t_1 \overset{red}{\to} s_1$.

Since $t_1 \in S^\infty$, there exists $S_2 \in \mathcal{C}$ such that $t_1 \in W$ for each $W \in \mathcal{C}$ with $W \geq S_2$, we may assume that $S_2 \geq S_1$ (In the opposite case you can take $S_2 = S_1$ anyway). Then it follows from (8) (ii) taking $R' = S_2$ and $t = t_1$ that:

(17) There exists $s_2 \in S_2$ such that $s_1 \overset{blue}{\to} s_2$, $s_2 \overset{T^{-1}}{\not\to} s_1$ and $y \overset{blue}{\to} s_2$. Again; taking $s' = s_2$ and $R = S_2$ it follows from (8) that there exists $t_2 \in S^\infty$ such that:

(18) $t_2 \overset{red}{\to} s_2. t_2 \in S^\infty$ implies that there exists $S_3 \in \mathcal{C}$ such that $t_2 \in W$ for each $W \in \mathcal{C}$ with $W \geq S_3$, and clearly we may assume $S_3 \geq S_2$ (As any $W \in \mathcal{C}$ with $W \geq S_3$ satisfies the same property as S_3). Thus, taking $R' = S_3$ and $t = t_2$ in (8) (ii) we obtain:

(19) There exists $s_3 \in S_3$ such that $s_2 \overset{blue}{\to} s_3$, $s_3 \overset{T^{-1}}{\not\to} s_2$ and $y \overset{blue}{\to} s_3$. Therefore: If we assume that for some $n \in \mathbb{N}$ we have that $\{t_1, \ldots, t_n\} \subseteq S^\infty$, $\{S_1, \ldots, S_{n+1}\} \subseteq \mathcal{C}$ with $S_1 \leq S_2 \leq \cdots \leq S_{n+1}$, $\{s_1, \ldots, s_{n+1}\} \subseteq V(D)$ are such that for each $i \in \{i, 2, \ldots, n\}$:

(20) (i) $t_i \overset{red}{\to} s_i$.

(ii) $\{t_i, s_{i+1}\} \subseteq S_{i+1}$, $s_i \overset{blue}{\to} s_{i+1}$, $s_{i+1} \overset{T^{-1}}{\not\to} s_i$ and $y \overset{blue}{\to} s_{i+1}$.

Then taking $s' = s_{n+1}$ and $R = S_{n+1}$ in (8) we have that there exists $t_{n+1} \in S^\infty$ such that:

(21) $t_{n+1} \overset{red}{\to} s_{n+1}. t_{n+1} \in S^\infty$ implies that there exists $S_{n+2} \in \mathcal{C}$ with $t_{n+1} \in W$ for each $W \in \mathcal{C}$, $W \geq S_{n+2}$, we may assume $S_{n+2} \geq S_{n+1}$; and from (8) (ii) (taking $s' = s_{n+1}$, $R = S_{n+1}$, $R' = S_{n+2}$ and $t = t_{n+1}$) the following:

(22) There exists $s_{n+2} \in S_{n+2}$ such that $s_{n+1} \overset{blue}{\to} s_{n+2}$, $s_{n+2} \overset{T^{-1}}{\not\to} s_{n+1}$ and $y \overset{blue}{\to} s_{n+2}$. Therefore we have a sequence of vertices of D, $(s_n)_{n \in \mathbb{N}}$ such that $s_{n+1} \overset{T^{-1}}{\to} s_{n+2}$ and $s_{n+2} \overset{T^{-1}}{\not\to} s_{n+1}$. Since T^{-1} is a right-pretransitive digraph, it follows from Lemma 2.1 that for each $k \in \mathbb{N}$, the sequence $(s_1, s_2, \ldots, s_k)$ is a directed path in T^{-1}; clearly this implies that $(s_n)_{n \in \mathbb{N}}$ is an infinite outward path in T^{-1}, a contradiction. So, $y \to S^\infty$ and then $S^\infty \in \mathcal{U}$. It follows from (5), (6) and (7) that S^∞ is an upper bound of $\mathcal{C}$. We have proved that any chain in $\mathcal{U}$ has an upper bound in $\mathcal{U}$, and so by Zorn's Lemma, $(\mathcal{U}, \leq)$ contains maximal elements. Let S be a maximal element of $(\mathcal{U}, \leq)$.

(23) S is a kernel of D.

(24) S is an independent set of D. Since $S \in \mathcal{U}$, S is an independent set of vertices of D.

(25) For each $x \in (V(D) - S)$ there exists an xS-arc. Suppose by contradiction that there exists $x \in (V(D) - S)$ such that $x \not\to S$. There exists a vertex $x_0 \in$

$V(D)$ such that $x_0 \not\to S$, and x_0 satisfies: $x_0 \xrightarrow{T} y$ and $y \not\to S$ imply $y \xrightarrow{T} x_0$, for all vertices $y \in V(D)$. This follows directly from Lemma 2.2 taking $D = T$ and $U = \{x \in V(D) \mid x \not\to S\}$ (notice that our assumption implies $U \neq \emptyset$). Let $P = \{s \in S \mid s \xrightarrow{blue} x_0\}$; we have the following assertions:

(26) $P \cup \{x_0\}$ is an independent set of vertices of D. P is an independent set (as $P \subseteq S$ and $S \in \mathcal{U}$); $x_0 \not\to P$ (because $x_0 \not\to S$ and $P \subseteq S$). From the definition of P we have $P \xrightarrow{blue} x_0$. Also $P \xrightarrow{red} x_0$ because $P \xrightarrow{red} x_0$ implies $S \xrightarrow{red} x_0$ and then $x_0 \to S$, a contradiction (recall $S \in \mathcal{U}$). (27) $P \cup \{x_0\} \in \mathcal{U}$. Suppose that $P \cup \{x_0\} \xrightarrow{red} y$; we will prove $y \to P \cup \{x_0\}$. In order to prove this we assume $y \not\to P$ and we will prove $y \to x_0$. We proceed by considering the two following cases:

Case a. $P \xrightarrow{red} y$. In this case we have $S \xrightarrow{red} y$ (as $P \subseteq S$) and then $y \to S$ (as $S \in \mathcal{U}$). Thus $y \to (S - P)$ (recall our assumption); and we have two possibilities: $y \xrightarrow{red} (S - P)$ or $y \xrightarrow{blue} (S - P)$. When $y \xrightarrow{red} (S - P)$, we consider $p \in P$ such that $p \xrightarrow{red} y$; and by applying that T is a left-pretransitive digraph we obtain $y \xrightarrow{T} p$ or $p \xrightarrow{T} (S - P)$. Since S is an independent set; we have $p \not\xrightarrow{T} (S - P)$ and then $y \xrightarrow{T} p$. This implies that $(p, y) \in \mathrm{Sym}(T) = \mathrm{Sym}(D)$; and thus $y \to P$, a contradiction to our assumption. So $y \not\xrightarrow{red} (S - P)$. When $y \xrightarrow{blue} (S - P)$ we consider $s \in (S - P)$ such that $y \xrightarrow{blue} s$. Since $s \xrightarrow{blue} x_0$ (recall the definition of P) and T^{-1} is a right-pretransitive digraph; we have $x_0 \xrightarrow{T^{-1}} s$ or $y \xrightarrow{T^{-1}} x_0$. If $x_0 \xrightarrow{T^{-1}} s$ then $(x_0, s) \in \mathrm{Sym}(T^{-1}) \subseteq \mathrm{Sym}(D)$, so $x_0 \to S$, a contradiction. If $y \xrightarrow{T^{-1}} x_0$, then $y \to x_0$ or $x_0 \to y$. In case $x_0 \to y$ we have the directed triangle of D, (x_0, y, s, x_0) which by hypothesis has two symmetrical arcs in D. Thus $(x_0, y) \in \mathrm{Sym}(D)$ (as $x_0 \not\to S$) and $y \to x_0$.

Case b. $x_0 \xrightarrow{red} y$. We analyze the two possibilities:

Subcase b.1. $y \not\to S$. In this case the definition of x_0 implies $y \xrightarrow{T} x_0$ and thus $(x_0, y) \in \mathrm{Sym}(T) \subseteq \mathrm{Sym}(D)$. So $y \to x_0$.

Subcase b.2. $y \to S$. In this case we take $s \in (S - P)$ such that $y \to s$ (recall $y \not\to P$). From the definition of P we have $s \xrightarrow{blue} x_0$ and then the directed triangle (x_0, y, s, x_0) is contained in D; which from the hypothesis has two symmetrical arcs in D (notice that $x_0 \to y$ otherwise $y \to x_0$ and we are done; also $s \to x_0$ because $x_0 \not\to S$). Since $x_0 \not\to S$ we have $(x_0, y) \in \mathrm{Sym}(D)$ and thus $y \to x_0$. We conclude that $P \cup \{x_0\} \in \mathcal{U}$.

(28) $S \leq P \cup \{x_0\}$. Let $s \in S$. When s belongs to P we are done. When $s \notin P$ we have $s \xrightarrow{blue} x_0$ and $x_0 \not\xrightarrow{blue} s$ (Recall the definitions of P and x_0 respectively); therefore $s \xrightarrow{T^{-1}} x_0$ and $x_0 \not\xrightarrow{T^{-1}} s$. Since $x_0 \notin S$ we conclude from (28) that $S < T \cup \{x_0\}$ contradicting the maximality of S. □

Notice that the hypothesis $\mathrm{Sym}(T) = \mathrm{Sym}(D)$ in Theorem 2.1 is used only to prove that any chain in $(\mathcal{U}, \leq)$ is upper bounded and for the rest of the proof we only need that $\mathrm{Sym}(T) \subseteq \mathrm{Sym}(D)$; so when G is a finite graph it suffices to ask that $\mathrm{Sym}(T) \subseteq \mathrm{Sym}(D)$ and the same proof as that of Theorem 2.1 works. So, we have the following result:

Theorem 2.2. *Let G be a finite graph and D an M-orientation of G. If there exists an orientation T of G that is a right-pretransitive digraph or a left-pretransitive digraph, with $\mathrm{Sym}(T) \subseteq \mathrm{Sym}(D)$. Then D has a kernel.*

Corollary 2.1. *Let G be a possibly infinite graph and D an M-orientation of G. If there exists a right (resp. left)-pretransitive orientation T of G such that T has no infinite outward path nor infinite inward path either and $\mathrm{Sym}(D) = \mathrm{Sym}(T)$. Then D is a kernel-perfect digraph.*

Corollary 2.2. *Let G be a finite graph and D an M-orientation of G. If there exists a right (resp. left)-pretransitive orientation T of G with $\mathrm{Sym}(T) \subseteq \mathrm{Sym}(D)$, then D is a kernel-perfect digraph.*

Corollary 2.3 (Champetier [6]). *Every M-orientation of a finite comparability graph G is kernel-perfect.*

Proof. If T is an asymmetrical transitive orientation of G, then clearly T satisfies the hypothesis of Corollary 2.2. $\square$

Remark 2.1. The hypothesis that T has no infinite outward path cannot be dropped in Theorem 2.1. Consider the graph G defined as follows: $V(G) = \{u_n \mid n \in \mathbb{N}\}$ and u_i is adjacent to u_j iff $i \neq j$. And consider the following orientation T of G; $(u_i, u_j) \in A(T)$ whenever $i < j$. Clearly T is transitive, and taking $D = T$ we have that D is an M-orientation (moreover T is asymmetrical) such that $\mathrm{Sym}(T) = \mathrm{Sym}(D)$; however D has no kernel.

Remark 2.2. The hypothesis that D be an M-orientation in Theorem 2.1 is tight.

Let G be the complete graph with set of vertices $V(G) = \{u, v, w, x\}$; D the following orientation; $A(D) = \{(u,v), (v,w), (w,x), (x,u), (u,w), (w,u), (v,x), (x,v)\}$ D is not an M-orientation and D has no kernel. And let T be the following orientation of G; $A(T) = \{(u,v), (w,v), (w,x), (u,x), (u,w), (w,u), (v,x), (x,v)\}$, clearly T is a right-pretransitive digraph, a left-pretransitive digraph and $\mathrm{Sym}(T) = \mathrm{Sym}(D)$.

Remark 2.3. In Theorem 2.2 the hypothesis $\mathrm{Sym}(T) \subseteq \mathrm{Sym}(D)$ cannot be dropped. Let G be an odd cycle of length at least 5; D the corresponding directed cycle (clearly D has no kernel and D is an M-orientation of G). We have the following right-pretransitive orientation of G: T with

$$A(T) = \{(v_1, v_2), (v_2, v_1)\}$$
$$\cup \{(v_{2i+1}, v_{2i}), (v_{2i+1}, v_{2i+2}) \mid i \in \{1, 2 \ldots, n\} \ (\mathrm{mod}\ 2n+1)\}$$

clearly $\mathrm{Sym}(T) \not\subseteq \mathrm{Sym}(D)$.

Remark 2.4. The following non perfect graph satisfies the hypothesis of Theorem 2.2. Let G be defined as follows:

$$V(G) = \{v_1, v_2, \ldots, v_{2n+1}\}$$
$$\cup \{u_1, u_2 \ldots, u_{2n-2}, u_{2n}, u_{2n+1}\}, u_{2n-1} = v_2.$$
$$E(G) = \{(v_i, v_{i+1}) \mid i \in \{1, \ldots, 2n+1\} (\bmod 2n+1)\}$$
$$\cup \{(u_i, u_{i+1}) \mid i \in \{1, 2, \ldots, 2n+1\} \ (\bmod 2n+1)\} \cup \{(v_1, u_{2n})\}.$$

Acknowledgments

The authors wish to thank the anonymous referee for many comments which improved the rewriting of this paper.

References

[1] C. Berge, *Les problèmes de coloration en théorie des graphes*, Publ. Inst. Statist. Univ. Paris **9** (1960), 123–160.

[2] C. Berge, Graphs, North-Holland, Amsterdam, New York, 1985.

[3] C. Berge and V. Chvátal, editors, Topics on perfect graphs, Ann. Discrete Mathematics (**21**), North-Holland Amsterdam, 1984.

[4] C. Berge and P. Duchet, *Recent problems and results about kernels in directed graphs*, Discrete Mathematics **86** (1990), 27–31.

[5] E. Boros and V. Gurvich, *Perfect graphs are kernel solvable*, Discrete Mathematics **159** (1996), 35–55.

[6] C. Champetier, *Kernels in some orientations of comparability graphs*, J. Combin. Theory Ser. B **47** (1989), 111–113.

[7] M. Chudnovsky, N. Robertson, P. Seymour, R. Thomas, The strong perfect graph theorem, Princeton University, 2002.

[8] P. Duchet, *Graphes noyau-parfaits*, Ann. Discrete Math. **9** (1980), 93–101.

[9] P. Duchet, *Classical perfect graphs*, an introduction with emphasis on triangulated and interval graphs, Ann. Discrete Math.**21** (1984),67–96.

[10] P. Duchet, *A sufficient condition for a digraph to be kernel-perfect*, J. Graph Theory II **1** (1987), 81–85.

[11] P. Duchet and H. Meyniel, *A note on kernel-critical graphs*, Discrete Mathematics, **33** (1981), 103–105.

[12] H. Galeana-Sánchez and V. Neumann-Lara, *On kernels and semikernels of digraphs*, Discrete Math. **48** (1984), 67–76.

[13] H. Galeana-Sánchez and R. Rojas-Monroy, *Kernels in pretransitive digraphs*, Discrete Math. **275** (2004), 129–136.

[14] Von Neumann and O. Morgenstern, Theory of games and economic behavior, Princeton University Press, Princeton, 1944.

[15] J. L. Ramírez-Alfonsín, B. A. Reed, Perfect graphs, Wiley Interscience, 2001.

[16] M. Richardson, *Solutions of irreflexive relations*, Ann. Math. **58** (2) (1953), 573.

[17] M. Richardson, *Extension theorems for solutions of irreflexive relations*, Proc. Nat. Acad. Sci. USA **39** (1953), 649.

[18] B. Sands, N. Sauer and R. Woodrow, *On monochromatic paths in edge coloured digraphs*, J. Combin. Theory Ser. B **33** (1982), 271–275.

Hortensia Galeana-Sánchez
Instituto de Matemáticas, UNAM
Ciudad Universitaria, Circuito Exterior
04510 México, D.F. México
e-mail: `hgaleana@matem.unam.mx`

Rocío Rojas-Monroy
Facultad de Ciencias
Universidad Autónoma del Estado de México
Instituto Literario No. 100, Centro
50000, Toluca, Edo. de México, México

Nonrepetitive Graph Coloring

Jarosław Grytczuk

To the memory of Claude Berge

Abstract. A coloring of the vertices of a graph G is *nonrepetitive* if no simple path in G looks like $a_1 a_2 \ldots a_n a_1 a_2 \ldots a_n$. The minimum number of colors needed for a graph G is denoted by $\pi(G)$. For instance, by the famous 1906 theorem of Thue, $\pi(G) = 3$ if G is a simple path with at least 4 vertices. This implies that $\pi(G) \leq 4$ if $\Delta(G) \leq 2$. But how large can $\pi(G)$ be for cubic graphs, k-trees, or planar graphs? This paper is a small survey of problems and results of the above type.

Mathematics Subject Classification (2000). Primary 05C38, 15A15; Secondary 05A15, 15A18.

Keywords. Graph coloring, Thue chromatic number, nonrepetitive sequence.

1. Introduction

A coloring f of the vertices of a graph G is *nonrepetitive* if there is no integer $n \geq 1$ and a simple path $v_1 v_2 \ldots v_{2n}$ in G such that $f(v_i) = f(v_{n+i})$ for all $i = 1, \ldots, n$. In other words, it is not possible to read a sequence of colors like $a_1 a_2 \ldots a_n a_1 a_2 \ldots a_n$ while tracing any path in G. This notion was introduced in [2] as a graph-theoretic variant of *nonrepetitive sequences* of Thue (see Section 2).

The minimum number of colors in a nonrepetitive coloring of G is denoted by $\pi(G)$ and is called the *Thue chromatic number* of a graph G. In fact, such a coloring must be proper in the usual sense, so, $\pi(G) \geq \chi(G)$. On the other hand, $\pi(G) \leq |V(G)|$, where $V(G)$ is the set of vertices of a graph G.

Unlike for most of chromatic graph invariants, determining $\pi(G)$ is a nontrivial task even for paths or cycles. It is not even clear *a priori* that $\pi(G)$ is bounded for paths. However, the celebrated theorem of Thue [24] asserts that there are arbitrarily long nonrepetitive sequences over just three symbols (see Section 2). This clearly implies that $\pi(P_n) = 3$ and $\pi(C_n) \leq 4$ for all $n \geq 4$.

This work was supported by KBN grant 1 P03A 017 27.

For other classes of graphs the situation is much less clear. For instance, it is not known how large the Thue chromatic number of a cubic graph can be, though it can be proved probabilistically that 108 colors suffice (see Section 3). Outerplanar graphs can be nonrepetitively colored using 12 colors, but we do not know if there are planar graphs of arbitrarily high Thue chromatic number (see Sections 4 and 5).

The present paper is intended as a short survey of problems and results of the above type. To make the presentation as self-contained as possible we include some proofs, as well as a glimpse of the history of this topic.

2. Chess, music, and sequences

During a game of chess the same position may appear several times on the board. Hence, in official rules of a chess tournament some convention must be adopted by which a draw can be claimed by either player. Suppose a draw could be claimed only if the same position appeared for the third time after the same sequence of moves. Is this condition sufficient to guarantee that no infinite chess play is possible?

The problem was considered by the famous Dutch chess master Max Euwe [13]. He discovered a peculiar recursive construction of an infinite binary sequence U without blocks of the form $xYxYx$, thereby providing a negative answer to the chess question. The sequence is defined as follows. Let $U_1 = 0$ and $U_{n+1} = U_n U'_n$, $n \geq 1$, where U'_n is the "negation" of U_n. Hence $U_2 = 01$, $U_3 = 0110$, $U_4 = 01101001$, and so on. Now an infinite sequence $U = u_1 u_2 \ldots$ is defined by taking $u_1 \ldots u_{2^{n-1}} = U_n$, for each $n \geq 1$. Correctness of this definition follows from the fact that U_n is a prefix of U_{n+1}.

Curiously, the same structure was independently introduced by Prouhet [19], Thue [25], Morse [18], Mahler [17] and Arshon [4], in different contexts of Number Theory and Dynamical Systems. In particular, Thue's fundamental work is considered the starting point of Combinatorics on Words, while Morse' ideas led to the birth of Symbolic Dynamics (cf. [1], [6], [7], [8], [9], [10], [11], [15], [16]).

However, the most surprising is the appearance of the sequence U in music. Around 1965 the famous Danish composer Per Nørgård discovered (or invented?) the sequence of integers defined by $N_{2n+1} = N_n + 1$ and $N_{2n} = -N_n$ for $n \geq 0$, with $N_0 = 0$ (see [1], [23]). In other words, the sequence is a *perfect shuffle* of the negative copy of itself with the shifted copy of itself:

$$\begin{array}{cccccccccc} 0 & & -1 & & 1 & & -2 & & -1 & & 0 & & 2 & & -3 & & 1 \\ & 1 & & 2 & & 0 & & 3 & & 2 & & 1 & & -1 & & 4 & & 0 \end{array} \ldots$$

Nørgård called it "the infinity series" and used it frequently in many of his compositions. For instance, in a mysterious symphony "Voyage into the Golden Screen" (1968), for chamber orchestra, or in "I Ching" (1982), for percussion solo.

It is not hard to show that the sequence U coincides with the infinity series reduced mod 2. We give now a proof of the "unending chess" property of the sequence U, first published by Thue in 1912.

Theorem 2.1. (Thue [25]) *The sequence U does not contain a block of the form $xYxYx$, where Y is a (possibly empty) binary word and x is a single letter.*

Proof. Clearly it is sufficient to prove the assertion for finite words U_n. So assume, for an induction argument, that U_n satisfies the property and that a bad block $B = 0Y0Y0$ appears in U_{n+1}. Color the odd positions of U_{n+1} blue and the even positions red. Then the blue subword coincides with U_n, while the red subword coincides with U'_n. For instance,

$$U_4 = 01101001 = \begin{matrix} 0 & & 1 & & 1 & & 0 & \\ & 1 & & 0 & & 0 & & 1 \end{matrix} \begin{matrix} = U_3 \text{ (blue subword)} \\ = U'_3 \text{ (red subword)} \end{matrix}.$$

We distinguish two cases with respect to the parity of the length of Y.

If the length of Y is odd then the three zeros of B are of the same color, say blue. But then the blue copy of U_n contains $0Y_{\text{blue}}0Y_{\text{blue}}0$, where Y_{blue} is the blue part of the word Y. This contradicts our inductive assumption on U_n. If the three zeros are red, then $0Y_{\text{red}}0Y_{\text{red}}0$ appears in U'_n. This also contradicts the inductive assumption, since then $(0Y_{\text{red}}0Y_{\text{red}}0)'$ appears in U_n.

Now suppose Y is of even length. Then the two utmost zeros are of the same color, say blue, while the middle zero is red. Denote $B = 0x_1 \ldots x_{2n}0y_1 \ldots y_{2n}0$. By the "perfect shuffle" property, in any word U_n, the next term after a blue zero is a red one, and the next term after a blue one is a red zero. Since the first zero is blue, x_1 must be a red one. So, $y_1 = 1$ (since $y_1 = x_1$) and it is blue (since the middle zero is red). Hence y_2 must be a red zero. Therefore, $x_2 = 0$ (since $x_2 = y_2$) and it is blue (since x_1 was red). Continuing this "ping-pong" we see that the first copy of Y in B must end with a blue zero. But this gives an impossible block of two zeros (the first blue and the second red) in the middle.

When the utmost zeros are red and the middle zero is blue, the argument is similar. This completes the proof. □

A finite or infinite sequence $S = s_1 s_2 \ldots$ is *nonrepetitive* if no two adjacent blocks in S are identical, that is, if there are no integers $k \geq 0$ and $n \geq 1$ such that $s_{k+i} = s_{k+n+i}$, for all $i = 1, \ldots, n$. In particular, S cannot contain blocks of the form aa, $abab$, $abcabc$, $abcdabcd$, etc.

It is easy to see that there are no nonrepetitive binary sequences with more than three terms. Curiously, if a third symbol is available then one can produce arbitrarily long nonrepetitive sequences. This fact, proved for the first time by Thue in 1906, can be deduced easily from the "unending chess" theorem.

Theorem 2.2. (Thue [24]) *There exists an infinite nonrepetitive sequence over three symbols $1, 2, 3$.*

Proof. Let $X = \{1, 4, 6, 7, \ldots\}$ be the set of positions occupied by zeros in the sequence U, and let $x_1 < x_2 < \cdots$ be the elements of X ordered increasingly.

Consider the sequence $T = t_1 t_2 \ldots$ defined by $t_n = x_{n+1} - x_n$, for each $n \geq 1$. Thus $T = 321312321231\ldots$, and clearly $t_i \in \{1,2,3\}$, for each $i \geq 1$.

Now suppose that for some $k,n \geq 1$, $t_{k+i} = t_{k+n+i}$, for all $i = 0,\ldots,n-1$. Then the triple x_k, x_{k+n}, x_{k+2n} forms a 3-term arithmetic progression of difference $t_k + \cdots + t_{k+n-1}$, and the set $\{x_{k+1},\ldots,x_{k+n-1}\}$ is a translated copy of the set $\{x_{k+n+1},\ldots,x_{k+2n-1}\}$. However this is equivalent to finding a block of the form $0Y0Y0$ in a sufficiently long word U_m. This contradicts Theorem 1. $\square$

Clearly, the above theorem provides a constructive nonrepetitive 3-coloring of any path P_n.

3. Graphs with bounded degree

Let $\pi(d)$ be the supremum of numbers $\pi(G)$, where G ranges over all graphs of maximum degree at most d. By Theorem 2 we know that $\pi(2) \leq 4$. It is not hard to check that in fact $\pi(2) = 4$, since the cycle C_5 demands four colors. Actually, as proved by Currie [12], $\pi(C_n) = 3$ for all $n \geq 3$, except $n = 5,7,9,10,14,17$.

For $d \geq 3$ the situation is less clear. In [2] it was proved that there are positive constants c_1, c_2 such that

$$c_1 \frac{d^2}{\ln d} \leq \pi(d) \leq c_2 d^2.$$

The proof is probabilistic and provides no explicit constructions. However, no other way of establishing at least the finiteness of $\pi(3)$ was found so far. Similarly for the lower bound; no explicit graphs of degree d and the Thue number of order $d^2/\ln d$ are known, though we know that almost every graph has this property.

We present here a slightly refined proof from [2], which shows that $\pi(d) \leq 36d^2$. The proof is based on the Lovász Local Lemma in the form stated below (see [3]). Recall that a *dependency graph* of random events $A_1,\ldots,A_n$ is any graph $D = (V,E)$ on the set of vertices $V = \{A_1,\ldots,A_n\}$, such that each event A_i is mutually independent of the events $\{A_j : A_i A_j \notin E\}$.

Lemma 3.1. (The local lemma) *Let $A_1,\ldots,A_n$ be events in any probability space with dependency graph $D = (V,E)$. Let $V = V_1 \cup \cdots \cup V_k$ be a partition such that all members of each part V_r have the same probability p_r. Suppose that the maximum number of vertices from V_s adjacent to a vertex from V_r is at most Δ_{rs}. If there are real numbers $0 \leq x_1,\ldots,x_k < 1$ such that $p_r \leq x_r \prod_{s=1}^{k}(1-x_s)^{\Delta_{rs}}$ then $\Pr(\bigcap_{i=1}^{n} \overline{A_i}) > 0$.*

Theorem 3.2. (Alon et al. [2]) *$\pi(G) \leq 36\Delta^2$, for every graph G of maximum degree at most Δ.*

Proof. Let G be a graph of maximum degree Δ. Consider a random coloring of the vertices of G with $N = 36\Delta^2$ colors. For each path P in G let A_P be the event that the first half of P is colored the same as the second. Let V_r be the set of all events A_P with P having $2r$ vertices. Clearly we have $p_r = N^{-r}$.

Now define a dependency graph so that A_P is adjacent to A_Q iff the paths P and Q have a common vertex. Since a fixed path with $2r$ vertices intersects at most $4rs\Delta^{2s}$ paths with $2s$ vertices in G, we may take $\Delta_{rs} = 4rs\Delta^{2s}$.

Next set $x_s = (5\Delta)^{-2s}$, and notice that $(1-x_s) \geq e^{-2x_s}$, as $x_s \leq 1/2$. Hence we get

$$x_r \prod_s (1-x_s)^{\Delta_{rs}} \geq (5\Delta)^{-2r} \prod_s e^{-8rs5^{-2s}} > (5\Delta)^{-2r} \exp\left(-2r \sum_{s=1}^{\infty} \frac{4s}{5^{2s}}\right).$$

Since the series $\sum_{s=1}^{\infty} \frac{4s}{5^{2s}}$ converges to $25/144$, we obtain

$$x_r \prod_s (1-x_s)^{\Delta_{rs}} \geq (5e^{25/144}\Delta)^{-2r} > (6\Delta)^{-2r} = p_r.$$

By the local lemma the proof is complete. □

4. Graphs of bounded treewidth

Another nontrivial class of graphs with bounded Thue chromatic number are graphs of bounded treewidth. This result was obtained independently by Kündgen and Pelsmajer [14] and Barát and Varjú [5].

Theorem 4.1. (Kündgen and Pelsmajer [14]) $\pi(G) \leq 4^k$, for any k-tree G.

For the ease of presentation we give here a proof of a weaker bound from [5]. Actually, we derive it from a slightly more general theorem relating nonrepetitive colorings to oriented colorings, which is implicitly proved in [5].

Let $\vec{G}$ be an orientation of a simple graph G. A *directed nonrepetitive coloring* of the vertices of $\vec{G}$ is defined similarly, by restricting the condition to directed paths. Denote by $\pi(\vec{G})$ the fewest number of colors in a directed nonrepetitive coloring of $\vec{G}$.

A coloring of the vertices of $\vec{G}$ is called a *proper oriented coloring* if each color class is an independent set in the underlying graph G, and all the edges between any two color classes are of the same direction. Let $\chi_o(\vec{G})$ denote the minimum number of colors in a proper oriented coloring of an oriented graph $\vec{G}$.

Finally recall that *fraternal orientation* of a graph is an orientation such that any two vertices with a common out-neighbor are adjacent.

Theorem 4.2. (Barát and Varjú [5]) *Let $\vec{G}$ be a fraternal orientation of a graph G. Then $\pi(G) \leq \pi(\vec{G})\chi_o(\vec{G})$.*

Proof. Let f be a directed nonrepetitive coloring of $\vec{G}$ and let g be a proper oriented coloring of $\vec{G}$. Consider a coloring h of the vertices of G defined by $h(v) = (f(v), g(v))$, for each vertex v of G. We will show that h is a nonrepetitive coloring of G. To this end suppose that a path $P = v_1 v_2 \ldots v_{2n}$ is a shortest path in G colored repetitively, that is, $h(v_i) = h(v_{n+i})$, for all $i = 1, \ldots, n$.

Now consider the related sequence of signs $S_P = s_1 s_2 \ldots s_{2n-1}$ defined by $s_i = +$, if v_i dominates v_{i+1} in $\vec{G}$, and $s_i = -$, otherwise. Since g is a proper oriented coloring, it follows that $s_i = s_{n+i}$ for all $i = 1, \ldots, n-1$. Since f is a directed nonrepetitive coloring, the sequence S_P cannot be constant. Hence there exists $1 < j \leq n$ such that v_j is a common out-neighbor of v_{j-1} and v_{j+1}.

Suppose first that $j < n$. Then $s_{j-1} \neq s_j$, say $s_{j-1} = +$ and $s_j = -$. In consequence, $s_{j-1+n} = +$ and $s_{j+n} = -$. This means that v_{j+n} is dominated by its neighbors on the path P. Since the orientation $\vec{G}$ is fraternal, both pairs of vertices $v_{j-1} v_{j+1}$ and $v_{j-1+n} v_{j+1+n}$ forms the edges of G. So removing the vertices v_j, v_{j+n} from the path P results in a shorter repetitive path, contrary to our assumption.

If $j = n$ then similarly the path $v_1 \ldots v_{n-1} v_{n+1} \ldots v_{2n-1}$ forms a shorter repetitive path. This completes the proof. $\square$

Let G be a k-tree and let $v_1, \ldots, v_n$ be a perfect elimination ordering of the vertices of G. This means that the vertices $v_1, \ldots, v_k$ induce a k-clique, and for each $i > k$, the neighbors of v_i with strictly smaller indices also induce a k-clique in G. Let $\vec{G}$ be an orientation of G such that v_i dominates v_j if $i < j$. Clearly, $\vec{G}$ is a fraternal orientation of G. Now using Theorem 2 it is not hard to prove that $\pi(\vec{G}) \leq 3^k$. An appropriate coloring by k-tuples $(x_1, \ldots, x_k)$ with $x_i \in \{1, 2, 3\}$ can be constructed as follows. Arrange the vertices of $\vec{G}$ into levels, in a tree-like fashion, and assign to the levels successive terms of a ternary nonrepetitive sequence. This fixes first coordinates of k-tuples. Then repeat the procedure inside each level. It is not hard to prove by induction that this coloring has the desired property.

On the other hand, by the theorem of Raspaud and Sopena [20], any orientation of a k-tree has a proper oriented coloring with at most $2^k(k+1)$ colors. Hence, we get the following.

Corollary 4.3. (Barát and Varjú [5]) $\pi(G) \leq 6^k(k+1)$, for any k-tree G.

The fact that $\pi(G)$ is bounded for graphs of bounded treewidth has further consequences for graphs with forbidden planar minor. The well-known theorem of Robertson and Seymour [21] asserts that graphs not containing a fixed planar graph as a minor has bounded treewidth. The bound on the treewidth improved later in [22] allows for the following corollary.

Corollary 4.4. Let H be planar graph with n vertices and m edges. Let G be any graph without a minor isomorphic to H. Then $\pi(G) \leq 4^{20^{2(2n+4m)^5}}$.

Therefore each minor-closed class of graphs with unbounded Thue chromatic number must contain the class of planar graphs.

5. Planar graphs

An intriguing open question is whether the Thue chromatic number is bounded for planar graphs. The results presented so far suggest that perhaps the following conjecture holds.

Conjecture 5.1. *There is a constant c such that $\pi(G) \leq c$, for each planar graph G.*

There are many intersecting sub-classes of planar graphs for which the conjecture is true and it is not so easy to guess an eventual counterexample. In the rest of this section we present several weaker versions of the problem, none of which has been completely decided for planar graphs.

5.1. Power-free colorings

We start with generalizing another property introduced by Thue. Let $k > 1$ be a fixed integer. A coloring f of the vertices of a graph G is k-*power-free* if there is no positive integer n and a path $v_1 v_2 \ldots v_{kn}$ in G such that $f(v_i) = f(v_{i+n}) = \cdots = f(v_{i+(k-1)n})$, for each $i = 1, \ldots, n$. Less formally, the condition says that no block is repeated k times in a row on any path in G.

Let $\pi_k(G)$ be the smallest number of colors in a k-power-free coloring of G. Clearly, $\pi_k(G) \leq \pi_m(G)$ if $k > m$ and $\pi_2(G) = \pi(G)$. Notice that for $k > 2$ a k-power-free coloring may not be proper.

Conjecture 5.2. *There exist constants k and c such that $\pi_k(G) \leq c$, for any planar graph G.*

Let us remark that it is not excluded that this seemingly weaker statement actually implies the previous conjecture. Indeed, the existence of nonrepetitive ternary sequences can be deduced from a weaker property for binary sequences, as we demonstrated in the proof of Theorem 2.

5.2. Are four colors enough?

Suppose for a while that Conjecture 10 is true. What is then the smallest value of c for which the conjecture holds with some, possibly huge k?

The following example shows that this minimum must be greater than 3. Let $G_1 = K_3$ and let G_n be a plane graph obtained from G_{n-1} by inserting a vertex inside each inner face of G_{n-1}, and joining it to the three vertices of the face. It is not hard to show that for every k there exists n such that in any 3-coloring of the vertices of G_n a monochromatic path with k vertices appears.

Conjecture 5.3. *There exists k such that $\pi_k(G) \leq 4$, for any planar graph G.*

This statement is hard to believe. In fact, we do not know if this is true even for outerplanar graphs, for which $\pi(G) \leq 12$ as proved in [5] and [14].

5.3. Induced paths

Brešar and Klavžar proposed another variant of nonrepetitive colorings. Consider a vertex coloring of a graph G in which only the *induced* paths are to be colored nonrepetitively. Denote by $\pi_{\text{ind}}(G)$ the *induced Thue chromatic* number of a graph G defined similarly as before. Clearly $\pi_{\text{ind}}(G) \leq \pi(G)$ for any graph G.

Conjecture 5.4. *There exists a constant c such that $\pi_{ind}(G) \leq c$, for any planar graph G.*

It seems plausible that $\pi_{\text{ind}}(G)$ should behave more like the classical chromatic number. For instance, a nice observation of Brešar shows that "perfect" graphs with respect to $\pi_{\text{ind}}(G)$ are exactly the P_4-free graphs. Indeed, let G be a π_{ind}-*perfect*, that is, $\pi_{\text{ind}}(H) = \omega(H)$ for any induced subgraph H of G, where $\omega(H)$ stands for the clique number of H. Since $\pi_{\text{ind}}(P_4) = 3$ and $\omega(P_4) = 2$, G cannot contain P_4 as an induced subgraph. On the other hand, any proper coloring of a P_4-free graph is certainly nonrepetitive on induced paths (we just don't care about paths with more than 3 vertices). The assertion follows since any P_4-free graph is perfect in the usual sense.

5.4. Subdivision

Let $S(G)$ be the set of all possible subdivisions of a graph G. By known properties of nonrepetitive sequences one can prove that for every graph G there is a graph $H \in S(G)$ such that $\pi(H) \leq 5$. Indeed, mark all vertices of G by symbol 5. Then subdivide each edge of G by one vertex and mark it by symbol 4. Next take $m = |E(G)|$ nonrepetitive sequences S_i of different lengths, constructed of symbols $1, 2, 3$. Assign the sequences S_i bijectively to subdivided edges, and turn each edge into a path $5S_i 4 S_i 5$. Now in a repetitive path P the symbols 4 and 5 must occupy the same positions in both halves of P. This is however impossible since the lengths of S_i are pairwise different.

Of course, no such result is possible in general if we bound the number of vertices subdividing an edge. How many vertices per edge one should add to a planar graph to make it 5-colorable in the sense of Thue?

Conjecture 5.5. *There are constants c, r such that each planar graph G has a subdivision H, with at most r vertices subdividing one edge, such that $\pi(H) \leq c$.*

Again, this looks easier than Conjecture 9, but it is not excluded that both statements are equivalent.

6. Metachromatic number

Similar problems can be considered in a more general setting. Suppose a large structure is given with adjacency relation on a collection of its substructures. We want to color the structure so as to distinguish the adjacent substructures, using as few colors as possible.

For instance, let G be a simple graph and let A, B be two connected subgraphs of G. The subgraphs A, B are said to be *adjacent* if their vertex sets are disjoint, and there is at least one edge uv in G with $u \in V(A)$ and $v \in V(B)$. Let f be a coloring of the vertex set of G. We say that the subgraphs A, B are *distinguished* by f if there is no color-preserving isomorphism between A and B. If f distinguishes every pair of adjacent connected subgraphs in G we call it a *metacoloring* of G. The minimum number of colors needed is the *metachromatic number* of a graph G, denoted by $\mathbb{M}(G)$.

Notice that for graphs of maximum degree at most 2, $\mathbb{M}(G) = \pi(G)$. For which other natural classes of graphs is $\mathbb{M}(G)$ bounded?

References

[1] J.-P. Allouche, J. Shallit, Automatic sequences. Theory, applications, generalizations, Cambridge University Press, Cambridge, 2003.

[2] N. Alon, J. Grytczuk, M. Hałuszczak, O. Riordan, Non-repetitive colorings of graphs, Random Struct. Alg. 21 (2002), 336–346.

[3] N. Alon, J.H. Spencer, The probabilistic method, Second Edition, John Wiley & Sons, Inc., New York, 2000.

[4] E.S. Arszon, Proof of the existence of n-valued infinite asymmetric sequences, Mat. Sb. 2 (44), (1939), 769–779. (In Russian).

[5] J. Barát, P. P. Varjú, Some results on square-free colorings of graphs, manuscript.

[6] D.R. Bean, A. Ehrenfeucht, G.F. McNulty, Avoidable patterns in strings of symbols, Pacific J. Math. 85 (1979), 261–294.

[7] J. Berstel, Axel Thue's work on repetitions in words; in P. Leroux, C. Reutenauer (eds.), Séries formelles et combinatoire algébrique Publications du LaCIM,, Université du Québec a Montréal, pp. 65–80, 1992.

[8] J. Berstel, Axel Thue's papers on repetitions in words: a translation, Publications du LaCIM, vol 20, Université du Québec a Montréal, 1995.

[9] J. Berstel, J. Karhumäki, Combinatorics on words – a tutorial, Bull. Eur. Assoc. Theor. Comput. Sci. EATCS No. 79 (2003), 178–228.

[10] R. Bott, Marston Morse and his mathematical works, Bull. Amer. Math. Soc. (New Series) 3 (1980), 907–950.

[11] Ch. Choffrut, J. Karhumäki, Combinatorics of Words, in: Handbook of Formal Languages, G. Rozenberg, A. Salomaa (Eds.) Springer-Verlag, Berlin Heidelberg, 1997, 329–438.

[12] J.D. Currie, There are ternary circular square-free words of length n for $n \geq 18$, Electron. J. Combin. 9 (2002) #N10, 7pp.

[13] M. Euwe, Mengentheoretische Betrachtungen über das Schachspiel, Proc. Konin. Akad. Wetenschappen Amsterdam 32 (1929), 633–642.

[14] A. Kündgen, M. J. Pelsmajer, Nonrepetitive colorings of graphs of bounded treewidth, manuscript.

[15] M. Lothaire, Combinatorics on Words, Addison-Wesley, Reading MA, 1983.

[16] M. Lothaire, Algebraic Combinatoric on Words, Cambridge, 2001.

[17] K. Mahler, On the translation properties of a simple class of arithmetical functions, J. Math. and Phys. 6 (1927), 158–163.

[18] M. Morse, A one-to-one representation of geodesics on a surface of negative curvature, Amer. J. Math. 43 (1921), 35–51.

[19] E. Prouhet, Memoire sur quelques relations entre les puissances des nombres, C.R. Acad. Sc. Paris 33 (1851), 31.

[20] A. Raspaud, E. Sopena, Good and semi-strong colorings of oriented planar graphs, Inform. Process. Lett. 51 (1994), 171–174.

[21] N. Robertson, P.D. Seymour, Graph minors V: Excluding a planar graph, J. Combin. Theory Ser. B 41 (1986), 92–114.

[22] N. Robertson, P. Seymour, R. Thomas, Quickly excluding a planar graph, J. Combin. Theory Ser. B 62 (1994), 323–348.

[23] N.J.A. Sloane, The On-line Encyclopedia of Integer Sequences, published electronically at http://akpublic.research.att.com/~njas/sequences/index.html

[24] A. Thue, Über unendliche Zeichenreihen, Norske Vid. Selsk. Skr., I Mat. Nat. Kl., Christiania, 7 (1906), 1–22.

[25] A. Thue, Über die gegenseitige Lage gleicher Teile gewisser Zeichenreihen, Norske Vid. Selsk. Skr., I Mat. Nat. Kl., Christiania, 1 (1912), 1–67.

Jarosław Grytczuk
Faculty of Mathematics, Informatics and Econometrics
University of Zielona Góra
PL-65-516 Zielona Góra, Poland
e-mail: J.Grytczuk@wmie.uz.zgora.pl

A Characterization of the 1-well-covered Graphs with no 4-cycles

B.L. Hartnell

> **Abstract.** The problem of determining which graphs have the property that every maximal independent set of vertices is also a maximum independent set was proposed in 1970 by M.D. Plummer who called such graphs well-covered. Whereas determining the independence number of an arbitrary graph is NP-complete, for a well-covered graph one can simply apply the greedy algorithm. A well-covered graph G is 1-well-covered if and only if, for every vertex v in G, $G - \{v\}$ is also well covered and has the same independence number. The notion of a 1-well-covered graph was introduced by J. Staples in her 1975 dissertation and was further investigated by M. Pinter in 1991 and later. In this note the 1-well-covered graphs with no 4-cycles are characterized.
>
> **Mathematics Subject Classification (2000).** 05C69.
>
> **Keywords.** Well-covered, maximal independent sets.

1. Introduction

A graph G is said to be well-covered if every maximal independent set of vertices has the same cardinality. These graphs were introduced by M.D. Plummer [16] in 1970. Although the recognition problem of well-covered graphs in general is Co-NP-complete [6, 19], it is polynomial for certain classes of graphs. For instance, well-covered graphs that are claw-free [22], have girth at least 5 [7], have neither 4-cycles nor 5-cycles (but 3-cycles are permitted) [8] or are chordal [18] are all recognizable in polynomial time (for further complexity issues see [4, 5, 22, 23]). The reader is referred to Plummer [17] for an excellent survey of the work on well-covered graphs and to a much lesser extent (but for more recent activity) to Hartnell [10].

In this investigation, we will be interested in a special class of well-covered graphs, called 1-well-covered. A well-covered graph G is 1-well-covered if and only

---

Research supported in part by NSERC of Canada.

if, for every vertex v in G, $G - \{v\}$ is also well-covered and has the same independence number. These graphs, also known as W_2 graphs, were introduced by J. Staples [20, 21] and were later investigated in more detail by M. Pinter [12, 13, 14, 15]. Using the characterization of cubic well-covered graphs [3] and girth 5 well-covered graphs [7], Pinter was able to determine the 1-well-covered ones in each class. He then extended this work to 4-regular 1-well-covered if these graphs were also planar and 3-connected [13] and to the planar 1-well-covered graphs of girth 4 [14]. In [15] he also provides several constructions to obtain infinite families of 1-well-covered graphs of girth 4. Here, we will address the question of what are the 1-well-covered graphs with no 4-cycles (but 3-cycles are allowed). In general, there is no known characterization of well-covered graphs without 4-cycles (for partial results see [9, 24]).

A few definitions and some notation are needed. For S a subset of vertices of a graph, $N[S]$ is the set S as well as all neighbors of vertices in S. A vertex will be called a stem if it has a neighbor of degree one (a leaf). A 3-cycle will be called basic if at least one of its vertices is of degree two. The following are defined in [7]. A 5-cycle is called basic if it contains no adjacent vertices of degree three or more. A vertex v in a well-covered graph G is said to be extendable if $G - \{v\}$ is also well-covered and has the same independence number as G. Hence every vertex is extendable in a 1-well-covered graph.

The following known results are required to establish the theorem of this paper.

Lemma 1. [1, 2, 7] *If G is a well-covered graph and I is an independent set of vertices in G, then $G - N[I]$ must also be well-covered.*

In a similar manner, we have the corresponding result for 1-well-covered graphs.

Lemma 2. [12, 13] *If G is a 1-well-covered graph and I is an independent set of vertices in G, then $G - N[I]$ is also a 1-well-covered graph.*

Lemma 3. [7] *Let G be a well-covered graph and v be an extendable vertex. Then there is no set I of independent vertices in G such that $G - N[I] \cong \{v\}$. That is, it is not possible to isolate v.*

Finally, the following result plays a crucial role in our investigation. The graph S_8 is shown in Figure 1.

Lemma 4. [11] *Let G be a well-covered graph without 4-cycles. Let v be a vertex in G satisfying the following two conditions:*

(i) *v is extendable.*
(ii) *v is not a stem and is not on a basic 3-cycle nor on a basic 5-cycle.*

Then v must be a vertex on an induced S_8.

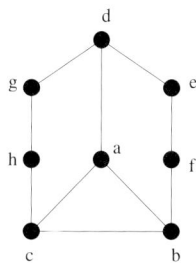

FIGURE 1. The graph S_8

2. The main result

Let F be a graph on $3t$ vertices such that F has no 4-cycles and such that the vertices can be partitioned into t subsets each of which induces a 3-cycle in F. Furthermore, at least two of the three vertices on this 3-cycle are of degree two in F (see Figure 2 for an illustration). Let $\mathcal{F}$ represent the family of all such graphs.

We shall show that if G is a 1-well-covered graph without 4-cycles, then G must, in fact, be K_2 or C_5 or a member of $\mathcal{F}$.

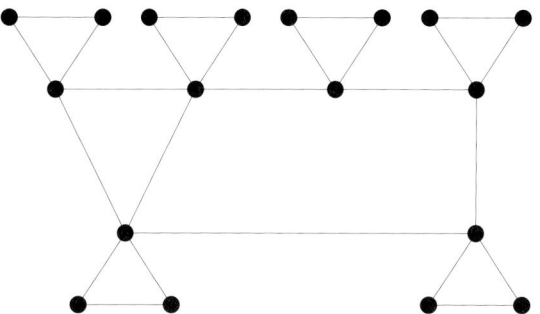

FIGURE 2. A graph in the family $\mathcal{F}$

Theorem 1. *Let G be a connected graph without 4-cycles. The graph G is 1-well-covered if and only if G is isomorphic to K_2, C_5 or a member of the family $\mathcal{F}$.*

Proof. We first consider the only if direction of the statement. Assume that the result does not hold. Let G be a connected graph without 4-cycles with as few vertices as possible such that G is 1-well-covered but is not K_2 nor C_5 nor a member of $\mathcal{F}$. Since every vertex of G is extendable, Lemma 4 ensures that there are limited possibilities for such a vertex. In particular, each vertex must either be a stem or on a basic 3-cycle or a basic 5-cycle or belong to an induced subgraph of G that is isomorphic to the graph S_8. We shall show that none of these cases can occur.

Assume some vertex of G is a stem. Let a neighbor of degree one of this stem be a and a neighbor other than a be called b (since G is not K_2 there must be such a vertex b). But then $G - N[b]$ contains the isolated vertex a which contradicts Lemma 3. Thus no vertex of G is a stem.

Next, assume some vertex is on a basic 5-cycle. Since G is not C_5, there must be a vertex, say a, of degree 3 or more on that 5-cycle. Let the 5-cycle be $abcde$ and v be a neighbor of a other than b or e. Since G has no 4-cycles, we observe that v and d are not adjacent. But then $G - N[\{v,d\}]$ contains the isolated vertex b which is impossible (Lemma 3). Hence no vertex of G is on a basic 5-cycle.

Now assume some vertex is on a basic 3-cycle. Note that there cannot be only one vertex, say a, of degree 2 on that 3-cycle, say abc. Say both b and c have neighbors, say $b*$ and $c*$ respectively, where these are not part of the 3-cycle. As there are no 4-cycles, $b*$ and $c*$ are neither adjacent nor equal. But then $G - N[\{b*,c*\}]$ contains the isolated vertex a which contradicts Lemma 3. Thus any basic 3-cycle must have two degree two vertices. But since G is not a member of $\mathcal{F}$ there must be some vertex not on such a basic 3-cycle.

Now select such a vertex of G that is not on a basic 3-cycle (with two degree two vertices) and label the S_8 that it belongs to as shown in Figure 1. Therefore, recalling that for any vertex w of G that $G - N[w]$ must be 1-well-covered (Lemma 2), it follows, since G is minimal, that each component of $G - N[w]$ is isomorphic to one of K_2, C_5 or a member of $\mathcal{F}$.

Consider $G - N[g]$ and the component containing the vertices a, b and c. This component is not K_2 nor C_5. Furthermore the vertices a, b and c form a 3-cycle where b is of degree at least three in that component and hence a and c must be of degree two in order for this component to belong to $\mathcal{F}$. Since there are no 4-cycles in the graph, g cannot be adjacent to a neighbor of a (other than d) nor to a neighbor of c (other than h). This implies that a and c must be of degree three in G itself. Similarly, in $G - N[e]$ the component containing a, b and c has c of degree three and thus b must be of degree two. As e is not adjacent to a neighbor of b (other than f) since there are no 4-cycles, b must be of degree three in G. Thus each of a, b and c is of degree 3 in G.

Now consider the component of $G - N[h]$ containing the 5-cycle with vertices a, b, f, e and d. Since $G - N[h]$ is isomorphic to one of K_2, C_5 or a member of $\mathcal{F}$, it follows that these must be all the vertices in that component. Hence d must be of degree three in G (since d and h cannot have another common neighbor besides g as there are no 4-cycles in G). Next consider the component of $G - N[c]$ containing d as well as g, e and f. Since d is of degree two in this component it cannot be part of a K_2 and so must belong a C_5 or to a K_3 as part of a graph in $\mathcal{F}$. As e and g are not adjacent it must, in fact, be a C_5. This forces e to be of degree two in G since c (being of degree three in G) shares no neighbor with e. But now observe that $G - N[\{b,g\}]$ has a component consisting of the single vertex e. This is a contradiction (see Lemma 3) as G is a 1-well-covered graph. This completes the only if part of the proof.

Noting that K_2, C_5 or a member of the family $\mathcal{F}$ is 1-well-covered the proof is complete. □

We conclude by observing that, given the characterization, it is easy to recognize in polynomial time if a graph is a 1-well-covered graph with no 4-cycles.

References

[1] S.R. Campbell, *Some results on cubic well-covered graphs*, Ph.D. thesis, Vanderbilt University, (1987).

[2] S.R. Campbell and M.D. Plummer, *On well-covered 3-polytopes*, Ars Combin. 25A (1988), 215–242.

[3] S.R. Campbell, M.N. Ellingham and G.F. Royle, *A characterization of well-covered cubic graphs*, J. Combin. Math. Combin. Comput. 13 (1993), 193–212.

[4] Y. Caro, *Subdivisions, parity and well-covered graphs*, J. Graph Theory 25 (1997), 85–94.

[5] Y. Caro, A. Sebö and M. Tarsi, *Recognizing greedy structures*, J. Algorithms 20 No. 1 (1996), 137–156.

[6] V. Chvátal and P.J. Slater, *A note on well-covered graphs*, Quo Vadis, Graph Theory?, Ann. Discrete Math. 55, North Holland, Amsterdam, (1993), 179–182.

[7] A. Finbow, B. Hartnell and R. Nowakowski, *A characterization of well-covered graphs of girth 5 or greater*, J. Combin. Theory Ser. B 57 (1993), 44–68.

[8] A. Finbow, B. Hartnell and R. Nowakowski, *A characterization of well-covered graphs that contain neither 4- nor 5-cycles*, J. Graph Theory 18 (1994), 713–721.

[9] S. Gasquoine, B. Hartnell, R. Nowakowski and C. Whitehead, *Techniques for constructing well-covered graphs with no 4-cycles*, J. Combin. Math. Combin. Comput. 17 (1995), 65–87.

[10] B.L. Hartnell, *Well-Covered Graphs*, J. Combin. Math. Combin. Comput. 29 (1999), 107–115.

[11] B.L. Hartnell, *On the local structure of well-covered graphs without 4-cycles*, Ars Combin. 45 (1997), 77–86.

[12] M. Pinter, W_2 *graphs and strongly well-covered graphs: two well-covered graph subclasses*, Ph.D. thesis, Vanderbilt University, (1991).

[13] M. Pinter, *Planar regular one-well-covered graphs*, Cong. Numer. 91 (1992), 159–187.

[14] M. Pinter, *A class of planar well-covered graphs with girth four*, J. Graph Theory 19 (1995), 69–81.

[15] M. Pinter, *A class of well-covered graphs with girth four*, Ars Combin. 45 (1997), 241–255.

[16] M.D. Plummer, *Some covering concepts in graphs*, J. Combin., Theory 8 (1970), 91–98.

[17] M.D. Plummer, *Well-covered graphs: a survey*, Quaestiones Math. 16 (1993), 253–287.

[18] E. Prisner, J. Topp and P.D. Vestergaard, *Well-covered simplicial, chordal and circular arc graphs*, J. Graph Theory 21 (1996), 113–119.

[19] R.S. Sankaranarayana and L.K. Stewart, *Complexity results for well-covered graphs*, Networks 22 (1992), 247–262.

[20] J. Staples, *On some subclasses of well-covered graphs*, Ph.D. thesis, Vanderbilt University, (1975).

[21] J. Staples, *On some subclasses of well-covered graphs*, J. Graph Theory 3 (1979), 197–204.

[22] D. Tankus and M. Tarsi, *Well-covered claw-free graphs*, J. Combin. Theory Ser. B 66 (1996), 293–302.

[23] D. Tankus and M. Tarsi, *The structure of well-covered graphs and the complexity of their recognition*, J. Combin. Theory Ser. B 69 (1997), 230–233.

[24] C.A. Whitehead, *A characterization of well-covered claw-free graphs containing no 4-cycles*, Ars Combin. 39 (1995), 189–198.

B.L. Hartnell
Saint Mary's University
Halifax, Canada B3H 3C3
e-mail: `bert.hartnell@smu.ca`

A Graph-theoretical Generalization of Berge's Analogue of the Erdős-Ko-Rado Theorem

A.J.W. Hilton and C.L. Spencer

Abstract. A family $\mathcal{A}$ of r-subsets of the vertex set $V(G)$ of a graph G is *intersecting* if any two of the r-subsets have a non-empty intersection. The graph G is r-*EKR* if a largest intersecting family $\mathcal{A}$ of independent r-subsets of $V(G)$ may be obtained by taking all independent r-subsets containing some particular vertex.

In this paper, we show that if G consists of one path P raised to the power $k_0 \geq 1$, and s cycles $_1C, _2C, \ldots, _sC$ raised to the powers $k_1, k_2, \ldots, k_s$ respectively, with

$$\min\left(\omega(_1C^{k_1}), \omega(_2C^{k_2}), \ldots, \omega(_sC^{k_s})\right) \geq \omega(P^{k_0}) \geq 2$$

where $\omega(H)$ denotes the clique number of H, and if G has an independent r-set (so r is not too large), then G is r-EKR. An intersecting family of the largest possible size may be found by taking all independent r-subsets of $V(G)$ containing one of the end-vertices of the path.

1. Introduction

We first discuss the Erdős-Ko-Rado theorem, Berge's analogue of it, and a recent further analogue due to Talbot. Then we show that all three can be presented in a unified way as being a property of some relevant graph. Then we give a much more general analogue, extending Berge's result.

1.1. The Erdős-Ko-Rado theorem

The Erdős-Ko-Rado (EKR) theorem [6] of 1961 states that if $\mathcal{A}$ is a family of r-subsets of $\{1, 2, \ldots, n\}$ with $r \leq n/2$ such that $\mathcal{A}$ is *intersecting* (that is $A_1, A_2 \in \mathcal{A} \Rightarrow A_1 \cap A_2 \neq \emptyset$), then $|\mathcal{A}| \leq \binom{n-1}{r-1}$. From the Hilton-Milner theorem [9] it follows that, except if $n = r/2$, the only way of obtaining the equality $|\mathcal{A}| = \binom{n-1}{r-1}$ is by taking all r-sets containing a common element (but, as Claude Berge observed to the first author, this fact can also be determined by a close examination of the original proof of the EKR theorem).

1.2. Berge's analogue of the EKR theorem

Let $X_1, X_2, \ldots, X_s$ be finite sets with $|X_i| = k_i$ $(1 \leq i \leq s)$ and $2 \leq k_1 \leq k_2 \leq \ldots \leq k_s$. In 1972 Berge considered the hypergraph, say H_0, with vertex set $X_1 \cup X_2 \cup \ldots \cup X_s$ and (hyper)edge set all $k_1, k_2, \ldots, k_s$ subsets $\{x_1, x_2, \ldots, x_s\}$ with $x_i \in X_i$ $(1 \leq i \leq s)$. The *chromatic index* $q(H)$ of a hypergraph H is the smallest number of colours needed to colour the edges of H so that no two edges with a vertex in common have the same colour. Berge [1] showed that

$$q(H_0) = k_2 k_3 \cdots k_s.$$

A corollary of this is the analogue of the EKR theorem mentioned in the title of this paper. This is that the greatest number of pairwise intersecting hyperedges in H is the same number, namely $k_2 k_3 \cdots k_s$; this number is clearly the greatest number of hyperedges containing a common vertex, i.e. the maximum degree in H. This corollary can be expressed in terms of integer sequences (e.g. [4], [5], [7], [8], [12], [14]) and in other formulations as well (e.g. [2], [11], [15]) and is a special case of Theorem 1.3.

1.3. Talbot's analogue of the EKR theorem

Very recently, in 2003, Talbot [15], investigating a problem of Holroyd [10], produced a further analogue of the EKR theorem. Considering the numbers $1, 2, \ldots, n$ in cyclic fashion, so that i and $i+1$ are adjacent $(1 \leq i \leq n-1)$ and n and 1 are adjacent, Talbot treated r-subsets of $\{1, 2, \ldots, n\}$ which are *separated*, that is no adjacent pair is in any separated r-subset. Talbot showed that if $\mathcal{A}$ is an intersecting family of separated r-subsets of $\{1, 2, \ldots, n\}$ then $|\mathcal{A}| \leq \binom{n-r-1}{r-1}$. He also characterized the families $\mathcal{A}$ for which there is equality here. Talbot's achievement in finding a proof of this was quite notable, as there seems to be no easy way of tackling this problem on the lines of the original proof [6], Katona's proof [13] or Daykin's proof [3], the three main proofs of the EKR theorem; Talbot's proof is more similar to the original proof than to the other two.

1.4. A unified viewpoint: r-EKR graphs

The EKR theorem, the corollary to Berge's theorem and Talbot's theorem can all be expressed in a very similar way in terms of graph theory. Let G be a given graph with n vertices and consider the *independent* (or stable) r-subsets of the vertex set $V(G)$ of G, that is, the r-subsets with no edge of G joining any pair of vertices. We look for an intersecting family of independent r-subsets. For the original EKR-theorem, we can take G to be the graph with $n \geq 2r$ vertices and no edges. For the corollary to Berge's theorem, we can take G to be the graph with r components, each a complete graph, the ith having order k_i. For Talbot's theorem we can take G to be an n-cycle.

We call a family $\mathcal{A}$ of independent r-subsets of $V(G)$, all containing the same vertex, say w, an r-*star*; the vertex w is called the *star centre*. We call a graph G r-EKR if some largest intersecting family of independent r-subsets of $V(G)$ is an r-star. We call G *strictly r-EKR* if every largest intersecting family of independent r-sets is an r-star. The EKR theorem, the corollary to Berge's theorem, and Talbot's theorem may all be expressed by saying that the relevant graph is r-EKR.

We mention that Talbot actually proved more, namely that the kth power of an n-cycle is r-EKR if $k \geq 1$, $r \geq 1$ and $n \geq r(k+1)$. He also showed exactly when C_n^k is strictly r-EKR.

Not all graphs need be r-EKR. For a much fuller discussion of this, see Holroyd and Talbot [12]. A simple example of a graph which is not r-EKR is provided, paradoxically, by the graph with n vertices and no edges when $n < 2r$. Then every r-set intersects every other r-set. Another simple example is provided by the graph G in Figure 1. This graph is not 3-EKR. A largest 3-star that can

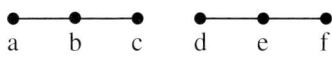

FIGURE 1

be obtained is clearly $\{acd, ace, acf, adf\}$, which has four members. Yet a largest intersecting family of independent 3-sets is $\{acd, ace, acf, adf, cdf\}$, which has five members.

We draw attention to the following interesting conjecture of Holroyd and Talbot. Let,

$$\mu(G) = \min\{|I| : I \text{ is a maximal independent subset of } V(G)\}.$$

Conjecture 1.1. *If $1 \leq r \leq \mu/2$, then G is r-EKR.*

1.5. Further extensions of Berge's analogue

Our main result, Theorem 1.3, generalizes Berge's theorem as well as a number of generalizations of Berge's theorem due to Gronau [8], Meyer [14], Deza and Frankl [4], Bollobás and Leader [2], culminating in the following theorem of Holroyd, Spencer and Talbot [11].

Theorem 1.2. *Let $t \geq r \geq 1$ and let G be a graph with t components, each being a complete graph of order at least two (the complete graphs not necessarily being of the same order). Then G is r-EKR, and a largest star may be found by taking the star centre to be a vertex in a complete graph of smallest order.*

The requirement in Theorem 1.2 that the components have order at least two is essential (apart from the fact (not observed by Holroyd, Spencer and Talbot) that we can permit one complete graph to be an isolated vertex.) In the extreme case, when all the components are isolated vertices, we are in the situation described in the EKR-theorem, where we needed the extra requirement that $r \leq t/2$ for G to be r-EKR.

1.6. Our further extension of Berge's Theorem

We let $\omega(G)$ be the clique number of a graph G, that is the largest order of a complete subgraph of G. Note that the formulae for the clique numbers of P_n^k and C_n^k, where P_n and C_n are the path and cycle respectively with n vertices, are

$$\omega(P_n^k) = \begin{cases} k+1 & \text{if } n \geq k+1, \\ n & \text{if } n \leq k, \end{cases}$$

and

$$\omega(C_n^k) = \begin{cases} k+1 & \text{if } n \geq 2k+2, \\ n & \text{if } n \leq 2k+1. \end{cases}$$

Our main result concerns a graph G consisting of cycles $_1C, _2C, \ldots, _sC$, raised to the powers $k_1, k_2, \ldots, k_s$ respectively and a path P raised to the power k_0. We let $c_i = |V(_iC)|$ and $p = |V(P)|$ and we let

$$\kappa_i = \begin{cases} \lfloor c_i/(k_i+1) \rfloor & \text{if } c_i \geq k_i+1, \\ 1 & \text{if } 2 \leq c_i \leq k_i+1. \end{cases}$$

We shall denote this graph G by $G(c_1^{k_1}, c_2^{k_2}, \ldots, c_s^{k_s}, p^{k_0})$. Our main result is:

Theorem 1.3. *Let $s \geq 0, p \geq 1$ and $c_i \geq 2$ ($1 \leq i \leq s$). Let*

$$\min\left(\omega(_1C^{k_1}), \omega(_2C^{k_2}), \ldots, \omega(_sC^{k_s})\right) \geq \max(\omega(P^{k_0}), 2), \tag{1}$$

and let

$$1 \leq r \leq \left(\sum_{i=1}^{s} \kappa_i\right) + \left\lceil \frac{p}{k_0+1} \right\rceil.$$

Then $G(c_1^{k_1}, c_2^{k_2}, \ldots, c_s^{k_s}, p^{k_0})$ is r-EKR. An r-star of maximum size may be obtained by taking all independent r-subsets of $V(G)$ containing one of the end vertices of the path P.

It is not hard to verify that Condition (1) is equivalent to the following Condition (2).

$$\min\left(\min_{1 \leq i \leq s}(k_i+1, c_i)\right) \geq \max\left(\min(k_0+1, p), 2\right). \tag{2}$$

Thus we have:

Lemma 1.4. *Conditions (1) and (2) are equivalent.*

In Theorem 1.3, we include K_2's as cycles (degenerate cycles!), so that the equation $\omega(_iC^{k_i}) = 2$ is permitted for any value of i, $1 \leq i \leq s$. The theorem remains true in this case, and the proof is considerably simplified. The theorem becomes untrue if we go further and include K_1's as (degenerate) cycles as well.

Graphs G consisting of powers of one path and several cycles may well be r-EKR even if $\omega(P^{k_0}) > \min_{1 \leq i \leq s} \omega(_iC^{k_i})$, but it is not clear to the authors where the star centre of a largest star might be.

A Generalization of Berge's Analogue of the EKR Theorem   229

The curious term $\max(\omega(P^{k_0}), 2)$ in Theorem 1.3 is there simply to take account of the fact that the cycles $_iC$ $(1 \leq i \leq s)$ all have to have length at least two, whereas the path can just have length 1.

Our proof of Theorem 1.3 is inspired by Talbot's clever proof. It takes Theorem 1.2 as its starting point.

1.7. Notation

Given the path P and the cycles $_1C, _2C, \ldots, _sC$, we let $p = |V(P)|$, $c_i = |V(_iC)|$, $\pi = c_1 + c_2 + \cdots + c_s$ and $n = p + c_1 + c_2 + \cdots + c_s (= p + \pi)$. We shall suppose that the vertices of $_iC$ are $c_1 + c_2 + \cdots + c_{i-1} + 1, \ldots, c_1 + c_2 + \cdots c_i$ and that the vertices $c_1 + c_2 + \cdots + c_{i-1} + j - 1$ and $c_1 + c_2 + \cdots + c_{i-1} + j$ are adjacent in $_iC$ $(1 \leq i \leq s, 2 \leq j \leq c_i)$ and that $c_1 + c_2 + \cdots + c_{i-1} + 1$ and $c_1 + c_2 + \cdots + c_i$ are adjacent in $_iC$. We shall suppose that the vertices of P are $\pi + 1, \pi + 2, \ldots, \pi + p(= n)$. The graph G described in Theorem 1.3 has cycles $_1C, _2C, \ldots _sC$ raised to the powers $k_1, k_2, \ldots, k_s$ respectively, and a path P raised to the power k_0; we shall suppose that G has vertex set $\{1, 2, \ldots, n\}$, and shall denote G by $G(c_1^{k_1}, c_2^{k_2}, \ldots, c_s^{k_s}, p^{k_0})$.

We let $\mathcal{I}^{(r)}$ or $\mathcal{I}^{(r)}(G)$ denote the set of all independent r-sets of G, and let $\mathcal{I}_a^{(r)}$ or $\mathcal{I}_a^{(r)}(G)$ denote the set of all independent r-sets of G containing some vertex $a \in V(G)$.

We shall use the letter a for an end vertex of the path P.

2. Proof of Theorem 1.3

The proof proceeds through a number of lemmas and sublemmas. Throughout $\mathcal{A}$ will be an intersecting family of independent r-subsets of

$$V(G(c_1^{k_1}, c_2^{k_2}, \ldots, c_s^{k_s}, p^{k_0})).$$

Lemma 2.1. *Theorem 1.3 is true for any graph which is the union of the k_0th power of a path and s cycles, where the ith cycle is raised to the power k_i, if*

(1) *the length of the path is at least 2 and at most $k_0 + 1$,*
(2) *for $1 \leq i \leq s$, the length of the ith cycle is at least 2 and at most $2k_i + 1$, and*
(3) *the clique number of the power of the path is not more than the smallest clique number of the powers of the cycles.*

Proof. In this case, the power of the path and the various powers of the cycles are cliques, and then Theorem 1.3 reduces to Theorem 1.2. □

Lemma 2.2. *Theorem 1.3 is true if $p = k_0 + 1$, $r = \sum_{i=1}^{r} \kappa_i + \lceil p/(k_0 + 1) \rceil$, and condition (1) is satisfied.*

Before proving Lemma 2.2, let us introduce another piece of terminology. Consider a bijection $\mu : \{1, 2, \ldots, n\} \to \{1, 2, \ldots, n\}$ given by:

$$\begin{aligned}
\mu(c_1 + c_2 + \cdots + c_{i-1} + j) &= c_1 + c_2 + \cdots + c_{i-1} + j + 1 & (1 \leq j < c_i, \\
& & 1 \leq i \leq s), \\
\mu(c_1 + c_2 + \cdots + c_i) &= c_1 + c_2 + \cdots + c_{i-1} + 1 & (1 \leq i \leq s), \\
\mu(\pi + j) &= \pi + j + 1 & (1 \leq j < p), \\
\mu(n) &= \pi + 1.
\end{aligned}$$

We call μ a *clockwise rotation*.

Proof of Lemma 2.2. Since $p = k_0+1$, $\lceil p/(k_0+1)\rceil = 1$. Let $r = \sum_{i=1}^{s} \kappa_i + 1$. Then r is the largest possible cardinality that an independent set can have; moreover any independent r-set must contain exactly one vertex of P.

By condition (2), $c_i \geq p = k_0 + 1$ ($1 \leq i \leq s$), so for any independent r-set A, the intersecting family $\mathcal{A}$ will contain at most one of $A, \mu(A), \mu^2(A), \ldots, \mu^{k_0}(A)$. Therefore $|\mathcal{A}| \leq |\mathcal{I}^{(r)}|/(k_0+1)$. But since $\mathcal{I}^{(r)} = \mathcal{I}^{(r)}_{\pi+1} \cup \cdots \cup \mathcal{I}^{(r)}_{\pi+p}$ and $\mathcal{I}^{(r)}_k \cap \mathcal{I}^{(r)}_l = \emptyset$ ($\pi+1 \leq k < l \leq \pi+p$), it follows that $|\mathcal{I}^{(r)}| = p|\mathcal{I}^{(r)}_{\pi+1}| = (k_0+1)|\mathcal{I}^{(r)}_{\pi+1}|$, so that $|\mathcal{A}| \leq |\mathcal{I}^{(r)}_{\pi+1}|$, which proves Lemma 2.2. $\square$

Lemma 2.3. *Theorem 1.3 is true if $|V(P)| = k_0 + 1$, $1 \leq r \leq \sum_{i=1}^{s} \kappa_i + 1$, and condition (1) is satisfied.*

Proof. In view of Lemma 2.1, we may assume that $_iC^{k_i}$ is not a complete graph for at least one i, $1 \leq i \leq s$. Without loss of generality, assume that $c_1 > \max(3, \omega(P^{k_0})) = \max(3, k_0 + 1)$. In particular, this implies that $c_1 \geq 4$ if $k_1 = 1$ and $c_1 \geq 2k_1 + 2$ if $k_1 \geq 2$ (since $_1C^{k_1}$ is a complete graph if $c_1 \leq 2k_1 + 1$). It also implies that $C_{c_1}^{k_1}$ contains a K_p. Notice that $C_{c_1-1}^{k_1}$ and $C_{c_1-k_1-1}^{k_1}$ also contain a K_p; this is obvious if neither of these is a complete graph, but if, for example, $c_1 = 2k_1 + 2$, then $K_{c_1-k_1-1}^{k_1}$ is a $K_{k_1+1} \supset K_{k_0+1} = K_p$.

We use induction on c_1 and, in particular, we shall assume that Lemma 2.3 is true for $c_1 - 1$ and $c_1 - 2$. Lemma 2.1 provides the base step for our induction hypothesis. In view of Lemma 2.2, we may assume that $r < \sum_{i=1}^{s} \kappa_i + 1$.

Define the function $f : \{1, 2, \ldots, n\} \to \{1, 2, \ldots, n-1\}$ by

$$f(j) = \begin{cases} 1 & \text{if } j = 1, \\ j - 1 & \text{if } 2 \leq j \leq n. \end{cases}$$

We shall need the following very easy sublemmas:

Sublemma 2.3.1. *If $\mathcal{G}$ is an intersecting family, then so is $f(\mathcal{G})$.*

Sublemma 2.3.2. *If A and B are independent r-subsets of $G(c_1^{k_1}, c_2^{k_2}, \ldots, c_s^{k_s}, p^{k_0})$ and $A \neq B$, then, for $1 \leq j \leq k_s$, $f^j(A) = f^j(B) \Rightarrow A \triangle B = \{c, d\}$ for some c, d with $1 \leq c < d \leq j + 1$.*

Consider the following partition of our intersecting family $\mathcal{A}$ of independent r-subsets of G:
$$\mathcal{A} = \mathcal{B} \cup \mathcal{C} \cup \left(\bigcup_{i=0}^{k_1} \mathcal{D}_i\right),$$
where
$$\mathcal{B} = \left\{A \in \mathcal{A} : 1 \notin A \text{ and } f(A) \in \mathcal{I}^{(r)}\left(G\left((c_1-1)^{k_1}, c_2^{k_2}, \ldots, c_s^{k_s}, p^{k_0}\right)\right)\right\},$$
$$\mathcal{C} = \left\{A \in \mathcal{A} : 1 \in A \text{ and } f(A) \in \mathcal{I}^{(r)}\left(G\left((c_1-1)^{k_1}, c_2^{k_2}, \ldots, c_s^{k_s}, p^{k_0}\right)\right)\right\},$$
$$\mathcal{D}_0 = \{A \in \mathcal{A} : 1, k_1+2 \in A\}$$
$$\mathcal{D}_i = \{A \in \mathcal{A} : c_1+1-i, k_1+2-i \in B\} \quad (1 \leq i \leq k_1).$$

Since $f(\mathcal{B}) \cup f(\mathcal{C}) = f(\mathcal{B} \cup \mathcal{C})$, and since, by Sublemma 2.3.1, $f(\mathcal{B} \cup \mathcal{C})$ is an intersecting family of independent r-subsets of $\mathcal{I}^{(r)}\left(G\left((c_1-1)^{k_1}, c_2^{k_2}, \ldots, c_s^{k_s}, p^{k_0}\right)\right)$, it follows by induction that
$$|f(\mathcal{B}) \cup f(\mathcal{C})| \leq \left|\mathcal{I}_a^{(r)}\left(G\left((c_1-1)^{k_1}, c_2^{k_2}, \ldots, c_s^{k_s}, p^{k_0}\right)\right)\right|. \tag{3}$$

It follows from Sublemma 2.3.2 with $j = 1$ that $|f(\mathcal{B})| = |\mathcal{B}|$ (as no set in $\mathcal{B}$ contains 1) and $|f(\mathcal{C})| = |\mathcal{C}|$ (as no set in $\mathcal{C}$ contains 2). Therefore $|\mathcal{B}| + |\mathcal{C}| = |f(\mathcal{B})| + |f(\mathcal{C})|$. Consequently
$$|\mathcal{B}| + |\mathcal{C}| = |f(\mathcal{B}) \cup f(\mathcal{C})| + |f(\mathcal{B}) \cap f(\mathcal{C})|.$$

Let
$$\mathcal{E} = f(\mathcal{B}) \cap f(\mathcal{C}).$$

Then, by (3),
$$|\mathcal{B}| + |\mathcal{C}| \leq |\mathcal{E}| + \left|\mathcal{I}_a^{(r)}\left(G\left((c_1-1)^{k_1}, c_2^{k_2}, \ldots, c_s^{k_s}, p^{k_0}\right)\right)\right|. \tag{4}$$

For any family $\mathcal{G}$ of sets, let $\mathcal{G} - \{1\} = \{G \setminus \{1\} : G \in \mathcal{G}\}$. Define,
$$\mathcal{F} = \left(f^{k_1-1}(\mathcal{E} - \{1\})\right) \cup \left(\bigcup_{i=0}^{k_1} \left(f^{k_1}(\mathcal{D}_i) - \{1\}\right)\right).$$

Note that if $E \in \mathcal{E}$ then $E = f(C)$ for some $C \in \mathcal{C}$, so $1 \in E$. Also note that if $D \in \mathcal{D}_i$ for some i, $0 \leq i \leq k_1$, then $1 \in f^{k_1}(D)$. Therefore $\mathcal{F}$ is a family of $(r-1)$-subsets. The family $\mathcal{F}$ has many further properties given in the following sublemmas.

Sublemma 2.3.3.

(1) $\mathcal{F}$ is a family of independent $(r-1)$-subsets of
$$V\left(G\left((c_1-k_1)^{k_1}, c_2^{k_2}, c_3^{k_3}, \ldots, c_s^{k_s}, p^{k_0}\right)\right).$$

(2) $f^{k_1}(\mathcal{D}_0 - \{1\}), f^{k_1}(\mathcal{D}_1 - \{1\}), \ldots, f^{k_1}(\mathcal{D}_k - \{1\})$ and $f^{k_1-1}(\mathcal{E} - \{1\})$ are pairwise disjoint families of sets.

(3) $\mathcal{F}$ is intersecting.

(4) $f(\mathcal{F})$ is a family of independent $(r-1)$-subsets of
$$V\left(G\left((c_1-k_1-1)^{k_1}, c_2^{k_2}, c_3^{k_3}, \ldots, c_s^{k_s}, p^{k_0}\right)\right).$$

Sublemma 2.3.4. With $p = k_0 + 1$,
$$\left|\mathcal{I}_a^{(r)}\left(G(c_1^{k_1}, c_2^{k_2}, \ldots, c_s^{k_s}, p^{k_0})\right)\right| = \left|\mathcal{I}_a^{(r)}\left(G\left((c_1-1)^{k_1}, c_2^{k_2}, \ldots, c_s^{k_s}, p^{k_0}\right)\right)\right|$$
$$+ \left|\mathcal{I}_a^{(r-1)}\left(G\left((c_1-k_1-1)^{k_1}, c_2^{k_2}, \ldots, c_s^{k_s}, p^{k_0}\right)\right)\right|$$

By Sublemma 2.3.2, f^{k_1} acts as an injective mapping on $\mathcal{D}_i$ ($0 \leq i \leq k_1$), so
$$|\mathcal{D}_i| = |f^{k_1}(\mathcal{D}_i)| \quad (0 \leq i \leq k_1). \tag{5}$$

By Sublemma 2.3.2 again, f^{k_1} also acts as an injective mapping on $\mathcal{C}$, so f^{k_1-1} acts injectively on $\mathcal{E}$, and so
$$|\mathcal{E}| = |f^{k_1-1}(\mathcal{E})|. \tag{6}$$

By Sublemma 2.3.3(2) it follows that $|\mathcal{F}| = |f^{k_1-1}(\mathcal{E})| + \sum_{i=0}^{k_1} |f^{k_1}(\mathcal{D}_i)|$. Therefore, by (5) and (6),
$$|\mathcal{F}| = |\mathcal{E}| + \sum_{i=0}^{k_1} |\mathcal{D}_i|. \tag{7}$$

As no set in $\mathcal{F}$ contains the vertex 1, the map $f : \mathcal{F} \to f(\mathcal{F})$ is bijective, so
$$|\mathcal{F}| = |f(\mathcal{F})|. \tag{8}$$

By Sublemma 2.3.3(3) $\mathcal{F}$ is intersecting, so by Sublemma 2.3.2, $f(\mathcal{F})$ is also intersecting. By Sublemma 2.3.3(4), $f(\mathcal{F})$ is a family of independent $(r-1)$-subsets of $V\left(G\left((c_1-k_1-1)^{k_1}, c_2^{k_2}, \ldots, c_s^{k_s}, p^{k_0}\right)\right)$. Therefore, by induction,
$$|f(\mathcal{F})| \leq \left|\mathcal{I}_a^{(r-1)}\left(G\left((c_1-k_1-1)^{k_1}, c_2^{k_2}, \ldots, c_s^{k_s}, p^{k_0}\right)\right)\right|. \tag{9}$$

Therefore, using (3), (7), (8) and (9),
$$\begin{aligned}
|\mathcal{A}| &= |\mathcal{B}| + |\mathcal{C}| + \sum_{i=0}^{k_1} |\mathcal{D}_i| \\
&= |f(\mathcal{B}) \cup f(\mathcal{C})| + |\mathcal{E}| + \sum_{i=0}^{k_1} |\mathcal{D}_i| \\
&= |f(\mathcal{B}) \cup f(\mathcal{C})| + |\mathcal{F}| \\
&= |f(\mathcal{B}) \cup f(\mathcal{C})| + |f(\mathcal{F})| \\
&\leq \left|\mathcal{I}_a^{(r)}\left(G\left((c_1-1)^{k_1}, c_2^{k_2}, \ldots, c_s^{k_s}, p^{k_0}\right)\right)\right| \\
&\quad + \left|\mathcal{I}_a^{(r-1)}\left(G\left((c_1-k_1-1)^{k_1}, c_2^{k_2}, \ldots, c_s^{k_s}, p^{k_0}\right)\right)\right|.
\end{aligned}$$

Therefore, by Sublemma 2.3.4,
$$|\mathcal{A}| \leq \left|\mathcal{I}_a^{(r)}\left(G\left(c_1^{k_1}, c_2^{k_2}, \ldots, c_s^{k_s}, p^{k_0}\right)\right)\right|.$$
Lemma 2.3 now follows by induction on c_1. □

Lemma 2.4. *Theorem 1.3 is true if $1 \leq |V(P)| \leq k_0 + 1$, $1 \leq r \leq \sum_{i=1}^{s} \kappa_i + \lceil p/(k_0+1) \rceil$, and Condition 1 is satisfied.*

Proof. From Lemma 2.3 we know that $G(c_1^{k_1}, c_2^{k_2}, \ldots, c_s^{k_s}, p^{k_0})$ is r-EKR if $|V(P)| = k_0 + 1$, $1 \leq r \leq (\sum_{i=1}^{s} \kappa_i) + 1$ and Condition (1) is satisfied, and that an r-star of maximum size can be found by taking all independent r-sets containing an endpoint a of P (in fact, since P^{k_0} is a complete graph, any vertex of P could be the centre of a suitable r-star).

If several vertices of P^{k_0} are removed leaving at least one vertex, say w, the number of independent r-sets centred on w remains unaltered. Lemma 2.4 follows. □

The rest of the proof of Theorem 1.3 bears a close resemblance to the proof of Lemma 2.3, and is similarly modelled on Talbot's proof of his separated sets result in [15]. We still need to show that, with the cycle powers fixed, we can "grow" the length of the path, P, to the required value p.

We argue by induction on p. The basis for the induction is provided by Lemma 2.4 which established Theorem 1.3 whenever $1 \leq r \leq (\sum_{i=1}^{s} \kappa_i) + 1$, Condition (1) is satisfied, and $1 \leq |V(P)| \leq k_0 + 1$. Recall that the vertices of P are labelled $\pi + 1, \pi + 2, \ldots, \pi + p$.

Consider the following partition of our intersecting family of $\mathcal{A}$ independent r-sets:
$$\mathcal{A} = \mathcal{Q} \cup \mathcal{R} \cup \mathcal{S}_0,$$
where
$$\mathcal{Q} = \left\{A \in \mathcal{A} : \pi + 1 \notin A \text{ and } g(A) \in \mathcal{I}_a^{(r)}\left(G(c_1^{k_1}, c_2^{k_2}, \ldots, c_s^{k_s}, (p-1)^{k_0})\right)\right\},$$
$$\mathcal{R} = \left\{A \in \mathcal{A} : \pi + 1 \in A \text{ and } g(A) \in \mathcal{I}_a^{(r)}\left(G(c_1^{k_1}, c_2^{k_2}, \ldots, c_s^{k_s}, (p-1)^{k_0})\right)\right\},$$
$$\mathcal{S}_0 = \{A \in \mathcal{A} : \pi + 1, \pi + k_0 + 2 \in A\}.$$
Define the function $g : \{1, 2, \ldots, n\} \to \{1, 2, \ldots, n-1\}$ by
$$g(j) = \begin{cases} j & \text{if } 1 \leq j \leq \pi + 1, \\ j - 1 & \text{if } \pi + 2 \leq j \leq \pi + p. \end{cases}$$
The analogues of Sublemmas 2.3.1 and 2.3.2 are:

Sublemma 2.4.1. *If $\mathcal{G}$ is an intersecting family, then so is $g(\mathcal{G})$.*

Sublemma 2.4.2. *If $\mathcal{A}$ and $\mathcal{B}$ are independent r-subsets of $G\left(c_1^{k_1}, c_2^{k_2}, \ldots, c_s^{k_s}, p^{k_0}\right)$ and $\mathcal{A} \neq \mathcal{B}$, then, for $1 \leq j \leq k_0$,*
$$g^j(\mathcal{A}) = g^j(\mathcal{B}) \Rightarrow \mathcal{A} \Delta \mathcal{B} = \{c, d\}$$
for some c, d with $\pi + 1 \leq c < d \leq \pi + 1 + j$.

Since $g(\mathcal{Q}) \cup g(\mathcal{R}) = g(\mathcal{Q} \cup \mathcal{R})$ and since, by Sublemma 2.4.1, $g(\mathcal{Q} \cup \mathcal{R})$ is an intersecting family of independent r-subsets of $\mathcal{I}_a^{(r)}\left(G\left(c_1^{k_1}, c_2^{k_2}, \ldots, c_s^{k_s}, (p-1)^{k_0}\right)\right)$, it follows by induction that

$$|g(\mathcal{Q}) \cup g(\mathcal{R})| \leq \left|\mathcal{I}_a^{(r)}\left(G\left(c_1^{k_1}, c_2^{k_2}, \ldots, c_s^{k_s}, (p-1)^{k_0}\right)\right)\right|. \tag{10}$$

It follows by Sublemma 2.4.2 with $j = 1$ that $|g(\mathcal{Q})| = |\mathcal{Q}|$ (as no set in $\mathcal{Q}$ contains $\pi + 1$) and that $|g(\mathcal{R})| = |\mathcal{R}|$ (as no set in $\mathcal{R}$ contains $\pi + 2$). Therefore $|\mathcal{Q}| + |\mathcal{R}| = |g(\mathcal{Q})| + |g(\mathcal{R})|$. Consequently

$$|\mathcal{Q}| + |\mathcal{R}| = |g(\mathcal{Q}) \cup g(\mathcal{R})| + |g(\mathcal{Q}) \cap g(\mathcal{R})|.$$

Let $\mathcal{T} = g(\mathcal{Q}) \cap g(\mathcal{R})$. Then, by (10),

$$|\mathcal{Q}| + |\mathcal{R}| \leq |\mathcal{T}| + \left|\mathcal{I}_a^{(r)}\left(G\left(c_1^{k_1}, c_2^{k_2}, \ldots, c_s^{k_s}, (p-1)^{k_0}\right)\right)\right|. \tag{11}$$

Define,

$$\mathcal{U} = \left(g^{k_0-1}(\mathcal{T} - \{\pi + 1\})\right) \cup \left(g^{k_0}(\mathcal{S}_0) - \{\pi + 1\}\right).$$

Note that if $T \in \mathcal{T}$ then $T = g(R)$ for some $R \in \mathcal{R}$, so $\pi + 1 \in T$. Also note that if $S \in \mathcal{S}_0$ then $\pi + 1 \in g^{k_0}(S)$. Therefore $\mathcal{U}$ is a family of $(r-1)$-subsets. The family $\mathcal{U}$ has the following further properties:

Sublemma 2.4.3.

(1) $\mathcal{U}$ *is a family of independent $(r-1)$-subsets of*
$$V\left(G\left(c_1^{k_1}, c_2^{k_2}, \ldots, c_s^{k_s}, (p-1)^{k_0}\right)\right),$$
(2) $g^{k_0}(\mathcal{S}_0 - \{\pi + 1\})$ *and* $g^{k_0-1}(\mathcal{T} - \{\pi + 1\})$ *are disjoint families of sets,*
(3) $\mathcal{U}$ *is intersecting,*
(4) $g(\mathcal{U})$ *is a family of intersecting $(r-1)$-subsets of*
$$V\left(G\left(c_1^{k_1}, c_2^{k_2}, \ldots, c_s^{k_s}, (p-k_0-1)^{k_0}\right)\right).$$

By Sublemma 2.4.2, g^{k_0} acts injectively on $\mathcal{S}_0$, so

$$|\mathcal{S}_0| = |g^{k_0}(\mathcal{S}_0)|. \tag{12}$$

Again, by Sublemma 2.4.2, g^{k_0} acts injectively on $\mathcal{R}$, and so g^{k_0-1} acts injectively on $\mathcal{T}$, and so

$$|\mathcal{T}| = \left|g^{k_0-1}(\mathcal{T})\right| \tag{13}$$

By Sublemma 2.4.3(2) it follows that

$$|\mathcal{U}| = \left|g^{k_0-1}(\mathcal{T})\right| + \left|g^{k_0}(\mathcal{S}_0)\right|.$$

Therefore, by (12) and (13),
$$|\mathcal{U}| = |\mathcal{T}| + |\mathcal{S}_0|. \tag{14}$$

As no set in $\mathcal{U}$ contains 1, the map $g : \mathcal{U} \to g(\mathcal{U})$ is injective, so
$$|\mathcal{U}| = |g(\mathcal{U})|. \tag{15}$$

By Sublemma 2.4.3(3), $\mathcal{U}$ is intersecting, so by Sublemma 2.4.1, $g(\mathcal{U})$ is also intersecting. By Sublemma 2.4.3(4), $g(\mathcal{U})$ is a family of independent $(r-1)$-subsets of $V\left(G\left(c_1^{k_1}, c_2^{k_2}, \ldots, c_s^{k_s}, (p-k_0-1)^{k_0}\right)\right)$. Therefore, by induction,
$$|g(\mathcal{U})| \leq \left|\mathcal{I}_a^{(r-1)}\left(G\left(c_1^{k_1}, c_2^{k_2}, \ldots, c_s^{k_s}, (p-k_0-1)^{k_0}\right)\right)\right|. \tag{16}$$

Therefore, using (10),(14),(15) and (16),
$$\begin{aligned}
|\mathcal{A}| &= |\mathcal{Q}| + |\mathcal{R}| + |\mathcal{S}_0| \\
&= |g(\mathcal{Q}) \cup g(\mathcal{R})| + |\mathcal{T}| + |\mathcal{S}_0| \\
&= |g(\mathcal{Q}) \cup g(\mathcal{R})| + |\mathcal{U}| \\
&= |g(\mathcal{Q}) \cup g(\mathcal{R})| + |g(\mathcal{U})| \\
&\leq \left|\mathcal{I}_a^{(r)}\left(G\left(c_1^{k_1}, c_2^{k_2}, \ldots, c_s^{k_s}, (p-1)^{k_0}\right)\right)\right| \\
&\quad + \left|\mathcal{I}_a^{(r-1)}\left(G\left(c_1^{k_1}, c_2^{k_2}, \ldots, c_s^{k_s}, (p-k_0-1)^{k_0}\right)\right)\right|.
\end{aligned}$$

We now need the following sublemma.

Sublemma 2.4.4. *If $p \geq k_0 + 2$ then*
$$\left|\mathcal{I}_a^{(r)}\left(G(c_1^{k_1}, c_2^{k_2}, \ldots, c_s^{k_s}, p^{k_0})\right)\right| = \left|\mathcal{I}_a^{(r)}\left(G\left(c_1^{k_1}, c_2^{k_2}, \ldots, c_s^{k_s}, (p-1)^{k_0}\right)\right)\right|$$
$$+ \left|\mathcal{I}_a^{(r-1)}\left(G\left(c_1^{k_1}, c_2^{k_2}, \ldots, c_s^{k_s}, (p-k_0-1)^{k_0}\right)\right)\right|.$$

Using this, it now follows that
$$|\mathcal{A}| \leq \left|\mathcal{I}_a^{(r)}\left(G\left(c_1^{k_1}, c_2^{k_2}, \ldots, c_s^{k_s}, p^{k_0}\right)\right)\right|.$$

Thus $\mathcal{G}$ is r-EKR. Theorem 1.3 now follows by induction on p.

3. Proofs of the lemmas

In this section we prove those lemmas used in the proof of Theorem 1.3 which still await a proof. We only give a proof of Sublemmas 2.3.1, 2.3.2, 2.3.3 and 2.3.4 because, for $1 \leq x \leq 4$, the proof of Sublemma 2.4.x is either virtually the same, or is a considerable simplification of the proof of Sublemma 2.3.x.

Proof of Sublemma 2.3.1. If $\mathcal{G}$ is an intersecting family and $A, B \in f(\mathcal{G})$, then there exist $C, D \in \mathcal{G}$ such that $A = f(C)$ and $B = f(D)$. Then $\emptyset \neq f(C \cap D) \subseteq f(C) \cap f(D) = A \cap B$. Thus $f(\mathcal{G})$ is intersecting. □

Proof of Sublemma 2.3.2. Let $A, B \in \mathcal{I}_a^{(r)}\left(G(c_1^{k_1}, c_2^{k_2}, \ldots, c_s^{k_s}, p^{k_0})\right)$ with $A \neq B$ but $f^j(A) = f^j(B)$ for some j, $1 \leq j \leq k_1$. If $2 \leq a \leq c_1 - j$ then $a \in f^j(A) \Leftrightarrow a + j \in A$. Hence

$$A \cap \{j+2, j+3, \ldots, c_1\} = B \cap \{j+2, j+3, \ldots, c_1\}.$$

So as $f^j(A) = f^j(B)$ but $A \neq B$, there exist $c, d \in \{1, 2, \ldots, j+1\}$ such that $c \in A$ and $d \in B$, say. But since $j \leq k_1$ we have that $A \cap \{1, 2, \ldots, j+1\} = \{c\}$ and $B \cap \{1, 2, \ldots, j+1\} = \{d\}$. Thus $A \triangle B = \{c, d\}$. □

Proof of Sublemma 2.3.3 (1). We have already that

$$\mathcal{F} = \left(f^{k_1-1}(\mathcal{E} - \{1\})\right) \cup \left(\bigcup_{i=0}^{k_1} \left(f^{k_i}(\mathcal{D}_i) - \{1\}\right)\right)$$

is a family of $(r-1)$-sets. We have to show that the sets are independent. There are three cases.

First consider the sets in $f^{k_1-1}(\mathcal{E} - \{1\})$. Let $H \in f^{k_1-1}(\mathcal{E})$. Then there exists $E \in \mathcal{E}$ such that $f^{k_1-1}(E) = H$, and, as $\mathcal{E} = f(\mathcal{B}) \cap f(\mathcal{C})$, there also exists $B \in \mathcal{B}$ and $C \in \mathcal{C}$ such that $f(B) = f(C) = E$. By Sublemma 2.3.2 with $j = 1$, we know that one of the sets B, C contains 1 and the other 2, and, by the definition of $\mathcal{B}$ and $\mathcal{C}$, we have that $1 \in C$, so $1 \in H$ and $2 \in B$. Moreover, $C \cap \{c_1 - k_1 + 1, \ldots, c_1\} = \emptyset$ so $E \cap \{c_1 - k_1, \ldots, c_1\} = \emptyset$. Therefore $H \cap \{c_1 - 2k_1 + 1, \ldots, c_1\} = \emptyset$. Since $2 \in B$ it follows that $E \cap \{2, \ldots, k_1 + 1\} = \emptyset$. It now follows that

$$H \cap (\{c_1 - 2k_1 + 1, \ldots, c_1 - k_1\} \cup \{1, 2\}) = \{1\}, \qquad (17)$$

and thence that $H - \{1\} \in f^{k_1-1}(\mathcal{E} - \{1\})$ is an independent $(r-1)$-subset of

$$V\left(H\left((c_1 - k_1^{k_1}), c_2^{k_2}, \ldots, c_s^{k_s}, p^{k_0}\right)\right).$$

Next suppose that $H \in f^{k_1}(\mathcal{D}_0)$. Then there exists $D \in \mathcal{D}_0$ such that $H = f^{k_1}(D)$, and, as $D \in \mathcal{D}_0$, $1, k_1 + 2 \in D$. Thus $1 \in H$ and

$$H \cap (\{c_1 - 2k_1 + 1, \ldots, c_1 - k_1\} \cup \{1, \ldots, k_1 + 2\}) = \{1, 2\} \qquad (18)$$

and so $H - \{1\}$ is an independent $(r-1)$-subset of

$$V\left(H\left((c_1 - k_1)^{k_1}, c_2^{k_2}, \ldots, c_s^{k_s}, p^{k_0}\right)\right).$$

Finally let $H \in f^{k_1}(\mathcal{D}_i)$ for some i, $1 \leq i \leq k_1$. Then there exists a $D \in \mathcal{D}_i$ with $H = f^{k_1}(D)$, and, as $D \in \mathcal{D}_i$, $c_1 + 1 - i, k_1 + 2 - i \in D$. Since $k_1 \geq i \geq 1$, $f^{k_1}(k_1 + 2 - i) = 1$, so $1 \in H$. Therefore

$$H \cap (\{c_1 - i - 2k_1 + 1, \ldots, c_1 - k_1\} \cup \{1, \ldots, k_1 + 2 - i\}) = \{c_1 - i - k_1 + 1, 1\}. \quad (19)$$

Hence $H - \{1\}$ is an independent $(r-1)$-subset of

$$V\left(H\left((c_1 - k_1)^{k_1}, c_2^{k_2}, \ldots, c_s^{k_s}, p^{k_0}\right)\right).$$

Sublemma 2.3.3(1) now follows from (17), (18) and (19). □

Proof of Sublemma 2.3.3 (2). To show that these families are pairwise disjoint, we consider how the members of each family intersect the set $\{c_1 - 2k_1 + 1, \ldots, c_1 - k_1\} \cup \{2\}$. Let $H \in f^{k_1-1}(\mathcal{E}) - \{1\}$. From (17) we have that

$$H \cap (\{c_1 - 2k_1 + 1, \ldots, c_1 - k_1\} \cup \{2\}) = \emptyset.$$

Next let $H \in f^{k_1}(\mathcal{D}_0)$. From (18) it follows that

$$H \cap (\{c_1 - 2k_1 + 1, \ldots, c_1 - k_1\} \cup \{2\}) = \{2\}.$$

Finally let $H \in f^{k_1}(\mathcal{D}_i) - \{1\}$ for some i, $1 \leq i \leq k_1$. From (19) we have

$$H \cap (\{c_1 - 2k_1 + 1, \ldots, c_1 - k_1\} \cup \{2\}) = \{c_1 - i - k_1 + 1\}.$$

Hence the families are pairwise disjoint. □

Proof of Sublemma 2.3.3 (3). Let $A, B \in \mathcal{F}$. First suppose that $A, B \in f^{k_1}(\mathcal{D}_i) - \{1\}$ for some i, $0 \leq i \leq k_1$. Then there exist $D', D'' \in \mathcal{D}_i$ with $A = f^{k_1}(D') - \{1\}$ and $B = f^{k_1}(D'') - \{1\}$. If $i = 0$ then $k+2 \in D' \cap D''$ and so $2 = f^{k_1}(k_1+2) \in A \cap B$. If $1 \leq i \leq k_1$ then $c_1 + 1 - i \in D' \cap D''$, so $c_1 + 1 - i - k_1 \in A \cap B$.

Now suppose that $A, B \in f^{k_1}(\mathcal{E}) - \{1\}$. Then there exist $E', E'' \in \mathcal{E}$ with $A = f^{k_1-1}(E') - \{1\}$ and $B = f^{k_1-1}(E'') - \{1\}$. As $E', E'' \in \mathcal{E} = f(\mathcal{B}) \cap f(\mathcal{C})$, it follows that there are $B_1 \in \mathcal{B}$ and $C_1 \in \mathcal{C}$ such that $f(B_1) = E'$ and $f(C_1) = E''$. As $B_1, C_1 \in \mathcal{A}$, we have that $B_1 \cap C_1 \neq \emptyset$. By the definitions of $\mathcal{B}$ and $\mathcal{C}$, and by Sublemma 2.3.2 with $j = 1$, we have that $1 \in C_1$ and $2 \in B_1$. It follows that, for some $j \geq k_1 + 3$, $j \in B_1 \cap C_1$. Therefore $3 \leq f^{k_1}(j) \in A \cap B$.

Next suppose that $0 \leq i < j \leq k_1$ and $A \in f^{k_1}(\mathcal{D}_i) - \{1\}$ and $B \in f^{k_1}(\mathcal{D}_j) - \{1\}$. In this case there exist $D' \in \mathcal{D}_i$ and $D'' \in \mathcal{D}_j$ with $A = f^{k_1}(D') - \{1\}$ and $B = f^{k_1}(D'') - \{1\}$. This implies that $D' \cap \{1, 2, \ldots, k_1 + 2\} = \{k_1 + 2 - j\}$ and $D'' \cap \{1, 2, \ldots, k_1 + 2\} = \{k_1 + 2 - i\}$, where $2 \leq k_1 + 2 - j < k_1 + 2 - i$. But $D', D'' \in \mathcal{A}$ which is intersecting, so $D' \cap D'' \neq \emptyset$. Therefore there is some $l \geq k_1 + 3$ with $l \in D' \cap D''$. Then $3 \leq f^{k_1}(l) \in A \cap B$.

Finally suppose that $A \in f^{k_1-1} - \{1\}$ and $B \in f^{k_1}(\mathcal{D}_i) - \{1\}$. In this case there exist $D \in \mathcal{D}_i$ and $E \in \mathcal{E}$ with $B = f^{k_1}(D) - \{1\}$ and $A = f^{k_1-1}(E) - \{1\}$, and, since $\mathcal{E} = f(\mathcal{B}) \cap f(\mathcal{C})$, there exist $B_1 \in \mathcal{B}$ and $C_1 \in \mathcal{C}$ such that $E = f(B_1) = f(C_1)$. From Sublemma 2.3.2 with $j = 1$ it follows that $1 \in C_1$ and $2 \in B_1$, so $B_1 \cap \{1, 2, \ldots, k_1 + 2\} = \{2\}$ and $C_1 \cap \{1, 2, \ldots, k_1 + 1\} = \{1\}$. Also, from the definition of $\mathcal{D}_i$ we have that

$$D \cap \{1, 2, \ldots, k_1 + 2\} = \begin{cases} \{1, k_1 + 2\} & \text{if } i = 0, \\ \{k_1 + 2 - i\} & \text{if } 1 \leq i \leq k_1. \end{cases}$$

As B_1, C_1, D are all elements of $\mathcal{A}$, we know that $B_1 \cap D$ and $C_1 \cap D$ are both non-empty. If $i = 0$ then there exists some $j \geq 2k_1 + 3$ such that $j \in B_1 \cap D$. Otherwise we have that $1 \leq i \leq k_1$, and in this case there exists some $j_1 \geq 2k_1 + 2 - i + 1 \geq k_1 + 3$ with $j \in C_1 \cap D$. Hence $3 \leq f^{k_1}(j) \in A \cap B$. □

Proof of Sublemma 2.3.3 (4). From Sublemma 2.3.3(1) we know that $\mathcal{F}$ is a family of independent $(r-1)$-subsets of $V\left(G\left((c_1-k_1)^{k_1}, c_2^{k_2}, \ldots, c_s^{k_s}, p^{k_0}\right)\right)$. Let $F \in \mathcal{F}$ and consider $f(\mathcal{F})$. Clearly $f(F)$ is an $(r-1)$-subset of

$$V\left(G\left((c_1-k_1-1)^{k_1}, c_2^{k_2}, \ldots, c_s^{k_s}, p^{k_0}\right)\right).$$

We just need to check that $f(\mathcal{F})$ is an independent set. Since F was an independent $(r-1)$-subset of $V\left(G\left((c_1-k_1)^{k_1}, c_2^{k_2}, \ldots, c_s^{k_s}, p^{k_0}\right)\right)$, the only way that $f(\mathcal{F})$ could fail to be an independent $(r-1)$-subset of

$$V\left(G\left((c_1-k_1-1)^{k_1}, c_2^{k_2}, \ldots, c_s^{k_s}, p^{k_0}\right)\right)$$

is if F contains one of the following pairs of elements:

$$(c_1-2k_1+1, 2), (c_1-2k_1+2, 3), \ldots, (c_1-k_1, k_1+1), (1, k_1+2).$$

The vertex 1 has been removed from every set in $\mathcal{F}$ so the last pair $(1, k_1+2)$ cannot be contained in $\mathcal{F}$. If $F \in f^{k_1-1}(\mathcal{E}) - \{1\}$ then by (17) (as the H there is in $f^{k-1}(\mathcal{E})$) it follows that

$$F \cap \{c_1-2k_1+1, \ldots, c_1-k_1\} = \emptyset.$$

This also follows from (18) if $F \in f^{k_1}(\mathcal{D}_0) - \{1\}$ (as the H in (18) is in $f^{k_1}(\mathcal{D}_0)$). It remains to check what happens if $F \in f^{k_1}(\mathcal{D}_i) - 1$ for some i, $1 \leq i \leq k_1$. In this case it follows from (19) that

$$F \cap (\{c_1-2k_1+1, \ldots, c_1-k_1\} \cup \{1, \ldots, k_1+2-i\}) = \{c_1-i-k_1+1\} \quad (20)$$

(as the H in (19) is in $f^{k_1}(\mathcal{D}_i)$). Note that all pairs of vertices in the list other than the excluded pair $(1, k_1+2)$ are of the form $(c_1-2k_1+j, j+1)$. Since we have $c_1-k_1-i+1 = c_1-2k_1+(k_1-i+1) \in F$ it follows from (20) that $(k_1-i+1)+1 = k_1-i+2 \notin F$. Thus F cannot contain any of the pairs of vertices in the list. □

Proof of Sublemma 2.3.4. To prove this we let

$$\mathcal{A} = \mathcal{I}_a^{(r)}\left(G\left(c_1^{k_1}, c_2^{k_2}, \ldots, c_s^{k_s}, p^{k_0}\right)\right)$$

and follow the line of reasoning in the induction step in the proof of Lemma 2.3. We may suppose here that the end-vertex a of the path is the vertex n.

From the definitions of $\mathcal{B}$ and $\mathcal{C}$ it follows that

$$f(\mathcal{B}) \cup f(\mathcal{C}) = \mathcal{I}_a^{(r)}\left(G\left((c_1-1)^{k_1}, c_2^{k_2}, \ldots, c_s^{k_s}, p^{k_0}\right)\right),$$

so that (3) holds with equality. We therefore have that

$$|\mathcal{B}| + |\mathcal{C}| = \left|\mathcal{I}_a^{(r)}\left(G\left((c_1-1)^{k_1}, c_2^{k_2}, \ldots, c_s^{k_s}, p^{k_0}\right)\right)\right| + |\mathcal{E}|$$

so that (4) holds with equality.

Since $\mathcal{A}$ is partitioned into $\mathcal{B}, \mathcal{C}, \mathcal{D}_0, \ldots, \mathcal{D}_k$, it follows that

$$\left|\mathcal{I}_a^{(r)}\left(G\left(c_1^{k_1}, c_2^{k_2}, \ldots, c_s^{k_s}, p^{k_0}\right)\right)\right|$$
$$-\left|\mathcal{I}_a^{(r)}\left(G\left((c_1-1)^{k_1}, c_2^{k_2}, \ldots, c_s^{k_s}, p^{k_0}\right)\right)\right| = |\mathcal{E}| + \sum_{i=0}^{k_1} |\mathcal{D}_i| \quad (21)$$
$$= |\mathcal{F}|, \text{ by (7)},$$
$$= |f(\mathcal{F})|, \text{ by (8)}.$$

From Sublemma 2.3.3(4) we know that $f(\mathcal{F})$ is a family of independent $(r-1)$-subsets of $V\left(G\left((c_1-k_1-1)^{k_1}, c_2^{k_2}, \ldots, c_s^{k_s}, p^{k_0}\right)\right)$. Since $f(\mathcal{F})$ is a subfamily of $f^{k_1+1}(\mathcal{A})$, it follows that every independent $(r-1)$-set in $f(\mathcal{F})$ contains the end-vertex of P^{k_0}, and thus in $\mathcal{I}_a^{(r-1)}\left(G\left((c_1-k_1-1)^{k_1}, c_2^{k_2}, \ldots, c_s^{k_s}, p^{k_0}\right)\right)$ it follows that

$$|f(\mathcal{F})| \leq \mathcal{I}_a^{(r-1)}\left(G\left((c_1-k_1-1)^{k_1}, c_2^{k_2}, \ldots, c_s^{k_s}, p^{k_0}\right)\right). \quad (22)$$

For a family $\mathcal{G}$ of subsets of $\{1, 2, \ldots, n\}$, let $\mathcal{G}+\{i\}$ denote the family $\{G \cup \{i\} : G \in \mathcal{G}\}$.

Now consider the "reverse" map $f^{-(k_1+1)} : \{1, 2, \ldots, n-k_1-1\} \to \{k_1+2, \ldots, n\}$ given by $f^{-(k_1+1)}(j) = k_1+1+j$. Under this map the independent $(r-1)$-sets in $\mathcal{I}_a^{(r-1)}\left(G\left((c_1-k_1-1)^{k_1}, c_2^{k_2}, \ldots, c_s^{k_s}, p^{k_0}\right)\right)$ are taken to $\mathcal{E}' \cup \sum_{i=0}^{k_1} \mathcal{D}'_i$, where $\mathcal{E}' \cup \{2\} \subseteq \mathcal{B}$, $\mathcal{E}' \cup \{1\} \subseteq \mathcal{C}$ (and $f(\mathcal{E}') \subseteq \mathcal{E} - \{1\}$) and $\mathcal{D}'_i + \{k_1+2-i\} \subseteq \mathcal{D}_i$ ($1 \leq i \leq k_1$) and $\mathcal{D}'_0 + \{1\} \subseteq \mathcal{D}_0$. Let us describe this in more detail. Consider an independent $(r-1)$-set S in $\mathcal{I}_a^{(r-1)}\left(G\left((c_1-k_1-1)^{k_1}, c_2^{k_2}, \ldots, c_s^{k_s}, p^{k_0}\right)\right)$.

(a) If S contains the vertex 1 (so does not contain any vertex in $\{c_1-2k_1, c_1-2k_1+1, \ldots, c_1-k_1-1\}$), then $f^{-(k_1+1)}(S)$ contains the vertex k_1+2 and does not contain any of the vertices in $\{c_1-k_1+1, \ldots, c_1\} \cup \{1, \ldots, k_1+1\}$. We let

$$\mathcal{D}'_0 = \left\{f^{-(k_1+1)}(S) : 1 \in S \text{ and} \right.$$
$$\left. S \in \mathcal{I}_a^{(r-1)}\left(G\left((c_1-k_1-1)^{k_1}, c_2^{k_2}, \ldots, c_s^{k_s}, p^{k_0}\right)\right)\right\};$$

then $\mathcal{D}'_0 \cup \{1\} \subseteq \mathcal{D}_0$.

(b) If S contains a vertex c_1-k_1-i for some i, $1 \leq i \leq k_1$ (and so does not contain any vertex in the set $\{c_1-k_1-i+1, \ldots, c_1-k_1-1\} \cup \{1, 2, \ldots, k_1+1-i\}$), then $f^{-(k_1+1)}(S)$ contains the vertex c_1-i+1 and does not contain any of the vertices in $\{c_1-i+2, \ldots, c_1\} \cup \{1, 2, \ldots, 2k_1+2-i\}$. For $1 \leq i \leq k_1$, we let

$$\mathcal{D}'_i = \left\{f^{-(k_1+1)}(S) : c_1-k_1-i \in S \text{ and} \right.$$
$$\left. S \in \mathcal{I}_a^{(r-1)}\left(G\left((c_1-k_1-1)^{k_1}, c_2^{k_2}, \ldots, c_s^{k_s}, p^{k_0}\right)\right)\right\};$$

then $\mathcal{D}'_i + \{k_1+2-i\} \subseteq \mathcal{D}_i$.

(c) If S contains no vertex from the set $\{c_1 - 2k_1, \ldots, c_1 - k_1 - 1\} \cup \{1\}$, then $f^{-(k_1+1)}(S)$ contains no vertex from the set $\{c_1 - k_1 + 1, \ldots, c_1\} \cup \{1, 2, \ldots, k_1 + 2\}$. We let

$$\mathcal{E}' = \left\{ f^{-(k_1+1)}(S) : S \cap (\{c_1 - 2k_1, \ldots, c_1 - k_1 - 1\} \cup \{1\}) = \emptyset \text{ and} \right.$$
$$\left. S \in \mathcal{I}_a^{(r-1)} \left(G \left((c_1 - k_1 - 1)^{k_1}, c_2^{k_2}, \ldots, c_s^{k_s}, p^{k_0} \right) \right) \right\}.$$

Then $\mathcal{E}' \cup \{2\} \subset \mathcal{B}$ and $\mathcal{E}' \cup \{1\} \subseteq \mathcal{C}$, so $\mathcal{E}' \subseteq \mathcal{E}$. It follows that

$$\left| \mathcal{I}_a^{(r-1)} \left(G \left((c_1 - k_1 - 1)^{k_1}, c_2^{k_2}, \ldots, c_s^{k_s}, p^{k_0} \right) \right) \right| \leq |\mathcal{E}'| + \sum_{i=0}^{k_1} |\mathcal{D}'_i|. \quad (23)$$

The family $\mathcal{E}' \cup \sum_{i=0}^{k_1} \mathcal{D}'_i$ has the properties that

$$|\mathcal{E}'| + \sum_{i=0}^{k_1} |\mathcal{D}'_i| \leq |\mathcal{E}| + \sum_{i=0}^{k_1} |\mathcal{D}_i| = |\mathcal{F}| = |f(\mathcal{F})|.$$

It therefore follows from (22) and (23) that

$$|f(\mathcal{F})| = \left| \mathcal{I}_a^{(r-1)} \left(G \left((c_1 - k_1 - 1)^{k_1}, c_2^{k_2}, \ldots, c_s^{k_s}, p^{k_0} \right) \right) \right|. \quad (24)$$

The equality we wish for now follows from (21) and (24). $\square$

4. Final remarks

There are a number of operations which can be used to obtain new r-EKR graphs from old. The first is described by the following lemma of Holroyd, Spencer and Talbot [11]. Let $N(v)$ denote the neighborhood of v, that is $N(v) = \{w : w \in V(G) \text{ and } vw \in E(G)\}$.

Lemma 4.1. *Let G be an r-EKR graph with a star centre v. If $S \subset N(v)$ then $G - S$ is also r-EKR with a star centre v.*

By applying this to the graph $G(c_1^{k_1}, c_2^{k_2}, \ldots, c_s^{k_s}, p^{k_0})$ in the case where $p \geq k_0 + 2$ and $S = N(n)$ (n being the end vertex of the path P), we obtain the following theorem.

Theorem 4.2. *Let $s \geq 0$, $p \geq 1$ and $c_i \geq 2$ ($1 \leq i \leq s$). Let G be a graph consisting of cycles $_1C, _2C, \ldots, _sC$ raised to the powers $k_1, k_2, \ldots, k_s$ respectively, a path P raised to the power k_0, and an isolated vertex. Let $c_i = |V(_iC)|$ ($1 \leq i \leq s$) and $p = |V(P)| + k_0 + 1$. Also let*

$$1 \leq r \leq \sum_{i=1}^{s} \left\lfloor \frac{c_i}{k_i + 1} \right\rfloor + \left\lceil \frac{p}{k_0 + 1} \right\rceil.$$

Let

$$\min \left(\omega(_1C^{k_1}), \omega(_2C^{k_2}), \ldots, \omega(_sC^{k_s}) \right) \geq k_0 + 1.$$

Then G is r-EKR with the isolated vertex w as star centre.

It is worth remarking that in a similar vein Holroyd, Spencer and Talbot in [11] showed that if G is a graph with q components being paths, cycles, complete graphs, and at least one isolated vertex, and if $q \geq 2r$, then G is r-EKR.

Finally we make two further comments.

(1) If $w \in N(v^\star)$, where $v^\star$ is a star centre of an r-EKR graph G, then it is clear that the addition of any edge wv produces a further graph that is r-EKR with star centre $v^\star$.
(2) If G is an r-EKR graph, then we can introduce a further vertex w and join it to each vertex of G, and by this means produce a further r-EKR graph. Conversely, if G is an r-EKR graph, and G contains a vertex w which is joined to all other vertices, then $G - w$ is also r-EKR.

References

[1] C. Berge, *Nombres de coloration de l'hypergraphe h-parti complet*, Hypergraph Seminar (Proc. First Working Sem., Ohio State Univ., Columbus, Ohio, 1972), Springer, Berlin, 1974, pp. 13–20. Lecture Notes in Math., Vol. 411.

[2] B. Bollobás and I. Leader, *An Erdős-Ko-Rado theorem for signed sets*, Comput. Math. Appl. **34** (1997), no. 11, 9–13, Graph theory in computer science, chemistry, and other fields (Las Cruces, NM, 1991).

[3] D.E. Daykin, *Erdős-Ko-Rado from Kruskal-Katona*, J. Combinatorial Theory Ser. A **17** (1974), 254–255.

[4] M. Deza and P. Frankl, *Erdős-Ko-Rado theorem – 22 years later*, SIAM J. Algebraic Discrete Methods **4** (1983), no. 4, 419–431.

[5] K. Engel and P. Frankl, *An Erdős-Ko-Rado theorem for integer sequences of given rank*, European J. Combin. **7** (1986), no. 3, 215–220.

[6] P. Erdős, Chao Ko, and R. Rado, *Intersection theorems for systems of finite sets*, Quart. J. Math. Oxford Ser. (2) **12** (1961), 313–320.

[7] P. Frankl and Z. Füredi, *The Erdős-Ko-Rado theorem for integer sequences*, SIAM J. Algebraic Discrete Methods **1** (1980), no. 4, 376–381.

[8] H.-D.O.F. Gronau, *More on the Erdős-Ko-Rado theorem for integer sequences*, J. Combin. Theory Ser. A **35** (1983), no. 3, 279–288.

[9] A.J.W. Hilton and E.C. Milner, *Some intersection theorems for systems of finite sets*, Quart. J. Math. Oxford Ser. (2) **18** (1967), 369–384.

[10] F.C. Holroyd, *Problem 338 (BCC16.25), Erdős-Ko-Rado at the court of King Arthur*, Discrete Math. **197/198** (1999), 812.

[11] F.C. Holroyd, C. Spencer, and J. Talbot, *Compression and Erdős-Ko-Rado graphs*, Discrete Math. (to appear).

[12] F.C. Holroyd and J. Talbot, *Graphs with the Erdős-Ko-Rado property*, Discrete Math. (to appear).

[13] G.O.H. Katona, *A simple proof of the Erdős-Chao Ko-Rado theorem*, J. Combinatorial Theory Ser. B **13** (1972), 183–184.

[14] J.C. Meyer, *Quelques problèmes concernant les cliques des hypergraphes h-complets et q-parti h-complets*, Hypergraph Seminar (Proc. First Working Sem., Ohio State Univ., Columbus, Ohio, 1972), Springer, Berlin, 1974, pp. 127–139. Lecture Notes in Math., Vol. 411.

[15] J. Talbot, *Intersecting families of separated sets*, J. London Math. Soc. (2) **68** (2003), no. 1, 37–51.

A.J.W. Hilton
Department of Mathematics
University of Reading
Whiteknights
Reading RG6 6AX, UK
e-mail: `a.j.w.hilton@rdg.ac.uk`

C.L. Spencer
Department of Mathematics
University of Reading
Whiteknights
Reading RG6 6AX, UK
e-mail: `c.l.spencer@rdg.ac.uk`

Independence Polynomials and the Unimodality Conjecture for Very Well-covered, Quasi-regularizable, and Perfect Graphs

Vadim E. Levit and Eugen Mandrescu

Abstract. If s_k denotes the number of stable sets of cardinality k in the graph G, then $I(G;x) = \sum_{k=0}^{\alpha} s_k x^k$ is the *independence polynomial* of G (Gutman, Harary, 1983), where $\alpha = \alpha(G)$ is the size of a maximum stable set in G. Alavi, Malde, Schwenk and Erdös (1987) conjectured that $I(T,x)$ is unimodal for every tree T, while, in general, they proved that for each permutation π of $\{1, 2, \ldots, \alpha\}$ there is a graph G with $\alpha(G) = \alpha$ such that $s_{\pi(1)} < s_{\pi(2)} < \cdots < s_{\pi(\alpha)}$. Brown, Dilcher and Nowakowski (2000) conjectured that $I(G;x)$ is unimodal for well-covered graphs. Michael and Traves (2003) provided examples of well-covered graphs with non-unimodal independence polynomials. They proposed the so-called *"roller-coaster"* conjecture: for a well-covered graph, the subsequence $(s_{\lceil \alpha/2 \rceil}, s_{\lceil \alpha/2 \rceil+1}, \ldots, s_\alpha)$ is unconstrained in the sense of Alavi *et al.* The conjecture of Brown *et al.* is still open for very well-covered graphs, and it is worth mentioning that, apart from K_1 and the chordless cycle C_7, connected well-covered graphs of girth ≥ 6 are very well covered (Finbow, Hartnell and Nowakowski, 1993).

In this paper we prove that $s_{\lceil (2\alpha-1)/3 \rceil} \geq \cdots \geq s_{\alpha-1} \geq s_\alpha$ are valid for (a) bipartite graphs; (b) quasi-regularizable graphs on 2α vertices.

In particular, we infer that these inequalities are true for (a) trees, thus doing a step in an attempt to prove the conjecture of Alavi *et al.*; (b) very well-covered graphs. Consequently, for the latter case, the unconstrained subsequence appearing in the roller-coaster conjecture can be shortened to $(s_{\lceil \alpha/2 \rceil}, s_{\lceil \alpha/2 \rceil+1}, \ldots, s_{\lceil (2\alpha-1)/3 \rceil})$. We also show that the independence polynomial of a very well-covered graph G is unimodal for $\alpha \leq 9$, and is log-concave whenever $\alpha \leq 5$.

Mathematics Subject Classification (2000). Primary 05C69, 05C17; Secondary 05A20, 11B83.

Keywords. Stable set, independence polynomial, unimodal sequence, quasi-regularizable graph, perfect graph, tree, very well-covered graph.

1. Introduction

Throughout this paper $G = (V, E)$ is a simple (i.e., a finite, undirected, loopless and without multiple edges) graph with vertex set $V = V(G)$ and edge set $E = E(G)$. If $X \subset V$, then $G[X]$ is the subgraph of G spanned by X. By $G - W$ we mean the subgraph $G[V - W]$, if $W \subset V(G)$. We also denote by $G - F$ the partial subgraph of G obtained by deleting the edges of F, for $F \subset E(G)$, which we abbreviate by $G - e$ whenever $F = \{e\}$.

A vertex v is *pendant* if its neighborhood $N(v) = \{u : u \in V, uv \in E\}$ contains only one vertex; an edge $e = uv$ is *pendant* if one of its endpoints is a pendant vertex.

$\overline{G}$ stands for the complement of G, while K_n, P_n, C_n denote respectively, the complete graph on $n \geq 1$ vertices, the chordless path on $n \geq 1$ vertices, and the chordless cycle on $n \geq 3$ vertices. As usual, a *tree* is an acyclic connected graph.

A set of pairwise non-adjacent vertices is called *stable*. If S is a stable set, then we denote $N(S) = \{v : N(v) \cap S \neq \emptyset\}$ and $N[S] = N(S) \cup S$. A stable set of maximum size will be referred to as a *maximum stable set* of G. The *stability number* of G, denoted by $\alpha(G)$, is the cardinality of a maximum stable set in G, and $\omega(G) = \alpha(\overline{G})$. A graph G is called *perfect* if $\chi(H) = \omega(H)$ for every induced subgraph H of G, where $\chi(H)$ denotes the chromatic number of H (Berge, [3]).

The *dis*joint union of the graphs G_1, G_2 is the graph $G = G_1 \cup G_2$ having as vertex set and edge set the disjoint unions of $V(G_1), V(G_2)$ and $E(G_1), E(G_2)$, respectively.

If G_1, G_2 are disjoint graphs, then their *Zykov sum* (Zykov, [40], [41]) is the graph $G_1 + G_2$ with

$$V(G_1 + G_2) = V(G_1) \cup V(G_2),$$
$$E(G_1 + G_2) = E(G_1) \cup E(G_2) \cup \{v_1 v_2 : v_1 \in V(G_1), v_2 \in V(G_2)\}.$$

In particular, $\cup nG$ and $+nG$ denote the disjoint union and Zykov sum, respectively, of $n > 1$ copies of the graph G.

According to Berge, a graph G is called *quasi-regularizable* if one can replace each edge of G with a non-negative integer number of parallel copies, so as to obtain a regular multigraph of degree $\neq 0$ (see [4], [5]).

Evidently, a disconnected quasi-regularizable graph has no isolated vertices. Moreover, a disconnected graph is quasi-regularizable if and only if each of its connected components spans a quasi-regularizable graph.

FIGURE 1. G_1 is a non-quasi-regularizable graph, while G_2 is a quasi-regularizable graph.

The following characterization of quasi-regularizable graphs, due to Berge, we shall use in the sequel.

Theorem 1.1. [4] *A graph G is quasi-regularizable if and only if $|S| \leq |N(S)|$ holds for any stable set S of G.*

Let s_k be the number of stable sets in G, of cardinality $k \in \{0, 1, \ldots, \alpha(G)\}$. The polynomial

$$I(G; x) = s_0 + s_1 x + s_2 x^2 + \cdots + s_\alpha x^\alpha, \ \alpha = \alpha(G),$$

is called the *independence polynomial* of G (Gutman and Harary, [19]). Various properties of this polynomial are presented in a number of papers, like [2], [6], [7], [8], [11], [15], [16], [17], [18], [20], [22], [29], [31], [38], [39].

A finite sequence of real numbers $(a_0, a_1, a_2, \ldots, a_n)$ is said to be:

- *unimodal* if there is some $k \in \{0, 1, \ldots, n\}$, called the *mode* of the sequence, such that $a_0 \leq \cdots \leq a_{k-1} \leq a_k \geq a_{k+1} \geq \cdots \geq a_n$;
- *log-concave* if $a_i^2 \geq a_{i-1} \cdot a_{i+1}$ holds for $i \in \{1, 2, \ldots, n-1\}$.

It is well known that every log-concave sequence of positive numbers is also unimodal.

A polynomial is called *unimodal (log-concave)* if the sequence of its coefficients is unimodal (log-concave), respectively. For instance, the independence polynomial $I(K_{1,3}; x) = 1 + 4x + 3x^2 + x^3$ is log-concave and hence unimodal, as well. However, the independence polynomial of $G = K_{24} + (K_3 \cup K_3 \cup K_4)$ is not unimodal, since $I(G; x) = 1 + 34x + 33x^2 + 36x^3$ (for other examples, see [1] and [30]). Moreover, Alavi et al. [1] proved that for any permutation π of $\{1, 2, \ldots, \alpha\}$ there is a graph G with $\alpha(G) = \alpha$ such that

$$s_{\pi(1)} < s_{\pi(2)} < \cdots < s_{\pi(\alpha)}.$$

Nevertheless, for trees, they stated the following conjecture.

Conjecture 1.2. [1] *The independence polynomial of a tree is unimodal.*

A graph G is called *well-covered* if all its maximal stable sets are of the same cardinality, (Plummer, [35], [36]). If, in addition, G has no isolated vertices and its order $|V(G)|$ equals $2\alpha(G)$, then G is *very well covered* (Favaron, [13]). Berge proved that every well-covered graph is quasi-regularizable (see [4], [5]).

By G^* we mean the graph obtained from G by appending a single pendant edge to each vertex of G (Dutton, Chandrasekharan, and Brigham, [12]).

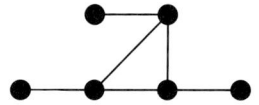

FIGURE 2. The graph K_3^*.

Let us notice that G^* is very well covered (see, for instance, [23]), and $\alpha(G^*)$ is equal to $|V(G)|$. Moreover, the following theorem shows that, under certain conditions, every well-covered graph is identical to G^* for some graph G.

Theorem 1.3. [14] *Let G be a connected graph of girth ≥ 6, which is isomorphic to neither C_7 nor K_1. Then G is well covered if and only if its pendant edges form a perfect matching.*

In other words, Theorem 1.3 shows that, apart from K_1 and C_7, connected well-covered graphs of girth ≥ 6 are very well covered. For example, a tree $T \neq K_1$ is well covered if and only if it is very well covered, in which case $T = G^*$ for some tree G (see also [37], [13], [24]).

In [6] it was conjectured that the independence polynomial of every well-covered graph G is unimodal. Michael and Traves [34] proved that this conjecture is true for $\alpha(G) \in \{1, 2, 3\}$, but false for $\alpha(G) \in \{4, 5, 6, 7\}$. A family of well-covered graphs with non-unimodal independence polynomials and stability numbers ≥ 8 is presented in [30]. However, the conjecture is still open for very well-covered graphs. In [25] and [26], unimodality of independence polynomials of some very well-covered graphs (e.g., $P_n^*, K_{1,n}^*$) was verified. To prove it for P_n^*, we showed that $I(P_n^*; x)$ is equal to the independence polynomial of a claw-free graph, and then we used the following theorem, due to Hamidoune.

Theorem 1.4. [21] *The independence polynomial of any claw-free graph is log-concave and hence unimodal.*

In [27] it was demonstrated that the polynomial $I(G^*; x)$ is unimodal for each graph G^* whose skeleton G has $\alpha(G) \leq 4$, while in [31] it was shown that $I(G^*; x)$ is log-concave whenever $\alpha(G) \leq 3$.

Michael and Traves formulated (and verified for well-covered graphs with stability numbers ≤ 7) the following so-called *"roller-coaster"* conjecture.

Conjecture 1.5. [34] *For any permutation π of the set $\{\lceil \alpha/2 \rceil, \lceil \alpha/2 \rceil + 1, \ldots, \alpha\}$, there exists a well-covered graph G, with $\alpha(G) = \alpha$, whose sequence $(s_0, s_1, \ldots, s_\alpha)$ satisfies*

$$s_{\pi(\lceil \alpha/2 \rceil)} < s_{\pi(\lceil \alpha/2 \rceil + 1)} < \cdots < s_{\pi(\alpha)}.$$

In [33], Matchett showed that this conjecture is true for well-covered graphs having stability numbers ≤ 11.

In this paper we prove that if G is a quasi-regularizable graph on $2\alpha(G)$ vertices, then

$$s_{\lceil (2\alpha(G)-1)/3 \rceil} \geq s_{\lceil (2\alpha(G)-1)/3 \rceil + 1} \geq \cdots \geq s_{\alpha(G)},$$

while if G is a perfect graph, then

$$s_{\lceil (\omega\alpha-1)/(\omega+1) \rceil} \geq s_{\lceil (\omega\alpha-1)/(\omega+1) \rceil + 1} \geq \cdots \geq s_\alpha, \text{ where } \alpha = \alpha(G), \omega = \omega(G).$$

We infer that for very well-covered graphs, the domain of the roller-coaster conjecture can be shortened to

$$\{\lceil \alpha/2 \rceil, \lceil \alpha/2 \rceil + 1, \ldots, \lceil (2\alpha - 1)/3 \rceil\}.$$

Moreover, we show that the independence polynomial of a very well-covered graph G is unimodal for $\alpha(G) \leq 9$, and log-concave whenever $\alpha(G) \leq 5$.

2. Very well-covered graphs

In [6] it was shown that any well-covered graph G on n vertices enjoys the following inequalities: $s_{k-1} \leq k \cdot s_k$ and $s_k \leq (n-k+1) \cdot s_{k-1}, 1 \leq k \leq \alpha(G)$.

Proposition 2.1. [34], [28] *If G is a well-covered graph with $\alpha(G) = \alpha$, then the following statements are true:*

(i) $(\alpha - k) \cdot s_k \leq (k+1) \cdot s_{k+1}$ *holds for $0 \leq k < \alpha$;*
(ii) $s_{k-1} \leq s_k$ *for every $1 \leq k \leq (\alpha+1)/2$.*

Notice that Proposition 2.1(i) can fail for non-well-covered graphs, e.g., the graph G_1 from Figure 3 has $\alpha(G_1) = 3$ and $(3-2)\cdot s_2 = 8 > 3 = (2+1)\cdot s_3$. However, there are non-well covered graphs satisfying Proposition 2.1(i), for instance, G_2 from Figure 3. Since $I(G_1; x) = 1 + 6x + 8x^2 + x^3$ and $I(G_2; x) = 1 + 5x + 4x^2$, we see that both G_1 and G_2 satisfy Proposition 2.1(ii). On the other hand, $K_{1,3}$ does not satisfy Proposition 2.1(ii), because $I(K_{1,3}; x) = 1 + 4x + 3x^2 + x^3$, $\alpha(K_{1,3}) = 3$, while $s_1 = 4 > 3 = s_2$.

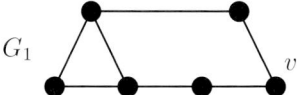
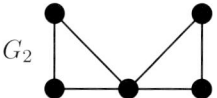

FIGURE 3. Non-well-covered graphs.

For a graph G of order n and having $\alpha(G) = \alpha$, we denote
$$\omega_{\alpha-k} = \max\{n - |N[S]| : S \text{ is a stable set with } |S| = k\}, \quad 0 \leq k \leq \alpha.$$

Clearly, $\omega_0 = 0$ and $\omega_\alpha = n$. While $\omega_1(G) \leq \omega(G)$, it is not necessary that $\omega_1(G)$ is equal to $\omega(G)$. For instance, the graph K_3^* (see Figure 2) has $\omega_1 = 2$ and $\omega(K_3^*) = 3$. It is worth mentioning that for every odd chordless cycle $C_{2n+1}, n \geq 2$, or even chordless path $P_{2n}, n \geq 2$, we have $\omega_1 = \omega$.

Lemma 2.2. *If G is a graph of order $n \geq 1$ with $\alpha(G) = \alpha$, then*
$$(k+1) \cdot s_{k+1} \leq \omega_{\alpha-k} \cdot s_k, \quad 0 \leq k < \alpha.$$
In particular, $\alpha \cdot s_\alpha \leq \omega_1 \cdot s_{\alpha-1} \leq \omega(G) \cdot s_{\alpha-1}$.

Proof. Let Ω_k be the set of all stable sets of k vertices in G. Define the bipartite graph $(\Omega_k \cup \Omega_{k+1}, \Psi)$ with partite sets Ω_k and Ω_{k+1} and an edge $X_k X_{k+1}$ in Ψ provided $X_k \subset X_{k+1}$. Now, since every stable set X_{k+1} has exactly $k+1$ subsets of cardinality k, we get $|\Psi| = (k+1)s_{k+1}$.

On the other hand, if $X \in \Omega_k$, then $X \cup \{v\} \in \Omega_{k+1}$ for every vertex v belonging to $V(G) - N[X]$, i.e., X has at most $w_{\alpha-k}$ neighbors in Ω_{k+1}. Hence, we get that
$$(k+1) \cdot s_{k+1} = |\Psi| \leq w_{\alpha-k} \cdot |\Omega_k| = w_{\alpha-k} \cdot s_k.$$
In particular, for $k = \alpha - 1$, we obtain $\alpha \cdot s_\alpha \leq w_1 \cdot s_{\alpha-1} \leq w(G) \cdot s_{\alpha-1}$. □

Let us remark that there are quasi-regularizable graphs with non-unimodal independence polynomials, e.g.,

(a) $G = K_{10} + \cup 6K_1$ is connected and has
$$I(G; x) = (1+x)^6 + 10x = 1 + \mathbf{16}x + 15x^2 + \mathbf{20}x^3 + 15x^4 + 6x^5 + x^6;$$

(b) $G = (K_{24} + \cup 6K_1) \cup (K_{25} + \cup 6K_1)$ is disconnected and has
$$\begin{aligned} I(G; x) &= \left((1+x)^6 + 24x\right)\left((1+x)^6 + 25x\right) \\ &= 1 + 61x + \mathbf{960}x^2 + 955x^3 + \mathbf{1475}x^4 + 1527x^5 \\ &\quad + 1218x^6 + 841x^7 + 495x^8 + 220x^9 + 66x^{10} + 12x^{11} + x^{12}. \end{aligned}$$

Proposition 2.3. *If G is a quasi-regularizable graph of order $n = 2\alpha(G) = 2\alpha$, then*
(i) $w_{\alpha-k} \leq 2(\alpha - k)$, $0 \leq k \leq \alpha$;
(ii) $(k+1) \cdot s_{k+1} \leq 2(\alpha - k) \cdot s_k$, $0 \leq k < \alpha$;
(iii) $s_{\lceil (2\alpha-1)/3 \rceil} \geq \cdots \geq s_{\alpha-1} \geq s_\alpha$.

Proof. (i) Let S be a stable set in G of size $k \geq 0$. From Theorem 1.1, it follows that $|S| \leq |N(S)|$, which implies $2 \cdot |S| \leq |S \cup N(S)| = |N[S]|$ and, hence,
$$2 \cdot (\alpha - k) = 2 \cdot (\alpha - |S|) \geq n - |N[S]|,$$
because $n = 2\alpha$. Consequently, we obtain $w_{\alpha-k} \leq 2(\alpha - k)$.
(ii) The result follows by combining Lemma 2.2 and part (i).
(iii) The fact that $(k+1) \cdot s_{k+1} \leq 2(\alpha - k) \cdot s_k$ implies that $s_{k+1} \leq s_k$ holds for $k + 1 \geq 2(\alpha - k)$, i.e., for $k \geq (2\alpha - 1)/3$. □

There are no quasi-regularizable graphs G of order $n > 2\alpha(G)$ that satisfy Proposition 2.3(i),(ii), since for $k = 0$, each of them demands $n \leq 2\alpha(G)$.

In addition, for the graphs G_1, G_2 in Figure 4, $I(G_1; x) = 1 + 6x + 8x^2$ and $I(G_2; x) = 1 + 8x + 19x^2 + 12x^3$ show that Proposition 2.3(iii) is sometimes, but not always, valid for a quasi-regularizable graph G on $n > 2\alpha(G)$ vertices. Notice that G_1 is also well covered, but not very well covered.

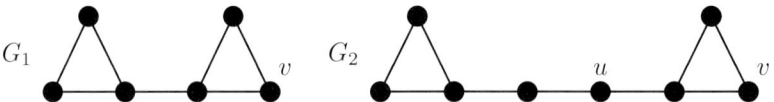

FIGURE 4. G_1, G_2 are quasi-regularizable graphs, but only G_1 is well covered.

The graph G in Figure 5 is very well covered and its independence polynomial $I(G;x) = 1 + 12x + 52x^2 + 110x^3 + 123x^4 + 70x^5 + 16x^6$ is not only unimodal but log-concave, as well.

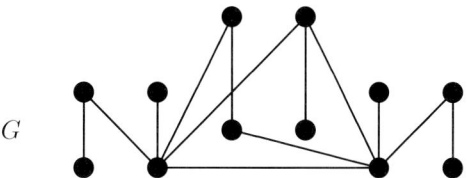

FIGURE 5. G is a very well-covered graph with a log-concave independence polynomial.

Theorem 2.4. *If G is a very well-covered graph of order $n \geq 2$ with $\alpha(G) = \alpha$, then:*
  (i) $(\alpha - k) \cdot s_k \leq (k+1) \cdot s_{k+1} \leq 2(\alpha - k) \cdot s_k, \quad 0 \leq k < \alpha;$
  (ii) $s_0 \leq s_1 \leq \cdots \leq s_{\lceil \alpha/2 \rceil}$ and $s_{\lceil (2\alpha-1)/3 \rceil} \geq \cdots \geq s_{\alpha-1} \geq s_\alpha;$
  (iii) *if $\alpha \geq 2$, then $s_{\alpha-2} \cdot s_\alpha \leq s_{\alpha-1}^2;$*
  (iv) *if $\alpha \leq 9$, then $I(G;x)$ is unimodal;*
  (v) *if $\alpha \leq 5$, then $I(G;x)$ is log-concave.*

Proof. (i) It follows from Proposition 2.1(i) and Proposition 2.3(ii), because each well-covered graph without isolated vertices is quasi-regularizable (see [4], [5]).

(ii) It is readily apparent from Proposition 2.1(ii) and Proposition 2.3(iii).

(iii) Taking $k = \alpha - 2$ in Proposition 2.1(i), we get $2 \cdot s_{\alpha-2} \leq (\alpha - 1) \cdot s_{\alpha-1}$, while substituting $k = \alpha - 1$ in part (i) assures that $\alpha \cdot s_\alpha \leq 2 \cdot s_{\alpha-1}$, which together lead to $2\alpha \cdot s_{\alpha-2} \cdot s_\alpha \leq 2(\alpha - 1) \cdot s_{\alpha-1}^2$ and, hence, $s_{\alpha-2} \cdot s_\alpha \leq s_{\alpha-1}^2$.

(iv) By part (ii), $s_0 \leq s_1 \leq \cdots \leq s_{\lceil \alpha/2 \rceil}$ and $s_{\lceil (2\alpha-1)/3 \rceil} \geq \cdots \geq s_{\alpha-1} \geq s_\alpha$. In addition, the fact that $\alpha(G) \leq 9$ ensures that $|\lceil \alpha/2 \rceil - \lceil (2\alpha - 1)/3 \rceil| \leq 1$.

(v) The result is trivial for $\alpha(G) = 1$.

Notice that
$$s_0 \cdot s_2 = |E(\overline{G})| \leq |V(G)|^2 = s_1^2$$
is true for every graph G with $\alpha(G) = \alpha \geq 2$. Moreover, $s_{\alpha-2} \cdot s_\alpha \leq s_{\alpha-1}^2$ holds by part (iii), since G is very well covered.

Now, it follows immediately that $I(G;x)$ is log-concave for $2 \leq \alpha(G) \leq 3$.

Assume that $\alpha(G) = 4$. Then part (i) implies that $3s_1 \leq 2s_2$ and $3s_3 \leq 4s_2$ and hence $s_1 \cdot s_3 \leq s_2^2$. Together with the inequalities $s_0 \cdot s_2 \leq s_1^2$ and $s_2 \cdot s_4 \leq s_3^2$ it follows that $I(G;x)$ is log-concave.

Suppose that $\alpha(G) = 5$. Then, taking $k \in \{1, 2, 3\}$ in part (i), we obtain that $4s_1 \leq 2s_2$, $3s_2 \leq 3s_3 \leq 6s_2$, and $4s_4 \leq 4s_3$, respectively. Consequently, $s_1 \cdot s_3 \leq s_2^2$ and $s_2 \cdot s_4 \leq s_3^2$. Therefore, $I(G;x)$ is log-concave, because $s_0 \cdot s_2 \leq s_1^2$ and $s_3 \cdot s_5 \leq s_4^2$ are true, as well. □

3. Perfect graphs

Lovász proved the theorem claiming that a graph G is perfect if and only if $|V(H)| \leq \alpha(H) \cdot \omega(H)$ holds for every induced subgraph H of G (see [32]).

Proposition 3.1. *If G is a perfect graph with $\alpha(G) = \alpha$ and $\omega = \omega(G)$, then*

$$s_{\lceil(\omega\alpha-1)/(\omega+1)\rceil} \geq \cdots \geq s_{\alpha-1} \geq s_\alpha.$$

Proof. Let S be a stable set in G of size $k \geq 0$. Then $H = G - N[S]$ is an induced subgraph of G and has $\alpha(H) \leq \alpha - k$. Therefore, by Lovász's theorem,

$$|V(H)| \leq \omega(H) \cdot \alpha(H) \leq \omega(H) \cdot (\alpha - k) \leq \omega \cdot (\alpha - k)$$

and, hence, $\omega_{\alpha-k} \leq \omega \cdot (\alpha - k)$. Further, according to Lemma 2.2, we obtain that

$$(k+1) \cdot s_{k+1} \leq \omega \cdot (\alpha - k) \cdot s_k, 0 \leq k < \alpha.$$

Now, $s_{k+1} \leq s_k$ is true while $k+1 \geq \omega \cdot (\alpha - k)$, i.e., for $k \geq (\omega\alpha - 1)/(\omega + 1)$. □

In fact, in Proposition 3.1 there is some k such that

$$\left\lceil \frac{\omega\alpha - 1}{\omega + 1} \right\rceil \leq k < \alpha$$

if and only if

$$\alpha - \frac{1+\alpha}{1+\omega} \leq \alpha - 1, \text{ i.e., } \alpha \geq \omega.$$

It is worth mentioning that, for general graphs, Lemma 2.2 assures that if a graph G satisfies $\omega(G) \leq \alpha = \alpha(G)$, then $s_\alpha \leq s_{\alpha-1}$. Let us notice that the converse assertion is not true, e.g., $\alpha(K_4 - e) = 2 < 3 = \omega(K_4 - e)$ and $I(K_4 - e; x) = 1 + 4x + x^2$, where by $K_4 - e$ we mean the graph obtained from K_4 by deleting one of its edges.

For non-perfect graphs, Proposition 3.1 is not necessarily false, for example, $I(C_7; x) = 1 + 7x + 14x^2 + 7x^3$. However, the graph $G = \cup 4C_5$ is not perfect, $\alpha(G) = 8, \omega(G) = 2$ and

$$\begin{aligned} I(\cup 4C_5; x) &= \left(1 + 5x + 5x^2\right)^4 = 1 + 20x + 170x^2 + 800x^3 + 2275x^4 \\ &+ 4000x^5 + \mathbf{4250}x^6 + 2500x^7 + 625x^8 \end{aligned}$$

is log-concave, but it does not satisfy Proposition 3.1, since

$$\lceil(\omega\alpha - 1)/(\omega + 1)\rceil = \lceil(2 \cdot 8 - 1)/(2 + 1)\rceil = 5 \text{ and } s_5 = 4000 < 4250 = s_6.$$

The validation of the Strong Perfect Graph Conjecture, due to Chudnovsky, Robertson, Seymour and Thomas, [9], [10], shows that the holes (i.e., $C_{2n+1}, n \geq 2$) and the antiholes (i.e., $\overline{C_{2n+1}}, n \geq 2$) are the only minimal imperfect graphs. Since both $C_{2n+1}, n \geq 2$, and $\overline{C_{2n+1}}, n \geq 2$, are claw-free graphs, we may infer that the polynomials $I(C_{2n+1}; x), I(\overline{C_{2n+1}}; x)$ are log-concave, according to Theorem 1.4.

However, there are imperfect graphs, whose independence polynomials are not unimodal, e.g., the disconnected graph $G = (K_{95} + \cup 4K_3) \cup C_5$ has

$$\begin{aligned} I(G;x) &= \left(1 + 107x + 54x^2 + 108x^3 + 81x^4\right)\left(1 + 5x + 5x^2\right) \\ &= 1 + 112x + 594x^2 + \mathbf{913}x^3 + 891x^4 + \mathbf{945}x^5 + 405x^6. \end{aligned}$$

Let $H = K_{97} + \cup 4K_3$, and G be the graph obtained from H by adding an edge that joins a vertex of K_{97} to a vertex of some C_5. Then G is a connected imperfect graph whose independence polynomial is not unimodal, because

$$\begin{aligned} I(G;x) &= \left(1 + 109x + 54x^2 + 108x^3 + 81x^4\right)\left(1 + 4x + 3x^2\right) \\ &\quad + x(1+2x)\left(1 + 108x + 54x^2 + 108x^3 + 81x^4\right) \\ &= 1 + 114x + 603x^2 + \mathbf{921}x^3 + 891x^4 + \mathbf{945}x^5 + 405x^6. \end{aligned}$$

Since every bipartite graph G is perfect and has $\omega(G) \leq 2$, we obtain the following result.

Corollary 3.2. *If G is a bipartite graph with $\alpha(G) = \alpha \geq 1$, then*
$$s_{\lceil (2\alpha-1)/3 \rceil} \geq \cdots \geq s_{\alpha-1} \geq s_\alpha.$$

In particular, we infer a similar result for trees, whose importance is significant vis-à-vis the conjecture of Alavi *et al.*

Corollary 3.3. *If T is a tree with $\alpha(T) = \alpha$, then*
$$s_{\lceil (2\alpha-1)/3 \rceil} \geq \cdots \geq s_{\alpha-1} \geq s_\alpha.$$

4. Conclusions

In this paper we prove that for very well-covered graphs the "chaotic interval"
$$(s_{\lceil \alpha/2 \rceil}, s_{\lceil \alpha/2 \rceil + 1}, \ldots, s_\alpha)$$
involved in the roller-coaster conjecture can be shortened to
$$(s_{\lceil \alpha/2 \rceil}, s_{\lceil \alpha/2 \rceil + 1}, \ldots, s_{\lceil (2\alpha-1)/3 \rceil}).$$
It seems that one can get even deeper results, by using more efficiently the power of the new defined parameters $\{w_k, 0 \leq k \leq \alpha\}$.

Based on our observations on log-concavity made in [29], [31], and this paper, we conclude with the two following conjectures sharpening the conjectures of Brown *et al.* and Alavi *et al.*, respectively.

Conjecture 4.1. $I(G;x)$ *is log-concave for any very well-covered graph G.*

Conjecture 4.2. $I(T;x)$ *is log-concave for any (well-covered) tree T.*

Acknowledgment

We would like to thank the anonymous referee for useful suggestions that helped us to improve the presentation of the paper.

References

[1] Y. Alavi, P.J. Malde, A.J. Schwenk and P. Erdös, *The vertex independence sequence of a graph is not constrained*, Congressus Numerantium **58** (1987) 15–23.

[2] J. L. Arocha, *Propriedades del polinomio independiente de un grafo*, Revista Ciencias Matematicas, vol. **V** (1984) 103–110.

[3] C. Berge, *Färbung von Graphen deren sämtliche bzw. deren ungerade Kreise starr sind (Zusammenfassung)*, Wiss. Z. Martin-Luther-Univ. Halle **10** (1961) 114–115.

[4] C. Berge, *Some common properties for regularizable graphs, edge-critical graphs and B-graphs*, in: Graph Theory and Algorithms, Lecture Notes in Computer Science **108** (1980) 108–123, Springer-Verlag, Berlin.

[5] C. Berge, *Some common properties for regularizable graphs, edge-critical graphs and B-graphs*, Annals of Discrete Mathematics **12** (1982) 31–44.

[6] J.I. Brown, K. Dilcher and R.J. Nowakowski, *Roots of independence polynomials of well-covered graphs*, Journal of Algebraic Combinatorics **11** (2000) 197–210.

[7] J.I. Brown, C.A. Hickman and R.J. Nowakowski, *On the location of roots of independence polynomials*, Journal of Algebraic Combinatorics **19** (2004) 273–282.

[8] J.I. Brown and R.J. Nowakowski, *Bounding the roots of independence polynomials*, Ars Combinatoria **58** (2001) 113–120.

[9] M. Chudnovsky, N. Robertson, P.D. Seymour and R. Thomas, *Progress on perfect graphs*, Mathematical Programming B **97** (2003) 405–422.

[10] M. Chudnovsky, N. Robertson, P.D. Seymour and R. Thomas, *The Strong Perfect Graph Theorem*, Annals of Mathematics (2004) (accepted).

[11] M. Chudnovsky and P. Seymour, *The roots of the stable set polynomial of a claw-free graph*, (2004) (submitted),
http://www.math.princeton.edu/~mchudnov/publications.html

[12] R. Dutton, N. Chandrasekharan and R. Brigham, *On the number of independent sets of nodes in a tree*, Fibonacci Quarterly **31** (1993) 98–104.

[13] O. Favaron, *Very well-covered graphs*, Discrete Mathematics **42** (1982) 177–187.

[14] A. Finbow, B. Hartnell and R.J. Nowakowski, *A characterization of well-covered graphs of girth 5 or greater*, Journal of Combinatorial Theory B **57** (1993) 44–68.

[15] D.C. Fischer and A.E. Solow, *Dependence polynomials*, Discrete Mathematics **82** (1990) 251–258.

[16] M. Goldwurm and M. Santini, *Clique polynomials have a unique root of smallest modulus*, Information Processing Letters **75** (2000) 127–132.

[17] I. Gutman, *Some analytical properties of independence and matching polynomials*, Match **28** (1992) 139–150.

[18] I. Gutman, *Some relations for the independence and matching polynomials and their chemical applications*, Bul. Acad. Serbe Sci. Arts **105** (1992) 39–49.

[19] I. Gutman and F. Harary, *Generalizations of the matching polynomial*, Utilitas Mathematica **24** (1983) 97–106.

[20] H. Hajiabolhassan and M.L. Mehrabadi, *On clique polynomials*, Australasian Journal of Combinatorics **18** (1998) 313–316.

[21] Y.O. Hamidoune, *On the number of independent k-sets in a claw-free graph*, Journal of Combinatorial Theory B **50** (1990) 241–244.

[22] C. Hoede and X. Li, *Clique polynomials and independent set polynomials of graphs*, Discrete Mathematics **125** (1994) 219–228.

[23] V.E. Levit and E. Mandrescu, *Well-covered and König-Egervàry graphs*, Congressus Numerantium **130** (1998) 209–218.

[24] V.E. Levit and E. Mandrescu, *Well-covered trees*, Congressus Numerantium **139** (1999) 101–112.

[25] V.E. Levit and E. Mandrescu, *On well-covered trees with unimodal independence polynomials*, Congressus Numerantium **159** (2002) 193–202.

[26] V.E. Levit and E. Mandrescu, *On unimodality of independence polynomials of some well-covered trees*, DMTCS 2003 (C. S. Calude *et al.* eds.), Lecture Notes in Computer Science, LNCS **2731**, Springer-Verlag (2003) 237–256.

[27] V.E. Levit and E. Mandrescu, *A family of well-covered graphs with unimodal independence polynomials*, Congressus Numerantium **165** (2003) 195–207.

[28] V.E. Levit and E. Mandrescu, *On the roots of independence polynomials of almost all very well-covered graphs*, Discrete Applied Mathematics (2005) (accepted).

[29] V.E. Levit and E. Mandrescu, *Graph products with log-concave independence polynomials*, WSEAS Transactions on Mathematics Issue **3**, vol. **3** (2004) 487–493.

[30] V.E. Levit and E. Mandrescu, *Independence polynomials of well-covered graphs: generic counterexamples for the unimodality conjecture*, European Journal of Combinatorics (2005) (accepted).

[31] V.E. Levit and E. Mandrescu, *Very well-covered graphs with log-concave independence polynomials*, Carpathian Journal of Mathematics **20** (2004) 73–80.

[32] L. Lovász, *A characterization of perfect graphs*,
Journal of Combinatorial Theory Series B **13** (1972) 95–98.

[33] P. Matchett, *Operations on well-covered graphs and the roller-coaster conjecture*, The Electronic Journal of Combinatorics **11** (2004) #R45

[34] T.S. Michael and W.N. Traves, *Independence sequences of well-covered graphs: non-unimodality and the roller-coaster conjecture*,
Graphs and Combinatorics **19** (2003) 403–411.

[35] M.D. Plummer, *Some covering concepts in graphs*, Journal of Combinatorial Theory **8** (1970) 91–98.

[36] M.D. Plummer, *Well-covered graphs: a survey*, Quaestiones Mathematicae **16** (1993) 253–287.

[37] G. Ravindra, *Well-covered graphs*, J. Combin. Inform. System Sci. **2** (1977) 20–21.

[38] D. Stevanovic, *Clique polynomials of threshold graphs*,
Univ. Beograd Publ. Elektrotehn. Fac., Ser. Mat. **8** (1997) 84–87.

[39] D. Stevanovic, *Graphs with palindromic independence polynomial*, Graph Theory Notes of New York XXXIV (1998) 31–36.

[40] A.A. Zykov, *On some properties of linear complexes*, Matematicheskij Sbornik **24** (1949) 163–188 (in Russian).

[41] A.A. Zykov, *Fundamentals of graph theory*, BCS Associates, Moscow, 1990.

Vadim E. Levit
Department of Computer Science
Holon Academic Institute of Technology
52 Golomb Str., P.O. Box 305
Holon 58102, Israel
e-mail: `levitv@hait.ac.il`

Eugen Mandrescu
Department of Computer Science
Holon Academic Institute of Technology
52 Golomb Str., P.O. Box 305
Holon 58102, Israel
e-mail: `eugen_m@hait.ac.il`

Precoloring Extension on Chordal Graphs

Dániel Marx

Abstract. In the precoloring extension problem (PrExt) we are given a graph with some of the vertices having preassigned colors and it has to be decided whether this coloring can be extended to a proper k-coloring of the whole graph. 1-PrExt is the special case where every color is assigned to at most one vertex in the precoloring. Answering an open question of Hujter and Tuza [7], we show that the 1-PrExt problem can be solved in polynomial time for chordal graphs.

Mathematics Subject Classification (2000). Primary 05C15; Secondary 05C85.

Keywords. Coloring, precoloring extension, chordal graph.

1. Introduction

In graph vertex coloring we have to assign colors to the vertices such that neighboring vertices receive different colors. Starting with [4] and [16], a generalization of coloring was investigated: in the *list coloring* problem each vertex can receive a color only from its list of available colors. A special case is the *precoloring extension* problem: a subset W of the vertices have a preassigned color and we have to extend this to a proper coloring of the whole graph, using only colors from a color set C. It can be viewed as a special case of list coloring: the list of a precolored vertex consists of a single color, while the list of every other vertex is C. A thorough survey on list coloring, precoloring extension, and list chromatic number can be found in [15].

Since vertex coloring is the special case when $W = \emptyset$, the precoloring extension problem is NP-complete in every class of graphs where vertex coloring is NP-complete. Therefore, we can hope to solve precoloring extension efficiently only on graphs that are easy to color. Biró, Hujter and Tuza [2, 6, 7] started a systematic study of precoloring extension in perfect graphs, where coloring can be done in polynomial time. It turns out that for some classes of perfect graphs, e.g., split graphs, complements of bipartite graphs, and cographs, the precoloring

---

Research is supported in part by grants OTKA 44733, 42559 and 42706 of the Hungarian National Science Fund.

extension problem can be solved in polynomial time. On the other hand, for some other classes like bipartite graphs, line graphs of bipartite graphs, and interval graphs, precoloring extension is NP-complete.

The d-PREXT problem is the restriction of the precoloring extension problem where every color is used at most d-times in the precoloring. It is easy to reduce PREXT to 1-PREXT: collapse the vertices precolored with the same color to a single vertex. Therefore, 1-PREXT is not easier than PREXT on classes of graphs that are closed for this operation. One can also show that 1-PREXT is NP-complete on bipartite graphs [6, 3].

However, there are cases where 1-PREXT is strictly easier than PREXT. For planar bipartite graphs, if the set of colors C contains only 3 colors, then PREXT is NP-complete [9], while 1-PREXT can be solved in polynomial time [11]. For interval graphs already 2-PREXT is NP-complete [2], but 1-PREXT can be solved in polynomial time [2]. For the special case of unit interval graphs PREXT remains NP-complete [10], and obviously 1-PREXT remains polynomial-time solvable.

Every chordal graph is perfect and interval graphs form a subset of chordal graphs (cf. [5]). Therefore, by [2], the 2-PREXT problem is NP-complete for chordal graphs. The complexity of 1-PREXT on chordal graphs is posed by Hujter and Tuza as an open question [7]. Here we show that 1-PREXT can be solved in polynomial time also for chordal graphs. The algorithm is a generalization of the method of [2] for interval graphs. As in [2], 1-PREXT is reduced to a network flow problem, but for chordal graphs a more elaborate construction is required than for interval graphs.

The paper is organized as follows. In Section 2 we review some known properties of chordal graphs. In Section 3 we define a set system that will be crucial in the analysis of the algorithm. The algorithm is presented in Section 4. In Section 5 we discuss some connections of the problem with matroid theory.

2. Tree decomposition

A graph is *chordal* if every cycle of length greater than 3 contains at least one chord, i.e., an edge connecting two vertices not adjacent in the cycle. Equivalently, a graph is chordal if and only if it does not contain a cycle of length greater than 3 as an induced subgraph. This section summarizes some well-known properties of chordal graphs. First, chordal graphs can be also characterized as the intersection graphs of subtrees of a tree (see e.g., [5]):

Theorem 2.1. *The following two statements are equivalent:*
1. $G(V, E)$ *is chordal.*
2. *There exists a tree $T(U, F)$ and a subtree $T_v \subseteq T$ for each $v \in V$ such that $u, v \in V$ are neighbors in $G(V, E)$ if and only if $T_u \cap T_v \neq \emptyset$.*

The tree T together with the subtrees T_v is called the *tree decomposition* of G. Given a chordal graph G, a tree decomposition can be found in polynomial time (see [5, 14]).

For clarity, we will use the word "vertex" when we refer to the graph $G(V, E)$, and "node" when referring to $T(U, F)$. We assume that T is a rooted tree with some root $r \in U$. For a node $x \in U$, let T^x be the subtree of T rooted at x. Consider those subtrees T_v that contain at least one node of T^x, denote by V_x the set of corresponding vertices v. The subgraph of G induced by V_x will be denoted by $G_x = G[V_x]$. For a node $x \in U$ of T, denote by K_x the union of v's where $x \in T_v$. Clearly, the vertices of K_x are in V_x, and they form a clique in G_x, since the corresponding trees intersect in T at node x. An important property of the tree decomposition is the following: for every node $x \in U$, the clique K_x separates $V_x \setminus K_x$ and $V \setminus V_x$. That is, among the vertices of V_x, only the vertices in K_x can be adjacent to $V \setminus V_x$.

Every inclusion-wise maximal clique of a chordal graph G is a clique K_x of the tree decomposition. This is a consequence of the fact that subtrees of a tree satisfy the Helly property (a family of sets is said to satisfy the Helly property if for each pairwise intersecting collection of sets from the family it follows that the sets in the collection have a common element). If K is a clique of G, then its vertices correspond to pairwise intersecting subtrees, hence by the Helly property, these trees have a common node x, implying $K \subseteq K_x$.

Since every chordal graph is perfect, the chromatic number of G equals its clique number, and it follows that G is k-colorable if and only if $|K_x| \le k$ for every node $x \in T$. Clearly, the precoloring can exist only if G is $|C|$-colorable, hence we assume in the following that $|K_x| \le |C|$ holds for every $x \in T$.

A tree decomposition will be called *nice* [8], if it satisfies the following additional requirements:

- Every node $x \in U$ has at most two children.
- If $x \in U$ has two children $y, z \in U$, then $K_x = K_y = K_z$ (x is a *join* node).
- If $x \in U$ has only one child $y \in U$, then either $K_x = K_y \cup \{v\}$ (x is the *add* node of v) or $K_x = K_y \setminus \{v\}$ (x is the *forget* node of v) for some $v \in V$.
- If $x \in U$ has no children, then K_x contains exactly one vertex (x is a *leaf* node).

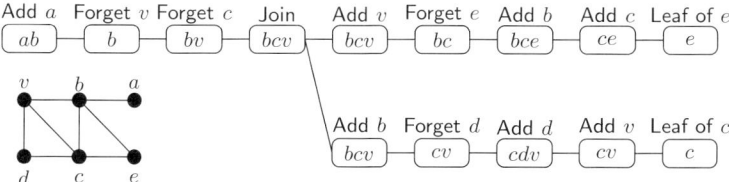

FIGURE 1. Nice tree decomposition of a chordal graph.

Figure 1 shows a nice tree decomposition. The node containing ab is the root of the tree; the children of each node appear to the right of the node in the figure.

It is easy to see that by splitting the nodes of the tree in an appropriate way, a tree decomposition of G can be transformed into a nice tree decomposition in polynomial time. If we go from the root towards the leaves, then the first node containing vertex v is the root of the subtree T_v, and the parent of this node is the forget node of v. The leaves of the subtree T_v are exactly the add nodes and leaf nodes of v. A vertex v can have multiple add nodes, but at most one forget node (the vertices in clique K_r of the root r have no forget nodes, but every other vertex has exactly one).

Given a graph G and a precolored set of vertices, we slightly modify the graph to obtain an even nicer tree decomposition. For each precolored vertex v, we add a clique K of $|C|-1$ new vertices to the graph, each vertex of K is connected to v; and we also add a new vertex v' that is connected to each vertex of K (but not to v). The precoloring of vertex v is removed and v' becomes a precolored vertex, the color of v is assigned to v'. It is easy to see that this transformation does not change the solvability of the instance: vertices v and v' receive the same color in every $|C|$-coloring of the new graph G' (since they are both connected to the same clique of $|C|-1$ vertices), thus a precoloring extension of G' induces a precoloring extension for G. Although the transformation increases the size of the graph and hence the size of the tree decomposition, it will be very useful, since now we can assume that the nice tree decomposition satisfies the following additional properties:

- If $x \in U$ is the add node of v, then v is not a precolored vertex.
- If $x \in U$ is a join node, then K_x does not contain precolored vertices.

We show how a nice tree decomposition T of G can be modified to obtain a nice tree decomposition T' of G' satisfying these two additional properties. Let $v_1, v_2, \ldots, v_{|C|-1}$ be the neighbors of v' in G'. Let x be an arbitrary node containing vertex v, let $K_x = \{v, w_1, w_2, \ldots, w_t\}$. Insert a new join node y between x and its parent. A new branch of the tree decomposition is attached to y: this branch will contain the subtrees representing the vertices $v', v_1, \ldots, v_{|C|-1}$. The new branch is a path, containing the following nodes (see Figure 2):

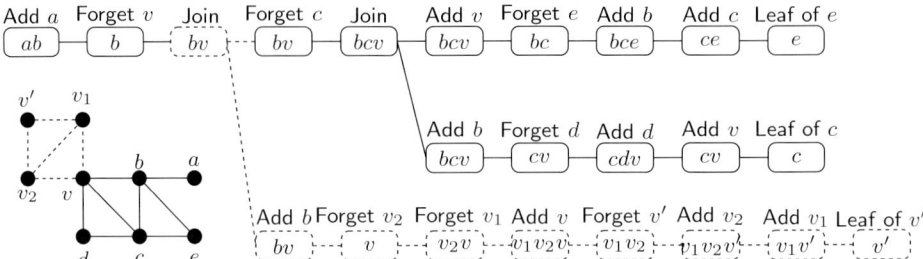

FIGURE 2. Nice tree decomposition of the graph shown on Figure 1, after adding the vertices v', v_1, v_2 to the graph. Dashed lines show the new parts of the tree decomposition.

- Leaf node containing v'.
- Add node of v_1, add node of v_2, ..., add node of $v_{|C|-1}$.
- Forget node of v'.
- Add node of v.
- Forget node of v_1, forget node of v_2, ..., forget node of $v_{|C|-1}$.
- Add node of w_1, add node of w_2, ..., add node of w_t.

It is clear that this modification results in a nice tree decomposition, and if we perform it for each precolored vertex v, then we obtain a decomposition of G'.

3. System of extensions

Let H be an induced subgraph of G, and let K be a clique of H. We define a set system $\mathscr{S}(H, K)$ over K that will play an important role in the analysis of the algorithm. Denote by $C_H \subseteq C$ those colors that the precoloring assigns to vertices in H. The set system $\mathscr{S}(H, K)$ is defined as follows:

Definition 3.1. For $S \subseteq K$, the set S is in $\mathscr{S}(H, K)$ if and only if there is a precoloring extension $\psi \colon V(H) \to C$ of subgraph H such that

- $\psi(v) \in C_H$ for every $v \in S$, and
- $\psi(v) \notin C_H$ for every $v \in K \setminus S$.

Thus the set system $\mathscr{S}(H, K)$ describes all the possible colorings that can appear on K in a precoloring extension of H, but this description only distinguishes between colors in C_H and colors not in C_H. In particular, the precoloring can be extended to H if and only if $\mathscr{S}(H, K)$ is not empty. If H contains no precolored vertices, but it can be colored with $|C|$ colors, then $\mathscr{S}(H, K)$ contains only the empty set.

The following observation bounds the possible size of a set in $\mathscr{S}(H, K)$:

Observation 3.2. *If $S \in \mathscr{S}(H, K)$, then*

$$|K| - |C \setminus C_H| \leq |S| \leq |C_H|$$

Proof. If $S \in \mathscr{S}(H, K)$, then there is a coloring ψ that assigns exactly $|S|$ colors from C_H to the vertices of K. Clearly, in ψ at most $|C_H|$ vertices of the clique K can receive colors from C_H, proving the upper bound. Coloring ψ assigns colors from $C \setminus C_H$ to the vertices in $K \setminus S$, hence $|C \setminus C_H| \geq |K| - |S|$, and the lower bound follows. □

The definition of this set system is somewhat technical, but it precisely captures the information necessary for solving the precoloring extension problem. Let K be a *clique separator* of G, that is, K is a clique such that its removal separates the graph into two or more components. Let $V \setminus K = V_1 \cup V_2$ be a partition of the remaining vertices such that there is no edge between V_1 and V_2 (that is, each of V_1 and V_2 contains one or more connected components of $V \setminus K$). Let $G_1 = G[V_1 \cup K]$ and $G_2 = G[V_2 \cup K]$. Assume that we have already extended the precoloring to

G_1 (coloring ψ_1) and to G_2 (coloring ψ_2). If $\psi_1(v) = \psi_2(v)$ for every vertex v of the clique K, then they can be merged to obtain a coloring of G. Therefore, G has a precoloring extension if and only if there is a precoloring extension ψ_1 of G_1, and a precoloring extension ψ_2 of G_2 such that they agree on K. This means that if we have the list of all possible colorings that a precoloring extension of G_1 can assign to K, then to decide if G has a precoloring extension this list is all the information required from the graph G_1. More formally, if we replace G_1 with a graph that has the same list of possible colorings on K, then this does not change the existence of a precoloring extension on G.

However, the following lemma shows that even less information is sufficient: we do not need the list of all possible colorings that can appear on clique K in a coloring of G_1, the set system $\mathscr{S}(G_1, K)$ is sufficient. More precisely, the set system $\mathscr{S}(G, K)$ can be constructed from $\mathscr{S}(G_1, K)$ and $\mathscr{S}(G_2, K)$, hence these two systems are sufficient to decide whether G has a precoloring extension.

Lemma 3.3. *Let K be a clique separator of $G(V_1 \cup K \cup V_2, E)$ containing no precolored vertices, let $G_1 = G[V_1 \cup K]$ and $G_2 = G[V_2 \cup K]$. A set $S \subseteq K$ is in $\mathscr{S}(G, K)$ if and only if $|S| \geq |K| - |C \setminus C_G|$ and S can be partitioned into disjoint sets $S_1 \in \mathscr{S}(G_1, K)$ and $S_2 \in \mathscr{S}(G_2, K)$.*

Proof. Assume first that $S \in \mathscr{S}(G, K)$ and let ψ be a coloring corresponding to the set S. Observation 3.2 implies that $|S| \geq |K| - |C \setminus C_G|$, as required. Coloring ψ induces a coloring ψ_i of G_i, let $S_i \in \mathscr{S}(G_i, K)$ be the set corresponding to ψ_i ($i = 1, 2$). Coloring ψ can assign three different types of colors to the vertices in K:

- If $\psi(v) \notin C_G$ (i.e., $\psi(v)$ is not used in the precoloring), then $v \notin S, S_1, S_2$.
- If the precoloring uses $\psi(v)$ in V_1, then $v \in (S \cap S_1) \setminus S_2$. (Since each color is used at most once in the precoloring, $\psi(v)$ cannot appear in V_2 on a precolored vertex.)
- If the precoloring uses $\psi(v)$ in V_2, then $v \in (S \cap S_2) \setminus S_1$.

Note that v cannot be a precolored vertex, hence the precoloring cannot use $\psi(v)$ in K. Therefore, S is the disjoint union of S_1 and S_2, as required.

Now assume that S can be partitioned into disjoint sets $S_1 \in \mathscr{S}(G_1, K)$ and $S_2 \in \mathscr{S}(G_2, K)$, let ψ_1 and ψ_2 be the two corresponding colorings. In general, ψ_1 and ψ_2 might be different on K, thus they cannot be combined to obtain a coloring of G. However, with some permutations of colors we modify the two colorings in such a way that they assign the same color to every vertex of K. Let C_1 (resp., C_2) be the colors of the precolored vertices in V_1 (resp., V_2). Notice that both ψ_1 and ψ_2 assign colors from $C \setminus C_1$ to S_2, (since S_1 and S_2 are disjoint). Modify coloring ψ_1: permute the colors of $C \setminus C_1$ such that $\psi_1(v) = \psi_2(v)$ holds for every $v \in S_2$ (this can be done since K is a clique, hence both ψ_1 and ψ_2 assign distinct colors to the vertices in S_2). Since the precolored vertices in V_1 have colors only from C_1, coloring ψ_1 remains a valid precoloring extension for G_1. Similarly, in coloring ψ_2, permute the colors of $C \setminus C_2$ such that $\psi_1(v) = \psi_2(v)$ for every $v \in S_1$. Now we have that ψ_1 and ψ_2 agree on S, there might be differences only on $K \setminus S$.

Moreover, ψ_1 uses only colors from $C \setminus C_1$ on $K \setminus S$, and ψ_2 uses colors only from $C \setminus C_2$ on this set. Now select a set $C' \subseteq C \setminus C_G$ such that $|C'| = |K \setminus S|$ (here we use the assumption $|S| \geq |K| - |C \setminus C_G|$, which implies that there are enough colors in $C \setminus C_G$). Permute again the colors of $C \setminus C_1$ in coloring ψ_1 such that ψ_1 assigns to $K \setminus S$ exactly the colors in C'. Similarly, permute the colors of $C \setminus C_2$ in coloring ψ_2 such that ψ_2 also uses C' on $K \setminus S$. Now the colorings ψ_1 and ψ_2 agree on K, hence we can combine them to obtain a coloring ψ of G. This coloring proves that $S = S_1 \cup S_2$ is in $\mathscr{S}(G, K)$, what we had to show. □

Lemma 3.3 implies that if we know the set systems $\mathscr{S}(G_1, K)$ and $\mathscr{S}(G_2, K)$, then the set system $\mathscr{S}(G, K)$ can be also determined. This suggests the following algorithm: for each node x of the tree decomposition, determine $\mathscr{S}(G_x, K_x)$. In principle, this can be done in a bottom-up fashion: the set system for node x can be determined from the systems of its children. Unfortunately, the size of $\mathscr{S}(G_x, K_x)$ can be exponential, thus it cannot be constructed explicitly during the algorithm. However, if G is a chordal graph, then these set systems have nice combinatorial structure that allows a compact representation. The main idea of the algorithm in Section 4 is to use network flows to represent the set systems $\mathscr{S}(G_x, K_x)$. As a consequence, it follows that $\mathscr{S}(G, K)$ has nice combinatorial structure: in Section 5 we show that if G is chordal and K is a clique of G, then $\mathscr{S}(G, K)$ is the projection of a matroid.

4. The algorithm

In this section we prove the main result of the paper:

Theorem 4.1. *1-PrExt can be solved in polynomial time for chordal graphs.*

Given an instance of the 1-PrExt problem, we construct a network flow problem that has a feasible flow if and only if there is a solution to 1-PrExt. We use the following variant of the flow problem. The *network* is a directed graph $D(U, A)$, each arc $e \in A$ has an integer capacity $c(e)$. The set of arcs entering (resp., leaving) node v will be denoted by $\delta^-(v)$ (resp., $\delta^+(v)$). The set of *sources* is $S \subseteq U$, and $T \subseteq U$ is the set of *terminals* in the network (we require $S \cap T = \emptyset$). Every source $v \in S$ produces exactly one unit amount of flow, and each terminal $v \in T$ has a capacity $w(v)$, it can consume up to $w(v)$ units. Formally, a *feasible flow* is a function $f : A \to \mathbb{Z}^+$ that satisfies $0 \leq f(e) \leq c(e)$ for every arc $e \in A$, and the following holds for every node $v \in U$:

- If $v \in S$, then $\sum_{e \in \delta^-(v)} f(e) - \sum_{e \in \delta^+(v)} f(e) = -1$.
- If $v \in T$, then $0 \leq \sum_{e \in \delta^-(v)} f(e) - \sum_{e \in \delta^+(v)} f(e) \leq w(v)$.
- If $v \in U \setminus (T \cup S)$, then $\sum_{e \in \delta^-(v)} f(e) = \sum_{e \in \delta^+(v)} f(e)$.

Using standard techniques, the existence of a feasible flow can be tested by a maximum flow algorithm. It is sufficient to add two new vertices s and t, an arc with capacity 1 from s to every vertex $v \in S$, and an arc with capacity $w(v)$ to t

from every vertex $v \in T$. Clearly, there is a feasible flow in the original network if and only if there is an $\overrightarrow{st}$ flow with value $|S|$ in the modified network. The maximum flow can be determined using at most $|S|$ iterations of the Edmonds-Karp augmenting path algorithm, hence the existence of a feasible flow in a network $D(U, A)$ can be tested in $O(|S||A|)$ time.

Given a chordal graph $G(V, E)$, its nice tree decomposition $T(U, F)$, $\{T_v \mid v \in V\}$, and the set of precolored vertices $W \subseteq V$, we construct a network as follows. Direct every edge of T towards the root r. For every $v \in V$ and for every $x \in T_v$ add a node x_v to the network. Denote by U_x the $|K_x|$ nodes corresponding to x. If the edge xy is in T_v, then connect $x_v \in U_x$ and $y_v \in U_y$ by an arc. If y is the child of x, then direct this arc from y_v to x_v. These new arcs $\overrightarrow{y_v x_v}$ have capacity 1, while the arcs $\overrightarrow{yx}$ of the tree T have capacity $|C| - |K_y|$ (recall that if the graph is $|C|$-colorable, then $|K_y| \le |C|$).

For each node $x \in T$, depending on the type of x, we do one of the following:

- If x is an add node of some vertex $v \notin W$, and y is the child of x, then add an arc $\overrightarrow{yx_v}$ to the network.
- If leaf node x contains some vertex $v \in W$, then add a new node x'_v to the network, add an arc $\overrightarrow{x'_v x_v}$ with capacity 1, and set x'_v to be a source.
- If x is a forget node of some vertex v (either in W or not), and y is the child of x, then add an arc $\overrightarrow{y_v x}$ to the network.

For join nodes and for leaf nodes containing vertices outside W we do nothing. Figure 3 sketches the construction for the different types of nodes.

So far there are no terminals in the network. The definition of the network is completed by adding terminals as follows. Here we define not only a single network, but several subnetworks that will be useful in the analysis of the algorithm. For every node $x \in U$ of the tree T, the network N_x contains only those nodes of the network that correspond to nodes in T^x (recall that T^x is the subtree of T rooted at x). Formally, the network N_x has the node set $T^x \cup \bigcup_{y \in T^x} U_y$, and the source nodes (if available) corresponding to the leaves of T^x. Moreover, in network N_x the nodes in U_x are set to be terminals with capacity 1, and node x is a terminal with capacity $|C| - |K_x|$. This completes the description of the network N_x.

Notice that there are sources only at the leaf nodes of precolored vertices. Therefore, the number of sources in network N_x is the same as the number of precolored vertices in V_x (recall that V_x is the set of those vertices v whose tree T_v has at least one node in T^x, and $G_x = G[V_x]$). We will denote by C_x the set of colors that appear on the precolored vertices of V_x. In network N_x, there are terminals only at x and U_x, these terminals must consume all the flow.

Observation 4.2. *The number of sources in N_x equals the number of precolored vertices in V_x, which is $|C_x|$. Consequently, in every feasible flow of N_x, the amount of flow consumed by the terminals at x and U_x is exactly $|C_x|$.* □

To prove Theorem 4.1, we show that the precoloring of G can be extended to the whole graph if and only if there is a feasible flow in N_r, where r is the

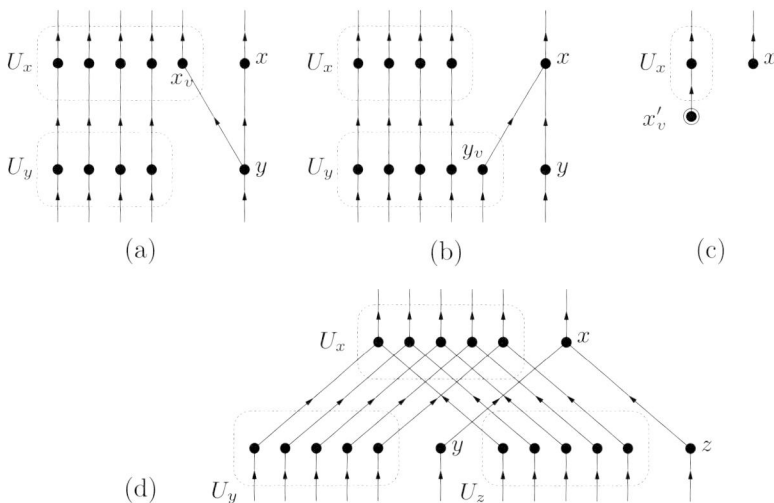

FIGURE 3. Construction of the network when node x is of the following types: (a) add node for $v \notin W$, (b) forget node for v (either in W or not), (c) leaf node for $v \in W$, (d) join node.

root of T. This gives a polynomial-time algorithm for 1-PrExt in chordal graphs, since constructing network N_r and finding a feasible flow in N_r can be done in polynomial time. The proof of this claim uses induction on the tree decomposition of the graph. For every node $x \in U$ of T, we prove the following more general statement: the network N_x has a feasible flow if and only if the precoloring of G can be extended to G_x.

More precisely, we show that the network N_x represents (in some well-defined sense) the set system $\mathscr{S}(G_x, K_x)$: every feasible flow corresponds to a set in the system. Therefore, N_x has no feasible flows if and only if $\mathscr{S}(G_x, K_x)$ is empty, or, equivalently, the precoloring cannot be extended to G_x.

We say that a feasible integer flow of N_x *represents the set* $S \subseteq K_x$ if for every $v \in S$, the terminal at x_v consumes one unit of flow, while for every $v \notin S$, there is no flow entering x_v. The following lemma establishes the connection between the constructed networks and the set systems $\mathscr{S}(G_x, K_x)$. The proof of this lemma completes the proof of Theorem 4.1, as it reduces the 1-PrExt problem to finding a feasible flow in N_r.

Lemma 4.3. *For an arbitrary node $x \in U$ of T, the network N_x has a feasible flow representing a set $S \subseteq K_x$ if and only if $S \in \mathscr{S}(G_x, K_x)$.*

Proof. The lemma is proved for every node x of T by a bottom-up induction on the tree T. After checking the lemma for the leaf nodes, we show that it is true for a node x assuming that it is true for the children of x. The proof is done separately for the different types of nodes. Verifying the lemma in every case is tedious, but it does not require any new ideas. The way the networks are constructed ensures that the set systems represented by the networks have the required properties.

Leaf node. For a leaf node x, the lemma is trivial: if the vertex v in K_x is precolored, then every flow of N_x represents $\{v\}$, otherwise N_x contains no sources, and every flow represents $\emptyset$.

Add node for $v \notin W$. Let x be an add node of $v \notin W$, and let y be the child of x. For every $S \in \mathscr{S}(G_x, K_x)$, it has to be shown that there is a feasible flow of N_x representing the set S. Assume first that $v \notin S$. Since $G_y = G_x \setminus v$ and $K_y = K_x \setminus \{v\}$, it follows that $S \in \mathscr{S}(G_y, K_y)$. Therefore, by the induction hypothesis, there is a flow f_y in N_y representing S. We modify this flow to obtain a flow f_x of N_x also representing S. For every $u \in S$, in flow f_y there is one unit of flow consumed by the terminal at y_u. To obtain flow f_x, direct this unit flow towards x_u, and consume it by the terminal at that node. Similarly, in the flow f_y, there is some amount of flow consumed by the terminal at y, direct this flow to x, and consume it by that terminal. By Observation 4.2, the amount of flow consumed at x is exactly $|C_x| - |S|$. Moreover, the lower bound of Observation 3.2 implies that this is at most $|C_x| - |K_x| + |C \setminus C_x| = |C| - |K_x| < |C| - |K_y|$, hence the capacity of the arc $\overrightarrow{yx}$ and the terminal at x is sufficient for the flow. Thus we obtained a feasible flow of N_x, and obviously it represents S.

We proceed similarly if $v \in S$. In this case $S \setminus \{v\} \in \mathscr{S}(G_y, K_y)$, thus N_y has a flow f_y representing $S \setminus \{v\}$. To obtain a flow f_x of N_x representing S, the flow consumed at y_u is directed to x_u, as in the previous paragraph. However, now we do not direct all the flow consumed at y to x, but we direct one unit amount through the arc $\overrightarrow{yx_v}$, and only the rest goes through arc $\overrightarrow{yx}$. Therefore, the amount of flow consumed by the terminal at x is one unit less than the flow consumed at y in flow f_y, hence the capacity of the terminal at x is sufficient. Clearly, this results in a flow f_x of N_x representing S, as required. The only thing to verify is that there is at least one unit of flow consumed at y in flow f_y. The flow f_y represents $S \setminus \{v\}$, and by Observation 4.2, the amount of flow consumed in $U_y \cup \{y\}$ is exactly $|C_y|$, hence the flow consumed at y is $|C_y| - |S| + 1$. Since v is not a precolored vertex, we have that $C_y = C_x$. We know that $S \in \mathscr{S}(G_x, K_x)$, therefore by the upper bound of Observation 3.2, $|C_y| - |S| + 1 \geq 1$, hence there is nonzero flow consumed at y in flow f_y.

Now assume that there is a flow f_x in N_x representing $S \subseteq K_x$, it has to be shown that $S \in \mathscr{S}(G_x, K_x)$. Let y be the child of x. Assume first that $v \notin S$, we show that N_y has a flow f_y in N_y representing S. To obtain this f_y, the flow f_x is modified the following way. For every vertex $x_w \in U_x$, where $w \neq v$, if there is flow on the arc $\overrightarrow{y_w x_w}$, then consume it by the terminal at y_w. Similarly, the flow on the arc $\overrightarrow{yx}$ can be consumed by the terminal at y (the capacity of the terminal at y equals the capacity of arc $\overrightarrow{yx}$). It is clear that these modifications result in a feasible flow for N_y that represents S. By the induction hypothesis, this means that $S \in \mathscr{S}(G_y, K_y)$, and there is a corresponding coloring ψ. Since v is the only vertex in $V_x \setminus V_y$, to prove $S \in \mathscr{S}(G_x, K_x)$ it is sufficient to show that coloring

ψ can be extended to v in such a way that v receives a color not in C_x. If there is no such extension, then this means that ψ uses every color of $C \setminus C_x$ on the neighbors of v, that is, on the clique K_y. By construction, ψ assigns exactly $|S|$ colors from C_x to the clique K_y, hence if every color of $C \setminus C_x$ is used on K_y, then $|K_y| = |C \setminus C_x| + |S|$. Therefore, $|K_x| = |K_y| + 1 = |C \setminus C_x| + |S| + 1$ and the capacity of the terminal at x is $|C| - |K_x| = |C_x| - |S| - 1$. However, in flow f_x of N_x that represents S, exactly $|C_x| - |S|$ unit of flow is consumed at x (Observation 4.2), a contradiction. Thus ψ can be extended to v, and $S \in \mathscr{S}(G_x, K_x)$ follows.

The case $v \in S$ can be handled similarly. If there is a flow f_x in N_x that represents S, then this is only possible if there is flow on the arc $\overrightarrow{yx_v}$. Therefore, by restricting the flow to N_y as in the previous paragraph, we can obtain a flow representing $S \setminus \{v\}$. (Notice that the capacity of the terminal at y equals the combined capacity of the terminals at x and x_v, hence it can consume the flow on the arcs $\overrightarrow{yx_v}$ and $\overrightarrow{yx}$.)

By the induction hypothesis, it follows that $S \setminus \{v\} \in \mathscr{S}(G_y, K_y)$, and there is a corresponding coloring ψ. Now it has to be shown that ψ can be extended to vertex v such that v receives a color from C_x. Coloring ψ assigns exactly $|S| - 1$ colors from C_x to K_y. The extension is not possible only in the case if every color of C_x is already used on K_y, that is, if $|C_x| = |S| - 1$. This would imply that in flow f_x of N_x, the amount of flow consumed at U_x is $|S| = |C_x| + 1$. However, by Observation 4.2, this is strictly larger than the number of sources in N_x, a contradiction.

Forget node for v (vertex v is either in W or not). Let x be the forget node of v, and let y be the child of x. Let $S \in \mathscr{S}(G_x, K_x)$. Since $G_x = G_y$, either S or $S \cup \{v\}$ is in $\mathscr{S}(G_y, K_y)$. In the first case, the flow f_y in N_y that represents S can be extended to a flow in N_x that also represents S. As before, the flow consumed at y_w is directed to x_w, and the flow consumed at y is directed to x. Recall that the capacity of the arc $\overrightarrow{yx}$ equals the capacity of the terminal at y, while the capacity of the terminal at x is strictly greater. Therefore, the resulting flow is feasible in N_x, and clearly it represents S. If $S \cup \{v\} \in \mathscr{S}(G_y, K_y)$, then we do the same, but the flow consumed at y_v is directed to x through the arc $\overrightarrow{y_v x}$. The resulting flow is feasible in N_x and represents S.

To prove the other direction, assume that N_x has a flow f_x representing $S \subseteq K_x$. Restrict this flow to N_y, that is, modify the flow such that the terminals at y and U_y consume all the flow. This results in a feasible flow f_y of N_y that represents S or $S \cup \{v\}$. Notice that the terminal at y has the same capacity as the arc $\overrightarrow{yx}$, hence this terminal can consume all the flow going through the arc. Therefore, by the induction hypothesis, either S or $S \cup \{v\}$ is in $\mathscr{S}(G_y, K_y)$, depending on whether there is flow consumed at y_v or not. In either case, $S \in \mathscr{S}(G_x, K_x)$ follows since $G_x = G_y$ and $K_x = K_y \setminus \{v\}$.

Join node. Let y and z be the two children of the join node x. Let $S \in \mathscr{S}(G_x, K_x)$. By Lemma 3.3 this means that

$$|S| \geq |K_x| - |C \setminus C_x| \qquad (4.1)$$

and S can be partitioned into disjoint sets $S_1 \in \mathscr{S}(G_y, K_y)$ and $S_2 \in \mathscr{S}(G_z, K_z)$. By the induction hypothesis, this implies that there are flows f_y, f_z in N_y and N_z that represent the sets S_1 and S_2, respectively. We combine these two flows to obtain a flow f_x of N_x that represents the set S. If there is flow consumed at a node $y_v \in U_y$ (resp., $z_v \in U_z$) in f_y (resp., f_z), then direct this flow on the arc $\overrightarrow{y_v x_v}$ (resp., $\overrightarrow{z_v x_v}$) to node x_v, and consume it there. The capacity of the terminal at x_v is only 1, but the disjointness of S_1 and S_2 implies that at most one unit of flow is directed to x_v. The flow consumed at node y and z is directed to x on the arc $\overrightarrow{yx}$, $\overrightarrow{zx}$, respectively. Since there are exactly $|S|$ units of flow consumed in U_x, therefore $|C_x| - |S|$ units of flow has to be consumed at x. By (4.1), this is at most $|C_x| - |K_x| + |C \setminus C_x| = |C| - |K_x|$, thus the capacity of the terminal at x is sufficient for consuming this flow. Therefore, we have obtained a flow f_x in network N_x that represents S.

Now assume that N_x has a flow f_x that represents S. Since the terminal at x has capacity at most $|C| - |K_x|$, and by Observation 4.2, the amount of flow consumed in $x \cup U_x$ is $|C_x|$, it follows that

$$|S| \geq |C_x| - (|C| - |K_x|) = |K_x| - |C \setminus C_x|. \tag{4.2}$$

If flow is consumed at a node $x_v \in U_x$, then the flow arrives to this node either from y_v or from z_v. Define the sets $S_1, S_2 \subseteq K_x$ such that $v \in S_1$ (resp., $v \in S_2$) if there is flow on arc $\overrightarrow{y_v x_v}$ (resp., $\overrightarrow{z_v x_v}$).

Based on the flow f_x of N_x representing S, we create a flow f_y of N_y that represents S_1 and a flow f_z of N_z that represents S_2. The flows f_y and f_z are constructed as follows. For every $y_v \in U_y$, if there is flow going through the arc $\overrightarrow{y_v x_v}$, then consume this flow at y_v, and similarly for the nodes $z_v \in U_z$. The flow on arcs $\overrightarrow{yx}$ and $\overrightarrow{zx}$ are consumed at y and z, respectively (the capacities of nodes x, y, and z are the same $|C| - |K_x| = |C| - |K_y| = |C| - |K_z|$). Clearly, flows f_y and f_z represent S_1 and S_2, respectively. By the induction hypothesis, the flows f_x and f_y imply that $S_1 \in \mathscr{S}(G_y, K_y)$ and $S_2 \in \mathscr{S}(G_z, K_z)$. Furthermore, it is clear that S_1 and S_2 are disjoint, and $S = S_1 \cup S_2$. Therefore, by Lemma 3.3 and Inequality (4.2), this proves that $S \in \mathscr{S}(G_x, K_x)$, as required. □

To determine the running time of the algorithm, we have to consider two main steps: the construction of the network and the solution of the flow problem. First of all, the transformation introduced at the end of Section 2 turns a graph $G_0(V_0, E_0)$ into a graph $G(V, E)$ with $|V| \leq |V_0||C|$. The tree decomposition of $G(V, E)$ can be constructed by first finding a perfect vertex elimination scheme [5, 14]. Based on this ordering of the vertices, one can build a tree $T(U, F)$ of size $O(|V|)$, and one subtree for each vertex of the graph. This tree decomposition can be found in time linear in the size of the output, that is, in $O(|V|^2)$ time. Converting $T(U, F)$ to a nice tree decomposition can introduce an increase of factor at most $|V|$, thus it can be done in $O(|V|^3)$ time. The network defined by the algorithm has size linear in the total size of the tree decomposition (size of $T(U, F)$ and the sum of the size of the subtrees), and clearly it can be constructed in linear time. Therefore, the constructed network has $O(|V|^3)$ nodes and $O(|V|^3)$ arcs, and the construction takes $O(|V|^3)$ time.

In a network with n nodes and m arcs, the maximum flow can be determined in $O(n^2m)$ or even in $O(n^3)$ time [1]. Moreover, it can be determined in $O(km)$ time if a flow with value k exists: the Edmonds-Karp algorithm produces such a flow after finding the first at most k augmenting paths (assuming that the capacities are integer). As discussed at the beginning of the section, the existence of a feasible flow can be tested by finding an s-t flow with value $|S|$, hence it can be done in $O(|S| \cdot |V|^3)$ time. By Observation 4.2, this is at most $O(|C| \cdot |V|^3) = O(|C|^4 \cdot |V_0|^3)$. We believe that the running time can be significantly improved by streamlining the algorithm. However, our aim was only to prove that the problem can be solved in polynomial time, thus we preferred ease of presentation over efficiently.

The algorithm described above determines whether a precoloring extension exists, but does not find a coloring. However, based on the feasible flow of network N_r, one can construct a precoloring extension of the graph. We have seen that the feasible flow of network N_x represents a set $S_x \in \mathscr{S}(G_x, K_x)$. Recursively for each $x \in U$, we compute a coloring ψ_x corresponding to S_x. For the leaf nodes this is trivial. Let x be an add node of vertex v, and let y be the child of x. To obtain ψ_x, coloring ψ_y has to be extended to v: if there is flow on $\overrightarrow{yx_v}$, then v has to receive a color from C_x, otherwise from $C \setminus C_x$. The construction ensures that there is always such a color not already used on the neighbors of v. If x is a forget node with child y, then ψ_x can be selected to be the same as ψ_y. Finally, assume that x is a join node with children y and z. By the way the network was constructed, S_y and S_z are disjoint, $S_x = S_y \cup S_z$, and $S \geq |K_x| - |C \setminus C_x|$. Therefore, the method described in the proof of Lemma 3.3 can be used to construct a coloring ψ_x of G_x that corresponds to $S_x \in \mathscr{S}(G_x, K_x)$.

5. Matroidal systems

The main idea of the algorithm in Section 4 is to represent the set system $\mathscr{S}(G, K)$ by a network flow. We have shown that for chordal graphs the set systems $\mathscr{S}(G_x, K_x)$ can be represented by network flows for every subgraph G_x and clique K_x given by the tree decomposition. It follows from this representation that the set systems have nice combinatorial structure:

Theorem 5.1. *Let $G(V, E)$ be a chordal graph, and let $W \subseteq V$ be a arbitrary set of precolored vertices such that every color of C is used at most once in the precoloring. If H is an induced subgraph of G, and K is a clique of H, then the set system $\mathscr{S}(H, K)$ is the projection of the basis set of a matroid.*

Theorem 5.1 will be proved at the end of this section. Recall that the a set system $\mathscr{B}$ is the basis set of a *matroid*, if it satisfies the following two conditions:
- Every set in $\mathscr{B}$ has the same size.
- For every $B_1, B_2 \in \mathscr{B}$ and $v \in B_1 \setminus B_2$, there is an element $u \in B_2 \setminus B_1$ such that $B_1 \cup \{u\} \setminus \{v\} \in \mathscr{B}$.

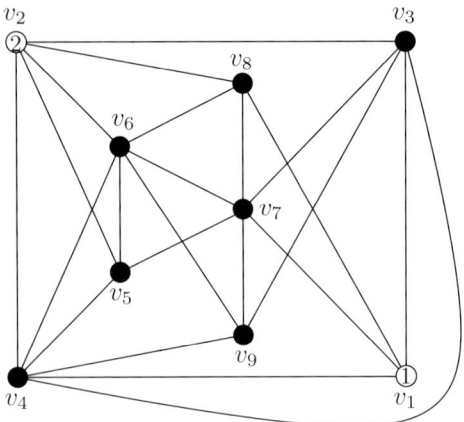

FIGURE 4. A non-chordal graph G and a clique $K = \{v_5, v_6, v_7\}$ such that $\mathscr{S}(G, K)$ is not the projection of a matroid ($|C| = 4$).

If $\mathscr{B}$ is a set system over X, then its *projection* to $Y \subseteq X$ is a set system over Y that contains $B' \subseteq Y$ if and only if there is a set $B \in \mathscr{B}$ with $B \cap Y = B'$. The projection of a matroid is always a so-called Δ-matroid [12], hence Theorem 5.1 also says that $\mathscr{S}(G, K)$ is a Δ-matroid. For further notions of matroid theory, the reader is referred to e.g., [13].

In general, if G is not chordal, then $\mathscr{S}(G, K)$ is not necessarily the projection of a matroid. Figure 4 shows a graph G with two precolored vertices v_1 and v_2. The graph is not chordal, since vertices v_1, v_4, v_2, v_8 induce a cycle of length 4. If we have only four colors, then G has four precoloring extensions: vertex v_3 can have only color 3 or 4, vertex v_9 can have only color 1 or 2, and setting the color of these two vertices forces a unique coloring for the rest of the graph. For example, if coloring ψ assigns color 3 to v_3, and color 1 to v_9, then $\psi(v_3) = 3$, $\psi(v_9) = 1$, $\psi(v_2) = 2$ imply $\psi(v_4) = 4$; $\psi(v_9) = 1$, $\psi(v_2) = 2$, $\psi(v_4) = 4$ imply $\psi(v_6) = 3$; $\psi(v_1) = 1$, $\psi(v_2) = 2$, $\psi(v_6) = 3$ imply $\psi(v_8) = 4$; and finally $\psi(v_5) = 1$ and $\psi(v_7) = 2$ follow in a similar fashion. Therefore, the clique $K = \{v_5, v_6, v_7\}$ receives one of the four colorings $(3, 1, 4)$, $(1, 3, 2)$, $(4, 1, 3)$, $(1, 4, 2)$ in every precoloring extension. Since $C_G = \{1, 2\}$, it follows that $\mathscr{S}(G, K) = \{\{v_6\}, \{v_5, v_7\}\}$, which cannot be the projection of a matroid (for example, it is not even a Δ-matroid).

The proof of Theorem 5.1 uses the following result of matroid theory. In a directed graph $D(U, A)$, we say that $Y \subseteq U$ can be *linked* onto $X \subseteq U$, if $|X| = |Y|$ and there are $|X|$ pairwise node disjoint paths from the nodes in X to the nodes in Y. The sets X and Y do not have to be disjoint, and the zero-length path consisting of a single node is also allowed. Hence X can be linked onto X in particular. The following theorem states that the graph G together with a set $X \subseteq U$ induces a matroid on the vertices of the graph (see e.g., [13]):

Theorem 5.2. *If $D(U, A)$ is a directed graph and $X \subseteq U$ is a fixed subset of nodes, then those subsets $Y \subseteq U$ that can be linked onto X form the bases of a matroid M over U.*

Considering the line graph of the directed graph, one can state an arc disjoint version of Theorem 5.2:

Theorem 5.3. *If $D(U, A)$ is a directed graph, $s \in U$ is a fixed vertex and r is a positive integer, then those r-element subsets $A' \subseteq A$ whose arcs can be reached from s by r pairwise arc disjoint paths form the bases of a matroid M over A.* □

To prove Theorem 5.1, we use the fact that $\mathscr{S}(G_x, K_x)$ can be represented by the network N_x (Lemma 4.3). Then Theorem 5.3 is used to show that the set system represented by a network is the projection of a matroid.

Proof (of Theorem 5.1). Clearly, it is sufficient to consider only the case when $H = G$, since every induced subgraph of a chordal graph is also chordal. Moreover, it can be assumed that K is a maximal (non-extendable) clique: if $K_1 \subseteq K_2$ are two cliques, then $\mathscr{S}(G, K_1)$ is the projection of $\mathscr{S}(G, K_2)$. Therefore, if $\mathscr{S}(G, K_2)$ is the projection of the basis set of a matroid, then this also follows for $\mathscr{S}(G, K_1)$. We have seen in Section 2 that given a tree decomposition $T(U, F)$, $\{T_v\}_{v \in V(G)}$ of the chordal graph G, every maximal clique of G is a clique K_x for some $x \in U$. Furthermore, since the choice of the root node of T is arbitrary, it can be assumed that x is the root, thus we have $G = G_x$ and $\mathscr{S}(G, K) = \mathscr{S}(G_x, K_x)$.

By Lemma 4.3, the sets in $\mathscr{S}(G_x, K_x)$ are exactly the sets represented by the feasible flows of the network N_x. Now, as described at the beginning of Section 4, add two new nodes s, t to the network, add an arc with unit capacity from s to every source, and for every terminal x, add an arc from x to t that has capacity equal to the capacity of x. Furthermore, replace every arc e having capacity $c(e)$ with $c(e)$ parallel arcs of unit capacity, clearly this does not change the problem. Call the resulting network N'_x. By Observation 4.2, the number of sources in N_x is $r = |C_x|$, hence every feasible flow of N_x corresponds to an $\overrightarrow{st}$ flow with value r in N'_x. Since every arc has unit capacity in N'_x, an integral $\overrightarrow{st}$ flow with value r corresponds to r arc disjoint paths from s to t. Now consider the matroid M given by Lemma 5.3. Denote by A_t the arcs incident to t, and let matroid M_t be the restriction of matroid M to A_t. Let $A'_t \subseteq A_t$ be those arcs of A_t that originate from some node $x_v \in U_x$ (and not from x). We claim that $\mathscr{S}(G, K_x)$ is isomorphic to the projection of M_t to A'_t (vertex $v \in K_x$ maps to arc $\overrightarrow{x_v t}$). By Lemma 4.3, if $S \in \mathscr{S}(G, K_x)$, then there is a feasible flow in N_x where flow is consumed only by those terminals of U_x that correspond to the elements in S. Based on this flow, one can find r arc disjoint $\overrightarrow{st}$ paths in N'_x, and it follows that the matroid M_t has a base whose intersection with A'_t is exactly S, hence S is in the projection of M_t to A'_t. It is easy to show the other direction as well: if S is in the projection of M_t, then there is a feasible flow of N_x where only the terminals corresponding to S consume flow in U_x. Thus by Lemma 4.3, $S \in \mathscr{S}(G, K_x)$, as required. □

Acknowledgments

I am grateful to Katalin Friedl for her useful comments on the paper.

References

[1] R.K. Ahuja, T.L. Magnanti, and J.B. Orlin. *Network flows: theory, algorithms, and applications*. Prentice Hall Inc., Englewood Cliffs, NJ, 1993.

[2] M. Biró, M. Hujter, and Zs. Tuza. Precoloring extension. I. Interval graphs. *Discrete Math.*, 100(1-3):267–279, 1992.

[3] H.L. Bodlaender, K. Jansen, and G.J. Woeginger. Scheduling with incompatible jobs. *Discrete Appl. Math.*, 55(3):219–232, 1994.

[4] P. Erdős, A.L. Rubin, and H. Taylor. Choosability in graphs. In *Proceedings of the West Coast Conference on Combinatorics, Graph Theory and Computing (Humboldt State Univ., Arcata, Calif., 1979)*, pages 125–157, Winnipeg, Man., 1980. Utilitas Math.

[5] M.C. Golumbic. *Algorithmic graph theory and perfect graphs*. Academic Press, New York, 1980.

[6] M. Hujter and Zs. Tuza. Precoloring extension. II. Graph classes related to bipartite graphs. *Acta Mathematica Universitatis Comenianae*, 62(1):1–11, 1993.

[7] M. Hujter and Zs. Tuza. Precoloring extension. III. Classes of perfect graphs. *Combin. Probab. Comput.*, 5(1):35–56, 1996.

[8] T. Kloks. *Treewidth*, volume 842 of *Lecture Notes in Computer Science*. Springer-Verlag, Berlin, 1994.

[9] J. Kratochvíl. Precoloring extension with fixed color bound. *Acta Mathematica Universitatis Comenianae*, 62(2):139–153, 1993.

[10] D. Marx. Precoloring extension on unit interval graphs. To appear in *Discrete Applied Mathematics*.

[11] B. Mohar, Zs. Tuza, and G. Woeginger, 1998. Manuscript.

[12] K. Murota. *Matrices and matroids for systems analysis*, volume 20 of *Algorithms and Combinatorics*. Springer-Verlag, Berlin, 2000.

[13] A. Recski. *Matroid theory and its applications in electric network theory and statics*, volume 6 of *Algorithms and Combinatorics*. Springer-Verlag, Berlin, New York and Akadémiai Kiadó, Budapest, 1989.

[14] D.J. Rose, R.E. Tarjan, and G.S. Lueker. Algorithmic aspects of vertex elimination on graphs. *SIAM J. Comput.*, 5(2):266–283, 1976.

[15] Zs. Tuza. Graph colorings with local constraints – a survey. *Discuss. Math. Graph Theory*, 17(2):161–228, 1997.

[16] V.G. Vizing. Coloring the vertices of a graph in prescribed colors. *Diskret. Analiz*, (29 Metody Diskret. Anal. v Teorii Kodov i Shem):3–10, 101, 1976.

Dániel Marx
Dept. of Computer Science and Information Theory
Budapest University of Technology and Economics
H-1521 Budapest, Hungary
e-mail: `dmarx@cs.bme.hu`

On the Enumeration of Bipartite Minimum Edge Colorings

Yasuko Matsui and Takeaki Uno

Abstract. For a bipartite graph $G = (V, E)$, an edge coloring of G is a coloring of the edges of G such that any two adjacent edges are colored in different colors. In this paper, we consider the problem of enumerating all edge colorings with the fewest number of colors. We propose a simple polynomial delay algorithm whose amortized time complexity is $O(|V|)$ per output, whereas the previous fastest algorithm took $O(|E|\log|V|)$ time per output. Although the delay of the algorithm is $O(|E||V|)$, the delay of our algorithm can be reduced to $O(|V|)$ by using a simple modification with a queue of polynomial size. We show an improvement to reduce the space complexity from $O(|V||E|)$ to $O(|E| + |V|)$. Furthermore, we obtain a lower bound $(|E| - |\hat{V}| + 1)\max\{2^{\Delta-3}, 2(|\hat{V}|/2 + 1)^{\Delta-3}/(\Delta-1)\}/\Delta$ of the number of edge colorings included in G, where Δ is the maximum degree and $\hat{V}$ is the set of vertices of the maximum degree.

Keywords. Enumeration, generation, listing, edge coloring, bipartite graph, algorithm, complexity, output polynomial.

1. Introduction

Enumeration problems and enumeration algorithms are quite fundamental in computer science. The subject has a long history, and many studies have been done [5, 7, 9, 14, 16, 18]. Enumeration has many applications in the other area of computer science, such as optimization, sampling, data mining, bioinformatics, and so on. For example, the basis of branch and bound algorithms is the enumeration, and many exact algorithms for NP-hard problems, which are actively studied in these 10 years, use enumeration algorithms. In data mining, the pattern mining algorithms, which finds all the patterns satisfying given constraints from a database, utilize the enumeration of candidate patterns[1]. Particularly, the recent increase of the power of computers supports the efficiency of enumeration approaches in practice.

A weak point of enumeration approach is that there are quite few generalized problems which contain many other enumeration problems as their special cases.

For example, in optimization, linear programming is a generalized problem, and it includes many other problems as its special cases, such as maximum matchings, network flow problems, assignment problems. In contrast, only few enumeration problems can be efficiently solved by other enumeration algorithms. For example, if we enumerate all spanning trees in a given graph G by enumerating all subtrees of G, it possibly takes exponential time for each spanning tree, since the number of subtrees is often exponentially larger than the number of spanning trees. If we want to reduce an enumeration problem A to an enumeration problem B, we have to preserve the structures with respect to all the solutions. In contrast to it, in optimization problems, we have to preserve only structures with respect to optimal solutions. Intuitively, this is one of difficulties of the reduction on enumeration problems.

One approach to handle enumeration problems efficiently is to develop fundamental techniques commonly applicable to many enumeration problems. For achieving polynomial algorithms, there are several techniques, such as divide and conquer (binary partition), backtracking, and reverse search. Here an enumeration algorithm is polynomial time if the computation time is polynomial in the input size and the output size of the input problem. For reducing the order of the time complexity of polynomial time algorithms, such as using the sparsity, data structures, and amortized analysis of the time complexity. There are quite many kinds of enumeration problems, thus it is important to develop and summarize efficient techniques. One of the big tasks in the research of enumeration algorithms is to clarify what kind of structures of the problems help to reduce the time complexity, and what kind of techniques can be applicable to the structures. In the literatures, we can see many efficient but simple enumeration algorithms for fundamental graph objects such as paths and cycles[11], spanning trees[7, 16], independent sets and cliques[6], and matchings[4], and fundamental geometrical objects such as vertices of polytopes[2], non-crossing spanning trees in plane[2], and floorplans[14].

In this paper, we consider the problem of enumerating all the minimum edge colorings of a given bipartite graph with multiple edges. Let $G = (V(= V_1 \cup V_2), E)$ be a bipartite graph with vertex set V and edge set E. An *edge coloring* of G is a coloring of all the edges of G such that no pair of adjacent edges is colored the same. An edge coloring with the minimum number of colors is called a *minimum edge coloring*. We simply denote a minimum edge coloring by an edge coloring if there is no confusion. We denote the maximum degree of G by $\Delta(G)$, and the set of vertices of maximum degree by $\hat{V}(G)$. If there is no confusion, we simply write them Δ and $\hat{V}$.

In 1916, König [8] proved that any minimum edge coloring of a bipartite graph G uses exactly Δ colors. Since no edges with the same color are adjacent, the set of edges with the same color forms a matching. Hence we can consider an edge coloring as a partition of E into Δ disjoint matchings. The enumeration problem of edge coloring considered here is to output all the ways of partitioning of E into disjoint Δ matchings.

An algorithm for solving the problem has been proposed by Matsui and Matsui [12, 13]. The amortized time complexity of the algorithm is $O(|E|\log|V|)$ per edge coloring, and the space complexity is $O(\Delta|E|)$. The current best time complexity algorithm for finding a minimum edge coloring in a given bipartite graph, which is proposed by Cole, et al., and Schrijver [3, 15], takes $O(|E|\log|V|)$ time, hence naturally we may think that $\Theta(|E|\log|V|)$ is a kind of lower bound for the time complexity. However, the structure of the set of edge coloring seems to have an advantage. For any edge coloring, we can generate another edge coloring by exchanging several edges of two matchings in it, along an alternating cycle. This implies that when we traverse the set of edge colorings, we basically need the time to exchange the edges along a cycle, which seems to be very short in average. Thus, naturally there is a question "Is there an algorithm for enumerating all minimum edge colorings in short time for each".

In this paper, we give a positive answer to the question. We propose a simple algorithm running in $O(|V|)$ time for each edge coloring. The space complexity is also reduced to $O(|E|+|V|)$. In detail, the delay[6] of the algorithm is $O(\Delta|E|\log|V|)$. The delay is the maximum computation time between two consecutive output. Actually, by the technique described in [17], we can reduce the delay to $O(|V|)$ by using $O(\Delta|E|)$ extra memory. We note that to output an edge coloring, our algorithm outputs the symmetric difference between an edge coloring and the next one instead of exact output. It reduces the computation time for output one edge coloring to $O(|V|)$ on average.

The main technique to reduce the time complexity is on the analysis of the time complexity. Actually, our algorithm is obtained by adding slight and simple modifications to the previous algorithm. It uses neither complicated data structures nor sophisticated algorithms. The modifications of the algorithm avoid the worst cases which make the time complexity of the previous algorithm tight. It is interesting that such kind of simple modifications reduces the time complexity so much. This is also an advantage from both theoretical and application viewpoints. Furthermore, as a corollary of the amortized analysis, we give $(|E|-|\hat{V}|+1)\max\{2^{\Delta-3}, 2(|\hat{V}|/2+1)^{\Delta-3}/(\Delta-1)\}/\Delta$ as a lower bound of the number of edge colorings included in G.

The organization of this paper is as follows. We explain the framework of our enumeration algorithm in Section 2. In Section 3, we analyze the time complexity, and show a lower bound of the number of edge colorings included in a graph. Finally, we explain a way to reduce the space complexity in Section 4.

2. Framework of algorithm for enumerating edge coloring

The basis of our enumeration algorithm is the same as that described in [12, 13]. We start the explanations with the definitions and properties. For a vertex set W, a matching covering all the vertices of W is called a *covering matching* for W. If no confusion can arise we will omit W. From König's theorem, any minimum edge coloring uses exactly Δ colors. This means that any vertex of $\hat{V}$ is incident to edges with all colors, and any matching which forms a minimum edge coloring is a covering matching for $\hat{V}$. Conversely, any covering matching M for $\hat{V}$ is included in at least one minimum edge coloring, since the removal of M from G is a graph with maximum degree $\Delta - 1$.

For an edge e, any edge coloring includes just one covering matching including e. Thus, edge colorings of G is partitioned into groups by the covering matchings including e. This observation immediately leads to the following algorithm.

ALGORITHM: BASIC_ALGORITHM $(G = (V, E), C)$
(BA1) **If** G is a matching **then output** C
(BA2) $e :=$ an edge of G
(BA3) **For each** covering matching M for $\hat{V}$ including e
 Call BASIC_ALGORITHM$((V, E \setminus M), C \cup \{M\})$

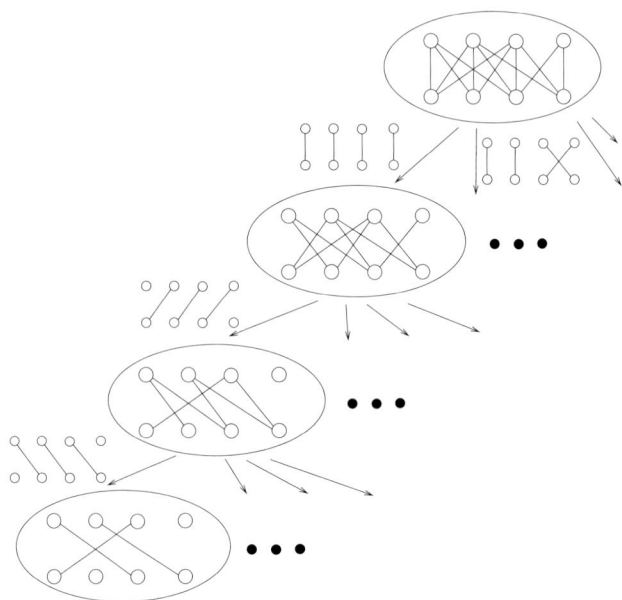

FIGURE 1. An example of enumeration of edge colorings. Each circle denotes a problem/subproblem, and arrows connect problems and their subproblems. Each arrow corresponds to a covering matching which is obtained from each subproblem.

Figure 1 shows an example of the execution of the algorithm. The enumeration of covering matchings in (BA3) can be done by using an algorithm proposed in [12, 13]. We explain their algorithm next. For conciseness, we modify their algorithm slightly.

For an edge $e = (u, v)$, let $G \setminus e = (V, E \setminus \{e\})$, and $G - \{u, v\}$ be the subgraph of G obtained by removing u, v, and all edges incident to either u or v. Here we consider the set of covering matchings for a vertex set W. Then, we can see that the set of the covering matchings not including e is equal to the set of covering matchings in $G \setminus e$. Similarly, we can see that the set of the covering matchings for $W \setminus \{u, v\}$ including e is equal to the set of matchings obtained by adding e to each covering matchings in $G - \{u, v\}$. Thus, this enumeration problem can be partitioned into two subproblems.

To partition the problem, we choose an edge e from the symmetric difference between two covering matchings M and M'. This ensures that both subproblems are non-empty, since either M or M' includes e. By augmenting the matching, we can find a covering matching which covers vertices in $\hat{V}$ We thus explain how to find a covering matching different from the given covering matching.

For the explanation, we define the following notation. Let Z be the set of isolated vertices in G. We recall that V_1 and V_2 are the partition of V so that G is a bipartite graph.

$\bar{U}_1$ is the set of vertices in $V_1 \setminus (Z \cup W)$ incident to an edge of M.
$\bar{U}_2$ is the set of vertices in $V_2 \setminus (Z \cup W)$ incident to an edge of M.
U_1 is the set of vertices in $V_1 \setminus (Z \cup W)$ incident to no edge of M.
U_2 is the set of vertices in $V_2 \setminus (Z \cup W)$ incident to no edge of M.

Let $D_1 = (G, M, W)$ be the directed graph obtained from G as follows:

1. remove all isolated vertices from G
2. orient the edges of M from V_1 to V_2, and the edges of $E \setminus M$ from V_2 to V_1
3. add a vertex s to V
4. add the arcs from s to each vertex of $\bar{U}_1 \cup U_2$, and
5. add the arcs from each vertex of $\bar{U}_2 \cup U_1$ to s.

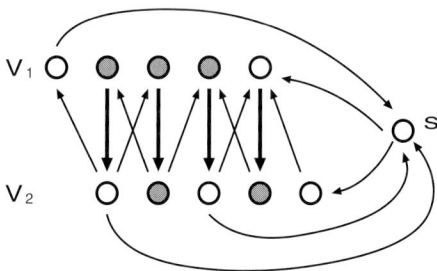

FIGURE 2. An instance of $D_1(G, M, W)$: Gray vertices are in W, and bold edges are in matching M.

For an example of $D_1(G, M, W)$, see Figure 2. For a directed cycle C of $D_1(G, M, W)$, we define $E(C)$ by the set of all edges corresponding to the arcs of C. From the rule of orienting edges, edges of M and edges not in M appear in $E(C)$ alternatively in any directed cycle. This is a common technique to find alternating paths and alternating cycles.

$D_1(G, M, W)$ is equivalent to the graph obtained by contracting the source and sink of the graph $G'(G, M)$ used in the **find-matching** algorithm of [13]. In [13], the following lemma is proved.

Lemma 1. [13] *The following two conditions hold.*
(1) *For any directed cycle C of $D_1(G, M, W)$, $M \triangle E(C)$ is a covering matching for $\hat{V}$.*
(2) *If $D_1(G, M, W)$ has no directed cycle, then M is the unique covering matching of G.* □

Since [13] is not a journal paper, we include a proof here.

Proof. For any two covering matchings M and M', the symmetric difference between them is composed of paths and cycles. Any cycle of those cycles forms a directed cycle in $D_1(G, M, W)$. Any path P of those paths forms a directed path P' of $D_1(G, M, W)$ whose end vertices are not in W. Moreover, since the end edges of P are not adjacent to the edges of M, one end vertex of P is in $\bar{U}_1 \cup U_2$ and the other is in $\bar{U}_2 \cup U_1$. Hence, by connecting s and the end vertices of P', we obtain a directed cycle.

Let C be a directed cycle of $D_1(G, M, W)$. If C does not include s, then $M \triangle E(C)$ is a covering matching, since the sets of vertices incident to the edges of matchings are the same for M and $M \triangle E(C)$. If C includes s, then $E(C)$ is a path connecting a vertex of $\bar{U}_1 \cup U_2$ and a vertex of $\bar{U}_2 \cup U_1$. Hence, $M \triangle E(C)$ is a matching. Since the end vertices of $E(C)$ are not in W, $M \triangle E(C)$ is a covering matching. □

A covering matching for $\hat{V}$ can be found in $O(|E| \log |V|)$ time by the algorithms of Cole et al., and Schrijver [3, 15]. Actually, the computation time can be also bounded by $O(|\hat{V}|^2 \Delta)$, by removing all the edges incident to no vertex of $\hat{V}$ in $O(|E|)$ time. Thus, the enumeration of covering matchings takes $O(|E| + |\hat{V}|^2 \Delta)$ time for the first covering matching, and $O(|E|)$ for each following covering matching.

Our algorithm is obtained by introducing three additional modifications to the basic algorithm. These modifications are the following. We note that a star is a graph such that all the edges are incident to a vertex.

(1) if the input graph is a star, then output the unique edge coloring directly
(2) choose an edge incident to not all the edges in (BA2)
(3) output by the difference from the previous output

The modified algorithm is described below. Note that the algorithm is composed of two procedures, the main part and the enumeration of covering matchings. The procedures are nested so that they call recursively each other. The algorithm memorize a graph with multiple edges by its underline graph with the multiplicity for each edge. Hence, the memory space never exceed $O(|V|^2)$. For outputting an edge coloring of a star, we output its underline graph and the multiplicities, and a message "each edge is a matching", instead of exact output. By this modification, the execution of (1) never take more than $O(|V|)$ time. In the next section we analyze the time complexity of the algorithm to bound it by $O(|V|)$ for each.

ALGORITHM: ENUM_EDGE_COLORING($G = (V, E)$:graph,
  Col:set of matchings to be edge colorings)
(1) **If** $\Delta(G) = 1$, **output** edge coloring $Col \cup \{E\}$
(2) **If** (G is a star) or (E is a matching) **then output** Col and the unique edge
  coloring of G // improvement (1)
(3) $(u, v) :=$ an edge not adjacent to all edges // improvement (2)
(4) $M :=$ a covering matching for $\hat{V}(G)$ of G including e
  // computed in $O(|\hat{V}|^2 \Delta)$ time
(5) **Call** ENUM_COVERING_MATCHING
  $(G - \{u, v\}, \hat{V}(G) \setminus \{u, v\}, M \setminus \{(u, v)\}, G, Col)$
  // enumerate covering matchings

ALGORITHM: ENUM_COVERING_MATCHING
  (H:graph, W:vertices to be covered,
  M:edge set to be a matching, G:original graph,
  Col:set of matching to be an edge coloring)
(6) **If** no directed cycle is in $D_1(H, M, W)$ **then**
  Call ENUM_EDGE_COLORING $((V, E \setminus M, Col \cup \{M\})$
  // generate recursive call when a new covering matching is found
(7) $C :=$ a directed cycle of $D_1(H, M, W)$; $M' := M \triangle E(C)$;
  $(u, v) :=$ an edge in $M \triangle M'$
(8) **If** $(u, v) \in M'$ **then swap** M and M'
  // now (u, v) is an edge in $M \setminus M'$
(9) **Call** ENUM_COVERING_MATCHING
  $(H - \{u, v\}), W \setminus \{u, v\}, M \setminus \{(u, v)\}, G, Col)$
  // enumerate covering matchings including (u, v)
(10) **Call** ENUM_COVERING_MATCHING $(H \setminus (u, v), W, M', G, Col)$
  // enumerate covering matchings not including (u, v)

3. Analysis of the time complexity

In this section, we now start with some definitions. We define an iteration of the enumeration algorithm of edge colorings by the set of operations to generate sub-problems for enumerating edge colorings from an input graph, i.e., the union of the iteration of ENUM_EDGE_COLORING inputting graph G and all the iterations of ENUM_COVERING_MATCHING which enumerate covering matchings of G. For each iteration x, we denote the input graph of x by $G_x = (V, E_x)$. Iterations generated by x are called the children of x. The depth of the recursion is up to Δ, and each iteration on the bottom of the recursion outputs an edge coloring.

Lemma 2. *For a graph G with $\Delta \geq 3$, a covering matching for $\hat{V}$ including an edge $e = (u, v)$ is unique if and only if all edges of G are adjacent to e.*

Proof. The 'if' part is obvious, thus we prove the 'only if' part by its contraposition. Let M be a covering matching for $\hat{V}$. We show that $D_1(G - \{u, v\}, M, \hat{V} \setminus \{u, v\})$ includes a directed cycle. From Lemma 1, this implies that G has a covering matching different from M. Since at least one edge is not adjacent to e, $G - \{u, v\}$ contains at least one edge. For any vertex w of $\hat{V} \setminus \{u, v\}$, w is incident to at most one of u and v. Hence, its degree in $G - \{u, v\}$ is no less than two, and w has at least one out-going arc in $D_1(G - \{u, v\}, M, \hat{V} \setminus \{u, v\})$. For any vertex w in $V \setminus \hat{V} \setminus \{u, v\}$, there is an arc (w, s) or (w, x) for some $x \neq u, v$ since (w, x) in M. Hence the out-degree of any non-isolated vertex of $D_1(G - \{u, v\}, M, \hat{V} \setminus \{u, v\})$ is at least one. Therefore, a depth-first search for the graph always reaches a vertex that has already been visited, and gives a directed cycle. □

If any edge of G_x is adjacent to all the other edges, then all the edges are incident to a vertex (note that G_x is bipartite), and G_x is a star. In this case the algorithm outputs the unique edge coloring in (1) and the iteration terminates, in $O(|V|)$ time. In the other case, the algorithm chooses an edge e so that at least one edge is not adjacent to e. From Lemma 2, at least two covering matchings includes e. Thus, we have the following corollary.

Corollary 1. *In algorithm* ENUM_EDGE_COLORING, *if an iteration has a child, the number of its children is at least two.* □

This corollary implies that the number of vertices in the enumeration tree is at most twice the number of leaves, which is the same as the number of edge colorings in G. To bound the number of iterations more, we give the following lemmas, where the first one is a standard result of graph theory.

Lemma 3. *For a directed graph $H = (V_H, A_H)$ in which each arc is included in a directed cycle, H contains at least $|A_H| - |V_H| + cc(H)$ directed cycles, where $cc(H)$ is the number of strongly connected components of H.*

Proof. If all the arcs are self-loops or $|V_H| = 1$, the statement holds. Assume that the statement holds if $|V_H| < k$, and we consider the case that $|V_H| = k$ and not all arcs of H are self-loops. Let C be a shortest directed cycle of H including no self-loop, and $|C|$ denote the number of arcs in C. Note that $|C| \geq 2$. Let $H' = (V_{H'}, A_{H'})$ be the graph obtained from H by removing all the arcs of C and contracting vertices of C into a vertex. Since $|V_{H'}| = |V_H| - |C| + 1 < k$, H' has at least

$$|A_{H'}| - |V_{H'}| + cc(H') = (|A_H| - |C|) - (|V_H| - |C| + 1) + cc(H)$$

directed cycles. Note that $cc(H) = cc(H')$. Since H has at least one more directed cycle (which is C) than H', H has at least

$$(|A_H| - |C|) - (|V_H| - |C| + 1) + cc(H) + 1 = |A_H| - |V_H| + cc(H)$$

directed cycles. □

Theorem 1. *Any graph G with $\Delta \geq 3$ and $|\hat{V}| \geq 3$ has at least $(|E| - |\hat{V}| + 1) \max\{2^{\Delta-3}, 2(|\hat{V}|/2 + 1)^{\Delta-3}/(\Delta-1)\}/\Delta$ edge colorings.*

Proof. Let Z be the set of isolated vertices in G, and M be a covering matching. In $D_1(G, M, \hat{V})$, any non-isolated vertex not in $\hat{V}$ is connected to s, hence its degree is at least 2, and the degree of s is $|V| - |Z| - |\hat{V}|$. Therefore, the number of arcs in $D_1(G, M, \hat{V})$ is at least $|E| + (|V| - |Z| - |\hat{V}|)$. For any edge e, an edge coloring of G includes both a covering matching which includes e and another covering matching which does not. This means that any arc of $D_1(G, M, \hat{V})$ is included in a directed cycle. From Lemma 3, $D_1(G, M, \hat{V})$ includes at least

$$|E| + (|V| - |Z| - |\hat{V}|) - (|V| + 1) + |Z| + 1 = |E| - |\hat{V}|$$

directed cycles. Thus, G includes at least $|E| - |\hat{V}| + 1$ covering matchings.

Let v be a vertex in $\hat{V}$ and F be the set of edges incident to v. Any covering matching includes an edge in F, hence there is an edge f in F such that f is included in at least $(|E| - |\hat{V}| + 1)/\Delta$ covering matchings. Thus, if $\Delta = 3$, G includes at least $(|E| - |\hat{V}| + 1)/3$ edge colorings.

We next consider the case $\Delta > 3$. Similar to the above, for any covering matching M, the graph $H = (V, E \setminus M)$ includes at least $(|E \setminus M| - |\hat{V}| + 1)/(\Delta - 1)$ covering matchings including an edge f, which is incident to a vertex in $\hat{V}(H)$. Since $|E| \geq \Delta|\hat{V}|/2$, we have

$$(|E \setminus M| - |\hat{V}| + 1)/(\Delta - 1) \geq ((\Delta - 3)|\hat{V}(H)|/2 + 1)/(\Delta - 1).$$

From Lemma 3, if $|\hat{V}| \geq 3$, there is an edge f incident to a vertex in $\hat{V}(H)$ included in at least two covering matchings. Therefore, by induction, G has at least

$$((|E| - |\hat{V}| + 1)/\Delta) \times \prod_{i=3}^{\Delta-1} \max\{2, ((i-1)|\hat{V}|/2 + 1)/i\}$$
$$\geq (|E| - |\hat{V}| + 1) \max\{2^{\Delta-3}, 2(|\hat{V}|/2 + 1)^{\Delta-3}/(\Delta-1)\}/\Delta$$

edge colorings. □

In particular, we can directly obtain the following corollary from the fact that $|E| \geq \Delta|\hat{V}|/2$.

Corollary 2. *Any graph G with $\Delta = 3$ has at least $|\hat{V}|/6$ edge colorings.*

From these lemmas and corollary, we obtain the following theorem.

Theorem 2. *The algorithm* ENUM_EDGE_COLORING *enumerates all minimum edge colorings of a bipartite graph $G = (V, E)$ with multiple edges in $O(|V|N)$ time, where N is the number of minimum edge colorings of G.*

Proof. For any child y obtained at iteration x, we have $|E_y| \geq |E_x| - |V|/2$, since E_y is obtained by removing a matching from E_x. An iteration x takes $O(|E_x| + |V| + |\hat{V}(G_x)|^2 \Delta(G_x))$ time and $O(|E_x| + |V|)$ time per child. By assigning a computation cost of $O(|E_x|+|V|)$ to each child, we suppose that the computation time $T(x)$ of x is

$$T(x) = \begin{cases} c \times |V| & \text{if } \Delta(G_x) < 3 \text{ or } G_x \text{ is a star} \\ c \times (|\hat{V}(G_x)|^2 \Delta(G_x) + |E_x| + |V|) & \text{otherwise} \end{cases}.$$

Here c is a constant, and $T(x)$ does not include the computation time for outputting the edge colorings. We have $|\hat{V}(G_x)| \leq |\hat{V}(G_y)|$, since any vertex of $\hat{V}(G_x)$ is the maximum degree in G_y. Hence, if $\Delta(G_x) > 3$ and G_x is not a star,

$$\begin{aligned} T(x) &= c \times (|\hat{V}(G_x)|^2 \Delta(G_x) + |E_x| + |V|) \\ &\leq c \times (|\hat{V}(G_y)|^2 (\Delta(G_y) + 1) + |E_y| + 1.5|V|). \end{aligned}$$

Hence, from Corollary 1, for any $4 \leq k \leq \Delta(G)$,

$$\sum_{x|\Delta(G_x)=k, G_x \text{ is not a star}} T(x)$$

$$\leq \sum_{y|\Delta(G_y)=k-1} c \times (|\hat{V}(G_y)|^2 (\Delta(G_y) + 1) + |E_y| + 1.5|V|)/2$$

$$\leq \sum_{y|\Delta(G_y)=k-1} 3T(y)/4.$$

Therefore, from Lemma 2,

$$\begin{aligned} \sum_{x|\Delta(G_x)\geq 3} T(x) &\leq \sum_{x|\Delta(G_x)=3} 4T(x) \\ &\leq \sum_{x|\Delta(G_x)=3} 48c(|\hat{V}(G_x)||V|) \\ &\leq 48cN|V| \end{aligned}$$

Next we consider the computation time for outputting the obtained edge colorings. Consider the recursive structure of our algorithm as a recursion tree. Our algorithm outputs the difference from the edge coloring output just before. From the edge coloring which is outputted just before, the size of the difference is at most $|V|/2$ times the number of edges in the recursion tree traced by the algorithm. Hence, the sum of these numbers of edges in the output is at most $|V|/2 \times 2(\#$ of iterations). Since the number of iterations is less than $2N$, the computation time for output is $O(|V|N)$ per output. We note that the algorithm outputs the unique edge coloring in a star in $O(|V|)$ time. □

4. Reducing delay

The delay is the maximum computation time between two consecutive outputs. An enumeration algorithm is said to be polynomial delay if its delay is polynomial of the input size[6]. If an enumeration algorithm is polynomial delay, it is output polynomial time, but an output polynomial time algorithm is not always polynomial delay. Thus, polynomial delay is a stronger result than output polynomial. We recall that an enumeration algorithm is output polynomial if the computation time is bounded by a polynomial of the input and output size. In this section, we carefully analyze the delay of our algorithm.

First, we show that the delay of covering matchings enumeration can be bounded by $O(|E|)$. In [10, 17], they claimed that if an enumeration algorithm outputs a solution in each iteration, then the delay can be 3 times the maximum computation time of an iteration. Algorithm ENUM_COVERING_MATCHING finds a new matching in each iteration, thus we can modify the algorithm to satisfy the condition. The idea in [10, 17] is to modify the algorithm so that the algorithm outputs a solution before executing recursive calls at odd levels of the recursion, and after the recursive calls at even levels. Then, we can see that at least one iteration of any consecutive three iterations must output a solution. Thus, delay is $O(|E|)$ without increasing neither time nor space complexity.

Next, we reduce the delay of the main algorithm. For an iteration I of an enumeration algorithm, let $Out(I)$ be the set of solutions output by the iterations which are descendants of I. Suppose that the delay of an enumeration algorithm A is D, and satisfies that for any iteration I, the amortized computation time taken by all descendants of I is T per solution in $Out(I)$. In [17], we can see that under these conditions, the delay can be reduced to $O(T)$ with using $O(D/T \times S)$ memory where S is the maximum size of output. The main idea is that we make a buffer and insert each solution into the buffer when the algorithm outputs it. The solutions in the buffer is extracted and output one by one with keeping the intermediate computation time equal to $6T$ unless the buffer is overflow. We can prove that after the buffer is once full, it never be empty. At the beginning of the enumeration, we do not extract solutions from the buffer until the first overflow of the buffer. Then, the delay is $6T = O(T)$.

On our algorithm, $T = O(|V|)$ and $S = |V|$. Since each iteration of both procedures takes $O(|E|)$ time, and the depth of the recursion is $O(\Delta)$, the delay is $O(\Delta|E|)$. According to the above, we can reduce the delay to $O(|V|)$ by using $O(\Delta|E|)$ memory.

Theorem 3. *Minimum edge colorings in a bipartite graph $G = (V, E)$ can be enumerated in $O(|V|)$ delay by using $O(\Delta|E|)$ memory, where Δ is the maximum degree in G.*

5. Reducing space complexity

In this section, we describe a way for reducing the space complexity. Since our enumeration algorithm is composed of two nested enumeration algorithms, which are for edge colorings and for covering matchings, the total depth of two recursions can be up to $\Theta(\Delta|E|)$. Note that here the iterations are different from the definition in Section 3. Here we consider that an iteration is the computation time in a recursive call except for the computation done in the further recursive calls generated in it.

Each recursive call requires $\Omega(1)$ memory, hence the total required memory is up to $\Theta(\Delta|E|)$. When ENUM_COVERING_MATCHING generates a subroutine call of ENUM_COVERING_MATCHING, $O(|E|+|V|)$ memory is required to store H in each level, hence the accumulated memory for storing H is up to $O(\Delta(|E|+|V|))$. These two parts are the bottle neck of the space complexity. Note that the total accumulated memory to store M and I is $O(|E|)$, since each time the algorithm executes, stored matchings $M \cup I$ compose a subset of an edge coloring. Note also that $\hat{V}$ can be computed from $\hat{V}$ and I in $O(|E|+|V|)$ time.

To improve these memory-consuming parts, we add the following two modifications to ENUM_COVERING_MATCHING. The first one is to use a loop instead of generating a recursive call with respect to $H \setminus e$. By this modification, a recursive call always adds an edge to I or a matching to C, hence the depth of the recursion is at most $|E|+|V|$.

The second modification is to use the minimum possible index edge e to partition the problem, in ENUM_COVERING_MATCHING. For a graph G, a matching M, and an index j, let $G(M,j) = (V,X)$ where

$$X = \{e_i \in E | i \geq j \text{ and } (e_i \text{ is not adjacent to any edge } e_h \in M, h < j)\}$$

For an example of $G(M,j)$, see Figure 3.

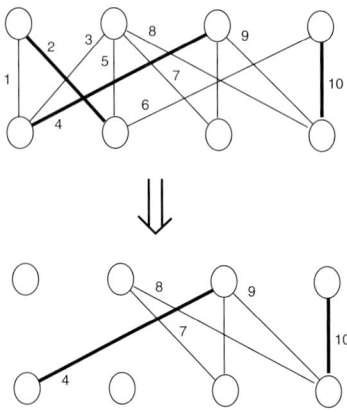

FIGURE 3. An example of G and $G(I \cup M, l)$. In the figure, the edges of $I \cup M$ are shown by the bold lines, and $l = 6$.

Suppose that an iteration of ENUM_COVERING_MATCHING inputs a graph H and a covering matching M. Let e_l be the minimum index edge of H. If $e_l \in E(C)$ for a directed cycle C of $D_1(H, M, \hat{V})$, the condition holds for both subproblems generated by e_l. If $e_l \notin E(C)$ for any directed cycle C of $D_1(H, M, \hat{V})$, the condition holds for H and $G(I \cup M, l)$. Therefore, by induction, at any iteration of ENUM_COVERING_MATCHING inputting $H, \hat{V}, M$, and I such that the minimum index edge of H is e_l, the set of covering matchings in H and that of $G(I \cup M, l)$ are equal. From this, we can see that a graph equivalent to H can be constructed from $G, I \cup M, l$ in $O(|E| + |V|)$ time.

By using these two modifications, the space complexity is reduced to $O(|E| + |V|)$. We now describe the algorithm with these two modifications.

ALGORITHM: ENUM_COVERING_MATCHING2 $(H, M, \hat{V})$
(1) **If** $D_1(H, M, \hat{V})$ includes no directed cycle **then output** M ; **return**
(2) $e_l (= (u_l, v_l)) :=$ the minimum index edge of H included
    in some directed cycle
(3) $H := G(M, l)$ // remove edges less than l and not included in M
(4) $C :=$ a directed cycle including e_l ; $M' := M \triangle E(C)$
(5) **If** $e_l \in M$ **then swap** M and M'
(6) **Call** ENUM_COVERING_MATCHING2
    $(H - \{u_l, v_l\}, M' \cup \{(u_l, v_l)\}, \hat{V} \setminus \{u_l, v_l\})$
         // H will be changed by the execution of the recursive call
(7) $H := G(M, l)$
(8) Remove e_l from H ; **go to** (1)

Step (1) outputs a covering matching if the covering matching of H is unique. An execution of (2) through (8) corresponds to an internal iteration which has some children in the enumeration tree. e_l is the edge to be used partitioning the problem, step (6) generates a recursive call for enumerating covering matchings including e_l, and step (8) corresponds to the recursive call for enumerating covering matchings not including e_l. Setting H to $G(M, l)$ in (3) is equivalent to removing the edges whose indices are less than l and not included in M. Since the recursive call in (6) changes H, H is not preserved after the termination of the recursive call. However, G is preserved in the execution[1], we can reconstruct H by setting H to $G(M, l)$. This is the key to save the memory for storing H during the execution of the recursive call. Step (8) removes e_l from H and go to the beginning. It corresponds to the recursive call for enumerating covering matchings not including e_l.

By using this algorithm, we obtain the following theorem.

Theorem 4. *The algorithm*

ENUM_EDGE_COLORING *with* ENUM_COVERING_MATCHING2

enumerates all minimum edge colorings of a bipartite graph $G = (V, E)$ with multiple edges in $O(|V|N)$ time and $O(|E| + |V|)$ space, where N is the number of minimum edge colorings of G. □

---
[1] In exact, G is changed by recursively calling ENUM_EDGE_COLORING, however it is reconstructed after the termination by adding the covering matching which is removed before the execution

6. Conclusion

We proposed an algorithm for enumerating all minimum edge colorings in a bipartite graph $G=(V,E)$ without using any sophisticated data structure or any sophisticated algorithm. The amortized time complexity of the algorithm is $O(|V|)$ per output. It improves the previous algorithm by a factor of $|E|\log|V|/|V|$. We also reduced the space complexity of the algorithm from $O(|E|\Delta)$ to $O(|E|+|V|)$. Although the delay of the algorithm is $O(\Delta|E|)$, we can reduce it to $O(|V|)$ by using a queue with $O(\Delta|E|)$ memory. We further give a lower bound $(|E|-|\hat{V}|+1)\max\{2^{\Delta-3}, 2(|\hat{V}|/2+1)^{\Delta-3}/(\Delta-1)\}/\Delta$ of the number of edge colorings included in G.

Acknowledgments

We would like to thank Professor Akihisa Tamura of Kyoto University and Professor Yoshiko T. Ikebe of Tokyo University of Science for their advice, and Professor David Avis of McGill University for his careful reading of our paper. This research is supported by Grant-in-Aid for scientific research from the ministry of education, science, sports and culture of Japan.

References

[1] R. Agrawal, H. Mannila, R. Srikant, H. Toivonen, and A.I. Verkamo, "Fast Discovery of Association Rules," *Advances in Knowledge Discovery and Data Mining*, MIT Press (1996), pp. 307–328, .

[2] D. Avis and K. Fukuda, "Reverse search for enumeration," *Discrete Applied Mathematics*, **65** (1996), pp. 21–46.

[3] R. Cole, K. Ost and S. Schirra, "Edge-Coloring Bipartite Multigraphs in $O(E\log D)$ Time," *Combinatorica* **21** (2001), pp. 5–12.

[4] K. Fukuda and T. Matsui, "Finding All the Perfect Matchings in Bipartite Graphs," *Applied Mathematics Letters* **7** (1994), pp. 15–18.

[5] L.A. Goldberg, "Efficient Algorithms for Listing Combinatorial Structures," *Cambridge University Press*, New York, (1993).

[6] D.S. Johnson, M. Yanakakis and C.H. Papadimitriou, "On Generating All Maximal Independent Sets," *Information Processing Letters*, **27** (1998), pp. 119–123.

[7] S. Kapoor, and H. Ramesh, "An Algorithm for Enumerating All Spanning Trees of a Directed Graph," *Algorithmica* **27** (2000), pp. 120–130.

[8] D. König, "Über Graphen und ihre Anwendung auf Determinantentheorie und Mengenlehre," *Math. Ann.* **77** (1916), pp. 453–465.

[9] D.L. Kreher and D.R. Stinson, "Combinatorial Algorithms," *CRC Press*, Boca Raton, (1998).

[10] S. Nakano, T. Uno, "Constant Time Generation of Trees with Specified Diameter," *Lecture Notes in Computer Science* **3353**, Springer-Verlag, (2004) pp. 33–45.

[11] R.C. Read and R.E. Tarjan, "Bounds on Backtrack Algorithms for Listing Cycles, Paths, and Spanning Trees," *Networks* **5**, 237–252 (1975).

[12] Y. Yoshida, Matsui and T. Matsui, "Finding All the Edge Colorings in Bipartite Graphs," *T.IEE. Japan 114-C* **4** (1994), pp. 444–449 (in Japanese).

[13] Y. Matsui, and T. Matsui, "Enumeration Algorithm for the Edge Coloring Problem on Bipartite Graphs," *Lecture Notes in Computer Science* **1120**, Springer-Verlag, (1996), pp. 18–26.

[14] S. Nakano, "Enumerating Floorplans with n Rooms," *Lecture Notes in Computer Science* **1350**, Springer-Verlag, (2001), pp. 107–115.

[15] A. Schrijver, "Bipartite Edge-Colouring in $O(\Delta m)$ Time," *SIAM Journal on Computing* **28** (1999), pp. 841–846.

[16] A. Shioura, A. Tamura, and T. Uno, "An Optimal Algorithm for Scanning All Spanning Trees of Undirected Graphs," *SIAM Journal on Computing* **26** (1997), pp. 678–692.

[17] Takeaki Uno, "Two General Methods to Reduce Delay and Change of Enumeration Algorithms," *NII Technical Report*, (2004), http://research.nii.ac.jp/TechReports/index.html

[18] H.S. Wilf, "Combinatorial Algorithms: An Update", SIAM, (1989).

Yasuko Matsui
Department of Mathematical Sciences
Faculty of Science
Tokai University
1117, Kitakaname, Hiratsuka-shi
Kanagawa, Japan
e-mail: `yasuko@ss.u-tokai.ac.jp`

Takeaki Uno
National Institute or Informatics
2-1-2 Hitotsubashi, Chiyoda-ku
Tokyo, 101-8430, Japan
e-mail: `uno@nii.jp`

Kempe Equivalence of Colorings

Bojan Mohar

Abstract. Several basic theorems about the chromatic number of graphs can be extended to results in which, in addition to the existence of a k-coloring, it is also shown that all k-colorings of the graph in question are Kempe equivalent. Here, it is also proved that for a planar graph with chromatic number less than k, all k-colorings are Kempe equivalent.

1. Introduction

Let G be a graph and $k \geq 1$ an integer. A vertex set $U \subseteq V(G)$ is *independent* if no two vertices of U are adjacent in G. A k-*coloring* of G is a partition of $V(G)$ in k independent sets $U_1, \ldots, U_k$, called *color classes*. If $v \in U_i$ ($i \in \{1, \ldots, k\}$), then v is said to have *color i*. Every k-coloring can be identified with a mapping $c : V(G) \to \{1, \ldots, k\}$ where $c(v)$ is the color of v. The *chromatic number* of G is denoted by $\chi(G)$.

Let $a, b \in \{1, \ldots, k\}$ be distinct colors. Denote by $G(a, b)$ the subgraph of G induced on vertices of color a or b. Every connected component K of $G(a, b)$ is called a *K-component* (short for *Kempe component*). By switching the colors a and b on K, a new coloring is obtained. This operation is called a *K-change* (short for *Kempe change*). Two k-colorings c_1, c_2 are K-*equivalent* (or K^k-*equivalent*), in symbols $c_1 \sim_k c_2$, if c_2 can be obtained from c_1 by a sequence of K-changes, possibly involving more than one pair of colors in successive K-changes.

Let $\mathcal{C}_k = \mathcal{C}_k(G)$ be the set of all k-colorings of G. The equivalence classes $\mathcal{C}_k / \sim_k$ are called the K^k-*classes* (or just K-*classes*). The number of K^k-classes of G is denoted by Kc(G, k).

K-changes have been introduced by Kempe in his false proof of the four color theorem. They have proved to be an utmost useful tool in graph coloring theory. It remains one of the basic and most powerful tools. The results of this paper show that some basic theorems about graph colorings can usually be turned into

---

Supported in part by the Ministry for Higher Education, Science and Technology of Slovenia, Research Program P1–0297.

stronger K-equivalence results where it can be proved that all k-colorings are K-equivalent. These results have been one of our motivations to study K-equivalence of colorings. Several such results have been published by Meyniel and Las Vergnas [9, 7] who proved, in particular, that all 5-colorings of a planar graph (respectively, a K_5-minor free graph) are K-equivalent. Fisk [5] proved that all 4-colorings of an Eulerian triangulation of the plane are K-equivalent. We extend these results by showing that in every planar graph G with chromatic number less than k, all k-colorings are K-equivalent (see Corollary 4.5).

The second motivation to study K-equivalence is the possibility to generate colorings either by using K-changes as a heuristic argument [1, 12], or with the goal of obtaining a random coloring by applying random walks and rapidly mixing Markov chains [16]. For instance, Vigoda [16] proved that the Markov chain, whose state space is $\mathcal{C}_k(G)$ and whose transitions correspond to K-changes, quickly converges to the stationary distribution if $k \geq \frac{11}{6}\Delta(G)$. On the other hand [8], there are bipartite graphs for which the Markov chain needs exponentially many steps to come close to the stationary distribution if $k = O(\Delta/\log \Delta)$. Later, Hayes and Vigoda [6] proved rapid mixing for $k > (1+\varepsilon)\Delta(G)$ for all $\varepsilon > 0$ assuming that G has girth more than 9 and $\Delta = \Omega(\log n)$. Dyer et al. [3] studied the same phenomenon on random graphs with expected average degree d, where d is a constant. Kempe change method has been successfully applied in some experiments [13] leading to new theoretical results. In relation to this, let us mention that K-changes appear in theoretical physics in the study of the Glauber dynamics for the hard-core lattice gas model at zero temperature. The related Wang-Swendsen-Kotecký dynamics [17, 18] uses K-changes to move from state to state. The question whether the associated Markov chain is ergodic is the same as asking if all colorings are K-equivalent. We refer to a survey by Sokal [14] for further details.

Finally, let us observe that Claude Berge, to whom we dedicate this paper, considered Kempe changes in some of his late papers, e.g., [2].

2. Basic results

The following result shows that the study of K^k-equivalence may be interesting also when k is much larger than the chromatic number of the graph. It also shows that it is possible that $\mathrm{Kc}(G, k-1) = 1$ and $\mathrm{Kc}(G, k) > 1$.

Proposition 2.1.

(a) Let G be a bipartite graph and $k \geq 2$ an integer. Then $\mathrm{Kc}(G, k) = 1$.
(b) For any integers $l \geq 3$ and $k > l$, there exists a graph G with chromatic number l such that $\mathrm{Kc}(G, l) = 1$ and $\mathrm{Kc}(G, k) > 1$.

Proof. (a) Clearly, any two 2-colorings are K^2-equivalent. Hence, it suffices to prove that every k-coloring of G is K-equivalent to a 2-coloring. This is easy to see and is left to the reader.

(b) Let G be the categorical product $K_l \times K_k$. Its vertices are pairs (i,j), $1 \le i \le l$, $1 \le j \le k$, and vertices (i,j) and (i',j') are adjacent if and only if $i \ne i'$ and $j \ne j'$. Let c be the l-coloring of G where $c((i,j)) = i$, and let c' be the k-coloring of G where $c'((i,j)) = j$. It is easy to see that c is the unique l-coloring of G, so $\chi(G) = l$ and $\mathrm{Kc}(G,l) = 1$. On the other hand, c' is not K-equivalent to any other k-coloring since all its 2-colored subgraphs $G(a,b)$ are connected. In particular, it is not K-equivalent to c, so $\mathrm{Kc}(G,k) > 1$. $\square$

There are other graphs with the same properties as in Proposition 2.1(b). They can be obtained from $K_l \times K_k$ by replacing every vertex (i,j) by an independent set $U(i,j)$ (of any size) and, for any two adjacent vertices (i,a) and (i',b) of $K_l \times K_k$ adding edges between $U(i,a)$ and $U(i',b)$ so that the subgraph induced on $\cup_{i=1}^{l}(U(i,a) \cup U(i,b))$ is connected. This construction describes the l-colorable graphs with a k-coloring which is not K-equivalent to any other k-coloring. More generally, it would be interesting to characterize l-colorable graphs ($l < k$) with a k-coloring (k large) which is not K-equivalent to any $(k-1)$-coloring. This problem was considered by Las Vergnas and Meyniel [7] who conjectured that such graphs contain the complete graph K_k as a minor.

Lemma 2.2. *Suppose that c_0 is a $(k-1)$-coloring of a graph G and that U is an independent vertex set of G. Then c_0 is K-equivalent in $\mathcal{C}_k(G)$ to a k-coloring of G, one of whose color classes is U.*

Proof. Let $U = \{u_1, \ldots, u_r\}$. For $i = 1, \ldots, r$, let c_i be the k-coloring of G that is obtained from c_{i-1} by recoloring the vertex u_i with color k. It is clear that the vertex u_i forms a K-component in c_{i-1} for colors k and $c_{i-1}(u_i) = c_0(u_i)$. Therefore, c_i is K-equivalent to c_{i-1}. This shows that c_r is a coloring that is K-equivalent to c_0, and one of its color classes is U. $\square$

Corollary 2.3. *Let k be an integer. Suppose that G is a graph such that every k-coloring of G is K-equivalent to some $(k-1)$-coloring. If U is an independent vertex set of G, then $\mathrm{Kc}(G,k) \le \mathrm{Kc}(G-U, k-1)$.*

Proof. For $i = 1, 2$, let c_i be a k-coloring of G. By assumption, c_i is K-equivalent to a $(k-1)$-coloring c_i'. By Lemma 2.2, c_i' is K-equivalent to a k-coloring c_i'', one of whose color classes is U. The restrictions of c_1'' and c_2'' to $G - U$ are $(k-1)$-colorings of $G - U$. If they are K^{k-1}-equivalent, then $c_1'' \sim_k c_2''$, and hence $c_1 \sim_k c_2$. This completes the proof. $\square$

A graph G is *d-degenerate* if every subgraph of G contains a vertex of degree $\le d$. Las Vergnas and Meyniel [7, Proposition 2.1] proved the following result, whose proof we include for completeness.

Proposition 2.4. *If G is a d-degenerate graph and $k > d$ is an integer, then $\mathrm{Kc}(G,k) = 1$.*

Proof. The proof is by induction on $|V(G)|$. The statement is clear if $G = K_1$. Otherwise, let v be a vertex of degree $\le d$, and let $G' = G - v$. Let c_1 and c be arbitrary k-colorings of G. By c_1' and c' we denote their restrictions to G'. By the induction hypothesis, c_1' is K-equivalent to c'. There is a sequence of K-changes, $c_1' \sim_k c_2' \sim_k \cdots \sim_k c_r' = c'$.

For $i = 2, \ldots, r$, let c_i be an extension of c_i' to G which is obtained as follows. There are two colors, say a_i and b_i, that are involved in the K-change yielding c_i' from c_{i-1}'. We assume that $a_i \neq c_{i-1}(v)$. Now we distinguish three cases. If $c_{i-1}(v) \neq b_i$, then we let $c_i(v) = c_{i-1}(v)$, and the same K-change as performed on G' shows that $c_i \sim_k c_{i-1}$. If $c_{i-1}(v) = b_i$ and v has precisely one neighbor u with $c_{i-1}'(u) = a_i$, then the K-components for a_i and b_i of the coloring c_{i-1} are the same as in G', except that the component containing u is extended by one vertex, namely v. Now we set $c_i(v) = c_{i-1}(v)$ if the K-change yielding c_i' does not involve u, and we set $c_i(v) = a_i$ if it does. In both cases, c_i is K-equivalent with c_{i-1}. Finally, suppose that $c_{i-1}(v) = b_i$ and v has more than one neighbor whose color in c_{i-1} is a_i. Then there is a color $b_i' \neq b_i$ that is not contained among the neighbors of v in c_{i-1}. Now we first make a K-change replacing $c_{i-1}(v)$ with b_i', and then another change as described above (since now the color of v is different from a_i and b_i). This gives the coloring c_i which is K-equivalent with c_{i-1} also in this case.

Finally, repeating the above changes, we see that c_1 is K-equivalent to c_r, an extension of c_r'. Note that c_r and c are the same except that they possibly disagree on v. Therefore, if $c_r \neq c$, another K-change can replace the color $c_r(v)$ by $c(v)$ since none of these colors appears in the neighborhood of v. This shows that $c_r \sim_k c$, and the proof is complete. □

An immediate corollary of Proposition 2.4 is:

Corollary 2.5. *Let Δ be the maximum degree of a graph G and let $k \geq \Delta + 1$ be an integer. Then $\mathrm{Kc}(G, k) = 1$. If G is connected and contains a vertex of degree $< \Delta$, then also $\mathrm{Kc}(G, \Delta) = 1$.*

We conjecture that the last statement of Corollary 2.5 can be extended to include all connected Δ-regular graphs with the exception of odd cycles and complete graphs.

The following proposition is left as an exercise.

Proposition 2.6. *Let G be a graph of order n, let α be the cardinality of a largest independent vertex set in G, and let $k \geq n - \alpha + 1$ be an integer. Then every k-coloring of G is K^k-equivalent to the k-coloring in which a fixed maximum independent set is a color class and every other color class is a single vertex. In particular, $\mathrm{Kc}(G, k) = 1$.*

3. Edge-colorings

Coloring the edges of a graph G is the same as coloring the vertices of its line graph $L(G)$. Vizing's Theorem states that the edges of a graph with maximum degree Δ can be colored with $\Delta + 1$ colors, i.e., $\chi'(G) = \chi(L(G)) \leq \Delta + 1$.

We prove:

Theorem 3.1. *Let Δ be the maximum degree of a graph G. If $k \geq \chi'(G) + 2$ is an integer, then $\mathrm{Kc}(L(G), k) = 1$.*

Proof. The proof is by induction on $\chi'(G)$. The case when $\chi'(G) \leq 2$ follows by Proposition 2.1(a), so assume that $\chi'(G) \geq 3$.

Let c be an arbitrary k-edge-coloring of G. First, we claim that c is K-equivalent to a $(k-1)$-edge-coloring. To prove this, we may assume that c has $m > 0$ edges of color k and that every k-edge-coloring which is K-equivalent to c has at least m edges of color k. The standard "fan" arguments of Vizing show that there is a sequence of K-changes which transforms c into an edge-coloring with $m - 1$ edges of color k, a contradiction. (Cf., e.g., [4] for details.) However, in order to make the proof self-contained, we repeat those arguments.

Let c and $m > 0$ be as above. We say that a color a is *missing* at a vertex v of G if no edge incident with v is colored a. Since $k \geq \Delta + 2$, at least two colors are missing at each vertex, and at least one of them is different from k. Clearly, if the same color a is missing at adjacent vertices u and v, then the change of the color of the edge uv to a represents a K-change.

Let $e = uv$ be an edge of color k. Suppose that color a_1 is missing at u and that a_0 is missing at v. If $a_0 = a_1$, then changing the color of e to a_0 is a K-change, yielding a coloring with $m - 1$ edges of color k, a contradiction. So, there is an edge vv_1 of color a_1. There is a color $a_2 \neq k$ which is missing at v_1. If a_2 is missing at v, we recolor vv_1 with a_2 and, as mentioned above, are henceforth able to get rid of the color k at e. Therefore, there is an edge vv_2 of color a_2.

Consider the K-component $K \subseteq G(a_0, a_2)$ at the vertex v_1. After the corresponding K-change, a_0 becomes missing at v_1. If a_0 is still missing at v, then we get a contradiction as above. Therefore, the corresponding K-change has changed the color a_2 at v to a_0, so $vv_2 \in E(K)$.

We shall now repeat the above procedure and henceforth have distinct edges $vv_1, \ldots, vv_r$ whose colors are $a_1, \ldots, a_r$ (respectively), and such that color a_i is missing at v_{i-1} for $i = 2, \ldots, r$. Moreover, the K-component in $G(a_0, a_i)$ at v_{i-1} is a path from v_{i-1} to v, whose last edge is $v_i v$. Having this situation, there is a color $a_{r+1} \neq k$ that is missing at v_r. If $a_{r+1} \notin \{a_0, \ldots, a_r\}$, then we consider the K-component $K \subseteq G(a_0, a_{r+1})$ at v_r. If this is a path ending at v, its last edge, call it $v_{r+1}v$, has color a_{r+1}, and we proceed with the next step. If K does not contain v, then after the K-change at K, a_0 is missing at v_r and at v. Now, we recolor vv_r with a_0, then we recolor vv_{r-1} with a_r, vv_{r-2} with a_{r-1}, $\ldots$, vv_1 with a_2. Finally, recolor e with a_1. All these recolorings were K-changes, so we have a coloring with $m - 1$ edges of color k, a contradiction.

From now on, we may assume that $a_{r+1} = a_j$, where $0 \leq j < r$. Let us consider $K \subseteq G(a_0, a_j)$ at v_r. The K-change at K makes a_0 missing at v_r. Since the component of $G(a_0, a_j)$ containing vv_j is a path from v_{j-1} to v_j and v, K does not contain vv_j. Therefore, a_0 is still missing at v. Now we conclude as above. This completes the proof of the claim.

By repeating the above arguments again if necessary, we conclude that c is K-equivalent to a $(\Delta+1)$-edge-coloring.

Fix an edge-coloring c_0 with the color partition $E(G) = M_1 \cup \cdots \cup M_r$, $r = \chi'(G)$. It suffices to prove that any $(\Delta+1)$-edge-coloring c of G is K-equivalent with c_0 in $\mathcal{C}_{\Delta+2}(L(G))$. By Lemma 2.2, c is K-equivalent to a $(\Delta+2)$-coloring c' whose first color class is M_r. Now, the proof is complete by applying induction on the graph $G - M_r$. □

It would be interesting to extend Theorem 3.1 to include $(\Delta+1)$-colorings as well. It is quite plausible that $\text{Kc}(L(G), \chi'(G) + 1)$, $\text{Kc}(L(G), \Delta + 2)$, or even $\text{Kc}(L(G), \Delta + 1)$ are always 1. Let us remark, however, that there are graphs for which $\text{Kc}(L(G), \chi'(G)) > 1$. Such examples are given after Theorem 3.3 below. We emphasize, specifically, the following interesting special case of the above speculations:

Conjecture 3.2. *If G is a graph with $\Delta(G) \leq 3$, then all its 4-edge-colorings are K-equivalent.*

The most challenging example, the Petersen graph, has been checked using computer by Drago Bokal (and it satisfies the conjecture).

We can say more if G is bipartite.

Theorem 3.3. *Let Δ be the maximum degree of a bipartite graph G. If $k \geq \Delta + 1$ is an integer, then $\text{Kc}(L(G), k) = 1$.*

Proof. The proof is the same as for Theorem 3.1 except that we need to show that every k-edge-coloring of G is K^k-equivalent to a Δ-edge-coloring. This is a standard exercise and is left to the reader. □

The complete bipartite graph $K_{p,p}$ (where p is a prime) has a p-edge-coloring in which any two color classes form a Hamiltonian cycle. This example shows that Theorem 3.3 cannot be extended to Δ-colorings, not even for complete bipartite graphs.

Problem 3.4. *For which cubic bipartite graphs is $\text{Kc}(L(G), 3) = 1$?*

A special case of this problem, when G is planar and 3-connected has been solved by Fisk. Let G be a 3-connected cubic planar bipartite graph. Its dual graph T is a 3-colorable triangulation of the plane. Fisk [5] proved (see Theorem 4.1 below) that any two 4-colorings of T are K-equivalent. If c_1, c_2 are 3-edge-colorings of G, they determine 4-colorings c_1^*, c_2^* (respectively) of T. It is easy to see that a K-change on 4-colorings of T corresponds to a sequence of one or more K-changes among the corresponding 3-edge-colorings in G. This implies that c_1 and c_2 are K-equivalent, and hence $\text{Kc}(L(G), 3) = 1$.

Let us observe that planarity is essential for the above examples since the graph $K_{3,3}$ has non-equivalent edge-colorings, $\text{Kc}(L(K_{3,3}), 3) = 2$.

4. Planar graphs

In [11], the author described an infinite class of "almost Eulerian" triangulations of the plane that have a special 4-coloring which is not K-equivalent to any other 4-coloring (and other 4-colorings exist). This shows that there are planar triangulations for which $\mathrm{Kc}(G,4) \geq 2$. By taking 3-sums of such graphs, we get planar triangulations with arbitrarily many equivalence classes of 4-colorings.

Meyniel [9] proved that $\mathrm{Kc}(G,5) = 1$ for every planar graph (and also $\mathrm{Kc}(G,k) = 1$ if $k \geq 6$, which follows by 5-degeneracy of planar graphs). In this section we prove a similar result for 4-colorings in the case when G is 3-colorable (cf. Theorem 4.4). A special case of this result, when G is a 3-colorable triangulation of the plane was proved by Fisk [5].

Theorem 4.1 (Fisk [5]). *Let G be a 3-colorable triangulation of the plane. Then $\mathrm{Kc}(G,4) = 1$.*

In order to extend Theorem 4.1, we shall need two auxiliary results.

Lemma 4.2. *Suppose that G is a subgraph of a graph $\tilde{G}$. Let $\tilde{c}_1, \tilde{c}_2$ be r-colorings of $\tilde{G}$. Denote by c_i the restriction of $\tilde{c}_i$ to G, $i = 1, 2$. If $\tilde{c}_1$ and $\tilde{c}_2$ are K^r-equivalent, then c_1 and c_2 are K^r-equivalent colorings of G.*

Proof. Any K-component in $\tilde{G}$ gives rise to one or more K-components in G, with respect to the induced coloring of G. This implies the lemma. □

A *near-triangulation* of the plane is a plane graph such that all its faces except the outer face are triangles.

Proposition 4.3. *Suppose that G is a planar graph with a facial cycle C. If c_1, c_2 are 4-colorings of G, then there is a near-triangulation T of the plane with the outer cycle C such that $T \cap G = C$ and there are 4-colorings c'_1, c'_2 of G which are K-equivalent to c_1 and c_2, respectively, such that they both can be extended to 4-colorings of $G \cup T$. Moreover, if the restriction of c_1 to C is a 3-coloring, then $c'_1 = c_1$, and c_1 can be extended to a 3-coloring of T.*

Proof. Let $C = v_1 v_2 \ldots v_k v_1$. The proof is by induction on k. If $k = 3$, then $T = C$, $c'_1 = c_1$, and $c'_2 = c_2$. Suppose now that $k \geq 4$. If there are indices i, j ($1 \leq i < j \leq k$) such that v_i and v_j are not consecutive vertices of C and such that $c_1(v_i) \neq c_1(v_j)$ and $c_2(v_i) \neq c_2(v_j)$, then we add the edge $v_i v_j$ inside C and apply induction on $C_1 = v_i v_{i+1} \ldots v_j v_i$ and $C_2 = v_j v_{j+1} \ldots v_k v_1 \ldots v_i v_j$.

More precisely, let $G_1 = G + v_i v_j$. By the induction hypothesis for the cycle C_1, there are sequences of K-changes in G_1 (and hence also in G, by Lemma 4.2) transforming c_1 into c_{11}, and transforming c_2 into c_{12}, respectively, and there is a near-triangulation T_1 with outer cycle C_1 such that c_{11} and c_{12} can be extended to colorings $\bar{c}_{11}$ and $\bar{c}_{12}$ of $G_1 \cup T_1$. Next, apply the induction hypothesis to $G_1 \cup T_1$ for the facial cycle C_2 and colorings $\bar{c}_{11}$ and $\bar{c}_{12}$. Let T_2 be the corresponding near-triangulation, and c'_{11}, c'_{12} the corresponding colorings of $G_1 \cup T_1$ that can be extended to $G_1 \cup T_1 \cup T_2$. By Lemma 4.2, the K-changes which produce c'_{11} and

c'_{12} from $\bar{c}_{11}$ and $\bar{c}_{12}$, respectively, can be made in G. All together, the restriction c'_l of the coloring c'_{1l} to G is K-equivalent to c_l in G ($l = 1, 2$). Clearly, c'_l has an extension to $G \cup T$, where $T = T_1 \cup T_2$. Therefore, T can be taken as the required near-triangulation for C.

If the restriction of c_1 to C is a 3-coloring, then $c_{11} = c_1$ and the restriction of c_1 to C_1 can be extended to a 3-coloring of T_1. Similarly, $c'_{11} = c_1$, and c_1 can be extended to a 3-coloring of T. This proves the "moreover" part of the proposition.

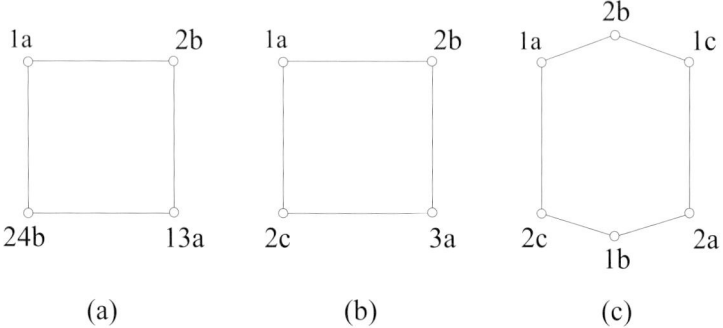

FIGURE 1. The special cases

Next, we show that vertices v_i, v_j exist unless one of the cases in Figure 1 occurs (where c_1 is represented by colors 1–4 and c_2 by colors a–d), up to permutations of colors, dihedral symmetries of C and up to changing the roles of c_1 and c_2. This is easy to see if $k = 4$. The details are left to the reader. If $k \geq 5$, we argue as follows. Suppose that v_i, v_j do not exist. We may assume that $c_1(v_1) = c_1(v_3) = 1$. Then $c_1(v_1) \neq c_1(v_4)$, so $c_2(v_1) = c_2(v_4) = a$ may be assumed. Suppose that $c_1(v_2) = 2$ and $c_2(v_2) = b$. Since $b = c_2(v_2) \neq c_2(v_4) = a$, we have $c_1(v_4) = c_1(v_2) = 2$. Now, $c_1(v_2) \neq c_1(v_5)$, so $c_2(v_5) = b$. Next, $c_2(v_5) \neq c_2(v_3)$ implies that $c_1(v_5) = c_1(v_3) = 1$. It follows, in particular, that $k \geq 6$. Similar conclusions as before imply that $c_1(v_6) = 2$ and $c_2(v_3) = c_2(v_6) = c$. If $k = 6$, this is the exceptional case of Figure 1(c). If $k \geq 7$, then we see that either v_2, v_7 or v_4, v_7 is the required pair v_i, v_j.

Let us consider the exceptional case shown in Figure 1(c). Let v_1 be the vertex with $c_1(v_1) = 1$ and $c_2(v_1) = a$ (upper-left). Try first a K-change of colors 1 and 3 at v_1. If this change gives rise to the same exception, there is a (1,3)-colored path P joining v_1 and v_3. Now, a K-change of colors 2 and 4 at v_2 changes c_1 into a coloring which does not fit Figure 1(c). None of these K-changes affects c_2, and we are done unless we do not want to change c_1 because of the "moreover" part. In that case we are allowed to use the fourth color in the extension of c_2, and we take T to be the near-triangulation with one interior point joined to all vertices on C.

Consider now the case of Figure 1(b). By symmetry, we may assume that c_1 is not allowed to be changed according to the "moreover" part of the proposition. Let v_1 be the vertex in the upper-left corner. By a K-change of colors a and d at v_1, or of b and c at v_2, we replace c_2 either by a 4-coloring which uses on C all four or only two of the colors. In each case, we can triangulate C by adding two adjacent vertices p, q such that p is adjacent to v_1, v_2, v_4, and q is adjacent to v_2, v_3, v_4.

The final case is the one shown in Figure 1(a). If we want $c'_2 = c_2$, then we let T be the near-triangulation consisting of C and a vertex of degree 4 inside. Then c_2 extends to a 3-coloring of T, and c_1 also extends to a 4-coloring with the exception of the case when all vertices on C have distinct colors. In the latter case, we can either K-change colors 1 and 3 at v_1, or change 2 and 4 at v_2, without affecting the colors at v_3 and v_4. The new coloring c'_1 extends to T.

Suppose, finally, that we want $c'_1 = c_1$. In this case, c_1 is a 3-coloring on C. Up to symmetries, we may assume either $c_1(v_3) = 1$ and $c_1(v_4) = 2$, or $c_1(v_3) = 3$ and $c_1(v_4) = 2$. In the first case we can take the same near-triangulation T as above (one interior vertex of degree 4). In the latter case, we take two interior vertices u_1, u_2 in T, where u_1 is adjacent to u_2, v_4, v_1, v_2 and u_2 is adjacent to u_1, v_2, v_3, v_4. This completes the proof. □

Theorem 4.4. *Let G be a 3-colorable planar graph. Then* $\mathrm{Kc}(G, 4) = 1$.

Proof. In order to be able to assume that G is 2-connected, we apply induction on the number of blocks of G. If $G = G_1 \cup G_2$, where $G_1 \cap G_2$ is either empty or a cutvertex v, we apply induction hypotheses on G_1 and G_2 but making sure that we never use a K-change on a K-component containing v. This is possible since a K-change using component K of $G_i(a, b)$ is the same as making K-changes on all components of $G_i(a, b)$ distinct from K.

From now on, we assume that G is 2-connected. Let c_1 be a 3-coloring of G. It suffices to see that every 4-coloring of G is K-equivalent to c_1.

Let c_2 be a 4-coloring of G. Let $C_1, \ldots, C_m$ be the facial cycles of G. Let $G_0 = G$, $c_1^0 = c_1$ and $c_2^0 = c_2$. For $i = 1, \ldots, m$, we apply Proposition 4.3 to the graph G_{i-1}, its facial cycle C_i and the colorings c_1^{i-1} and c_2^{i-1}. We conclude that there is a near-triangulation T_i with outer cycle C_i such that $G_{i-1} \cap T_i = C_i$. Let $G_i = G_{i-1} \cup T_i$. By Proposition 4.3, c_1^{i-1} can be extended to a 3-coloring c_1^i of G_i, and c_2^{i-1} is K-equivalent in G_{i-1} to a 4-coloring that has an extension c_2^i to G_i.

The final graph G_m is a triangulation with the 3-coloring c_1^m. By Theorem 4.1, c_2^m is K-equivalent to c_1^m. By successively applying Lemma 4.2 to $G_{m-1} \subseteq G_m$, etc. up until $G_0 \subseteq G_1$, we conclude that $c_2^{m-1} \sim_k c_1^{m-1}$ in G_{m-1}, etc., until finally concluding that $c_1 = c_1^0 \sim_k c_2^0 = c_2$ in $G_0 = G$. □

It is worth mentioning that there exist 3-colorable planar graphs G with $\mathrm{Kc}(G, 3) \geq 2$. An infinite family of such examples can be obtained as follows. In [11], a family of planar triangulations T is constructed for which a special 4-coloring exists for which no nontrivial K-change exists. Let T^* be the dual cubic graph, and let L be its line graph. It is well-known that every 4-coloring of a triangulation gives rise to a 3-edge-coloring of its dual. The special 4-coloring of T therefore determines a (vertex) 3-coloring of L. The property of the 4-coloring of T implies that the 3-coloring of L is not K-equivalent with any other 3-coloring of L. Since T admits other 4-colorings, also L admits other 3-colorings, hence $\mathrm{Kc}(L, 3) \geq 2$.

Theorem 4.4 combined with Proposition 2.1(a) and the aforementioned result of Meyniel [9] yield:

Corollary 4.5. *Let G be a planar graph and $k > \chi(G)$ and integer. Then*
$$\mathrm{Kc}(G, k) = 1.$$

A planar graph G may have 4-colorings which are not K-equivalent. However, if G is "almost 3-colorable", this is not likely to happen.

Problem 4.6. *Suppose that G is a 4-critical planar graph. Is it possible that G has two 4-colorings that are not K-equivalent to each other?*

5. Some further open problems

In the preceding sections, we have exposed several open problems about K-classes of graph colorings. Two further questions are presented below.

Meyniel [10] proved that a graph, in which every odd cycle of length 5 or more has at least two chords, is perfect. He conjectured that for every such graph G and every integer $k \geq \chi(G)$, $\mathrm{Kc}(G, k) = 1$. He proved that every k-coloring of G is K-equivalent to some $\chi(G)$-coloring. The last property does not hold for arbitrary perfect graphs.

Let G be a triangulation of some orientable surface, and let c be a 4-coloring of G. Let t_1^+ be the number of facial 3-cycles whose coloring (in the clockwise order around the face) is 234, and let t_1^- be the number of facial 3-cycles whose coloring is 432. Let $d(c) = |t_1^+ - t_1^-|$. The number $d(c)$ turns out to be invariant on permutations of the colors, and it is called the *degree* of the coloring c [5]. The degree, in particular its parity has been studied by Tutte who also observed that this is Kempe invariant, i.e., all colorings within the same K-class have the same parity.

One can also define two colorings to be *close* if they have at least one color class in common. Two colorings are *similar* if there is a sequence of colorings, starting with one and ending with the other, such that any two consecutive colorings in this sequence are close. The parity of the degree is constant on close colorings. It is not difficult to find triangulations of the plane with two similarity classes of 4-colorings. However, in all such examples known, colorings in different similarity classes have different parity of the degree. Tutte [15] asked if it is possible to have non-similar 4-colorings of a planar triangulation whose degrees have the same parity.

References

[1] U. Baumann, Über das Erzeugen minimaler Kantenfärbungen von Graphen, in Graphentheorie und ihre Anwendungen, Päd. Hochsch. Dresden, Dresden, 1988, pp. 5–8.

[2] C. Berge, A new color change to improve the coloring of a graph, Discrete Appl. Math. **24** (1989) 25–28.

[3] M. Dyer, A. Flaxman, A. Frieze, and E. Vigoda, Randomly coloring sparse random graphs with fewer colors than the maximum degree, to appear in Random Structures and Algorithms.

[4] A. Ehrenfeucht, V. Faber, H.A. Kierstead, A new method of proving theorems on chromatic index, Discrete Math. **52** (1984) 159–164.

[5] S. Fisk, Geometric coloring theory, Adv. Math. **24** (1977) 298–340.

[6] T.P. Hayes, E. Vigoda, A non-Markovian coupling for randomly sampling colorings, in Proc. 44th Ann. IEEE Symp. on Found. of Comp. Sci., 2003, pp. 618–627.

[7] M. Las Vergnas, H. Meyniel, Kempe classes and the Hadwiger conjecture, J. Combin. Theory Ser. B **31** (1981) 95–104.

[8] T. Łuczak, E. Vigoda, Torpid mixing of the Wang-Swendsen-Kotecký algorithm for sampling colorings, J. Discrete Algorithms **3** (2005) 92–100.

[9] H. Meyniel, Les 5-colorations d'un graphe planaire forment une classe de commutation unique, J. Combin. Theory Ser. B **24** (1978) 251–257.

[10] H. Meyniel, The graphs whose odd cycles have at least two chords, in Topics on perfect graphs, Eds. C. Berge and V. Chvátal, Ann. Discrete Math. **21** (1984) 115–119.

[11] B. Mohar, Akempic triangulations with 4 odd vertices, Discrete Math. **54** (1985) 23–29.

[12] C.A. Morgenstern, H.D. Shapiro, Heuristics for rapidly four-coloring large planar graphs, Algorithmica **6** (1991) 869–891.

[13] T. Sibley, S. Wagon, Rhombic Penrose tilings can be 3-colored, Amer. Math. Monthly **107** (2000) 251–253.

[14] A.D. Sokal, A personal list of unsolved problems concerning lattice gasses and antiferromagnetic Potts models, Markov Processes and Related Fields **7** (2001) 21–38.

[15] W.T. Tutte, Invited address at the Workshop in Combinatorics and Discrete Structures — In honour of Prof. W.T. Tutte, UNICAMP, Campinas, Brazil, August 2–4, 1999.

[16] E. Vigoda, Improved bounds for sampling colorings, J. Math. Phys. **41** (2000) 1555–1569.

[17] J.-S. Wang, R.H. Swendsen, and R. Kotecký, Antiferromagnetic Potts models, Phys. Rev. Lett. **63** (1989) 109–112.

[18] J.-S. Wang, R.H. Swendsen, and R. Kotecký, Three-state antiferromagnetic Potts models: A Monte Carlo study, Phys. Rev. B **42** (1990) 2465–2474.

Bojan Mohar
Department of Mathematics, University of Ljubljana
1000 Ljubljana, Slovenia
e-mail: `bojan.mohar@uni-lj.si`

Acyclic 4-choosability of Planar Graphs with Girth at Least 5

Mickaël Montassier

Abstract. A proper vertex coloring of a graph $G = (V, E)$ is acyclic if G contains no bicolored cycle. A graph G is L-list colorable, for a given list assignment $L = \{L(v) : v \in V\}$, if there exists a proper coloring c of G such that $c(v) \in L(v)$ for all $v \in V$. If G is L-list colorable for every list assignment with $|L(v)| \geq k$ for all $v \in V$, then G is called k-choosable. A graph is said to be acyclically k-choosable if these L-list colorings can be chosen to be acyclic. In this paper, we prove that if G is planar with girth $g \geq 5$, then G is acyclically 4-choosable. This improves the result of Borodin, Kostochka and Woodall [BKW99] concerning the acyclic chromatic number of planar graphs with girth at least 5.

1. Introduction

An acyclic coloring of a graph G is a proper coloring of the vertices of G such that the graph induced by every union of two color classes is a forest. We denote by $\chi_a(G)$ the minimum number of colors in an acyclic coloring of G. Acyclic colorings were introduced by Grünbaum in [Grü73]. For planar graphs, Grünbaum conjectured that $\chi_a(G) \leq 5$ and proved $\chi_a(G) \leq 9$. After many studies (Mitchem [Mit74], Albertson and Berman [AB77], Kostochka [Kos76]), Borodin proved that $\chi_a(G) \leq 5$ [Bor79]. This bound is best possible since there exist planar graphs which are not acyclically colorable with four colors [Grü73, KM76]. In 1999, Borodin, Kostochka and Woodall improved this bound for planar graphs with large girth:

Theorem 1. [BKW99]
1. If G is planar with girth $g \geq 5$, then $\chi_a(G) \leq 4$.
2. If G is planar with girth $g \geq 7$, then $\chi_a(G) \leq 3$.

A graph G is L-list colorable, for a given list assignment $L = \{L(v) : v \in V(G)\}$, if there exists a coloring c of the vertices such that $c(v) \in L(v)$ and $c(v) \neq c(u)$ if

u and v are adjacent in G. If G is L-list colorable for every list assignment with $|L(v)| \geq k$ for all $v \in V(G)$, then G is called k-choosable. We denote by $\chi_l(G)$ the smallest integer k such that G is k-choosable. In [Tho94], Thomassen proved that every planar graph is 5-choosable (i.e., $\chi_l(G) \leq 5$) and Voigt proved that there are planar graphs which are not 4-choosable [Voi93]. In the following, we are interested in the acyclic choosability of graphs (the L-list colorings can be chosen to be acyclic). In [BFDFK+02], the following theorem is proved and the next conjecture is given:

Theorem 2. [BFDFK+02] *Every planar graph is acyclically 7-choosable.*

This means that for any given list assignment L such that $\forall v \in V$, $|L(v)| = 7$, we can choose for each vertex v a color in $L(v)$ such that the obtained coloring of G is acyclic.

Conjecture 1. [BFDFK+02] *Every planar graph is acyclically 5-choosable.*

Conjecture 1 is very strong, since it implies the result of Borodin [Bor79]. Moreover, we know that Borodin's proof is tough.

In [MOR05], we study the acyclic choosability of graph with bounded maximum average degree. The maximum average degree, $\mathrm{Mad}(G)$, of the graph G is defined as

$$\mathrm{Mad}(G) = \max\{2|E(H)|/|V(H)|, H \subset G\}.$$

Theorem 3. [MOR05]
1. *Every graph G with $\mathrm{Mad}(G) < \frac{8}{3}$ is acyclically 3-choosable.*
2. *Every graph G with $\mathrm{Mad}(G) < \frac{19}{6}$ is acyclically 4-choosable.*
3. *Every graph G with $\mathrm{Mad}(G) < \frac{24}{7}$ is acyclically 5-choosable.*

The following are the immediate consequences of Theorem 3.

Corollary 1. [MOR05]
1. *Every planar graph with girth at least 8 is acyclically 3-choosable.*
2. *Every planar graph with girth at least 6 is acyclically 4-choosable.*
3. *Every planar graph with girth at least 5 is acyclically 5-choosable.*

Our main result is Theorem 4; it improves Theorem 1.1 and Corollary 1.

Theorem 4. *Every planar graph with girth at least 5 is acyclically 4-choosable.*

The proof of Theorem 1 is based on the method of reducible configurations and a discharging procedure. The proof of Theorem 4 has a similar structure and uses the techniques developed in [BKW99]. Let H be a counterexample to Theorem 4 having the minimum number of vertices and edges. First, by minimality of H, we prove that H does not contain some configurations. Finally, we apply a discharging procedure in order to obtain a contradiction with Euler's formula.

In the following, a k-vertex (resp. $\geq k$-vertex, $\leq k$-vertex) is a vertex of degree k (resp. $\geq k$, $\leq k$). A $k(l)$-vertex is a k-vertex adjacent to at least l 2-vertices. An i,j-path is a bicolored path with colors i and j. An r-cycle (resp. $\leq r$-cycle, $\geq r$-cycle) is a cycle with length r (resp. $\leq r$, $\geq r$).

2. Proof of Theorem 4

2.1. Reducible configurations

Lemma 1. *The minimum counterexample H satisfies the following:*

1. *There is no 1-vertices.*
2. *No 2-vertex is adjacent to a 2-vertex or 3-vertex.*
3. *There are no $d(d)$-vertices $(2 \leq d \leq 15)$, no $d(d-1)$-vertices $(2 \leq d \leq 9)$ and no $d(d-2)$-vertices $(3 \leq d \leq 4)$.*
4. *If w is a 5(3)-vertex, then the three 2-vertices occur consecutively in cyclic order round w, and both of the two faces between consecutive 2-vertices are $^{>}5$-faces.*
5. *If a 5(2)-vertex is adjacent to three 3-vertices, then it is incident to at least one $^{>}5$-face.*
6. *A 5(3) or 6(4)-vertex is not adjacent to any 3-vertices.*

Proof.
1. Suppose that H contains a 1-vertex v. By minimality of H, the graph $H' = H \setminus \{v\}$ is acyclically 4-choosable: so, for any list assignment L, there exists an acyclic proper coloring c with $\forall u \in V(H'), c(u) \in L(u)$. It is easy to extend the coloring c to H by choosing a color for v in its list different from the color of its neighbor. The obtained coloring is proper and acyclic and $\forall v \in V(H), c(v) \in L(v)$. The contradiction completes the proof.
2. follows immediately from 3.

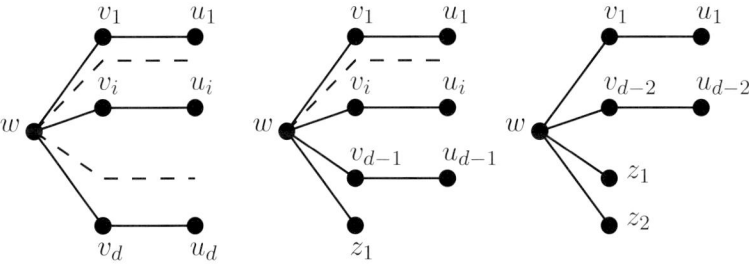

FIGURE 1.

3. Suppose that H contains a $d(d)$-vertex w adjacent to d 2-vertices $v_1, \ldots, v_d$ (see Figure 1). Each vertex v_i is adjacent to w and to another vertex u_i, $1 \leq i \leq d$. By minimality of H, the graph $H' = H \setminus \{w\}$ is acyclically 4-choosable. Hence, for any list assignment $L = \{L(v) : v \in V\}$, there exists an acyclic coloring c of H' with $c(v) \in L(v)$. Now, we show that we can extend c to H. Since $2 \leq d \leq 15$, there exists a color j of $L(w)$ that appears on at most three of $u_1, \ldots, u_d$. Set $c(w) = j$. Now, we give distinct proper colors to the v_i such that $c(u_i) = j$ and any proper color to the other u_i. The obtained coloring is proper and acyclic and $\forall v \in V(H), c(v) \in L(v)$.

Suppose that H contains a $d(d-1)$-vertex w adjacent to $d-1$ 2-vertices $v_1,\ldots,v_{d-1}$ and to another vertex z_1. Each vertex v_i is adjacent to w and to another vertex u_i, $1 \le i \le d-1$. By minimality of H, the graph $H' = H \setminus \{w\}$ is acyclically 4-choosable. Since $2 \le d \le 9$, there exists a color j of $L(w)$ distinct from $c(z_1)$ that appears on at most two of $u_1,\ldots,u_{d-1}$. Set $c(w) = j$. Now, we give distinct proper colors different from $c(z_1)$ to the v_i such that $c(u_i) = j$ and any proper color to the other v_i.

Finally, suppose that H contains a $d(d-2)$-vertex w adjacent to $d-2$ 2-vertices $v_1,\ldots,v_{d-2}$ and to two other vertices z_1,z_2. Each vertex v_i is adjacent to w and to another vertex u_i, $1 \le i \le d-2$. By minimality of H, the graph $H' = H \setminus \{v_1\}$ is acyclically 4-choosable. If $c(w) \ne c(u_1)$, we give any proper color to v_1. Hence, we assume that $c(w) = c(u_1)$. If $c(u_2) \ne c(w)$ (or if u_2 does not exist), then we color v_1 with $c(v_1) \in L(v_1) \setminus \{c(w), c(z_1), c(z_2)\}$. Suppose now that $c(w) = c(u_1) = c(u_2)$. We consider two cases: if $c(z_1) = c(z_2)$, we color v_1 with $c(v_1) \in L(v_1) \setminus \{c(w), c(z_1), c(v_2)\}$; if $c(z_1) \ne c(z_2)$, we erase the color of v_2, we modify the color of w and choose $c(w) \in L(w) \setminus \{c(z_1), c(z_2), c(u_1)\}$, then we give any proper color to v_i, $i = 1, 2$.

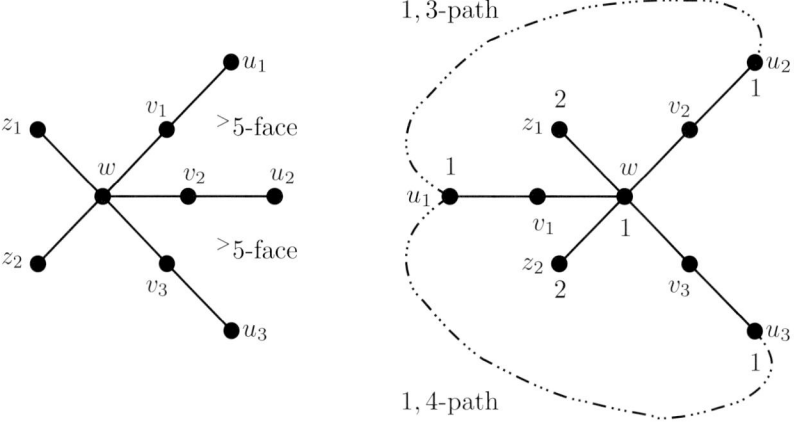

FIGURE 2.

4. Suppose that H contains a 5(3)-vertex w. The vertex w is adjacent to three 2-vertices v_1, v_2, v_3 and to two other vertices z_1, z_2 (see Figure 2). Each 2-vertex v_i is adjacent to w and to another vertex u_i. By minimality of H, the graph $H' = H \setminus \{v_1\}$ is acyclically 4-choosable. If $c(w) \ne c(u_1)$, we color v_1 with any proper color. Assume now that $c(u_1) = c(w)$.

First, observe that a face between two 2-vertices is necessarily a $>$5-face. If not, suppose w.l.o.g. that v_1 and v_2 occur consecutively in cyclic order round w and the face with boundary $u_1 u_2 v_2 w v_1$ is a 5-face. Since u_1 and u_2 are adjacent, then $c(u_1) \ne c(u_2)$. If $c(u_3) \ne c(u_1)$, then we color v_1

with $c(v_1) \in L(v_1) \setminus \{c(w), c(z_1), c(z_2)\}$. So, suppose that $c(u_3) = c(u_1)$. If we cannot color v_1, this implies that w.l.o.g. $L(v_1) = \{1,2,3,4\}$, $c(w) = c(u_1) = c(u_3) = 1$, $c(z_1) = 2$, $c(z_2) = 3$, and $c(v_3) = 4$. Now, we erase the colors of v_2 and v_3; we recolor w with a color different from 1, 2, 3. We color then v_1 and v_3 with any proper color and v_2 with a proper color if $c(w) \neq c(u_2)$ and with a color different from $c(w), c(z_1), c(z_2)$ otherwise.

Now, we consider the case $c(w) = c(u_1) = c(u_2) = c(u_3)$ and w.l.o.g. suppose $c(w) = 1$ (the case where w, u_1, u_2, u_3 do not have the same color can be reduced with the previous reasoning). Observe that if $c(z_1) \neq c(z_2)$, we erase the colors of v_2, v_3, we modify the color of w by choosing $c(w) \in L(w) \setminus \{1, c(z_1), c(z_2)\}$ and we give any proper color to v_i, $i = 1,2,3$. W.l.o.g., we assume that $c(z_1) = c(z_2) = 2$. If the v_i are not consecutive in cyclic order round w, assume that v_1 is between z_1 and z_2. $L(v_1)$ contains 1 and 2; otherwise we color v_1 with $c(v_1) \in L(v_1) \setminus \{1, 2, c(v_2), c(v_3)\}$ (there remains at least one color). We assume w.l.o.g. that $L(v_1) = \{1,2,3,4\}$. If we cannot color v_1, this implies that there exists a 1,4-path and a 1,3-path between u_1 and u_2, u_3. Now, we erase the colors of v_2, v_3. We modify the color of w with a color different from 1 and 2: we cannot create a bicolored cycle between z_1, z_2 (since there exist 1,4-path and 1,3-path). Finally, we give any proper color to v_i, $i = 1,2,3$.

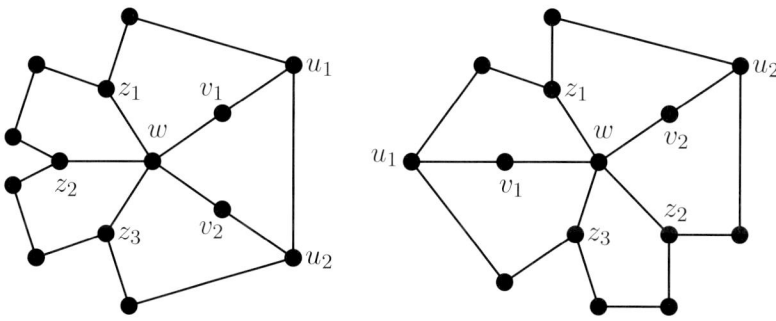

FIGURE 3.

5. Suppose that H contains a 5(2)-vertex w adjacent to two 2-vertices v_1 and v_2 and three 3-vertices z_1, z_2, z_3. Assume that w is incident to five 5-faces. By minimality of H, the graph $H' = H \setminus \{v_1\}$ is acyclically 4-choosable. If $c(w) \neq c(u_1)$, we give any proper color to v_1. Suppose that $c(w) = c(u_1)$, w.l.o.g. $c(w) = c(u_1) = 1$. Note that at most three different $1,j$-paths of length > 1 can start from w and go through z_1, z_2, z_3, v_2 (see Figure 3); this means that, if $1 \notin L(v_1)$, then at least one color remains to color v_1. W.l.o.g., we suppose that $L(v_1) = \{1,2,3,4\}$. If $c(u_2) = 1$, then v_1 and v_2 are not consecutive in the cyclic order round w, and at most one of z_1, z_2, z_3 has an outer neighbor colored 1: this contradicts the existence of the three

1,2-, 1,3-, 1,4-paths and we can color v_1. Hence, suppose that $c(u_2) \neq 1$. If we cannot color v_1, this implies that z_1, z_2, z_3 are colored with 2, 3, 4 and each has an outer neighbor colored 1. We erase the color of v_2. Observe that $L(w) = \{1,2,3,4\}$, otherwise we modify w with a color different from 1,2,3,4 and give any proper color to v_i. There exists a color $j \in L(w) \setminus \{1, c(u_2)\}$ that occurs on at most one of the outer neighbors of z_1, z_2, z_3, say z_l. We color w with j and give any proper color to v_1, v_2, z_l.

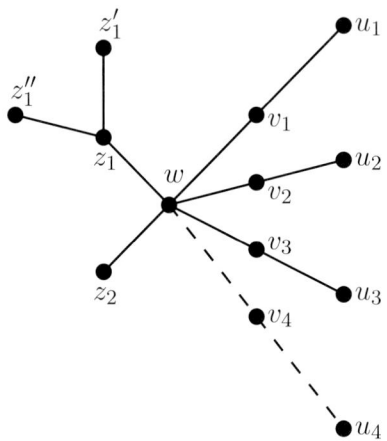

FIGURE 4.

6. Suppose that H contains a 5(3)-vertex w adjacent to three 2-vertices v_1, v_2, v_3 (adjacent each to another vertex u_i), a 3-vertex z_1 (adjacent to z_1' and z_1'') and another vertex z_2. By minimality of H, the graph $H' = H \setminus \{v_1\}$ is acyclically 4-choosable. If $c(u_1) \neq c(w)$, we give any proper color to v_1. Now, we assume that $c(u_1) = c(w) = 1$. If $c(z_1) \neq c(z_2)$, we erase the colors of v_2, v_3 and we modify the color of w by choosing a color different from $c(z_1), c(z_2)$ that occurs on at most one of u_1, u_2, u_3. Then, we give a proper color different from $c(z_1), c(z_2)$ to v_i such that $c(u_i) = c(w)$ and proper color to the others. Hence, we assume that $c(z_1) = c(z_2) = 2$. Observe that $L(v_1)$ contains 1 and 2; otherwise, we can color v_1. We assume w.l.o.g. that $L(v_1) = \{1, 2, 3, 4\}$. If we cannot color v_1 this implies that $c(u_1) = c(u_2) = c(u_3) = 1$, $c(v_2) = 3, c(v_3) = 4$. If $c(z_1') \neq c(z_1'')$, we recolor z_1 with a color different from $2, c(z_1'), c(z_1'')$. Now, $c(z_1) \neq c(z_2)$). We erase then the colors of v_2, v_3, we color w with a color different from $1, 2, c(z_1)$ and we give any proper colors to the v_i. Hence, $c(z_1') = c(z_1'')$. We modify the color of w with a color different from $c(z_1'), 2, 1$ and give any proper colors to v_1, v_2, v_3. So, H does not contain a 5(3)-vertex adjacent to a 3-vertex.

Suppose that H contains a 6(4)-vertex w adjacent to four 2-vertices v_1, v_2, v_3, v_4, a 3-vertex z_1 (adjacent to z_1' and z_1'') and another vertex z_2. By minimality of H, the graph $H' = H \setminus \{v_1\}$ is acyclically 4-choosable. W.l.o.g assume that $c(w) = c(u_1) = 1$. Consider the number of u_i colored with 1:

6.1. Suppose that $c(w) = c(u_1) = 1$ and $c(u_2), c(u_3), c(u_4)$ are different from 1. We color v_1 with $c(v_1) \in L(v_1) \setminus \{c(w), c(z_1), c(z_2)\}$.

6.2. Suppose that $c(w) = c(u_1) = c(u_2) = 1$ and $c(u_3), c(u_4) \neq 1$. Observe that if $c(z_1) = c(z_2)$, we can color v_1 with $c(v_1) \in L(v_1) \setminus \{1, c(z_1), c(v_2)\}$; thus we may assume that $c(z_1) \neq c(z_2)$. Now, assume that $c(u_3) \neq c(u_4)$. We recolor w with a color different from $1, c(z_1), c(z_2)$; since $c(u_3) \neq c(u_4)$, there exists at most one pair w, u_i with $c(w) = c(u_i)$, we color the corresponding v_i with a color different from $c(w), c(z_1), c(z_2)$ and give any proper colors to the other v_j. Finally, assume that $c(u_3) = c(u_4)$, say $c(u_3) = 2$. Now, $c(z_1) \neq 1, 2$ and $c(z_2) \neq 1, 2$; otherwise we can recolor w with a color different from $1, 2, c(z_1), c(z_2)$ and give any proper colors to the v_i, $i = 1, 2, 3, 4$. Say $c(z_1) = 3$ and $c(z_2) = 4$. By the same observation, $L(w) = \{1, 2, 3, 4\}$. Remark that at least one of z'_1, z''_1 is colored with 1; otherwise we can color v_1 with a color different from $1, c(v_2), c(z_2)$. Recolor w with 2 and the same observation shows that at least one of z'_1, z''_1 is colored with 2. So, $\{c(z'_1), c(z''_1)\} = \{1, 2\}$. Now, recolor w with 3 and give proper colors to z_1 and v_i, $i = 1, 2, 3, 4$.

6.3. Suppose that $c(w) = c(u_1) = c(u_2) = c(u_3) = 1$ and $c(u_4) \neq 1$. If $c(z_1) \neq c(z_2)$, we recolor w with a color different from $1, c(z_1), c(z_2)$; we give any proper colors to v_1, v_2, v_3 and we color v_4 with a proper color if $c(w) \neq c(u_4)$ and with a color different from $c(w), c(z_1), c(z_2)$ otherwise. So, $c(z_1) = c(z_2)$. If $c(z'_1) \neq c(z''_1)$, we recolor z_1 with a color different from $c(z_2), c(z'_1), c(z''_1)$ and we are in the previous case. If $c(z'_1) = c(z''_1)$, we recolor w with a color different from $1, c(z_1), c(z'_1)$ and we are in an earlier case.

6.4. Suppose that $c(w) = c(u_1) = c(u_2) = c(u_3) = c(u_4) = 1$. If $c(z_1) \neq c(z_2)$, we recolor w with a color different from $1, c(z_1), c(z_2)$ and give proper colors to v_i. Assume now that $c(z_1) = c(z_2)$. If $c(z'_1) = c(z''_1)$, we recolor w with a color different from $1, c(z_1), c(z'_1)$ and give proper colors to v_i. If $c(z'_1) \neq c(z''_1)$, we recolor z_1 with a color different from $c(z'_1), c(z''_1), c(z_2)$ and we are in an earlier case. □

We call a vertex *weak* if it is a 2- or 3-vertex or a 4-vertex adjacent to both a 2-vertex and a 3-vertex.

Lemma 2. *Each 3-vertex is adjacent to at most one weak vertex.*

Proof. Observe that by Lemma 1.3, a 3-vertex is not adjacent to a 2-vertex. Let w be a 3-vertex adjacent to x, y, z where x, y are weak, with degree 3 or 4. Let the outer neighbors of x be x_1, x_2 and if $d(x) = 4$, x_3 where $d(x_3) = 2$ and the other neighbor of x_3 is x'_3. So is it for y (see Figure 5).

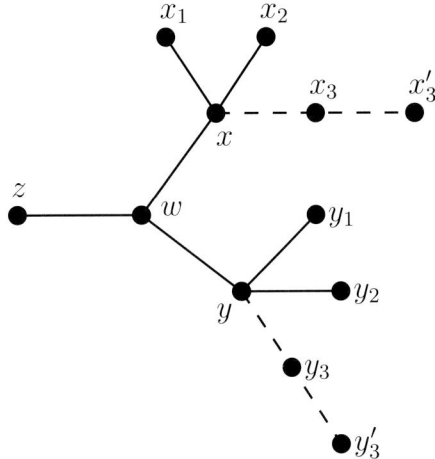

Figure 5.

By minimality of H, the graph $H' = H \setminus \{w, x_3, y_3\}$ is acyclically 4-choosable. We have to consider the different cases:

1. The vertices x, y, z have all distinct colors, $c(x) \neq c(y) \neq c(z) \neq c(x)$. We color w with $c(w) \in L(w) \setminus \{c(x), c(y), c(z)\}$. If $c(x) = c(x'_3)$, we color x_3 with $c(x_3) \in L(x_3) \setminus \{c(x), c(x_1), c(x_2)\}$ and with any proper color otherwise (color y_3 similarly).
2. Exactly two vertices of x, y, z have the same color.
   2.1. Suppose that $c(x) = c(y) = 2$ and $c(z) = 1$.
   First, we show that $L(w)$ contains 1 and 2. Suppose that $L(w)$ does not contain 1 and 2; w.l.o.g. we assume that $L(w) = \{3, 4, 5, 6\}$. Then, if $c(x) = c(x'_3)$, we color x_3 with $c(x_3) \in L(x_3) \setminus \{c(x), c(x_1), c(x_2)\}$; if not, we color x_3 with a proper color (deal analogously with y_3); finally, we color w with $c(w) \in L(w) \setminus \{c(x_1), c(x_2), c(x_3)\}$.
   Assume now that $L(w)$ contains 1 and not 2: $L(w) = \{1, 3, 4, 5\}$. Hence, if $c(x) = c(x'_3)$, we color x_3 with $c(x_3) \in L(x_3) \setminus \{c(x), c(x_1), c(x_2)\}$; if not, we color x_3 with a proper color (deal analogously with y_3). If we cannot color w, this implies that $\{c(x_1), c(x_2), c(x_3)\} = \{3, 4, 5\}$ and $c(x'_3) = 2$. So, we color w with $c(x_3)$ and we change the color of x. If we cannot change the color of x, this implies that $L(x) = \{2, 3, 4, 5\}$. In this case, we color x with $c(x_3)$ and we give a proper color to x_3 and w. The case "$L(w)$ contains 2" may be dealt in the same as "$L(w)$ contains 1".
   So, $L(w)$ contains 1 and 2: $L(w) = \{1, 2, 3, 4\}$.

First assume that there is no 2, 3-path between x and y (going through x_1 or x_2 and y_1 or y_2; w, x_3, y_3 are not colored). We color w with 3. Now, if we cannot color x_3, this implies that $c(x) = c(x_3') = 2$, $L(x_3) = \{2, 3, c(x_1), c(x_2)\}$ with $c(x_1) \neq c(x_2)$, and $c(x_1), c(x_2) \neq 3$. By the same way, if we cannot color y_3, this implies that $c(y) = c(y_3') = 2$, $L(y_3) = \{2, 3, c(y_1), c(y_2)\}$ with $c(y_1) \neq c(y_2)$, and $c(y_1), c(y_2) \neq 3$. W.l.o.g. suppose that we cannot color x_3. We recolor x with a color different from $2, c(x_1), c(x_2)$. If this new color is different from 1, then the colors of x, y, z are all distinct and we obtain case 1. Now, if this new color is equal to 1, we color x_3 with 3 (we recall that $c(x_1), c(x_2) \neq 3$, $c(x) = 1$, and $c(x_3') = 2$). And finally, we color y_3 with a color different from $2, c(y_1), c(y_2)$ if $c(y) = c(y_3')$ and with any proper color otherwise. We can do the same if there is no 2, 4-path connecting x and y; hence we may suppose that both paths exist and $c(x_1) = 3, c(x_2) = 4$, $\{c(y_1), c(y_2)\} = \{3, 4\}$.

Observe that $L(x) = L(y) = \{1, 2, 3, 4\}$; otherwise, we can recolor x (or y) with a color different from $1, 2, 3, 4$ and we obtain case 1. Now either the 2, 3-path (completed to a cycle through w) separates x_2 from z or the 2, 4-path (similarly completed) separates x_1 from z. Suppose the former. Consequently, there is no 1, 4-path connecting x_2 to z. We color w with 4, x with 1. Finally, we color x_3 (resp. y_3) with any proper color if $c(x) \neq c(x_3')$ (resp. $c(y) \neq c(y_3')$) and with a color different from $1, 3, 4$ (resp. $2, 3, 4$) otherwise.

2.2. Suppose that $c(x) = c(z) = 1$ and $c(y) = 2$. If $c(x_1) \neq c(x_2)$, we change $c(x)$ to get case 1 or 2. Now, if $c(x_1) = c(x_2)$, we color w with $c(w) \in L(w) \setminus \{1, 2, c(x_1)\}$, x_3 with $c(x_3) \in L(x_3) \setminus \{c(x), c(x_1), c(w)\}$ if $c(x) = c(x_3')$ or a proper color otherwise and y_3 with $c(y_3) \in L(y_3) \setminus \{c(y), c(y_1), c(y_2)\}$ if $c(y) = c(y_3')$ or a proper color otherwise.

3. The vertices x, y, z have the same color, w.l.o.g. assume that $c(x) = c(y) = c(z) = 1$. If $c(x_1) \neq c(x_2)$ or $c(y_1) \neq c(y_2)$, we change $c(x)$ or $c(y)$ to get case 1 or 2. Now, suppose that $c(x_1) = c(x_2)$ and $c(y_1) = c(y_2)$. We color w with $c(w) \in L(w) \setminus \{1, c(x_1), c(y_1)\}$ and x_3 with $c(x_3) \in L(x_3) \setminus \{1, c(x_1), c(w)\}$ if $c(x) = c(x_3')$ or a proper color otherwise (we deal analogously for y). □

Lemma 3. *Every connected planar graph satisfies:*

$$\sum_{v \in V}(3d(v) - 10) + \sum_{f \in F}(2r(f) - 10) = -20$$

where $d(v)$ denotes the degree of the vertex v and $r(f)$ the length of the face f.

Proof. We can rewrite the Euler's formula $n - m + f = 2$ in the form $(6m - 10n) + (4m - 10f) = -20$. Now, $\sum_{v \in V} d(v) = \sum_{f \in F} r(f) = 2m$. □

2.2. Discharging procedure

We complete the proof with a discharging procedure. First, we assign to each vertex v a charge $\omega(v)$ such that $\omega(v) = 3d(v) - 10$ and to each face f a charge $\omega(f)$ such that $\omega(f) = 2r(f) - 10$. Then we apply the following rules:

Rule 1. Each 2-vertex receives 2 from each adjacent vertex.

Rule 2. Each 3-vertex receives $\frac{1}{2}$ from each adjacent non weak vertex.

Rule 3. Each $\geq$6-face f with bounding cycle $v_1 v_2 \ldots v_{r(f)} v_1$ gives $\frac{1}{2}$ to each vertex v_i for which $d(v_{i-1}) \leq 3$ and $d(v_{i+1}) \leq 3$ (index modulo $r(f)$).

Let ω^* be the new charge after the discharging process. Clearly

$$\sum_{v \in V(H)} \omega^*(v) + \sum_{f \in F(H)} \omega^*(f) = \sum_{v \in V(H)} \omega(v) + \sum_{f \in F(H)} \omega(f) = -20 \quad (1)$$

by Lemma 3. We will prove that $\omega^*(v) \geq 0$ for every vertex v and $\omega^*(f) \geq 0$ for every face f, and this contradiction with (1) will complete the proof.

Let f be a face. Since a 2-vertex is not adjacent to a 2- or 3-vertex (Lemma 1.2) and since a 3-vertex is adjacent to at most one 3-vertex (Lemma 2), the boundary of f cannot contain three consecutive vertices with degree less or equal to 3; so, f gives at most $\frac{1}{2} r(f) \frac{1}{2}$; hence, $\omega^*(f) \geq 2 \cdot r(f) - 10 - \frac{1}{4} \cdot r(f) \geq 0$ if $r(f) \geq 6$.

Let v be a k-vertex.

- If $k = 2$, $\omega(v) = -4$ and $\omega^*(v) = -4 + 2 \cdot 2 = 0$ by *Rule 1*.
- If $k = 3$, $\omega(v) = -1$ and $\omega^*(v) \geq -1 + 1 = 0$ by Lemma 2 and *Rule 2*.
- If $k = 4$, $\omega(v) = 2$. If v is adjacent to a 2-vertex, then v is not adjacent to other 2-vertices by Lemma 1.3, and v gives nothing to 3-vertices by *Rule 2*, since if v is adjacent to a 3-vertex then v is weak; so, it gives 2. If however v is not adjacent to a 2-vertex then v gives $\frac{1}{2}$ to each adjacent 3-vertex (at most $4 \cdot \frac{1}{2}$). In either case, $\omega^*(v) \geq 0$.
- If $k = 5$, $\omega(v) = 5$. If v is adjacent to at most one 2-vertex, it is easy to see that $\omega^*(v) \geq 5 - 2 - 4 \cdot \frac{1}{2} \geq 1$. If v is adjacent to two 2-vertices and to at most two 3-vertices, then $\omega^*(v) \geq 5 - 2 \cdot 2 - 2 \cdot \frac{1}{2} \geq 0$. If v is adjacent to two 2-vertices and three 3-vertices, then v receives $\frac{1}{2}$ from an incident $\geq$6-face (Lemma 1.5) and $\omega^*(v) \geq 5 - 2 \cdot 2 - 3 \cdot \frac{1}{2} + \frac{1}{2} \geq 0$. Finally, if v is adjacent to three 2-vertices, then v is not adjacent to 3-vertices (Lemma 1.6) and v is incident to at least two $\geq$6-faces which gives each $\frac{1}{2}$ to v by Lemma 1.4 and *Rule 3*. Hence, $\omega^*(v) \geq 0$.
- If $k = 6$, $\omega(v) = 8$. By Lemmas 1.3 and 1.6, v gives at most 8, either to four 2-vertices, or to at most three 2-vertices and three 3-vertices; hence, $\omega^*(v) \geq 0$.
- If $k = 7$, $\omega(v) = 11$. By Lemma 1.3, v gives to at most five 2-vertices and two 3-vertices; hence, $\omega^*(v) \geq 0$.
- If $k = 8$, $\omega(v) = 14$. By Lemma 1.3, v gives to at most six 2-vertices and two 3-vertices; hence, $\omega^*(v) \geq 0$.

- If $k = 9$, $\omega(v) = 17$. By Lemma 1.3, v gives to at most seven 2-vertices and two 3-vertices; hence, $\omega^*(v) \geq 0$.
- If $k \geq 10$, $\omega(v) = 3 \cdot k - 10$. The vertex v gives at most $2 \cdot k$; hence, $\omega^*(v) \geq 0$.

This contradiction with (1) completes the proof of Theorem 4.

3. Concluding remarks

In this paper, we give a sufficient condition for a planar graph to be acyclically 4-choosable: if its girth is at least 5.

Recently, we gives some new sufficient conditions for a planar graph to be acyclically 4-choosable:

Theorem 5. [MRW05] *Let G be a planar graph without 4-cycles and 5-cycles. The graph G is acyclically 4-choosable if G furthermore satisfies one of the following conditions: G does not contain (1) 6-cycles; (2) 7-cycles; (3) intersecting triangles.*

We conclude with a challenging problem regarding the acyclic choosability of planar graphs:

Problem 1. *Is acyclically 5-choosable every planar triangle-free graph?*

Acknowledgement

We would like to thank the referees for their fruitful comments.

References

[AB77]   M.O. Albertson and D.M. Berman. Every planar graph has an acyclic 7-coloring. *Israel J. Math.*, (28):169–174, 1977.

[BFDFK+02]   O.V. Borodin, D.G. Fon-Der Flaass, A.V. Kostochka, A. Raspaud, and E. Sopena. Acyclic list 7-coloring of planar graphs. *J. Graph Theory*, 40(2):83–90, 2002.

[BKW99]   O.V. Borodin, A.V. Kostochka, and D.R. Woodall. Acyclic colourings of planar graphs with large girth. *J. London Math. Soc.*, 1999.

[Bor79]   O.V. Borodin. On acyclic coloring of planar graphs. *Discrete Math.*, 25:211–236, 1979.

[Grü73]   B. Grünbaum. Acyclic colorings of planar graphs. *Israel J. Math.*, 14:390–408, 1973.

[KM76]   A.V. Kostochka and L.S. Mel'nikov. Note to the paper of Grünbaum on acyclic colorings. *Discrete Math.*, 14:403–406, 1976.

[Kos76]   A.V. Kostochka. Acyclic 6-coloring of planar graphs. *Discretny analys.*, (28):40–56, 1976. In Russian.

[Mit74]   J. Mitchem. Every planar graph has an acyclic 8-coloring. *Duke Math. J.*, (41):177–181, 1974.

[MOR05]   M. Montassier, P. Ochem, and A. Raspaud. On the acyclic choosability of graphs. *Journal of Graph Theory*, 2005. To appear.

[MRW05]   M. Montassier, A. Raspaud, and W. Wang. Acyclic 4-choosability of planar graphs without cycles of specific lengths. Technical report, LaBRI, 2005.

[Tho94] C. Thomassen. Every planar graph is 5-choosable. *J. Combin. Theory Ser. B*, 62:180–181, 1994.
[Voi93] M. Voigt. List colourings of planar graphs. *Discrete Mathematics*, 120:215–219, 1993.

Mickaël Montassier
LaBRI UMR CNRS 5800
Université Bordeaux 1
F-33405 Talence Cedex, France
e-mail: `montassi@labri.fr`

Graph Theory
Trends in Mathematics, 311–325
© 2006 Birkhäuser Verlag Basel/Switzerland

Automorphism Groups of Circulant Graphs – a Survey

Joy Morris

Abstract. A circulant (di)graph is a (di)graph on n vertices that admits a cyclic automorphism of order n. This paper provides a survey of the work that has been done on finding the automorphism groups of circulant (di)graphs, including the generalisation in which the arcs of the (di)graph have been assigned colours that are invariant under the aforementioned cyclic automorphism.

Mathematics Subject Classification (2000). 05C25.

Keywords. Circulant graphs, automorphism groups, algorithms.

1. Introduction

The aim of this paper is to provide a history and overview of work that has been done on finding the automorphism groups of circulant graphs. We will focus on structural theorems about these automorphism groups, and on efficient algorithms based on these theorems, that can be used to determine the automorphism group of certain classes of circulant graphs.

We must begin this discussion by defining the terms that will be central to the topic. In what follows, we sometimes refer simply to "graphs" or to "digraphs," but all of our definitions, and many of the results on automorphism groups, can be generalised to the case of "colour digraphs:" that is, digraphs whose arcs not only have directions, but colours.

Definition 1.1. *Two graphs $X = X(V, E)$ and $Y = Y(V', E')$ are said to be* **isomorphic** *if there is a bijective mapping ϕ from the vertex set V to the vertex set V' such that $(u, v) \in E$ if and only if $(\phi(u), \phi(v)) \in E'$. The mapping ϕ is called an* **isomorphism**. *We denote the fact that X and Y are isomorphic by $X \cong Y$.*

The author gratefully acknowledges support from the National Science and Engineering Research Council of Canada (NSERC).

That is, an isomorphism between two graphs is a bijection on the vertices that preserves edges and nonedges. In the case of digraphs, an isomorphism must also preserve the directions assigned to the arcs, and in the case of colour (di)graphs, the colours must also be preserved.

This definition has the following special case:

Definition 1.2. *An* **automorphism** *of a (colour) (di)graph is an isomorphism from the (colour) (di)graph to itself.*

When we put all of the automorphisms of a (colour) (di)graph together, the result is a group:

Definition 1.3. *The set of all automorphisms of a graph X forms a group, denoted* Aut(X), *the* **automorphism group** *of X.*

Now that we know what automorphism groups of graphs are, we must define circulant graphs.

Definition 1.4. *A* **circulant graph** $X(n; S)$ *is a Cayley graph on $\mathbb{Z}_n$. That is, it is a graph whose vertices are labelled $\{0, 1, \ldots, n-1\}$, with two vertices labelled i and j adjacent iff $i - j \pmod{n} \in S$, where $S \subset \mathbb{Z}_n$ has $S = -S$ and $0 \notin S$.*

For a circulant digraph, the condition that $S = -S$ is removed. For a colour circulant (di)graph, each element of S also has an associated (not necessarily distinct) colour, which is assigned to every edge (or arc) whose existence is a consequence of that element of S.

With these basic definitions in hand, we can explore the history of the question: what do the automorphism groups of circulant graphs look like, and how can we find them?

Although we define most of the terms used in this paper, for permutation group theoretic terms that are not defined, the reader is referred to Wielandt's book on permutation groups [43], which has recently come back into print in his collected works [44]. Another good source is Dixon and Mortimer's book [12]. Throughout this paper, the symmetric group on n points is denoted by $\mathcal{S}_n$.

2. History

In 1936, König [28] asked the following question: "When can a given abstract group be interpreted as the group of a graph and if this is the case, how can the corresponding graph be constructed? This same question could be asked for directed graphs." (This quote is from the English translation published by Birkhäuser in 1990). By the group of a graph, he is referring to the automorphism group.

This question was answered by Frucht, in 1938 [19]. The answer was yes; in fact, it went further: there are infinitely many such graphs for any group G.

By no means was this the end of the matter. One method of construction involved the creation of graphs on large numbers of vertices, that encoded the colour information from the Cayley colour digraph of a group, into structured subgraphs. This construction led to graphs that had, in general, many more vertices than the order of the group.

One major area of research that spun off from this, was the search for "graphical regular representations" of a particular group G: that is, graphs whose automorphism group is isomorphic to G, and whose number of vertices is equal to the order of G. This is not a topic that we will pursue further in this paper, however.

A related question was presumably also considered by mathematicians. That is, given a particular representation of permutation group G, is there a graph X for which $\operatorname{Aut}(X) \cong G$ as permutation groups? We know of no reference for this question prior to 1974, when it appears in [3].

The answer to this question did not prove to be so straightforward. For example, if $\rho = (0\ 1\ 2 \ldots n)$, no graph has $\langle \rho \rangle$ as its automorphism group, because the reflection that maps a to $-a$ for every $0 \leq a \leq n$ will always be an automorphism of any such graph. Another way of explaining this, is that the dihedral group acting on n elements has the same orbits on unordered pairs (i, j) as $\mathbb{Z}_n$ has. This shows that there is not always a graph X whose automorphism group is a particular representation of a permutation group.

Although it is not our aim to discuss this version of the question, we will give here the most significant results that have been obtained on it.

The first result is due to Hemminger [22], in 1967.

Theorem 2.1. *Let G be a transitive Abelian permutation group that is abstractly isomorphic to $\mathbb{Z}_{n_1} \times \cdots \times \mathbb{Z}_{n_k}$. Then if the number of factors in the direct product of order 2 is not 2, 3, or 4, and the number of factors of order 3 is not 2, there is a directed graph whose automorphism group is isomorphic to G.*

In 1981, Godsil [21] proved the following result.

Theorem 2.2. *Let G be a finite permutation group. A necessary condition for the Cayley graph $X = X(G; S)$ to have G as its automorphism group, is that the subgroup of G that fixes some vertex of X, is isomorphic to the automorphisms of G that fix S set-wise.*

He also proved that this condition is sufficient for many p-groups, and obtained necessary and sufficient conditions for the occurrence of the dihedral groups of order 2^k, and of certain Frobenius groups, as the full automorphism groups of vertex-transitive graphs and digraphs.

In 1989, a paper by Zelikovskij [45] appeared in Russian. We are unaware of a translation, so can only report the main result as stated in the English summary: that for every finite Abelian permutation group G whose order is relatively prime to 30, the paper provides necessary and sufficient conditions for the existence of a simple graph whose automorphism group is isomorphic to G. Note that the facts that Zelikovskij produces a simple graph and apparently does not require transitivity, represent improvements over Hemminger's result, although he does not cover all of the orders that Hemminger does.

In 1999, Peisert proved [37] the following result.

Theorem 2.3. *If two permutation groups each have representations as automorphism groups of graphs, then the direct product of these representations, will also have a graph for which it is isomorphic to the automorphism group, unless the two original groups are isomorphic as permutation groups, transitive, and have a unique graph for which they are the automorphism group (up to isomorphism).*

The approach to this question that we will follow in this paper, is to ask what is the automorphism group of a given graph X? Again, the answer is not easy, so we will limit our consideration to circulant graphs.

3. Algorithms for finding automorphism groups

As one focus of this paper will be finding efficient algorithms for calculating the automorphism group of a graph, it is necessary to spend some time considering what makes an algorithm "efficient" in this regard.

There is, after all, a very straightforward algorithm that is guaranteed to find the automorphism group of any graph on n vertices: simply consider every possible permutation in $\mathcal{S}_n$, and include those that turn out to be automorphisms of our graph. As $\mathcal{S}_n$ has order $n!$, this algorithm is exponential in n.

Depending on our goal, it may not be possible to do any better than this. Specifically, if our goal is to list every automorphism in the automorphism group, then as there may be as many as $n!$ elements, generating the list in anything less than exponential time will not be feasible in general.

If the number of prime factors of n is bounded with a sufficiently low bound, then we actually can do more than this. In such cases, the number of subgroups of $\mathcal{S}_n$ that can be automorphism groups of circulant graphs may be sufficiently small that we can choose the automorphism group of our graph from a short list. We will not be listing all of its elements, but we may be able to describe its structure precisely without the use of generating sets, as for example by saying it is isomorphic to $\mathcal{S}_p \times \mathcal{S}_q$.

However, when the number of prime factors is unbounded, we lose this ability to explicitly describe the automorphism group. Thus, when we consider algorithms intended to determine the automorphism group of a circulant graph on a number of vertices whose number of factors is unbounded, what we are looking for is not an explicit listing of all of the elements of the group, but the provision of a generating set for the group.

Each of the algorithms that we provide below runs in polynomial time in the number of vertices of the graph, and provides either a generating set for the automorphism group of the graph, or (where possible) a precise structural description of the automorphism group.

4. Circulant graphs on a prime number of vertices

In 1973, Alspach proved the following result on the automorphism groups of circulant graphs on a prime number of vertices.

Theorem 4.1. [2] *Let p be prime. If $S = \emptyset$ or $S = \mathbb{Z}_p^*$, then $\mathrm{Aut}(X) = \mathcal{S}_p$; otherwise, $\mathrm{Aut}(X) = \{T_{a,b} : a \in E(S), b \in \mathbb{Z}_p\}$, where $T_{a,b}(v_i) = v_{ai+b}$, and $E(S)$ is the largest even-order subgroup of $\mathbb{Z}_p^*$ such that S is a union of cosets of $E(S)$.*

Notice that since $S = -S$, S must be a union of cosets of $\{1, -1\}$, so $E(S)$ can always be found.

This leads to the following algorithm for finding the automorphism group of such a graph.

Algorithm for finding Aut(X):
1. If $S = \emptyset$ or $S = \mathbb{Z}_p^*$, Aut(X)= $\mathcal{S}_p$.
2. For each even-order subgroup H of $\mathbb{Z}_p^*$, verify whether S is a union of cosets of H.
3. Since $\mathbb{Z}_p^*$ is cyclic, there is one H of every order dividing $p-1$. Set $E(S)$ to be the largest H that satisfies (2).
4. If Aut(X)$\neq \mathcal{S}_p$, Aut(X)= $\{T_{a,b} : a \in E(S), b \in \mathbb{Z}_p\}$.

Proof of Alspach's theorem relies on the following theorem by Burnside.

Theorem 4.2. [9] *If G is a transitive group acting on a prime number p of elements, then either G is doubly transitive or $G = \{T_{a,b} : a \in H < \mathbb{Z}_p^*, b \in \mathbb{Z}_p\}$.*

Although the algorithm given above may be the most natural way to create an algorithm from the statement of Alspach's theorem, we will re-state the algorithm slightly differently. The purpose of this, is to provide a closer parallel to the algorithms that we will be constructing subsequently, to cover other possible numbers of vertices.

Alternate algorithm:
1. Find A, the set of all multipliers $a \in \mathbb{Z}_p^*$ for which $aS = S$.
2. If $A = \mathbb{Z}_p^*$, then Aut(X)= $\mathcal{S}_p$.
3. Otherwise, Aut(X)= $\{T_{a,b} : a \in A, b \in \mathbb{Z}_p\}$.

Before presenting any generalisations of this algorithm, the concept of wreath products will be required.

5. Wreath products

Although we will present a formal definition of the wreath product of two graphs in a moment, we will first give a description which may make the formal presentation easier to follow. If we are taking a wreath product of two (di)graphs, X and Y, we replace every vertex of X by a copy of the (di)graph Y. Between two copies of Y, we include all edges (or all arcs in a particular direction) if there was an edge (or an arc in the appropriate direction) between the corresponding vertices of X.

Now for the formal definition.

Definition 5.1. *The wreath product of the (di)graph X with the (di)graph Y, denoted $X \wr Y$, is defined in the following way.*

The vertices of $X \wr Y$ are the ordered pairs (x, y) where x is a vertex of X and y is a vertex of Y. There is an arc (or edge) from the vertex (x_1, y_1) to the vertex (x_2, y_2) if and only if one of the following holds:

1. $x_1 = x_2$ *and* (y_1, y_2) *is an arc (or edge) of Y; or*
2. (x_1, x_2) *is an arc (or edge) of X.*

We say two sets of vertices A and B are **wreathed** if either ab is an edge (or arc) for every $a \in A$ and every $b \in B$, or ab is a nonedge (or nonarc) for every $a \in A$ and every $b \in B$.

In graph theory, what we have called the "wreath" product is also often called the "lexicographic" product, and has also been called the "composition" of graphs. The notion of a wreath product is also defined on groups; in fact, the term "wreath product" comes from group theory. Although it can be defined on abstract groups, we will only be considering permutation groups, where the definition is simpler, so it is this definition that we provide.

Definition 5.2. *The wreath product of two permutation groups, H and K, acting on sets U and V respectively, is the group of all permutations f of $U \times V$ for which there exist $h \in H$ and an element k_u of K for each $u \in U$ such that*

$$f((u, v)) = (h(u), k_{h(u)}(v))$$

for all $(u, v) \in U \times V$. It is written $H \wr K$.

It is easy to verify that

$$\mathrm{Aut}(X) \wr \mathrm{Aut}(Y) \leq \mathrm{Aut}(X \wr Y)$$

is true for any graphs X and Y; in fact, it is often the case that equality holds.

6. Circulant graphs on pq vertices, or p^n vertices

The following theorem, proven by Klin and Pöschel in 1978, characterises the automorphism groups of circulant graphs on pq vertices, where p and q are distinct primes.

Theorem 6.1. [26] *If G is the automorphism group of a circulant graph on pq vertices, then G is one of:*

1. S_{pq};
2. $A_1 \wr A_2$, *or* $A_2 \wr A_1$, *where A_1 and A_2 are automorphism groups of circulant graphs on p and q vertices, respectively;*
3. $S_p \times A_2$, *or* $A_1 \times S_q$, *where A_1 and A_2 are automorphism groups of circulant graphs on p and q vertices, respectively; or,*
4. $\{T_{a,b} : a \in A \leq \mathbb{Z}_n^*, b \in \mathbb{Z}_n\}$ *(a subgroup of the holomorph).*

The following algorithm can be constructed from this characterisation, to determine the automorphism group of a graph on pq vertices.

Algorithm:
1. If $S = \emptyset$ or $S = \mathbb{Z}_{pq} - \{0\}$, then $\text{Aut}(X) = \mathcal{S}_{pq}$. END.
2. If $(v_0 \; v_p \; v_{2p} \ldots v_{(q-1)p}) \in \text{Aut}(X)$, then $\text{Aut}(X) = A_1 \wr A_2$, where A_1 is the automorphism group of the induced subgraph of X on the vertices $\{v_0, v_q, \ldots, v_{(p-1)q}\}$ and A_2 is the automorphism group of the induced subgraph of X on the vertices $\{v_0, v_p, \ldots, v_{(q-1)p}\}$. Use the previous algorithm (for finding the automorphism group of a circulant graph on a prime number of vertices) to find A_1 and A_2. END.
3. Repeat (2) with the roles of p and q reversed.
4. Let A be the group of all multipliers a in $\mathbb{Z}_{pq}^*$ for which $aS = S$.
5. Define E_p by $E_p = \{T_{a,b} : a \in A, a \equiv 1 \pmod{p}, b \in q\mathbb{Z}_p\}$; if $E_p \cong \text{AGL}(1,p)$ then $\text{Aut}(X) = \mathcal{S}_p \times A_2$, where A_2 is as in step (2). Use the previous algorithm to find A_2. END.
6. Repeat (4) with the roles of p and q reversed, and A_1 taking the role of A_2. END.
7. Otherwise, $\text{Aut}(X) = \{T_{a,b} : a \in A, b \in \mathbb{Z}_{pq}\}$.

The full automorphism group of circulant graphs on p^n vertices has been determined, first by Klin and Pöschel, and later independently by Dobson; both results are unpublished.

The full result is technical, but the following nice result gets much of the way.

Theorem 6.2. [13, 27] *A circulant graph on p^n vertices is either a wreath product, or its automorphism group has a normal Sylow p-subgroup.*

7. The square-free case

There is no known characterisation for the automorphism groups of circulant graphs in the general square-free case that is as straightforward as the results we have described above, when the number of vertices is p or pq and p, q are distinct primes.

However, Dobson and Morris [15] did prove the following structural theorem about the automorphism groups of circulant graphs of arbitrary square-free order. The definition of a group being 2-closed is quite technical, but the important thing to know about 2-closed groups is that the automorphism group of a vertex-transitive graph or digraph is always 2-closed. The terms used in the second point of this theorem are defined later in this paper, in the section "A Strategy," for readers who are interested.

Theorem 7.1. *Let mk be a square-free integer and $G \leq \mathcal{S}_{mk}$ be 2-closed and contain a regular cyclic subgroup, $\langle \rho \rangle$. Then one of the following is true:*
1. $G = G_1 \cap G_2$, *where* $G_1 = \mathcal{S}_r \wr H_1$ *and* $G_2 = H_2 \wr \mathcal{S}_k$, *where* H_1 *is a 2-closed group of degree mk/r, H_2 is a 2-closed group of order m, and $r | m$; or*

2. *there exists a complete block system $\mathcal{B}$ of G consisting of m blocks of size k, and there exists $H \triangleleft G$ such that H is transitive, 2-closed, and $\langle \rho \rangle \leq H = H_1 \times H_2$ (with the canonical action), where $H_1 \leq \mathcal{S}_m$ is 2-closed and $H_2 \leq \mathcal{S}_k$ is 2-closed and primitive.*

Unlike the previous structural results by Alspach, and by Klin and Pöschel, this clearly does not provide a clear, short list of groups from among which the automorphism group of a circulant graph on any square-free number of vertices must be chosen. However, it is sufficient to allow Dobson and Morris, in a subsequent paper [16], to construct the following algorithm that will determine the automorphism group of any such graph.

General algorithm (n square-free)
Inputs: The number n of vertices of X, and the connection set $S \subseteq \mathbb{Z}_n$.

1. Let A be the group of all multipliers a in $\mathbb{Z}_n^*$ for which $aS = S$.
2. For each prime divisor p of n, define E_p by $E_p = \{T_{a,b} : a \in A, a \equiv 1 \pmod{p}, b \in \frac{n}{p}\mathbb{Z}_p\}$; if $E_p \cong \mathrm{AGL}(1,p)$ then define $E_p = \mathcal{S}_p$.
3. Let $p_1, \ldots, p_t$ be all primes such that $E_p = \mathcal{S}_p$.
   (a) For each pair of distinct primes p_i, p_j, if the transposition that switches $n/p_i + kp_ip_j$ with $n/p_j + kp_ip_j$ for every $0 \leq k \leq n/p_ip_j$ and fixes all other vertices is an automorphism of X, then $E_{p_ip_j} = \mathcal{S}_{p_ip_j}$. Otherwise, $E_{p_ip_j} = E_{p_i} E_{p_j}$.
   (b) Define a relation R on $\{p_1, \ldots, p_t\}$ by $p_i R p_j$ iff there exists a sequence of primes $p_{k_1}, \ldots, p_{k_s}$ where $p_i = p_{k_1}, p_j = p_{k_s}$ such that $E_{\mathrm{lcm}(p_{k_l}, p_{k_{l+1}})} = \mathcal{S}_{\mathrm{lcm}(p_{k_l}, p_{k_{l+1}})}$. This is an equivalence relation. For each equivalence class $\mathcal{E}_i$, let $m_i = \Pi_{j \in \mathcal{E}_i} p_j$, then let $E_{m_i} = \mathcal{S}_{m_i}$.
   (c) For any divisor m of n,
   $$E_m = \Pi_\mathcal{E} \mathcal{S}_{\gcd(m_i,m)} \Pi_{p | \gcd(n/p_1 \ldots p_t, m)} E_p.$$
4. For each composite divisor m of n, let $A_m = \{a \in A : a \equiv 1 \pmod{\frac{n}{m}}\}$, and let $A_n = A$. Define $E'_m = \langle E_m, A_m \rangle$.
5. Let $G = E'_n$. For each complete block system of E'_n, $\mathcal{B}$, consisting of n/k blocks of size k, do:
   (a) For each complete block system of E'_n, $\mathcal{D}$, consisting of n/kk' blocks of size kk', determine whether or not
   $$\rho^{n/k}|_{D_0},$$
   the mapping that acts as $\rho^{n/k}$ on the vertices of D_0, and fixes all other vertices, is an automorphism of X. If it is, let our new G be the group generated by the old G together with every $E'_k|_D$, where $D \in \mathcal{D}$; if not, leave G unchanged.
6. G is the automorphism group of X.

To avoid some technical details, we have oversimplified some of the notation in this algorithm; the astute reader may observe, for example, that if we define $E_p = \mathcal{S}_p$, then E_p is acting on p vertices, and subsequent products involving

E_p ($E_m = E_p E_q$, etc.,) may not be properly defined. These issues are properly addressed in the paper that presents the algorithm, but as the intent of this paper was only to give the flavour of the algorithm, the technical details seemed likely to unnecessarily complicate our presentation.

In the same paper, Dobson and Morris prove that this algorithm runs in polynomial time on the number of vertices of the graph.

8. A strategy

In this section, we outline a general strategy that is often used to obtain structural results about circulant graphs.

Definition 8.1. *Let V be a set, and G a permutation group acting on the elements of V. The subset $B \subseteq V$ is a G-**block** if for every $g \in G$, either $g(B) = B$, or $g(B) \cap B = \emptyset$.*

In some cases, the group G is clear from the context and we simply refer to B as a block.

It is a simple matter to realise that if B is a G-block, then for any $g \in G$, $g(B)$ will also be a G-block. Also, intersections of G-blocks remain G-blocks.

Let G be a transitive permutation group, and let B be a G-block. Then, as noted above, $\{g(B) : g \in G\}$ is a set of blocks that (since G is transitive) partition the set V. We call this set the **complete block system** of G generated by the block B.

Notice that any singleton in V, and the entire set V, are always G-blocks.

The **size** of the block B is the cardinality of the set B. A block B is **nontrivial** if the size of B is neither 1 nor the cardinality of V.

Definition 8.2. *The transitive permutation group G is said to be **imprimitive** if G admits nontrivial blocks. If G is transitive but not imprimitive, then G is said to be **primitive**.*

The notion of transitivity for permutation groups can be generalised.

Definition 8.3. *The permutation group G acting on the set V is k-**transitive** if given any two k-tuples $(v_1, \ldots, v_k)$ and $(u_1, \ldots, u_k)$ with $v_1, \ldots, v_k, u_1, \ldots, u_k \in V$, there exists some $g \in G$ such that $g(v_i) = u_i$ for $1 \leq i \leq k$.*

In particular, we often say that a 2-transitive group is doubly transitive.

Definition 8.4. *The abstract group G is a **Burnside group** if every primitive permutation group containing the regular representation of G as a transitive subgroup is doubly transitive.*

Burnside gave the first example of such a group, hence the name. This is extremely useful in the theory of circulant graphs, due to the following theorem.

Theorem 8.5 (Theorem 25.3, [43]). *Every cyclic group of composite order is a Burnside group.*

In particular, if n is composite, the automorphism group of a circulant graph of order n is either doubly transitive or imprimitive.

Corollary 8.6. *For any circulant graph $X = X(n; S)$, one of the following holds:*
1. $\text{Aut}(X) = \mathcal{S}_n$;
2. $\text{Aut}(X)$ *is imprimitive; or*
3. n *is prime.*

This can be used repeatedly to show that the minimal blocks of a circulant graph must either have prime size, or the induced subgraph on these vertices is complete or empty.

We can also use this to determine additional information about how $\text{Aut}(X)$ acts upon any blocks - once again, this action must be imprimitive or doubly transitive unless there are a prime number of blocks.

9. Related problems

In this section, we discuss a number of problems that are closely related to finding the automorphism group of a circulant graph.

9.1. The Cayley Isomorphism problem for circulants

One of the most-studied problems related to finding the automorphism group of circulant graphs, is the Cayley Isomorphism, or CI, problem.

Definition 9.1. *A circulant graph $X = X(n; S)$ is said to have the* **Cayley Isomorphism (CI) property** *if whenever $Y = Y(n; S')$ is isomorphic to X, there is some $a \in \mathbb{Z}_n^*$ for which $aS = S'$.*

The cyclic group of order n is said to have the Cayley Isomorphism (CI) property if every circulant graph $X(n; S)$ has the CI property.

The CI problem, of course, is to determine which graphs (or which groups) have the CI property. When applied to digraphs, the CI property is referred to as the DCI property (for "directed Cayley Isomorphism" property).

Theorem 9.2. [6, 7] *A circulant graph X on n vertices has the CI property if and only if any two n-cycles in $\text{Aut}(X)$ are conjugate in $\text{Aut}(X)$.*

So, knowing that a graph has the CI-property gives us significant information about its automorphism group.

We give a brief history of the major results on this problem.

In 1967, Àdàm conjectured [1] that all cyclic groups are CI-groups. Elspas and Turner proved in 1970 that this was not the case [18], and that, in fact, $\mathbb{Z}_{p^2}$ is not CI for $p \geq 5$. There were also many positive results on this conjecture, however. First, in 1967, Turner proved that $\mathbb{Z}_p$ is CI [41]. In a computer search, McKay [33] found that $\mathbb{Z}_n$ is CI, when $n \leq 37$ and $n \neq 16, 24, 25, 27, 36$. In 1977, Babai [7] proved that $\mathbb{Z}_{2p}$ is CI, which was generalised in 1979 by Alspach and Parsons [6], who proved that $\mathbb{Z}_{pq}$ is CI. In 1983, Godsil [20] proved that $\mathbb{Z}_{4p}$ is

CI. Muzychuk completed the work in 1997 [35, 36] by showing that $\mathbb{Z}_n$ is DCI if and only if $n \in \{k, 2k, 4k\}$ where k is odd and squarefree; $\mathbb{Z}_n$ is CI if and only if $n \in \{8, 9, 18\}$ or $\mathbb{Z}_n$ is DCI.

This problem has also been studied for particular families of graphs; Huang and Meng proved in 1996 [23], that $X(n; S)$ has the CI property if S is a minimal generating set for $\mathbb{Z}_n$. In 1977, Toida conjectured that if $S \subseteq \mathbb{Z}_n^*$ then $X(n; S)$ is CI [40]. This conjecture was proven by Klin, Pöschel and Muzychuk [25], and independently by Dobson and Morris [14].

The special case where the connection set, S, is small, has also been studied. In 1977, Toida [40] proved that if $|S| \le 3$, $X(n; S)$ is CI. In 1988, Sun [39] was the first to prove that if $|S| = 4$, $X(n; S)$ is CI, although others later also proved this result. In 1995, Li proved that if $|S| = 5$, $X(n; S)$ is CI [29].

This is by no means intended as a full history of work that has been done on the CI problem; for further information, the reader is referred to Li's survey of the problem [30].

9.2. Edge-transitivity, arc-transitivity, 2-arc-transitivity

Definition 9.3. *A k-arc is a list $v_1, \ldots, v_k$ of vertices for which any two sequential vertices are adjacent, and any 3 sequential vertices are distinct.*

This definition allows us to consider graphs whose automorphism groups are transitive on the set of k-arcs of the graph, for various values of k; for our purposes, we will consider only $k = 1$ and $k = 2$.

The following theorem was proven by Chao [10] in 1971; the proof was simplified by Berggren [8] in 1972 and further simplified by Alspach in Theorem 1.16 of [4].

Theorem 9.4. *A circulant graph on p vertices is edge-transitive iff $S = \emptyset$ or S is a coset of an even order subgroup $H \le \mathbb{Z}_p^*$.*

Notice that in the case of circulant graphs, the reflection is an automorphism, so arc-transitivity is equivalent to edge-transitivity.

This result, and those that follow, provide significant information about how the automorphism group of a graph can act on the edge set of the graph. They are therefore related to finding the automorphism group, although they do not directly provide much information about the automorphism group of an arbitrary circulant graph.

The above theorem was extended to classifications of arc-transitive graphs on pq vertices, for any distinct primes p and q, in papers by Cheng and Oxley [11], who classified the graphs on $2p$ vertices; Wang and Xu [42], who classified the graphs on $3p$ vertices; and Praeger, Wang and Xu [38], who completed the classification.

In 2001, the following classification was obtained of arc-transitive circulants on a square-free number of vertices, by Li, Marušič and Morris [31].

Theorem 9.5. *If X is an arc-transitive circulant graph of square-free order n, then one of the following holds:*

1. $X = K_n$;
2. $\mathrm{Aut}(X)$ *contains a cyclic regular normal subgroup; or*
3. $X = Y \wr \bar{K}_b$, *or* $X = Y \wr \bar{K}_b - bY$, *where* $n = mb$, *and Y is an arc-transitive circulant of order m.*

Another approach has been taken to classifying edge-transitive circulant graphs, using the more stringent condition that the complement must also be edge-transitive. The following result was proven by Zhang in 1996 [46].

Theorem 9.6. *If G and $\overline{G}$ are both edge-transitive circulants, then G is one of: mK_n, $\overline{mK_n}$, or a self-complementary Paley graph (n is prime, $1 \pmod 4$) and $S = \{a^2 : a \in \mathbb{Z}_n^*\}$).*

The property of 2-arc-transitivity is stronger than arc-transitivity, but accordingly tells us more about the action of the automorphism group. Circulant graphs that are 2-arc-transitive have been fully classified by Alspach, Conder, Marušič and Xu [5], as follows.

Theorem 9.7. *A connected 2-arc-transitive circulant graph is one of:*

1. K_n *(exactly 2-arc-transitive);*
2. $K_{n/2,n/2}$ *(exactly 3-arc-transitive);*
3. $K_{n/2,n/2}$ *minus a 1-factor, $n \geq 10$, $n/2$ odd (exactly 2-arc-transitive);*
4. C_n *(k-arc-transitive for every $k \geq 0$).*

9.3. Other regular subgroups in Aut(X)

The automorphism group of any circulant graph will have a cyclic subgroup that acts regularly on the vertices of the graph. (A permutation group on a set V is said to act regularly if for any pair of points in V, there is exactly one permutation in the group that maps one to the other.) Sometimes, the automorphism group of a graph may have multiple, nonisomorphic, regular subgroups. This is of interest from a different perspective, because it means that the graph in question can be represented as a Cayley graph on some noncyclic group, besides being a circulant graph. From our perspective, knowing the regular subgroups of the automorphism group may be useful in determining the automorphism group.

There are only a few results of note on this topic. The first result was proven by Joseph in the special case $n = p^2$ [24], and extended by Morris [34] to all prime powers.

Theorem 9.8. *If $n = p^e$, X is a circulant graph on n vertices, and $\mathrm{Aut}(X)$ contains a regular subgroup that is not cyclic, then X is isomorphic to a wreath product of smaller circulant graphs.*

Recently, Marušič and Morris [32] proved the following results.

Theorem 9.9. *Let $X = X(n; S)$ be a circulant graph, and $\mathbb{Z}_n^*(S)$ be the subgroup of $\mathbb{Z}_n^*$ that fixes S set-wise. Then if $\gcd(n, |\mathbb{Z}_n^*(S)|) > 1$, the automorphism group of X has a noncyclic regular subgroup.*

In fact, if p is any prime divisor of $\gcd(n, |\mathbb{Z}_n^*(S)|)$, then the automorphism group of X contains a regular subgroup that is isomorphic to $\mathbb{Z}_p \ltimes \mathbb{Z}_{n/p}$.

They also show that the converse is not true in general; that is, there may be noncyclic regular subgroups in the automorphism group even if this greatest common divisor is 1.

In the same paper, they prove the following result. The condition that we must have a normal circulant means that the regular cyclic group must be normal in the automorphism group of the graph.

Theorem 9.10. *Let X be a normal circulant graph of order n, n not divisible by 4. Then if the automorphism group of X has a noncyclic regular subgroup, that group must be metacyclic, generated by two cyclic subgroups whose orders are relatively prime.*

10. Concluding remarks

Efficient algorithms have been found for determining the automorphism group of a circulant graph, when the number of vertices is any product of distinct primes. In the case where the number of vertices is a prime power, structural theorems about the automorphism groups exist (in unpublished form), but no algorithms have been constructed. No work has been done on combining these into results that may hold when the number of vertices is divisible by at least two distinct primes, but is not square-free. There is, therefore, a great deal of room for more results on this problem.

Unfortunately, in the results that we have seen, the complexity (of the proofs) seems to be growing. The result on circulant graphs on a square-free number of vertices is the culmination of two densely-packed, long papers; the first provides a structural theorem, and the second the algorithm. The result that deals with circulant graphs whose number of vertices is a prime power seems to be unpublished to this point, largely because of its length and complexity. It may be that new techniques will need to be developed before the problem can be completed in its full generality.

Of course, circulant graphs are just one very small step towards answering the general question with which we began, of finding the automorphism group of any graph. In full generality, this problem seems unmanageable, but it may be that for some other classes of graphs, the solution turns out to be feasible, or even easy.

References

[1] A. Àdàm, *Research problem 2–10.* J. Combin. Theory **2** (1967), 309.
[2] B. Alspach, *Point-symmetric graphs and digraphs of prime order and transitive permutation groups of prime degree.* J. Combinatorial Theory Ser. B **15** (1973), 12–17.

[3] B. Alspach, *On constructing the graphs with a given permutation group as their group*. Proc. Fifth Southeastern Conf. Combin., Graph Theory and Computing (1974), 187–208.

[4] B. Alspach, *Isomorphism and Cayley graphs on abelian groups*. Graph Symmetry, G. Hahn and G. Sabidussi, eds., Kluwer, 1997, 1–22.

[5] B. Alspach, M. Conder, D. Marušič and M.-Y. Xu, *A classification of 2-arc-transitive circulants*. J. Algebraic Combin. **5** (1996), 83–86.

[6] B. Alspach and T.D. Parsons, *Isomorphism of circulant graphs and digraphs*. Discrete Math. **25** (1979), 97–108.

[7] L. Babai, *Isomorphism problem for a class of point-symmetric structures*. Acta Math. Sci. Acad. Hung. **29** (1977), 329–336.

[8] J.L. Berggren, *An algebraic characterization of symmetric graphs with a prime number of vertices*. Bull. Austral. Math. Soc. **7** (1972), 131–134.

[9] W. Burnside, *On some properties of groups of odd order*. J. London Math. Soc. **33** (1901), 162–185.

[10] C.Y. Chao, *On the classification of symmetric graphs with a prime number of vertices*. Trans. Amer. Math. Soc. **158** (1971), 247–256.

[11] Y. Cheng and J. Oxley, *On weakly symmetric graphs of order twice a prime*. J. Combin. Theory Ser. B **42** (1987), 196–211.

[12] J.D. Dixon and B. Mortimer, *Permutation Groups*, Springer-Verlag, Graduate Texts in Mathematics, **163**, 1996.

[13] E. Dobson, *An extension of a classical result of Burnside*. Unpublished manuscript.

[14] E. Dobson and J. Morris, *Toida's conjecture is true*. Electronic J. Combin. **9(1)** (2002), R35.

[15] E. Dobson and J. Morris, *On automorphism groups of circulant digraphs of square-free order*. Discrete Math. **299** (2005), 79–98.

[16] E. Dobson and J. Morris, *Automorphism groups of circulant digraphs of square-free order*. In preparation.

[17] E. Dobson and D. Witte, *Transitive permutation groups of prime-squared degree*. J. Algebraic Combin. **16** (2002), 43–69.

[18] B. Elspas and J. Turner, *Graphs with circulant adjacency matrices*. J. Combin. Theory **9** (1970), 297–307.

[19] R. Frucht, *Herstellung von Graphen mit vorgegebener abstrakter Gruppe*. Compositio Math. **6** (1938), 239–250.

[20] C.D. Godsil, *On Cayley graph isomorphisms*. Ars Combin. **15** (1983), 231–246.

[21] C.D. Godsil, *On the full automorphism group of a graph*. Combinatorica **1** (1981), 243–256.

[22] R.L. Hemminger, *Directed graphs with transitive abelian groups*. Amer. Math. Monthly **74** (1967), 1233–1234.

[23] Q.X. Huang and J.X. Meng, *On the isomorphisms and automorphism groups of circulants*. Graphs Combin. **12** (1996), 179–187.

[24] A. Joseph, *The isomorphism problem for Cayley digraphs on groups of prime-squared order*. Discrete Math. **141** (1995), 173–183.

[25] M.H. Klin, M. Muzychuk and R. Pöschel, *The isomorphism problem for circulant graphs via Schur ring theory*. Codes and Association Schemes, American Math. Society, 2001.

[26] M.H. Klin and R. Pöschel, *The König problem, the isomorphism problem for cyclic graphs and the method of Schur*. Proceedings of the Inter. Coll. on Algebraic methods in graph theory, Szeged 1978, Coll. Mat. Soc. János Bolyai **27**.
[27] M.H. Klin and R. Pöschel, *The isomorphism problem for circulant digraphs with p^n vertices*. Unpublished manuscript (1980).
[28] D. König, *Theorie der Endlichen und Unendlichen Graphen*, Akademische Verlagsgesellschaft, 1936.
[29] C.H. Li, *Isomorphisms and classification of Cayley graphs of small valencies on finite abelian groups*. Australas. J. Combin. **12** (1995), 3–14.
[30] C.H. Li, *On isomorphisms of finite Cayley graphs – a survey*. Discrete Math. **256** (2002), 301–334.
[31] C.H. Li, D. Marušič and J. Morris, *Classifying arc-transitive circulants of square-free order*. J. Algebraic Combin. **14** (2001), 145–151.
[32] D. Marušič and J. Morris, *Normal circulant graphs with noncyclic regular subgroups*. J. Graph Theory **50** (2005), 13–24.
[33] B.D. McKay, unpublished computer search for cyclic CI-groups.
[34] J. Morris, *Isomorphic Cayley graphs on nonisomorphic groups*. J. Graph Theory **31** (1999), 345–362.
[35] M. Muzychuk, *Àdàm's conjecture is true in the square-free case*. J. Combin. Theory A **72** (1995), 118–134.
[36] M. Muzychuk, *On Àdàm's conjecture for circulant graphs*. Discrete Math. **167/168** (1997), 497–510.
[37] W. Peisert, *Direct product and uniqueness of automorphism groups of graphs*. Discrete Math. **207** (1999), 189–197.
[38] C.E. Praeger, R.J. Wang and M.-Y. Xu, *Symmetric graphs of order a product of two distinct primes*. J. Combin. Theory B **58** (1993), 299–318.
[39] L. Sun, *Isomorphisms of circulant graphs*. Chinese Ann. Math. **9A** (1988), 259–266.
[40] S. Toida, *A note on Àdàm's conjecture*. J. Combin. Theory B **23** (1977), 239–246.
[41] J. Turner, *Point-symmetric graphs with a prime number of points*. J. Combin. Theory **3** (1967), 136–145.
[42] R.J. Wang and M.-Y. Xu, *A classification of symmetric graphs of order $3p$*. J. Combin. Theory B **58** (1993), 197–216
[43] H. Wielandt, *Finite Permutation Groups*, Academic Press, 1964.
[44] H. Wielandt, *Mathematische Werke/Mathematical works. Vol. 1. Group theory*, edited and with a preface by Bertram Huppert and Hans Schneider, Walter de Gruyter & Co., 1994.
[45] A.Z. Zelikovskij, *Knig's problem for Abelian permutation groups*. Izv. Akad. Nauk BSSR, Ser. Fiz.-Mat. Nauk **5** (1989), 34–39.
[46] H. Zhang, *On edge transitive circulant graphs*. Tokyo J. Math. **19** (1996), 51–55

Joy Morris
Department of Mathematics and Computer Science
University of Lethbridge
Lethbridge, AB. T1K 3M4. Canada
e-mail: `joy@cs.uleth.ca`

Hypo-matchings in Directed Graphs

Gyula Pap

Abstract. We give a common generalization of results on hypo-matchings given in a sequence of papers by G. Cornuéjols, D. Hartvigsen and W. Pulleyblank in [2, 3, 4, 5] and results on even factors given by W.H. Cunningham and J.F. Geelen in [7] and by the author and L. Szegő in [13].

1. Introduction

The main result of this paper is a common generalization of the hypo-matching formula and the even factor formula.

The maximum hypo-matching problem is the following. Given an undirected graph G and a family $\mathcal{F}$ of factor-critical (hypo-matchable) subgraphs of G. A hypo-matching in G is a subgraph the components of which are some members of $\mathcal{F}$ and some components isomorphic to K_2. The problem is to maximize the number of nodes covered by a hypo-matching, for results on this problem, see papers of G. Cornuéjols, D. Hartvigsen and W. Pulleyblank [2, 3, 4, 5].

The maximum even factor problem is given in directed graphs. An even factor is the arc set M of a subgraph the weak components of which are directed cycles of even length and directed paths of arbitrary length. The problem is to maximize the cardinality of M. This problem can only be solved for a class of directed graphs called *odd-cycle-symmetric* – a directed graph is odd-cycle-symmetric if each directed odd cycle is symmetric, i.e., its arcs also exist in the opposite direction. W.H. Cunningham proposed this problem in [6] as a generalization of the optimum path-matching problem, the algebraic method of Cunningham and J.F. Geelen [7, 8] may be extended to even factors. The author and L. Szegő gave a simplified min-max formula in [13].

We present a theorem generalizing results in both topics, let us make some remarks here on the method of proof. In [16] M. Loebl and S. Poljak gave an elegant

---

The author is supported by the Egerváry Research Group of the Hungarian Academy of Sciences. Supported by European MCRTN Adonet, Contract Grant No. 504438. Research supported by the Hungarian National Foundation for Scientific Research Grant, OTKA T037547.

proof of the hypo-matching formula, which uses the Edmonds-Gallai decomposition of G (see [9, 10]). In [13], the author and Szegő used a non-constructive inductive method ("divide-and-conquer") to show the maximum even factor formula, where they also described an Edmonds-Gallai-type decomposition for even factors. It would be desirable to use this decomposition for the elegant method of Loebl and Poljak, but there is an obstacle. For hypo-matchings in undirected graphs the inclusion $D_G^{\mathcal{F}} \subseteq D_G$ holds, using widespread notation for the Edmonds-Gallai-type decomposition in graph G. However, the analogue inclusion does not hold for the concept in this paper for hypo-matchings in directed graphs.

The "divide-and-conquer" method of proof could only be extended to prove a special case of Theorem 3.1, see [14].

The proof of this paper uses the constructive method first presented in [15] by the author to solve the maximum even factor problem algorithmically. In fact, the proof of the paper is also constructive in the same sense as those of results on hypo-matchings. That is, given a digraph D and $\mathcal{H}$ as in Definition 2.1, and suppose we have an oracle for testing $\mathcal{H}$-criticality of subgraphs, we can solve the maximum $\mathcal{H}$-matching problem in polynomial time.

2. Definitions

Consider a digraph $D = (V, A)$ where we allow loops or parallel arcs. A *cycle (path)* is the arc-set of a closed (unclosed) directed walk without repetition of arcs or nodes. The emptyset is regarded as the zero-length path, any node may be regarded as the start and end of a zero-length path. A loop-arc gives a one-arc cycle. We call an arc $e = uv \in A$ *symmetric* (in D) if there is an arc $f = vu \in A$, otherwise e is *asymmetric* (in D). A cycle or a path is *even* (*odd*) if it consists of an even (odd) number of arcs. A cycle is *asymmetric* (in D) if it has at least one asymmetric arc. Loop-arcs are regarded as symmetric. The weakly connected components of a digraph (i.e., connected components in the undirected sense) are called *weak components*, for short. A *path-cycle-matching* is the arc-set of a subgraph the weak components of which are directed cycles and directed paths.

An undirected graph $G = (V, E)$ is called *factor-critical* if for all $v \in V$ the graph $G - v$ has a perfect matching. (For a survey on matching theory, see L. Lovász and M.D. Plummer [12].) A digraph is called *symmetric-critical* if all its arcs are symmetric and the underlying undirected graph is factor-critical. Let $\mathcal{H}$ be a (possibly empty) family of symmetric-critical subgraphs in D. For example, single node subgraphs may be put in $\mathcal{H}$ as they are symmetric-critical.

Definition 2.1. *An $\mathcal{H}$-matching in $D = (V, A)$ is a subset M of A so that for each weak component (V_0, M_0) of the digraph (V, M)*

1. M_0 *is a path, or*
2. M_0 *is an even cycle, or*
3. M_0 *is an asymmetric odd cycle, or*
4. (V_0, M_0) *is a member of $\mathcal{H}$.*

The size(M) is defined as the number of arcs in all these paths and cycles plus the number of nodes covered by members of $\mathcal{H}$ used in M. We say M uses a member H of $\mathcal{H}$ if H is a weak component in (V, M). Let $\nu^{\mathcal{H}}(D)$ denote the maximum size of an $\mathcal{H}$-matching in D.

Some further definitions, notation: for a set $X \subseteq V$ let $\Gamma_D^+(X) := \{x \in V - X : \exists y \in X, yx \in A\}$ and $\varrho_D(X) = |\{uv \in A : u \in V - X, v \in X\}|$ and $\delta_D(X) = |\{uv \in A : u \in X, v \in V - X\}|$. For $uv \in A$ we call v the *head* of arc uv, while u is called the *tail* of arc uv; uv *leaves* u and *enters* v.

A set $X \subseteq V$ induces the subgraph $D[X] = (X, A[X])$ where $A[X] = \{uv \in A : u, v \in X\}$. $D - X := D[V - X]$. Also, for a set $M \subseteq A$ we define the set of arcs in M induced in X as $M[X] := \{uv \in M : u, v \in X\}$. For a set $U \subseteq V$ we denote the *contracted graph* by D/U having node set $V/U = V - U + \{U\}$ and arc-set A/U given by deleting the arcs in $A[U]$ and identifying the nodes in U by $\{U\}$ (we will use contractions only if $D[U]$ is connected, thus this definition is equivalent with the usual contraction of the arcs in $A[U]$). Consider an induced subgraph $D[U]$, let $\mathcal{H}_U$ be the family of members of $\mathcal{H}$ that are subgraphs of $D[U]$. To make the notation simpler, an $\mathcal{H}_U$-matching in $D[U]$ will be called an $\mathcal{H}$-matching in $D[U]$.

For an $\mathcal{H}$-matching M let $V_\mathcal{H}(M)$ denote the set of nodes covered by a member of $\mathcal{H}$ used in M. Let $V^+(M) := \{v \in V : \delta_M(v) = 1\} \cup V_\mathcal{H}(M)$ and $V^-(M) := \{v \in V : \varrho_M(v) = 1\} \cup V_\mathcal{H}(M)$, hence $|V^+(M)| = |V^-(M)| = \text{size}(M)$. We say the set $V - V^-(M)$ consists of the M-*source-nodes* and $V - V^+(M)$ consists of the M-*sink-nodes*. M is defined to be *perfect* if it size$(M) = |V|$, or equivalently it has no sink-nodes or source-nodes.

Consider an induced subgraph $D[U]$ which is symmetric-critical. Let $D[U]$ be called $\mathcal{H}$-*critical* if there is no perfect $\mathcal{H}$-matching in $D[U]$. Here we remark that, a symmetric-critical subgraph is not necessarily induced, but only induced symmetric-critical subgraphs may be called $\mathcal{H}$-critical, by definition. (This definition of $\mathcal{H}$-critical subgraphs makes also sense because of the following observation. If some digraph has a symmetric-critical spanning subgraph and has an asymmetric arc, then there exists a perfect $\mathcal{H}$-matching – this follows from Claim 3.5.)

A set $S \subseteq V$ is called a *source-component* in D if $D[S]$ is strongly connected and $\varrho_D(S) = 0$. Let $\sigma_\mathcal{H}(D[X])$ denote the number of those source-components in $D[X]$ which are $\mathcal{H}$-critical.

The *deficiency* of an $\mathcal{H}$-matching M is $\text{def}_{D,\mathcal{H}} M := |V| - \text{size}(M)$, which is non-negative, of course. The *deficiency* of a set $X \subseteq V$ is defined by $\text{def}_{D,\mathcal{H}} X := \sigma_\mathcal{H}(D[X]) - |\Gamma_D^+(X)|$. Let us use the notation $\tau_{D,\mathcal{H}}(X) := |V| + |\Gamma_D^+(X)| - \sigma_\mathcal{H}(D[X]) = |V| - \text{def}_{D,\mathcal{H}} X$. Define $\tau^\mathcal{H}(D) := \min_{X \subseteq V} \tau_{D,\mathcal{H}}(X)$.

To prove the main theorem 3.1, in the next section we will use the following well-known statements about factor-critical undirected graphs.

Lemma 2.2. *Consider an undirected graph $G = (V, E)$. If for some $U \subseteq V$ the induced subgraph $G[U]$ is factor-critical and G/U is factor-critical, then G is factor-critical, too.* □

Lemma 2.3. *Suppose we are given a factor-critical undirected graph $G = (V, E)$ and nodes $s, t \in V$. Then there is an even path P from s to t and a perfect matching M in $G - V(P)$.* □

3. A min-max formula

In this section we prove the following theorem, the main result of the paper.

Theorem 3.1. *If $D = (V, A)$ is a digraph and $\mathcal{H}$ is a family of symmetric-critical subgraphs of D, then*
$$\nu^{\mathcal{H}}(D) = \tau^{\mathcal{H}}(D). \tag{1}$$

The easy part of the proof is to see that the left hand side in (1) is at most the right hand side. This follows from the following lemma.

Lemma 3.2. *For any $\mathcal{H}$-matching M and any set $X \subseteq V$ we have $|X - V^+(M)| \geq \sigma_{\mathcal{H}}(D[X]) - |\Gamma_D^+(X)|$.*

Proof. First we show that we may assume without loss of generality that M does not use any member H of $\mathcal{H}$ with $V(H) \cap X \neq \emptyset$, $V(H) - X \neq \emptyset$. Otherwise, if M uses some H split by X, then consider a node v in $V(H) \cap \Gamma_D^+(X)$, which is non-empty since H is strongly connected. Since H is symmetric-critical, there exists a perfect matching in the underlying undirected graph of $H - v$. We construct another $\mathcal{H}$-matching M' by replacing H in M by a subset of $A(H)$ of $|V(H)| - 1$ arcs which is the union of node-disjoint two-arc cycles covering $V(H) - v$: these two-arc cycles are constructed from a perfect matching in the underlying undirected graph of $H - v$. Then $X - V^+(M') = X - V^+(M)$.

Consider the set $M[X]$ of arcs of M induced in X. By the above assumption, $M[X]$ is an $\mathcal{H}$-matching. Consider an $\mathcal{H}$-critical source-component S in $D[X]$. Since S is a source-component, $M[S]$ is an $\mathcal{H}$-matching, too. There is no perfect $\mathcal{H}$-matching in S, hence $S - V^-(M[S]) \neq \emptyset$. Since S is a source-component, $S - V^-(M[S]) = S - V^-(M[X])$. Thus $|X - V^+(M[X])| = |X - V^-(M[X])| \geq \sigma_{\mathcal{H}}(D[X])$.

Each node a in $(X - V^+(M[X])) - (X - V^+(M))$ is covered by an arc $ab \in M$ with $b \in \Gamma^+(X)$. This arc ab can only be an arc on a cycle or a path of M, hence $|(X - V^+(M[X])) - (X - V^+(M))| \leq |\Gamma^+(X)|$. □

Definition 3.3. *A set $X \subseteq V$ is a verifying set for an $\mathcal{H}$-matching M if $\text{size}(M) = \tau_{D,\mathcal{H}}(X)$, or equivalently $\text{def}_{D,\mathcal{H}} M = \text{def}_{D,\mathcal{H}} X$.*

Lemma 3.2 implies
$$\text{size}(M) = |V^+(M) \cap X| + |V^+(M) - X|$$
$$\leq |X| - \sigma_{\mathcal{H}}(D[X]) + |\Gamma_D^+(X)| + |V - X| = \tau_{D,\mathcal{H}}(X),$$
so we can easily see the following "slackness" type condition.

Claim 3.4. *If M is an $\mathcal{H}$-matching with a verifying set X, then $V - X \subseteq V^+(M)$.*

The following claim is straightforward from Lemma 2.3.

Claim 3.5. *Suppose $D = (V, A)$ is symmetric-critical, and $s, t \in V$ are two not necessarily distinct nodes. Then there is an $\mathcal{H}$-matching M_{st} for which*

1. *$\text{size}(M_{st}) = |V| - 1$,*
2. *M_{st} consists of two-arc cycles and an even $s - t$ path P_{st} (in case of $s = t$ this path has length zero).* □

Claim 3.6. *Consider symmetric-critical induced subgraph $D[U]$, suppose J induces a symmetric-critical subgraph in D/U with $\{U\} \in J$. Let R be the pre-image of J in D. Then $D[R]$ is symmetric-critical, or there is a perfect $\mathcal{H}$-matching in $D[R]$ (or both).*

Proof. By Lemma 2.2, the underlying undirected graph of $D[R]$ is factor-critical, so either $D[R]$ is symmetric-critical, or there is an asymmetric arc $ab \in A[R]$. Then ab is not induced in U, so the images a' and b' in D/U are distinct. By Claim 3.5, in D/U there is a node-disjoint family of two-arc cycles and a $b' - a'$ path internally covering J, the union of these is denoted by N'. Then by Claim 3.5, there is an expansion N in D which internally covers R by even cycles and a $b - a$ path. So $N + ab$ covers R by even cycles and one asymmetric cycle. □

Definition 3.7. *Suppose $D[U]$ is a symmetric-critical induced subgraph in D, let $D' := D/U$. We construct a family $\mathcal{H}'$ in D' as follows. We put a symmetric-critical induced subgraph $D'[Q']$ into $\mathcal{H}'$ if and only if the pre-image Q of Q' induces a subgraph $D[Q]$ which is not symmetric-critical, or a symmetric-critical subgraph $D[Q]$ which is not $\mathcal{H}$-critical.*

By Claim 3.6 we make the following observation on this definition.

Observation 3.8. *If $D'[J] \in \mathcal{H}'$ and R is the pre-image of J in D, then there is a perfect $\mathcal{H}$-matching in $D[R]$.*

Definition 3.9. *Suppose $D[U]$ is symmetric-critical. We say an $\mathcal{H}$-matching N fits U if $\delta_N(U) = 0$, $\varrho_N(U) \leq 1$, and $\text{size}(N[U]) = |U| - 1$.*

Notice, it follows from $\delta_N(U) = 0$ that N does not use any member H of $\mathcal{H}$ split by U. Thus $N[U]$ is an $\mathcal{H}$-matching, so $\text{size}(N[U])$ also makes sense. If N fits U, then N/U will be an $\mathcal{H}'$-matching in D/U with $\text{size}(N) - \text{size}(N/U) = |U| - 1$. The following Lemma shows the key property of the definition of "N fitting U" which will be used later in an inductive fashion.

Lemma 3.10. *Consider an $\mathcal{H}$-matching N in D which fits U. Define $D' = D/U$ and $N' = N/U$. For $\mathcal{H}'$, use Definition 3.7.*

1. *If N is a maximum $\mathcal{H}$-matching, then N' is a maximum $\mathcal{H}'$-matching.*
2. *If X' is a verifying set for N', then the pre-image X is a verifying set for N.*

We prove this lemma by showing Claim 3.11 and Claim 3.12 which imply the first and second assertions, respectively.

Claim 3.11. *Suppose $D[U]$ is symmetric-critical and M' is an $\mathcal{H}'$-matching in $D' = D/U$. Then*

1. *there is an $\mathcal{H}$-matching M in D of size $\text{size}(M') + |U| - 1$,*
2. $\nu^{\mathcal{H}}(U) \geq \nu^{\mathcal{H}'}(D') + |U| - 1$.

Proof. The second statement follows from the first, we need to prove the first. We construct M as the pre-image of M', except for the weak component of (V', M') incident with $\{U\}$.

If this component is a member $D'[J]$ of $\mathcal{H}'$, then by Observation 3.8 there is a perfect $\mathcal{H}$-matching in J's pre-image R. We add this perfect $\mathcal{H}$-matching in $D[R]$ to the pre-images of the other components of (V', M').

If this component is a cycle or a path, say Z', then Z' contains at most one arc entering $\{U\}$, and at most one arc leaving $\{U\}$. So the pre-image Z of the arcs in Z' contains at most one arc entering U, say $s's$, otherwise choose $s \in U$ arbitrarily. Similarly, Z contains at most one arc leaving U, say tt', otherwise choose $t \in U$ arbitrarily. Let M_{st} be the $\mathcal{H}$-matching in $D[U]$ from Claim 3.5. Adding M_{st} to the pre-image of M', we get an $\mathcal{H}$-matching M in D size $\text{size}(M') + |U| - 1$. $\square$

Claim 3.12. *Suppose $D[U]$ is symmetric-critical and N is an $\mathcal{H}$-matching in D fitting U. Define $D' = D/U$ and $N' = N/U$. If X' is a verifying set in D' for N', then $\{U\} \in X'$ and the pre-image $X := X' - \{U\} \cup U$ is a verifying set for N.*

Proof. Since X' is a verifying set for N', N' is a maximum $\mathcal{H}'$-matching. By definition $\delta_{N'}(\{U\}) = 0$, thus by Claim 3.4 we get that $\{U\} \in X'$. Consider an $\mathcal{H}'$-critical source-component Q' in $D'[X']$. If $\{U\} \notin Q'$, then Q' is an $\mathcal{H}$-critical source-component in $D[X]$. If $\{U\} \in Q'$ then we claim that $Q := Q' - \{U\} \cup U$ induces an $\mathcal{H}$-critical source-component in $D[X]$, which can be seen as follows: $D[Q]$ is a source component in $D[X]$ – here we use that $D'[Q']$ and $D[U]$ are strongly connected, hence so is $D[Q]$. Moreover, by Definition 3.7 and Claim 3.6, $D[Q]$ is $\mathcal{H}$-critical.

Thus $D[X]$ has at least as many $\mathcal{H}$-critical source-components as the number $\mathcal{H}'$-critical source-components of $D'[X']$, that is $\sigma_{\mathcal{H}}(D[X]) \geq \sigma_{\mathcal{H}'}(D'[X'])$. Furthermore $\{U\} \in X'$ implies $\Gamma_D^+(X) = \Gamma_{D'}^+(X')$.

$$\text{size}(N) = \text{size}(N') + |U| - 1 = |V/U| + |\Gamma_{D'}^+(X')| - \sigma_{\mathcal{H}'}(D'[X']) + |U| - 1$$
$$\geq |V| + |\Gamma_D^+(X)| - \sigma_{\mathcal{H}}(D[X]) \geq \text{size}(N).$$
$\square$

The following definition presents the auxiliary graph which will be used as a tool in the proof to find N fitting U.

Definition 3.13. *Let M be a fixed $\mathcal{H}$-matching in $D = (V, A)$, let $\{H_1, H_2, \cdots, H_m\}$ be the family of symmetric-critical subgraphs used by M. Let $D^* = (V^*, A^*)$ be the graph with each $V(H_i)$ contracted to a single node $\{H_i\}$, a loop l_i added on this* node, moreover a loop is added on a node $a \in V^*$ if $\{a\} \in \mathcal{H}$ (these nodes a are called pseudonodes). M^* is defined by replacing H_i by l_i, i.e.,

$$V^* := V/V(H_1)/V(H_2)/\cdots/V(H_m),$$
$$A^* := A/V(H_1)/V(H_2)/\cdots/V(H_m)$$
$$+ \{l_1, l_2, \cdots, l_m\} + \{f : f \text{ a loop on a pseudonode }\},$$
$$M^* := M - H_1 - H_2 - \cdots - H_m + l_1 + l_2 + \cdots + l_m$$

So M^* is a path-cycle-matching with $|M^*| = \text{size}(M) - |V| + |V^*|$. Let $K^+ = K^+(M^*) := V^* - V^+(M^*)$ be the set of sink-nodes for M^*. A sequence $W = (v_0, e_0, v_1, e_1, \ldots, v_{n-1}, e_{n-1}, v_n)$ is called an M^*-alternating walk if

1. $v_0 \in K^+$ and $v_i \in V^*$,
2. if i is even then $e_i = v_i v_{i+1} \in A^*$,
3. if i is odd then $e_i = v_{i+1} v_i \in M^*$.

Here n is called the length of W, v_0 is the first node of W and v_n is the last node of W. W is called even/odd by the parity of its length. Let $A^*(W) = \{e_i : 0 \leq i \leq n-1\}$ denote the set of arcs in W. A node v_i with i even/odd is called an even/odd node of W.

Definition 3.14. An M^*-alternating walk $W = (v_0, e_0, v_1, e_1, \ldots, v_{n-1}, e_{n-1}, v_n)$ is called special if its even nodes are pairwise distinct, and its odd nodes are pairwise distinct. The starting segment of a walk $(v_0, e_0, v_1, e_1, \ldots, v_n)$ of length k is $(v_0, e_0, v_1, e_1, \ldots, v_k)$. Notice, the starting segment of a special M^*-alternating walk is a special M^*-alternating walk, too.

Claim 3.15. If for some nodes $w, z \in V^*$ there is an even M^*-alternating walk from w to z, then there is a special even M^*-alternating walk from w to z, too.

Proof. If an M^*-alternating walk W is not special, then $v_i = v_j$ for some $i < j$, $i \equiv j \mod 2$. A shorter M^*-alternating walk is constructed by deleting the section from v_i to v_j. So a shortest even M^*-alternating walk from w to z must be special. □

Claim 3.16. For a special even M^*-alternating walk W, $M^* \Delta A^*(W)$ is a path-cycle-matching in D^*.

Proof. Path-cycle-matchings are exactly those arc-sets having in- and out-degree at most one in any node. A special M^*-alternating walk has the property that it traverses any arc at most once. The in-degree of nodes is only inflicted for nodes v_i with i odd. The symmetric difference $M^* \Delta A^*(W)$ is constructed in such a way that we replace an arc in M with head v_i (i odd) by another arc with head v_i. So the in-degree of nodes does not change. Similar reasoning shows that the out-degree of nodes does not change, except for v_0 and v_n. There the out-degree is 0 and 1 in M, and is 1 and 0 in $M^* \Delta A^*(W)$, respectively. □

A cycle which is symmetric in D^* is called D^*-*symmetric*, for short. Let us call a D^*-symmetric odd cycle C in D^* *feasible* if the pre-image of $V^*(C)$ is not $\mathcal{H}$-critical.

Claim 3.17. *If N^* is a path-cycle-matching in D^* such that all D^*-symmetric odd cycles in N^* are feasible, then there is an $\mathcal{H}$-matching N in D of size $|N^*|+|V|-|V^*|$ such that the contraction of N gives N^*.*

Proof. Notice, by Claim 3.5 for an even cycle C in D^*, there is a node-disjoint family of one even cycle and some two-arc cycles in D partitioning the pre-image of $V^*(C)$. For an asymmetric odd cycle C in D^*, there is a node-disjoint family of one asymmetric odd cycle and some two-arc cycles in D partitioning the pre-image of $V^*(C)$. For a path P in D^* there is a family of one path and some two-arc cycles in D partitioning the pre-image of $V^*(P)$. By Claim 3.6, if C is a feasible symmetric odd cycle, then there is a perfect $\mathcal{H}$-matching in the pre-image of $V^*(C)$. □

Definition 3.18. *Let $L = L(D, M) \subseteq V^*$ be the set of nodes $v \in V^*$ for which there exists an even alternating walk with last node v.*

Notice, a node in K^+ sets up a zero-length alternating walk, thus $K^+ \subseteq L$. So if M is not a perfect $\mathcal{H}$-matching, then L will be non-empty.

Proof of Theorem 3.1. We prove Theorem 3.1 by induction on $|V| + |A|$. Let M be a maximum $\mathcal{H}$-matching. By Lemma 3.2 it is enough to present a verifying set for M. Let D^* be the contracted graph defined in 3.13.

Case I. Suppose there is an arc $ab = e \in A^*$ with $a \in L = L(D, M)$ and $b \in V^* - V^-(M^*)$. (Arc ab may be a loop!) In this case we will find a maximum $\mathcal{H}$-matching N which fits a symmetric-critical subgraph, the proof will be completed using Theorem 3.10.

By Claim 3.15 there is a special even M^*-alternating walk W with last node a. Suppose $q = v_0 \in K^+$ is the first node, i.e.,

$$W = (q = v_0, e_0, v_1, e_1, \ldots, v_{n-1}, e_{n-1}, v_n = a).$$

Let $n = 2l$ be the length of W, let W_i be the starting segments of W of length $2i$ (for $i = 0, \ldots, l$). Of course, each W_i is a special even M^*-alternating walk. By Claim 3.16 $M_i^* := M^* \Delta A^*(W_i)$ are path-cycle-matchings in D^*. Notice, the only D^*-symmetric odd cycles in $M_0^* = M^*$ are the loops l_i, which are are feasible. From this fact we will only use later that D^*-symmetric odd cycles in M_0^* are feasible. It is easy to see that for $i \leq n - 1$

$$M_{i+1}^* = M_i^* + v_{2i}v_{2i+1} - v_{2i+2}v_{2i+1}, \tag{2}$$

i.e., M_{i+1}^* is obtained from M_i^* by replacing an arc entering v_{2i+1} by a different arc entering v_{2i+1}.

Subcase Ia. Suppose each D^*-symmetric odd cycle in M_l^* is feasible. It is easy to see that $M_l^* + ab$ is a path-cycle-matching. If each D^*-symmetric odd cycle in $M_l^* + ab$ is feasible, then by Claim 3.17 one could construct an $\mathcal{H}$-matching larger than M. So $M_l^* + ab$ has a unique D^*-symmetric odd cycle C which is not feasible, and we have $ab \in C$. Let U be the pre-image of $V^*(C)$. Then $D[U]$ is symmetric-critical since C is not feasible. By Claim 3.17 applied for $N^* = M_l^*$ there is an $\mathcal{H}$-matching N of size $\text{size}(M)$ which fits U.

Subcase Ib. Suppose there is a D^*-symmetric odd cycle in M_l^* which is not feasible. Consider the smallest index $0 \le i < l$ for which M_{i+1}^* has a D^*-symmetric odd cycle which is not feasible. So each D^*-symmetric odd cycle in M_i^* is feasible. Then by (2) there is a unique D^*-symmetric odd cycle C in M_{i+1}^* which is not feasible, and we have $v_{2i}v_{2i+1} \in C$. Let U be the pre-image of $V^*(C)$. By Claim 3.17 applied for $N^* = M_l^*$ there is an $\mathcal{H}$-matching N of size $\text{size}(M)$ which fits U.

In both subcases we have a maximum $\mathcal{H}$-matching N which fits U. By part 1 of Lemma 3.10 $N' = N/U$ is a maximum $\mathcal{H}'$-matching in $D' = D/U$. By induction, there is a verifying set X' for N' in D'. By part 2 of Lemma 3.10 there is a verifying set for N in D, which completes the proof in case I.

Case II. Suppose there is no arc $ab = e \in A^*$ with $a \in L = L(D, M)$ and $b \in V^* - V^-(M^*)$. Let $X \subseteq V$ be the pre-image of L, we will prove that X is a verifying set for M. Let $M_1 := M[X] = \{vz \in M : v \in X, z \in X\}$, $M_2 := \{vz \in M : v \in X, z \in V - X\}$ and $M_3 := \{vz \in M : v \in V - X\}$. Let S be the set of nodes v in L for which there is no arc uv with $u \in L$ – these are exactly the source-nodes in $D^*[L]$ without a loop. Notice, by definition, the pre-images of the nodes in S are $\mathcal{H}$-critical source-components in $D[X]$.

Claim 3.19. *In Case II we have* $\text{size}(M_3) = |V| - |X|$, $\text{size}(M_2) = |\Gamma_D^+(X)|$ *and* $\text{size}(M_1) = |X| - |S|$.

Proof. The sizes defined in the claim make sense because M_1, M_2, M_3 are $\mathcal{H}$-matchings, since M uses no H split by X.

The first equality follows from $K^+ \subseteq L$.

Consider a node b in $\Gamma_D^+(X)$. We claim that b is covered by a cycle or a path in M. First suppose for contradiction that $b \in V(H_i)$ for some $H_i \in \mathcal{H}$ used by M. Then $\{H_i\} \in \Gamma_{D^*}^+(L)$, so there is an arc $a\{H_i\} \in A^*$ with $a \in L$. By the definition of L there is an even M^*-alternating walk W with last node a. The extension of W by arcs $a\{H_i\}$ and l_i gives an even M^*-alternating walk with last node $\{H_i\}$, thus $\{H_i\} \in L$, a contradiction. Hence a cycle or a path in M covers b. Then $b \in V^*$, i.e., b is not a contracted node. So $b \in \Gamma_{D^*}^+(L)$, and there is an arc $ab \in A^*$ with $a \in L$. From the assumption of case II we get that $b \in V^-(M^*)$, i.e., there must be an arc $cb \in M^*$. By the definition of L there is an even M^*-alternating walk W with last node a. The extension of W by arcs ab and cb gives an even M^*-alternating walk with last node c, thus $c \in L$. But $b \in \Gamma_{D^*}^+(L)$, so $b \ne c$. Hence arc cb is not equal to any loop l_i. Then c is also a non-contracted node in V^*, so $c \in X$, $cb \in A$. Thus, each node b in $\Gamma_D^+(X)$ is covered by an arc cb of a cycle or a path in M leaving X, this proves $\text{size}(M_2) = |\Gamma_D^+(X)|$.

For the third equality, consider a node $b \in L - S$. Then there is an arc $ab \in A^*$ with $a \in L$. From the assumption of case II we get that $b \in V^-(M^*)$, i.e., there must be an arc $cb \in M^*$. (Here cb may be one of the loops l_i.) There is an even M^*-alternating walk W with last node a. The extension of W by arcs ab and cb gives an even M^*-alternating walk with last node c, thus $c \in L$. We get that each node in $b \in L - S$ is covered by an arc $cb \in M^*$. This implies the third equality. □

By definition, there is no $H \in \mathcal{H}$ used by M split by X, which explains the first part of the following calculation, completing the proof of Theorem 3.1.

$$\text{size}(M) = \text{size}(M_1) + \text{size}(M_2) + \text{size}(M_3)$$
$$= |V| + |\Gamma_D^+(L)| - |S| \geq |V| + |\Gamma_D^+(L)| - \sigma(D[L]) \geq \text{size}(M). \quad \square\square$$

Acknowledgment

The author is grateful to the referee who pointed out several inaccuracies.

References

[1] C. Berge, *Sur le couplage maximum d'un graph*, Comptes Rendus Hebdomadaires des Séances de l'Académie des Sciences **247** (1958) 258–259.

[2] G. Cornuéjols and D. Hartvigsen, *An extension of matching theory*, Journal of Combinatorial Theory Ser. B **40** (1986) 285–296.

[3] G. Cornuéjols, D. Hartvigsen and W. Pulleyblank, *Packing subgraphs in a graph*, Op. Res. Letters, (1982) 139–143.

[4] G. Cornuéjols and W. Pulleyblank, *A matching problem with side conditions*, Discrete Mathematics **29** (1980) 135–159.

[5] G. Cornuéjols and W. Pulleyblank, *Critical graphs, matchings and tours or a hierarchy of the travelling salesman problem*, Combinatorica **3(1)** (1983) 35–52.

[6] W.H. Cunningham, *Matching, Matroids and Extensions*, Math. Program. Ser. B **91** (2002) 3, 515–542.

[7] W.H. Cunningham and J.F. Geelen, *The Optimal Path-Matching Problem*, Combinatorica, **17/3** (1997), 315–336.

[8] W.H. Cunningham and J.F. Geelen, *Combinatorial Algorithms for Path-Matching*, manuscript, (2000).

[9] J. Edmonds, *Paths, trees, and flowers*, Canadian Journal of Mathematics **17** (1965) 449–467.

[10] T. Gallai, *Maximale Systeme unabhängiger Kanten*, A Magyar Tudományos Akadémia Matematika Kutatóintézetének Közleményei **9** (1964) 401–413.

[11] A. Frank and L. Szegő, *A Note on the Path-Matching Formula*, J. of Graph Theory **41/2**, (2002) 110–119.

[12] L. Lovász and M.D. Plummer, Matching Theory, Akadémiai Kiadó, Budapest, 1986.

[13] G. Pap and L. Szegő, *On the Maximum Even Factor in Weakly Symmetric Graphs*, Journal of Combinatorial Theory Ser. B, **91/2** (2004) 201–213.

[14] G. Pap and L. Szegő, *On factorizations of directed graphs by cycles*, EGRES Technical Report, TR-2004-01.

[15] G. Pap, *Alternating paths revisited I: even factors*, EGRES Technical Report, TR-2004-18.

[16] M. Loebl and S. Poljak, *Efficient Subgraph Packing*, Journal of Combinatorial Theory Ser. B, 59 (1993), 106–121.

[17] W.T. Tutte, *The factorization of linear graphs*, The Journal of the London Mathematical Society **22**, (1947) 107–111.

Gyula Pap
Dept. of Operations Research
Eötvös University
Pázmány Péter sétány 1/C
Budapest, Hungary H-1117.
e-mail: `gyuszko@cs.elte.hu`

On Reed's Conjecture about ω, Δ and χ

Bert Randerath and Ingo Schiermeyer

Abstract. For a given graph G, the clique number $\omega(G)$, the chromatic number $\chi(G)$ and the maximum degree $\Delta(G)$ satisfy $\omega(G) \leq \chi(G) \leq \Delta(G)+1$. Brooks showed that complete graphs and odd cycles are the only graphs attaining the upper bound $\Delta(G)+1$. Reed conjectured $\chi(G) \leq \lceil \frac{\Delta+1+\omega}{2} \rceil$. In this paper we will present some partial solutions for this conjecture.

Mathematics Subject Classification (2000). 05C15.

Keywords. Brooks Theorem, colouring, chromatic number.

1. Introduction

We refer to [16] for terminology and notation not defined here and consider finite, simple and undirected graphs only. A *k-colouring* of a graph G is an assignment of k different colours to the vertices of G such that adjacent vertices receive different colours. The minimum cardinality k for which G has a k-colouring is called the *chromatic number* of G and is denoted by $\chi(G)$ or briefly χ if no ambiguity can arise.

An obvious lower bound for χ is the size of a largest clique in a graph G. This number is called the *clique number* of G and denoted by $\omega(G)$ or briefly ω. Unfortunately, the computations of χ and ω are both NP-hard.

By a classical result of Erdős [14] we know that the difference $\chi(G) - \omega(G)$ can be arbitrarily large. On the other hand, the graphs for which χ attains the lower bound ω form a graph class of great variety, even if we impose the equality to all induced subgraphs of a graph. A graph G is called *perfect* if the chromatic number $\chi(H)$ equals the clique number $\omega(H)$ for every induced subgraph H of G. More than four decades ago Berge [2] introduced the concept of perfect graphs motivated by Shannon's notion of the zero-error capacity of a graph which has been applied in Shannon's work on communication theory.

Parts of this research were performed within the RIP program (Research in Pairs) at the Mathematisches Forschungsinstitut Oberwolfach. Hospitality and financial support are gratefully acknowledged.

Berge [3] conjectured that a graph G is perfect if and only if neither G nor its complement $\bar{G}$ contains an induced odd cycle of order at least five. In honor of Berge the graphs defined by the righthand side of the conjecture are known as *Berge graphs*. This famous longstanding conjecture known as *Strong Perfect Graph Conjecture* has recently been solved by Chudnovsky, Robertson, Seymour and Thomas [8] (see also [12] and[13]). Polynomial time recognition algorithms for Berge graphs have recently be announced by Chudnovsky and Seymour and Cornuéjols, Liu and Vušković (see [9], [10], [13]).

Upper bounds for χ can be obtained by studying the degrees of the vertices of a graph G. In particular, we are interested in the *maximum degree* of G, which is denoted by $\Delta(G)$ or simply Δ. Obviously, the chromatic number of G is at most $\Delta + 1$. In fact, there is a simple recursive greedy algorithm for colouring G with at most $\Delta + 1$ colours. Having coloured $G - v$, we just colour the vertex v of G with one of the colours not appearing on any of the at most Δ neighbors of v.

Hence, for a given graph G, the clique number $\omega(G)$, the chromatic number $\chi(G)$ and the maximum degree $\Delta(G)$ satisfy

$$\omega(G) \leq \chi(G) \leq \Delta(G) + 1.$$

In 1941 Brooks [6] determined for connected graphs G the families of graphs attaining the upper bound $\Delta(G) + 1$, namely complete graphs and odd cycles. This characterization leads to an improvement of the upper bound.

Theorem 1.1. [6] *If a connected graph $G = (V, E)$ is neither complete nor an odd cycle, then G has a $\Delta(G)$-colouring.*

Based on Lovász algorithmic proof [22] of Brooks Theorem it is possible to design a linear time algorithm (see for instance [1] for an implementation in time $O(|V| + |E|)$).

In this work we will strengthen the upper bound $\Delta + 1$ for χ by considering additional parameters in terms of vertex degrees and the clique number. It is important to mention that all proofs in this work for upper bounds lead to polynomial time algorithms attaining these improved bounds.

2. ω, Δ and χ

We start this section with a well-known corollary of Brooks' Theorem.

Corollary 2.1. *If a graph of maximum degree $\Delta \geq 3$ has no clique containing more than Δ vertices, then its chromatic number χ is at most Δ.*

Reed [27] (see also [24]) believes that this corollary is just the *tip of the iceberg*. He conjectured that the chromatic number is bounded by the average of the trivial upper and lower bound.

Conjecture 2.2. For any graph G of maximum degree Δ and clique number ω,

$$\chi(G) \leq \lceil \frac{\Delta + 1 + \omega}{2} \rceil.$$

The Chvátal graph [11], the smallest 4-regular, triangle-free graph of order 12 with chromatic number 4, shows that the rounding up in this conjecture is necessary. The next conjecture, a variation of Conjecture 2.2 omitting rounding up, is likewise due to Reed [27].

Conjecture 2.3. For any graph G of maximum degree $\Delta \geq 3$,

$$\chi(G) \leq \frac{2(\Delta+1)}{3} + \frac{\omega}{3}.$$

Conjecture 2.2 is obviously true for $\omega \in \{\Delta, \Delta+1\}$ and $\omega \in \{\Delta-2, \Delta-1\}$ by Brooks' Theorem. For completeness we add the special case, where $\omega = 2$, of Conjecture 2.2.

Conjecture 2.4. Any triangle-free graph G satisfies $\chi(G) \leq \frac{\Delta(G)}{2} + 2$.

Asymptotically even a smaller upper bound is valid as shown by Johannson [19] and independently by Kim [20]. In fact, they proved that there is a constant c such that if $\omega = 2$ for a graph G, then

$$\chi(G) \leq \frac{c\Delta(G)}{\ln \Delta(G)}.$$

In [27] it was observed that Conjecture 2.2 is also valid for all graphs $G = (V, E)$ with $\Delta(G) = |V| - 1$. The main result in [27] asserts that if Δ is sufficiently large and ω is sufficiently close to Δ, then Conjecture 2.2 holds.

Theorem 2.5. ([27]) *There is a constant Δ_0 such that for $\Delta \geq \Delta_0$ and if G is a graph of maximum degree Δ having clique number ω with $\omega \geq \lfloor (1 - \frac{1}{70000000})\Delta \rfloor$, then $\chi(G) \leq \frac{\Delta+1+\omega}{2}$.*

A related result due to Reed [28] thereby mainly proving a conjecture of Beutelspacher and Hering [4] asserts that for sufficiently large Δ, any graph with maximum degree at most Δ and no cliques of size Δ has a $\Delta - 1$ colouring.

We shall prove now that Conjecture 2.2 holds for a given ω, if Δ is sufficiently large.

Theorem 2.6. *For every $k \geq 3$ there is a constant c_k such that if G is a graph of order n with clique number ω and maximum degree $\Delta \geq \frac{2n}{k} + c_k \omega^{k-1}$, then $\chi(G) \leq \frac{\Delta+1+\omega}{2}$.*

The proof of our main result will make use of Ramsey numbers. For given integers $p, q \geq 1$ the *Ramsey number* $r(K_p, K_q)$ is the smallest integer n such that any graph G of order n contains a clique of size p or an independent set of size q. For example $r(K_3, K_3) = 6$ solves the simple party problem asking for the minimal number of people required such that at least three persons pairwise know each other or pairwise don't know each other. An upper bound for the Ramsey number $r(K_p, K_q)$ has been obtained by Erdős and Szekeres [15].

Theorem 2.7. ([15]) *For $p, q \geq 3$ the inequality*
$$r(K_p, K_q) \leq r(K_{p-1}, K_q) + r(K_p, K_{q-1})$$
holds, with strict inequality if both $r(K_{p-1}, K_q)$ and $r(K_p, K_{q-1})$ have even parity. Moreover, the inequality implies
$$r(K_p, K_q) \leq \binom{p+q-2}{q-1}.$$

Now we are able to start with the proof of Theorem 2.6.

Proof. We first iteratively choose k independent vertices. This is possible as long as there are at least $r(K_{\omega+1}, K_k)$ vertices remaining.

Therefore, we can find $\lceil \frac{n-(r(K_{\omega+1},K_k)-1)}{k} \rceil$ disjoint subsets of k independent vertices. We colour the k vertices of each subset with a single colour. All remaining $n - k \lceil \frac{n-r(K_{\omega+1},K_k)+1}{k} \rceil$ vertices can be coloured differently. Hence by applying the last theorem we obtain,
$$\chi(G) \leq n - (k-1)\lceil \frac{n-r(K_{\omega+1},K_k)+1}{k} \rceil \leq \frac{n-k+1}{k} + \frac{k-1}{k}\binom{\omega+k-1}{k-1}.$$

Now we compare this bound for the chromatic number with the desired bound
$$\frac{n-k+1}{k} + \frac{k-1}{k}\binom{\omega+k-1}{k-1} \leq \frac{\Delta+1+\omega}{2}$$
$$\Leftrightarrow \Delta \geq \frac{2n}{k} + \frac{2k-2}{k}\binom{\omega+k-1}{k-1} - \omega - 1 - \frac{2k-2}{k}.$$

For every $k \geq 3$ there exists a constant c_k such that
$$c_k \omega^{k-1} \geq \frac{2k-2}{k}\binom{\omega+k-1}{k-1} - \omega - 1 - \frac{2k-2}{k}.$$

Hence, $\chi(G) \leq \frac{\Delta+1+\omega}{2}$ for all sufficiently large Δ satisfying $\Delta \geq \frac{2n}{k} + c_k \omega^{k-1}$. □

Remark 2.8. For every k a vertex colouring with at most $\frac{\Delta+1+\omega}{2}$ colours can be found in time $O(n^{k+1})$ by an algorithm indicated in the proof given above.

The approach above can be used to show that Conjecture 2.2 holds for all graphs G with maximum degree $\Delta(G) = n - k$ if the independence number of G satisfies $\alpha(G) \geq k + 1$.

Theorem 2.9. *Let G be a graph with maximum degree $\Delta = n - k$ for some $k \geq 1$, independence number α and clique number ω. If G satisfies $\alpha \geq k + 1$, then $\chi(G) \leq \frac{\Delta+1+\omega}{2}$.*

Proof. First choose an independent set I containing $k+1$ vertices and colour these vertices with one colour. Now let $H = G - I$ and compute a maximum matching M in $\overline{H}$. If M is missing p vertices of $\overline{H}$, then it has size $\frac{n-k-1-p}{2}$. Hence $\omega(G) \geq \omega(H) \geq p$. Now choose one colour for every vertex not contained in the matching M and one colour for every pair of vertices of a matching edge. This colouring of G satisfies

$$\chi(G) \le 1 + p + \frac{n-k-1-p}{2} = \frac{n-k+p+1}{2} \le \frac{\Delta+\omega+1}{2}. \qquad \square$$

Therefore, to prove Conjecture 2.2 for all graphs with $\Delta(G) = n - k$ for some fixed $k \ge 2$, it remains to prove it for all graphs with $\Delta(G) = n - k$ and $2 \le \alpha(G) \le k$.

We will now show that Conjecture 2.2 holds for all graphs with maximum degree $n - 4 \le \Delta(G) \le n - 1$.

Theorem 2.10. *Let G be a graph with maximum degree $\Delta = n - k$ for some k with $1 \le k \le 4$. Then $\chi(G) \le \lceil \frac{\Delta+1+\omega}{2} \rceil$.*

Proof. We first compute a maximum matching M in $\overline{G}$. Let $q = |M|$ and $u_i w_i \in E(\overline{G})$ for $1 \le i \le q$ be the edges of M. Then $H = G[V(G) - V(M)]$ is complete. Set $n = p + 2q$ and let $V(H) = \{v_1, v_2, \ldots, v_p\}$. We now consider the following fact.

(F1) There are no triples of integers i, j, t with $1 \le i < j \le p$ and $1 \le t \le q$ such that $v_i u_t, v_j w_t \notin E(G)$, since otherwise $M - u_t w_t \cup \{v_i u_t, v_j w_t\}$ would be a matching of larger cardinality in $\overline{G}$.

As remarked earlier, the conjecture is easily verified for all ω with $\Delta - 2 \le \omega \le \Delta + 1$. Hence we may assume $\omega \le \Delta - 3$.

If $\omega \ge n - 2q + k - 2$, then $\lceil \frac{(n-k)+(n-2q+k-2)+1}{2} \rceil = n - q \ge \chi(G)$. Hence we may assume $p = n - 2q \le \omega \le n - 2q + k - 3$.

If $p = 0$, then $q = \frac{n}{2}$. Hence $\chi(G) \le q = \frac{n}{2} \le \lceil \frac{(n-4)+\omega+1}{2} \rceil$, since $\omega \ge 2$ and n is even.

If $p = 1$, then $q = \frac{n-1}{2}$. Hence for all ω, Δ with $\Delta + \omega \ge n - 1$ we have $\chi(G) \le q + 1 = \frac{n+1}{2} \le \lceil \frac{\Delta+\omega+1}{2} \rceil$, since n is odd. If $\Delta + \omega \le n - 2$, then $\omega = 2$ and $\Delta = n - 4$. Let u be a vertex of maximum degree $\Delta = d(u) = n - 4$. Then u can receive colour 1 and all vertices in $N(u)$ colour 2. Now the remaining vertices in $V(G) - (\{u\} \cup N(u))$ can be coloured with colours 1 and 3. Hence $\chi(G) \le 3 < \frac{n-1}{2} \le \lceil \frac{\Delta+\omega+1}{2} \rceil$, since $2 \le \omega \le \Delta - 3 = n - 7$ implies $n \ge 9$. Hence we may assume $p \ge 2$.

If there are integers i, t with $1 \le i \le p$ and $1 \le t \le q$ such that $v_i u_t, v_i w_t \notin E(G)$, then u_t and w_t can receive the same colour as v_i. Then $\chi(G) \le p + (q - 1) = n - q - 1 = \lceil \frac{(n-4)+(n-2q)+1}{2} \rceil \le \lceil \frac{\Delta+\omega+1}{2} \rceil$. Hence, using (F1), we may assume that $u_i v \in E(G)$ for all $v \in V(K_p)$ and $1 \le i \le q$. Therefore, $p + 1 = n - 2q + 1 \le \omega \le n - 2q + k - 3$.

Thus $k = 4$ and $\omega = n - 2q + 1$.

Now $u_i u_j \notin E(G)$ for $1 \le i < j \le p$, since $\omega = p + 1$. Since $q \le \alpha \le k = 4$ and $\omega = n - 2q + 1 \le \Delta - 3 = n - 7$, we conclude that $p = n - 8$ and $q = 4$. Let $F = G[\{w_1, w_2, w_3, w_4\}]$. If F is bipartite, then $\chi(G) \le p + 1 + 2$. If F is not bipartite, then F contains a K_3, say with vertices w_1, w_2, w_3.

Because of $\omega = p+1$, there are no two vertices $w_i, w_j \in V(K_3), 1 \leq i < j \leq 3$, such that $w_i v, w_j v \in E(G)$ for all $v \in V(K_p)$, and there are no $p-1$ vertices of $V(K_p)$, say $v_1, v_2, \ldots, v_{p-1}$, such that $v_i w_j \in E(G)$ for $1 \leq i \leq p-1$ and $1 \leq j \leq 3$.

Hence there are integers i_1, i_2, i_3, i_4 with $1 \leq i_1 < i_2 \leq 3$ and $1 \leq i_3 < i_4 \leq p$ such that $w_{i_1} v_{i_3}, w_{i_2} v_{i_4} \notin E(G)$. Then w_{i_1} and w_{i_2} can receive the colours of v_{i_3} and v_{i_4}, respectively. Further u_{i_1} and u_{i_2} can receive the same colour, say the one of u_{i_1}, since $u_{i_1} u_{i_2} \notin E(G)$. Hence $\chi(G) \leq p+1+2 = n-5 = \lceil \frac{(n-4)+(n-7)+1}{2} \rceil$. □

In the book of Jensen and Toft [18] (Problem 4.6, p. 83) the problem of improving Brooks' Theorem for the class of triangle-free graphs is stated or, more generally provided that the graph contains no K_{r+1} (cf. also [26]). The problem has its origin in a paper of Vizing [29]. Besides the already mentioned asymptotic result of Johannson and Kim the best known improvement of Brooks' Theorem in terms of the maximal degree for the class of triangle-free or, more generally K_{r+1}-free graphs is due to Borodin, Kostochka [5], Catlin [7], Kostochka [21]. They proved that if $3 \leq r \leq \Delta(G)$ and G contains no K_{r+1}, then $\chi(G) \leq \frac{r}{r+1}(\Delta(G)+2)$. Kostochka [21] proved $\chi(G) \leq 2/3(\Delta(G)+3)$ for every triangle-free graph G. The remaining authors independently proved that $\chi(G) \leq 3/4(\Delta(G)+2)$ for every triangle-free graph G. For the class of triangle-free graphs Brooks' Theorem can be restated in terms of forbidden induced subgraphs, since triangle-free graphs G satisfy $G[N_G[x]] \cong K_{1,d_G(x)}$ for every vertex x of G.

Theorem 2.11 ([6]). (Triangle-free version of Brooks' Theorem)
Let G be a triangle-free and $K_{1,r+1}$-free graph. Then G is r-colourable unless G is isomorphic to an odd cycle or a complete graph with at most two vertices.

The following theorem will extend this triangle-free version of Brooks' Theorem. An r-sunshade (with $r \geq 3$) is a star $K_{1,r}$ with one branch subdivided once. The 3-sunshade is sometimes called *chair* and the 4-sunshade *cross*. In [25] we obtained the following result.

Theorem 2.12. ([25]) Let G be a connected, triangle-free and r-sunshade-free graph with $r \geq 3$, which is not an odd cycle. Then

(a) G is r-colourable;
(b) G is bipartite, if $\Delta(G) \geq 2r-3$;
(c) G is $(r-1)$-colourable, if $r = 3, 4$ or if $\Delta(G) \leq r-1$.

Problem 2.13. Let G be the class of all connected, triangle-free and r-sunshade-free graphs with $5 \leq r \leq \Delta(G) \leq 2r-4$. Does there exist an r-chromatic member $G^* \in G$?

Using Kostochka's result that $\chi(G) \leq 2/3(\Delta(G)+3)$ for every triangle-free graph G, it is not very difficult for $r \geq 9$ to reduce Problem 2.13 to the range $3/2(r-3) \leq \Delta(G) \leq 2r-4$. If Conjecture 2.4 is true then it is not very difficult to reduce Problem 2.13 to the range $2r-5 \leq \Delta(G) \leq 2r-4$, which seems to be tractable. Moreover, an affirmative answer to Conjecture 2.4 would imply that there exists no 5-regular, 5-chromatic or 6-regular, 6-chromatic triangle-free graph. These negative results would settle the remaining cases of Grünbaum's girth problem ([17], see also [18]).

Acknowledgement

We thank the referee for some helpful comments and suggestions.

References

[1] B. Baetz and D.R. Wood, *Brooks' Vertex Colouring Theorem in Linear Time*, TR CS-AAG-2001-05, Basser Dep. Comput. Sci., Univ. Sydney, (2001) 4 pages.

[2] C. Berge, *Les problèms de coloration en théorie des graphes*, Publ. Inst. Statist. Univ. Paris 9 (1960), 123–160.

[3] C. Berge, *Perfect graphs*, in: Six papers on graph theory, Indian Statistical Institute, Calcutta (1963), 1–21.

[4] A. Beutelsbacher and P.R. Hering, *Minimal graphs for which the chromatic number equals the maximal degree*, Ars Combin. 18 (1984), 201–216.

[5] O.V. Borodin and A.V. Kostochka, *On an upper bound of a graph's chromatic number, depending on the graph's degree and density*, J. Combin. Theory Ser. B 23 (1977), 247–250.

[6] R.L. Brooks, *On colouring the nodes of a network*, Proc. Cambridge Phil. Soc. 37 (1941) 194–197.

[7] P.A. Catlin, *A bound on the chromatic number of a graph*, Discrete Math. 22 (1978), 81–83.

[8] M. Chudnovsky, N. Robertson, P. Seymour and R. Thomas, *The strong perfect graph theorem*, to appear in Annals of Mathematics.

[9] M. Chudnovsky, G. Cornuéjols, X. Liu, P. Seymour and K. Vušković, *Recognizing Berge Graphs*, Combinatorica 25 (2005), 143–187.

[10] M. Chudnovsky, G. Cornuéjols, X. Liu, P. Seymour and K. Vušković, *Cleaning for Bergeness,* manuscript (2002), 13 pages.

[11] V. Chvátal, *The smallest triangle-free, 4-chromatic, 4-regular graph*, J. Combin. Theory 9, (1970), 93–94.

[12] G. Cornuéjols, *The Strong Perfect Graph Conjecture*, Proc. Intern. Congress of Math. III, Invited Lecture Beijing (2002), 547–559.

[13] G. Cornuéjols, *The Strong Perfect Graph Theorem*, Optima 70 (2003), 2–6.

[14] P. Erdős, *Graph theory and probability*, Canad. J. Math. 11 (1959), 34–38.

[15] P. Erdős and G. Szekeres, *A combinatorial problem in geometry*, Composito Math. 2 (1935), 463–470.

[16] J.L. Gross and J. Yellen, *Handbook of Graph Theory*, CRC Press, (2004).

[17] B. Grünbaum, *A problem in graph coloring*, Amer. Math. Monthly 77 (1970) 1088–1092.

[18] T.R. Jensen and B. Toft, *Graph colouring problems*, Wiley, New York (1995).

[19] A.R. Johansson, *Asymptotic choice number for triangle-free graphs*, Preprint DIMACS, (1996).

[20] J.H. Kim, *On Brooks' theorem for sparse graphs*, Combin. Prob. Comput. 4 (1995), 97–132.

[21] A.V. Kostochka, *A modification of a Catlin's algorithm*, Methods and Programs of Solutions Optimization Problems on Graphs and Networks 2 (1982), 75–79, (Russian).

[22] L. Lovász, *Three short proofs in graph theory*, J. Combin. Theory Ser. B 19 (1975), 269–271.

[23] S.E. Markossyan, G.S. Gasparyan and B.A. Reed, *β-Perfect Graphs*, J. Combin. Theory Ser. B 67 (1996), 1–11.

[24] M. Molloy and B. Reed, eds.,*Graph Colourings and the Probabilistc Method*, Algorithms and Combinatorics 23, Springer-Verlag Berlin (2002).

[25] B. Randerath and I. Schiermeyer, *A note on Brooks' theorem for triangle-free graphs*, Australas. J. Combin. 26, (2002), 3–9.

[26] B. Randerath and I. Schiermeyer, *Vertex colouring and forbidden subgraphs – a survey*, Graphs and Combinatorics 20 (1), (2004), 1–40.

[27] B.A. Reed, ω, Δ and χ, J. Graph Theory 27 (4), (1998), 177–212.

[28] B.A. Reed, *A Strengthening of Brooks' Theorem,* J. Combin. Theory Ser. B 76 (2), (1999), 136–149.

[29] V.G. Vizing, *Some unsolved problems in graph theory (in Russian)*, Uspekhi Mat. Nauk 23, (1968), 117–134, [Russian] English translation in Russian Math. Surveys 23, 125–141.

Bert Randerath
Institut für Informatik
Universität zu Köln
D-50969 Köln, Germany
e-mail: `randerath@informatik.uni-koeln.de`

Ingo Schiermeyer
Institut für Diskrete Mathematik und Algebra
Technische Universität Bergakademie Freiberg
D-09596 Freiberg, Germany
e-mail: `schierme@tu-freiberg.de`

On the Generalization of the Matroid Parity Problem

András Recski* and Jácint Szabó**

Abstract. Let $T_1, T_2, \ldots, T_t$ be disjoint k-element sets and let their union be denoted by S. Let $\mathcal{A}$ be a subset of $\{0, 1, 2, \ldots, k\}$. For an integer $0 \leq c \leq t$, a subset $X \subseteq S$ is called $(\geq c)$-*legal* if $|X \cap T_i| \in \mathcal{A}$ holds for at least c subscripts and it is called c-*legal* if $|X \cap T_i| \in \mathcal{A}$ holds for exactly c subscripts. Let $\mathcal{M}$ be a matroid on S. In this paper we study problems like "Does there exist a $(\geq c)$-legal (or c-legal) independent set of given cardinality in $\mathcal{M}$?" Observe that if $\mathcal{A} = \{0, k\}$ and $c = t$ then both problems reduce to the matroid k-parity problem (in particular, to the classical matroid parity problem for $k = 2$). The problems have some motivations from engineering applications and are also related to the more recent theory of jump systems.

Mathematics Subject Classification (2000). Primary 05B35; Secondary 90C27, 05C70.

Keywords. Matroid parity.

1. Preliminaries

For brevity, we denote the set $\{0, 1, \ldots, k\}$ by $[k]$. The problems stated in the abstract are trivial in the case when $\mathcal{A} = \emptyset$ and when $\mathcal{A} = [k]$. Hence, in what follows, we suppose that $\mathcal{A}$ is a proper subset of $[k]$ and we denote its complement by $\mathcal{B}$. Throughout, α and β denote the smallest elements of $\mathcal{A}$ and $\mathcal{B}$, respectively.

The two questions (existence of a $(\geq c)$-legal or a c-legal subset) may or may not be accompanied with a cardinality constraint. If the cardinality of the independent set may be arbitrary, we shall speak about the *weak version* of the problem while the *strong version* means that legal independent sets of a given size p are requested.

---

* Research is supported by OTKA grants T 37547, T 42559 and TS 44733.
** The author is a member of the Egerváry Research Group (EGRES). Research is supported by OTKA grants T 037547 and TS 049788, by European MCRTN Adonet, Contract Grant No. 504438 and by the Egerváry Research Group of the Hungarian Academy of Sciences.

The complexity of the problems may also depend on the way how the matroid $\mathcal{M}$ is given. For example, the strong version of the existence of a t-legal set for $k = 2$ and $\mathcal{A} = \{0, 2\}$ (i.e., the classical matroid parity problem) is in **P**, i.e., is of polynomial time complexity if $\mathcal{M}$ is a *linear* matroid, and a representation with subspaces of a vector space over a field is given [Lo81], but is non-polynomial if $\mathcal{M}$ is given by an independence oracle [JeKo, Lo81]. Here we concern complexity issues so if $\mathcal{M}$ is represented over the field K then we tacitly assume that it is possible to perform elementary computations with the elements of K.

Hence for each prescription $\mathcal{A} \subseteq [k]$ we have the problem in 8 variations. Namely, given a matroid $\mathcal{M}$ by an independence oracle (or by a linear representation if $\mathcal{M}$ is linear) with ground set $S = T_1 \dot\cup \cdots \dot\cup T_t$, $|T_i| = k$, together with integers c and p, decide if there exists a ($\geq c$)-legal (or c-legal) independent set of arbitrary size (or of size p).

In what follows, full solution of six variations will be presented. If the matroid is given with a linear representation then we have only partial results for the strong version.

Remark 1.1. Although the questions concerning ($\geq c$)-legal or c-legal sets are clearly related, we shall consider them separately. Note that for a given $\mathcal{A}$, the complexity of the ($\geq c$)-legal version is at most t times the complexity of the c-legal version, hence if the latter is polynomial so is the former. The other implication is certainly false, see Statements 3.1 and 3.2 below for an example.

Similarly, for a given $\mathcal{A}$, the complexity of the weak version is at most $|S|$ times the complexity of the strong version.

If E is a set of edges in a graph then the set of vertices covered by E is denoted by $V(E)$. We will need the following observation.

Lemma 1.2. *If $\mathcal{M}$ is a linear matroid given by its representation on the ground set S and E is a set of undirected edges on the vertex set S, then we can find in polynomial time a matching $M \subseteq E$ of maximum size with the property that $V(M)$ is independent in $\mathcal{M}$.*

Proof. For each $v \in S$ replace v by $\deg_E(v)$ parallel elements such that E becomes a perfect matching on the new ground set S'. Note that the representation of the new matroid $\mathcal{M}'$ can easily be produced. For all new edges $uv = e \in E$ let $U_e = \{u, v\}$. By the matroid parity algorithm of Lovász [Lo81] we can find in polynomial time an independent set of $\mathcal{M}'$ of maximum size which is the union of some sets U_e. Such an independent set clearly corresponds to a matching M with the required property. □

Finally, throughout the paper, we define the matroid $\mathcal{N}$ to be the partitional matroid [Re74] on S with $I \subseteq S$ independent if and only if $|I \cap T_i| \leq 1$ holds for all i.

2. The "weak version" of the problems

In this section we study the complexity of the problem formulated in Section 1 without the cardinality constraint. Recall that $\alpha = \min \mathcal{A}$ and $\beta = \min \mathcal{B}$. Note that the weak version is equivalent to the following: given a matroid $\mathcal{M}$ and an integer c, decide if there exists an independent set I of $\mathcal{M}$ with at least c (or exactly c in the c-legal case) subsets T_i with $|I \cap T_i| = \alpha$ and at most (or exactly) $t - c$ subsets T_i with $|I \cap T_i| = \beta$. Hence if $\alpha \geq 1$, both the c-legal and the $(\geq c)$-legal versions are equivalent to deciding if there exists an independent set I of $\mathcal{M}$ intersecting exactly c of the T_i's each in exactly α elements.

Hence our problem is closely related to the *matroid l-parity problem*, which is the following. Given a matroid $\mathcal{M}'$ on ground set $S' = U_1 \dot\cup \ldots \dot\cup U_t$ with $|U_i| = l$, determine the maximum size of an independent set I of $\mathcal{M}'$ with the property that for all $1 \leq i \leq t$ either $U_i \subseteq I$ or $U_i \cap I = \emptyset$ holds. $\mathcal{M}'$ may be given either by an independence oracle, or by a linear representation if it is linear. The case $l = 2$ is the classical matroid parity problem. Now we show that the matroid l-parity problem reduces to the weak version of our problem with respect to an arbitrary prescription $\mathcal{A}$ with $l = \alpha \geq 1$. First, add as a direct sum $t \cdot (k - l)$ loops to $\mathcal{M}'$ resulting in the matroid $\mathcal{M}$ on ground set S, and let the partition $S = T_1 \dot\cup \ldots \dot\cup T_t$, $|T_i| = k$ such that $U_i \subseteq T_i$ holds for $1 \leq i \leq t$. The independence oracle (or the representation) of $\mathcal{M}$ is immediate. Moreover, the largest c for which there exists a c-legal (or $(\geq c)$-legal) independent set in $\mathcal{M}$ with respect to $\mathcal{A}$ implies the answer to this instance of the matroid l-parity problem.

Here we cite the following results. Lovász [Lo81] proved that the matroid parity problem is polynomial for linearly represented matroids, and is of exponential order for matroids given by an independence oracle (proved also in [JeKo]). On the other hand, the matroid α-parity problem for $\alpha \geq 3$, even if restricted to graphic matroids, is known [We] to contain an **NP**-complete problem (determine whether a directed Hamiltonian path with given initial and terminal points exists in a directed graph).

These considerations will be used in proving the following statements.

Statement 2.1. *For matroids given by an independence oracle the problem concerning $(\geq c)$-legal sets is in* **P** *if and only if $\alpha \leq 1$.*

Proof. First, if $\alpha = 0$ then $\emptyset$ is always a $(\geq c)$-legal independent set. If $\alpha = 1$ then a $(\geq c)$-legal subset (of arbitrary size) which is independent in $\mathcal{M}$ exists if and only if there exists a common independent set of $\mathcal{M}$ and $\mathcal{N}$ with cardinality c, with the matroid $\mathcal{N}$ defined in the end of Section 1. This condition can be checked in polynomial time for any pair of matroids given by independence oracles. Finally, by our above observation, the case $\alpha \geq 2$ is not in **P**, because the matroid α-parity problem can be reduced to it. $\square$

Statement 2.2. *For represented linear matroids the problem concerning $(\geq c)$-legal sets is in* **P** *if $\alpha \leq 2$ and is* **NP**-*complete for $\alpha \geq 3$.*

Proof. For the case $\alpha \leq 1$ we refer to the proof of Statement 2.1.

In case $\alpha = 2$ we can make the following reduction to the matroid parity problem. Add as a direct sum $2t$ coloops v'_i, v''_i, $1 \leq i \leq t$, to $\mathcal{M}$ resulting in a matroid $\mathcal{M}'$ on ground set S'. Note that the representation of $\mathcal{M}'$ can easily be produced in linear time. Next we define a graph on vertex set S' with edge set $E = \{v'_i v''_i : 1 \leq i \leq t\} \cup \{v'_i s, v''_i s : s \in T_i, 1 \leq i \leq t\}$. By Lemma 1.2 we can find in polynomial time a matching $M \subseteq E$ of maximum size with the property that $V(M)$ is independent in $\mathcal{M}'$. We can assume that $v'_i, v''_i \in V(M)$ for all $1 \leq i \leq t$ so it is easy to see that $|M| \geq c + t$ if and only if $\mathcal{M}$ has an independent set intersecting at least c of the T_i's each in exactly 2 elements. In other words, if and only if it has a $(\geq c)$-legal independent set.

Finally, by our above observation, the case $\alpha \geq 3$ is **NP**-complete because the matroid α-parity problem for represented matroids can be reduced to it. □

As for the c-legal version, note that a matroid has a c-legal independent set with respect to the prescription $\mathcal{A}$ if and only if it has a $(t-c)$-legal independent set with respect to the prescription $\mathcal{B}$. So the roles of $\mathcal{A}$ and $\mathcal{B}$ are symmetric, hence we may assume that $\alpha \geq 1$. Above we showed that in the case $\alpha \geq 1$ the c-legal and the $(\geq c)$-legal versions are equivalent in the weak version so the above statements imply the following results.

Statement 2.3. *For matroids given by an independence oracle the problem concerning c-legal sets is in* **P** *if and only if* $\max(\alpha, \beta) = 1$.

Statement 2.4. *For represented linear matroids the problem concerning c-legal sets is in* **P** *if* $\max(\alpha, \beta) \leq 2$ *and is* **NP**-*complete for* $\max(\alpha, \beta) \geq 3$.

3. The "strong version" for matroids given by an oracle

In this section we consider the problem of deciding if there exists a $(\geq c)$-legal (or c-legal) independent set of cardinality p in the matroid $\mathcal{M}$, given the integers c, p and the matroid $\mathcal{M}$ by an independence oracle. Note that by truncating this problem is equivalent to finding a $(\geq c)$-legal (or c-legal) base of $\mathcal{M}$. Moreover, by dualizing, this problem is equivalent to finding a $(\geq c)$-legal (or c-legal) base with respect to $\mathcal{A}^R = \{a : k - a \in \mathcal{A}\}$. The independence oracles for truncating and dualizing are straightforward to produce.

Statement 3.1. *For matroids given by an independence oracle the problem concerning $(\geq c)$-legal sets of cardinality p is in* **P** *if and only if* $\{1, 2, \ldots, k-1\} \subseteq \mathcal{A}$.

Proof. An integer q is called a *gap* of $\mathcal{A}$ if $q \notin \mathcal{A}$ yet neither $\mathcal{A} \cap \{0, 1, \ldots, q-1\}$ nor $\mathcal{A} \cap \{q+1, q+2, \ldots, k\}$ is empty. In [Re83] it was proved that if $\mathcal{A}$ contains at least one gap then the above problem is not in **P** even in the special case $c = t$.

We can reduce the matroid α-parity problem to the case $\alpha \geq 2$ in exactly the same way as we did in the beginning of Section 2. Namely, extend U_i to T_i by adding $k - \alpha$ loops as a direct sum for all $1 \leq i \leq t$. Now, finding the largest c

for which there exists a ($\geq c$)-legal independent set of cardinality $c \cdot \alpha$ solves the matroid α-parity problem. This is of exponential order for matroids given by an independence oracle when $\alpha \geq 2$ so this holds for our problem, too. Hence – by dualizing – we get that the case $\min \mathcal{A}^R \geq 2$ is of exponential order as well. So $\{1, 2, \ldots, k-1\} \subseteq \mathcal{A}$ is necessary for the polynomial solvability of this problem.

As for the sufficiency, we consider only the case $k \geq 2$, otherwise being trivial. For a base B of $\mathcal{M}$ let

$$s^+(B) = |\{i : T_i \subseteq B\}|,$$
$$s_-(B) = |\{i : T_i \cap B = \emptyset\}|$$

and $s_-^+(B) = s^+(B) + s_-(B)$. Moreover, let $s^+(\mathcal{M})$, $s_-(\mathcal{M})$ and $s_-^+(\mathcal{M})$ resp., denote the minima of the above values taken over all bases of $\mathcal{M}$. Note that one can determine $s_-(\mathcal{M})$ and $s^+(\mathcal{M})$ in polynomial time: $t - s_-(\mathcal{M})$ is just the maximum cardinality common independent set of $\mathcal{M}$ and $\mathcal{N}$ (defined in the end of Section 1), while $s^+(\mathcal{M}) = s_-(\mathcal{M}^*)$.

Omitting the reference to the matroid, one can observe that $s_-^+ \geq s_- + s^+$. In fact, equality holds here. Though the linking property for polymatroids implies this, we include a short proof. Let B_0 be a base with $s_-(B_0) = s_-$ and let B be a base with $s^+(B) = s^+$ such that $B \cap B_0$ is maximal. We show that $s_-^+(B) = s_- + s^+$, proving the equality. Otherwise $s_-(B) > s_-$ so $B \cap T_i = \emptyset$, $B_0 \cap T_i \neq \emptyset$ holds for some T_i. For $u \in B_0 \cap T_i$ there exists $v \in B \setminus B_0$ such that $B + u - v$ is a base as well. Now $s^+(B) \leq s^+$ and $B \cap B_0$ increased, a contradiction. Hence $s_-^+ = s_- + s^+$ can be determined in polynomial time as well.

First, let $\mathcal{A} = \{1, 2, \ldots, k\}$ (recall, that $\mathcal{A}$ is assumed to be a proper subset of $[k]$). Now $\mathcal{M}$ has a ($\geq c$)-legal base if and only if $s_-(\mathcal{M}) \leq t - c$. For $\mathcal{A} = \{0, 1, \ldots, k-1\}$ replace this condition with $s^+(\mathcal{M}) \leq t - c$ and for $\mathcal{A} = \{1, 2, \ldots, k-1\}$ with $s_-^+(\mathcal{M}) \leq t - c$. □

Statement 3.2. *For matroids given by an independence oracle the problem concerning c-legal sets of cardinality p is in* **P** *if and only if $k = 1$.*

Proof. By Remark 1.1 all the polynomially solvable cases satisfy $\{1, 2, \ldots, k-1\} \subseteq \mathcal{A}$. On the other hand, as we observed before Statement 2.3, the roles of $\mathcal{A}$ and $\mathcal{B}$ are symmetric in the c-legal version. So even $\{1, 2, \ldots, k-1\} \subseteq \mathcal{B}$ must hold in a polynomial case. This can happen only if $k = 1$, when the problem is trivial. □

4. The "strong version" for represented matroids

4.1. Necessary conditions for the solvability

Statement 4.1. *Deciding the existence of a ($\geq c$)-legal independent set of size p is* **NP**-*complete for linearly represented matroids unless*
  1. *$\mathcal{A}$ contains no adjacent gaps and*
  2. *$\mathcal{A}$ intersects $\{0, 1, 2\}$ and $\{k-2, k-1, k\}$.*

Proof. If $\mathcal{A}$ contains adjacent gaps then even the special case $c = t$ is **NP**-complete, see [Re83].

Suppose that $\{0, 1, 2\} \cap \mathcal{A} = \emptyset$, i.e., $\alpha \geq 3$. As usual, we make a reduction of the matroid α-parity problem: extend U_i to T_i by adding $k - \alpha$ loops as a direct sum for all $1 \leq i \leq t$. Now, finding the largest c for which there exists a $(\geq c)$-legal independent set of cardinality $c \cdot \alpha$ solves the matroid α-parity problem, which is **NP**-complete for linearly represented matroids when $\alpha \geq 3$.

Suppose now that $\{k - 2, k - 1, k\} \cap \mathcal{A} = \emptyset$. So $\alpha' \geq 3$ with the notation $\alpha' = k - \max \mathcal{A}$. We reduce the matroid α'-parity problem of the matroid $\mathcal{M}'$ to this problem: extend U_i to T_i by adding $k - \alpha'$ coloops as a direct sum for all $1 \leq i \leq t$ resulting in the new matroid $\mathcal{M}$. A linear representation of $\mathcal{M}$ is straightforward to produce. Now there exists a $(\geq c)$-legal independent set of size $(t - c) \cdot \alpha' + t(k - \alpha')$ in $\mathcal{M}$ if and only if $\mathcal{M}'$ has an independent set consisting of $t - c$ of the U_i's. Hence finding the smallest such c solves the matroid α'-parity problem. □

As we noted before Statement 2.3, in the c-legal version the roles of $\mathcal{A}$ and $\mathcal{B}$ are symmetric, hence we get the following corollary.

Corollary 4.2. *Deciding the existence of a c-legal independent set of size p is* **NP**-*complete for linearly represented matroids unless*

1. *$\mathcal{A}$ and $\mathcal{B}$ contain no adjacent gaps (hence both are jump systems [BoCu, Lo97]) and*
2. *both $\mathcal{A}$ and $\mathcal{B}$ intersect $\{0, 1, 2\}$ and $\{k - 2, k - 1, k\}$.*

For example, if $k = 3$ then $\mathcal{A} = \{0, 3\}$ and $\mathcal{A} = \{1, 2\}$ are excluded by 1. and $\mathcal{A} = \{0\}$, $\mathcal{A} = \{3\}$, $\mathcal{A} = \{0, 1, 2\}$ and $\mathcal{A} = \{1, 2, 3\}$ are excluded by 2.

4.2. Sufficient conditions for the solvability

Statement 4.3. *If $\mathcal{A}$ is the set of even integers then it is polynomial to decide whether a $(\geq c)$-legal independent set of size p exists in a matroid represented over the reals, given the integers p and c.*

Proof. We may assume that $p \equiv t - c \pmod{2}$. Let us extend the underlying set S of the matroid $\mathcal{M}$ by t new elements $s_1, s_2, \ldots, s_t$, and define $\mathcal{M}'$ on this extended set S' as the direct sum of $\mathcal{M}$ and the uniform matroid $\mathcal{U}_{t, t-c}$. A linear representation of $\mathcal{M}'$ over the reals can easily be produced. Next we define a graph on vertex set S' with edge set

$$E = \{s_i s_j : 1 \leq i < j \leq t\} \cup \{v s_i : v \in T_i, 1 \leq i \leq t\} \cup \{vv' : v, v' \in T_i, 1 \leq i \leq t\}.$$

By Lemma 1.2 we can find in polynomial time a matching $M \subseteq E$ of maximum size with the property that $V(M)$ is independent in $\mathcal{M}'$. Let us call such a matching *independent*.

Now we prove that the existence of a $(\geq c)$-legal independent set of size p is equivalent to the existence of an independent matching covering at least $p + t - c$ elements. One direction is clear using that $p \equiv t - c \pmod{2}$. So suppose that

$M \subseteq E$ is an independent matching covering at least $p + t - c$ elements. Since $\mathcal{U}_{t,t-c}$ is a direct component in $\mathcal{M}'$ we may assume that M covers exactly $t - c$ of the elements s_i. Hence deleting the edges of M incident to some s_i results in a ($\geq c$)-legal independent set I of $\mathcal{M}$ such that $|I| - p \geq 0$ is even. If $|I| > p$ then it is possible to delete either two elements from I contained in one subset T_i or two elements from two subsets T_i, T_j with $|I \cap T_i|$, $|I \cap T_j|$ odd. In this way we do not decrease the number of subscripts i with $|I \cap T_i|$ even. Hence there exists a ($\geq c$)-legal independent set of size p.

By Lemma 1.2, the existence of an independent matching covering at least $p + t - c$ elements can be decided in polynomial time by Lovász' matroid parity algorithm. $\square$

We do not know the complexity of the remaining cases, they may include both polynomial and **NP**-complete problems. We mention that deciding the existence of a c-legal (or ($\geq c$)-legal) *base* is much easier. This would be the case if $p = r(\mathcal{M})$ or if we could represent the p-truncation of the given matroid over some field in polynomial time (the complexity of such a representation is unknown). In this case we could decide the existence of a c-legal base for $\mathcal{A} = \{\text{even numbers}\}$ (and hence for $\mathcal{A} = \{\text{odd numbers}\}$) and for $\mathcal{A} = \{0,1\}$, $k = 3$. We do not go into details.

It may be interesting to show a graph theoretic problem, which is similar in flavor. Both the degree-sequences of the subgraphs of an undirected graph, and the vectors $(|I \cap T_1|, \ldots, |I \cap T_t|)$ for all independent sets I of a matroid with ground set $S = T_1 \dot\cup \ldots \dot\cup T_t$ form a jump system [BoCu]. Hence it is tempting to formulate the related problem for degree-sequences: given an undirected graph G, an integer c and a degree-prescription $\mathcal{A} \subseteq \mathbb{N}$, does there exist a subgraph $F \subseteq E(G)$ such that $\deg_F(v) \in \mathcal{A}$ holds for at least (or exactly) c vertices v? Besides some trivially polynomial cases (e.g. when $\mathcal{A}$ consists of the odd numbers, which is an instance of the degree prescribed factor problem introduced by Lovász, see [Lo72]), one can prove by a reduction to the set cover problem that the ($\geq c$)-legal case is **NP**-complete even for the simple prescription $\mathcal{A} = \{1\}$ [Be].

References

[Be] A. BERNÁTH, Private communication. (2004)

[BoCu] A. BOUCHET AND W.H. CUNNINGHAM, Delta-matroids, jump systems, and bisubmodular polyhedra. *SIAM J. Disc. Math.* (1995) **8** 17–32.

[JeKo] P.M. JENSEN AND B. KORTE, Complexity of matroid property algorithms. *SIAM J. Computing* (1982) **11** 184–190.

[Lo72] L. LOVÁSZ, The factorization of graphs. II. *Acta Math. Acad. Sci. Hungar.* (1972) **23** 223–246.

[Lo81] L. LOVÁSZ, The matroid matching problem. *Coll. Math. Soc. J. Bolyai* (1981) **25** 495–517.

[Lo97] L. LOVÁSZ, The membership problem in jump systems. *J. Combinatorial Theory Ser. B.* (1997) **70** 45–66.

[Re74] A. RECSKI, On partitional matroids with applications. *Coll. Math. Soc. J. Bolyai* (1974) **10** Vol. 3, 1169–1179.

[Re83] A. RECSKI, On the generalization of the matroid parity and the matroid partition problems, with applications. *Annals of Discrete Mathematics* (1983) **17** 567–573.

[We] D.J.A. WELSH, *Matroid Theory.* Academic Press, London. 1976

András Recski
Dept. of Computer Science and Information Theory
and Center for Applied Mathematics and Computational Physics
Budapest University of Technology and Economics
H-1521 Budapest, Hungary
e-mail: `recski@cs.bme.hu`

Jácint Szabó
Dept. of Operations Research
Eötvös University
Pázmány P. s. 1/C
H-1117 Budapest, Hungary
e-mail: `jacint@elte.hu`

Reconstruction of a Rank 3 Oriented Matroids from its Rank 2 Signed Circuits

Ilda P.F. da Silva

Abstract. We consider the problem of reconstructing oriented rank 3 matroids from its family of rank 2 signed circuits. We prove that the oriented matroid of affine dependencies of the set of points of a $m \times n$ rectangular lattice can be reconstructed from its rank 2 circuits.

Keywords. Orthogonality of signed sets, oriented matroid, rectangular lattice.

1. Introduction

It is well known that a simple matroid of rank 3 is determined by its rank 2 circuits. For oriented matroids this result is not true in general. In fact, a rank 3 matroid over an n-element set with no rank 2 circuits is the uniform matroid $U_{n,3}$ and, asymptotically, it has more then $2^{\frac{n^2}{8}}$ non-isomorphic orientations, [5] [3].

Which matroids have this property? Are there any? Is this property a property of the matroid or of the orientation?

In the next paragraphs we introduce and play two "orthogonality games" which give a quick first answer to these questions, namely Proposition 1.1 says that this property is an invariant of the orientation class of the matroid.

The main result of the paper is Theorem 2.1 which says that for every $m, n \geq 3$ the oriented matroid of affine dependencies of the set $B(m,n) := \{(p,q) \in \mathbb{N} : 1 \leq p \leq m,\ 1 \leq q \leq n\}$ is determined by its rank 2 signed circuits.

The proof of Theorem 2.1. is constructive: we combine orthogonality between signed circuits and cocircuits with elimination for modular pairs of cocircuits to obtain from the rank 2 circuits, the cocircuit signature of the oriented matroid. Determining convenient modular pairs of cocircuits leading the signature of a new cocircuit involves geometric properties of the integer lattice namely Pick's Theorem (see, for instance, [1]).

We start by recalling the notion of orthogonality of signed sets as introduced by R. Bland and M. Las Vergnas in [4] so that any reader may enjoy playing the "orthogonality games". The interested reader is referred to [3] for details on oriented matroids.

1.1. Orthogonality of signed sets

A *signed subset of a (finite) set* E is a map $X : E \longrightarrow \{-1, 0, 1\}$. The preimages of $-1, 0, 1$ are denoted, respectively, X^-, X^0, X^+. The *support* of X is the subset $\underline{X} := X^+ \cup X^- = E \setminus X^0$ and X is identified with the ordered pair $X = (X^+, X^-)$ of subsets of E.

The opposite signed subset $-X : E \longrightarrow \{-1, 0, 1\}$, defined by $(-X)(e) = -X(e)$ is identified with the ordered pair $-X = (X^-, X^+)$.

If X is a signed subset of E then the restriction of X to a subset F of E is the signed subset $X(F) = (X^+ \cap F, X^- \cap F)$ of F.

Two *signed subsets* X, Y of E are *orthogonal* if the following condition is satisfied:

(O) $$X(\underline{X} \cap \underline{Y}) \neq \pm Y(\underline{X} \cap \underline{Y})$$

Two *families* $\mathcal{C}$ *and* $\mathcal{D}$ *of signed subsets of* E are *orthogonal* if the following condition is satisfied:

(OF) $$\forall X \in \mathcal{C}, Y \in \mathcal{D}, \ X \text{ and } Y \text{ are orthogonal}$$

In concrete examples signed subsets $X = (X^+, X^-)$ of a set E are described by fixing a total order on E and identifying X with the word $X = e_{i_1}^{\epsilon_1} \ldots e_{i_k}^{\epsilon_k}$ with $i_1 < \cdots < i_k$ and $\epsilon_j = +$ if $e_j \in X^+$, $\epsilon_j = -$ if $e_j \in X^-$.

Example. Consider $E = \{1, 2, 3, 4\}$ and the signed subsets

$$X = 1^+ 2^+ 3^- (\Leftrightarrow X^+ = \{1, 2\}, \ X^- = \{3\} \ X^0 = \{4\} \Leftrightarrow X = (\{1, 2\}, \{3\})),$$

$$Y = 1^+ 2^- 4^- \text{ and } Z = 1^- 3^+.$$

Then X, Y are orthogonal but X, Z and Y, Z are not orthogonal.

1.2. Orthogonality games

Game 1. Consider the set E of points of $\mathbb{R}^2$ represented in Figure 1.

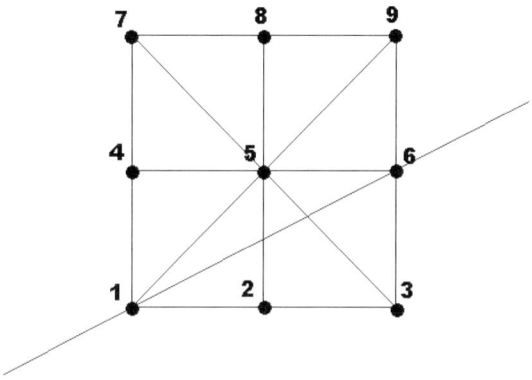

FIGURE 1

Let $\mathcal{C}_2$ be the family of signed subsets of E defined by:
$\mathcal{C}_2 := \{\mathbf{i}^+\mathbf{j}^-\mathbf{k}^+ : \mathbf{i}, \mathbf{j}, \mathbf{k} \in E$ such that $\mathbf{j}$ lies in the interior of the line segment $\mathbf{ik}\}$, i.e.,
$\mathcal{C}_2 = \{\mathbf{1}^+\mathbf{2}^-\mathbf{3}^+, \ \mathbf{1}^+\mathbf{4}^-\mathbf{7}^+, \ \mathbf{1}^+\mathbf{5}^-\mathbf{9}^+, \ \mathbf{2}^+\mathbf{5}^-\mathbf{8}^+, \ \mathbf{3}^+\mathbf{5}^-\mathbf{7}^+, \ \mathbf{3}^+\mathbf{6}^-\mathbf{9}^+, \ \mathbf{4}^+\mathbf{5}^-\mathbf{6}^+, \mathbf{7}^+\mathbf{8}^-\mathbf{9}^+\}.$

Objective:
1) Determine a signed subset X of E orthogonal to $\mathcal{C}_2$ with $X^0 = \{\mathbf{1}, \mathbf{6}\}$.
2) Play the same game as in 1) replacing X^0 by any set of points of E on a straight line $\mathbf{s}$ containing at least two points of E (there are 20 such lines, but symmetry helps).

Solutions of Game 1. There is exactly one pair of opposite signed subsets of E in the required conditions:

$$X = \mathbf{2}^-\mathbf{3}^-\mathbf{4}^+\mathbf{5}^+\mathbf{7}^+\mathbf{8}^+\mathbf{9}^+, \quad -X = \mathbf{2}^+\mathbf{3}^+\mathbf{4}^-\mathbf{5}^-\mathbf{7}^-\mathbf{8}^-\mathbf{9}^-.$$

This conclusion is obtained arguing in the following way.

Let $X = (X^+, X^-)$ denote a signed subset of E satisfying the required conditions. X must be orthogonal to $\mathbf{3}^+\mathbf{6}^-\mathbf{9}^+ \in \mathcal{C}_2$ implying that $\mathbf{3}$ and $\mathbf{9}$ must have different signs in X. Assume that $\mathbf{3} \in X^-$ and $\mathbf{9} \in X^+$. Then orthogonality with the signed subsets $\mathbf{1}^+\mathbf{5}^-\mathbf{9}^+, \mathbf{1}^+\mathbf{2}^-\mathbf{3}^+ \in \mathcal{C}_2$ implies $\mathbf{5}, \mathbf{9} \in X^+$ and $\mathbf{2}, \mathbf{3} \in X^-$. The signs of $\mathbf{4}, \mathbf{7}$ and $\mathbf{8}$ in X are now fixed by orthogonality of X with the signed subsets $\mathbf{1}^+\mathbf{4}^-\mathbf{9}^+, \mathbf{3}^+\mathbf{5}^-\mathbf{7}^+$ and $\mathbf{2}^+\mathbf{5}^-\mathbf{8}^+$ implying that $X = \mathbf{2}^-\mathbf{3}^-\mathbf{4}^+\mathbf{5}^+\mathbf{7}^+\mathbf{8}^+\mathbf{9}^+$.

Note that if the assumption $\mathbf{3} \in X^-$ and $\mathbf{9} \in X^+$ is replaced by $\mathbf{3} \in X^+$ and $\mathbf{9} \in X^-$ the same steps would lead us to the opposite vector $-X = \mathbf{2}^+\mathbf{3}^+\mathbf{4}^-\mathbf{5}^-\mathbf{7}^-\mathbf{8}^-\mathbf{9}^-$. Both X and $-X$ are orthogonal to the family $\mathcal{C}_2$. Remark that the partition $X^+ \uplus X^-$ of $E \setminus X^0$ determined by both signed sets describes the partition of the points of $E \setminus X^0$ into the points that are on each side of the straight line spanned by $X^0 = \{\mathbf{1}, \mathbf{6}\}$.

2) For each one of the 20 straight lines there is exactly one pair of opposite signed sets orthogonal to $\mathcal{C}_2$. In all the cases the solution corresponds to the partition of the set E into the open half-spaces determined by the straight line.

Game 2. Play the same game now on the "board" E' of points of $\mathbb{R}^2$ represented in Figure 2 (see next page).

As in the previous case consider $\mathcal{C}'_2$ the family of signed subsets of E' defined by:
$\mathcal{C}'_2 := \{\mathbf{i}^+\mathbf{j}^-\mathbf{k}^+ : \mathbf{i}, \mathbf{j}, \mathbf{k} \in E'$ such that $\mathbf{j}$ lies in the interior of the line segment $\mathbf{ik}\}$.

1) Determine a signed subset X of E' orthogonal to $\mathcal{C}'_2$ such that $X^0 = \{\mathbf{1}, \mathbf{6}\}$.
2) Play the same game as in 1) replacing X^0 by any the set of points of E' on a straight line $\mathbf{s}$ containing at least two points of E' (there are 20 lines).

Solutions of Game 2.
1) There are 4 possible solutions, two pairs of opposite signed subsets: X_1 and $-X_1$, X_2 and $-X_2$ with

$$\mathbf{X_1} = \mathbf{2}^-\mathbf{3}^-\mathbf{4}^+\mathbf{5}^-\mathbf{7}^+\mathbf{8}^-\mathbf{9}^- \text{ and } \mathbf{X_2} = \mathbf{2}^-\mathbf{3}^-\mathbf{4}^+\mathbf{5}^-\mathbf{7}^+\mathbf{8}^+\mathbf{9}^-.$$

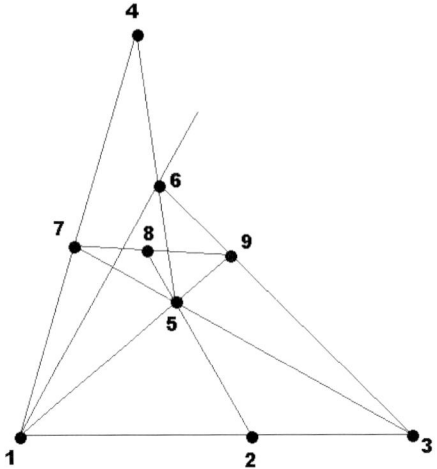

FIGURE 2

2) For $X^0 = \{\mathbf{2,4}\}, \{\mathbf{2,9}\}$ or $\{\mathbf{6,8}\}$ there are 3 pairs of opposite signed sets as possible solutions. For $X^0 = \{\mathbf{1,6}\}, \{\mathbf{1,8}\}, \{\mathbf{2,5,8}\}, \{\mathbf{2,6}\}, \{\mathbf{4,5,6}\}, \{\mathbf{4,8}\}, \{\mathbf{4,9}\}$ there are two pairs of signed sets as possible solutions. For the remaining 10 lines there is a unique pair of signed sets as possible solutions.

1.3. The general context of the games and results

In both cases the game consisted in giving a set E of points in $\mathbb{R}^2$ (the board) together with the family $\mathcal{C}_2$ of (signed) rank 2 circuits of the oriented matroid $Aff(E)$, the oriented matroid of affine dependencies of E over $\mathbb{R}$. The purpose of the game consisted in finding signatures of a (all) cocircuit(s) orthogonal to the family $\mathcal{C}_2$.

The general questions underlying this game are the following:

Question 1. *Let $M = M(E)$ be a simple (no loops nor parallel elements) oriented matroid of rank 3 over a set E. Denote by $\mathcal{C}_2$ the family of signed circuits of rank 2 of M. Characterize those oriented matroids (of rank 3) which satisfy the following property:*

(C2) $\qquad\qquad\qquad \mathcal{C}_2$ *determines the oriented matroid.*

The unicity of the solution for the 20 lines of E in Game 1 shows that the matroid of affine dependencies of E over $\mathbb{R}$ has the property (C2).

Question 2. *Is (C2) a property of the underlying matroid or of the orientation?*

The analysis of Games 1 and 2 answers this question in the following way: *Property (C2) is a property of the oriented matroid and not of the underlying matroid.*

In fact: The oriented matroids $Aff(E)$ and $Aff(E')$ determined by the sets E and E' of Figure 1 and 2 have the same underlying matroid but, while the family of signed cocircuits of $Aff(E)$ (and therefore the oriented matroid) can be recovered from $\mathcal{C}_2$, the oriented matroid $Aff(E')$ cannot be reconstructed from $\mathcal{C}_2'$.

In order to prove that $Aff(E')$ cannot be recovered from $\mathcal{C}_2'$ consider the set E^* of points of $\mathbb{R}^2$ obtained from E' replacing **8** by a point $\mathbf{8_1}$ on the interior of the line segment **79** but on the other side of the straight line **16** and then replacing **2** by $\mathbf{2_1}$ the intersection point of lines $\mathbf{58_1}$ and **13**. The oriented matroids $Aff(E')$ and $Aff(E^*)$ have the same underlying matroid, the same family $\mathcal{C}_2'$ of circuits of rank 2 but while the pair of signed cocircuits of $Aff(E)$ complementary of line $\{\mathbf{1,6}\}$ is the pair of signed subsets $X_1, -X_1$ with $X_1 = \mathbf{2^- 3^- 4^+ 5^- 7^+ 8^- 9^-}$ (pair of solutions of Game 2.1), the pair of signed cocircuits of $Aff(E^*)$ complementary of line $\{\mathbf{1,6}\}$ is the pair of signed subsets $X_2, -X_2$ with $X_2 = \mathbf{2_1^- 3^- 4^+ 5^- 7^+ 8_1^+ 9^-}$ (the other pair of solutions of Game 2.1).

The next proposition shows that Property $(C2)$ is an invariant of the class of orientations of a matroid.

Proposition 1.1. *Let $M(E)$ be an oriented matroid satisfying property $(C2)$ then for all $A \subseteq E$ the reorientation $_{-A}M$ obtained from A reversing signs on A also has property $(C2)$*

Proof. The proof is by contradiction: assume that $M = M(E)$ is an oriented matroid satisfying property $(C2)$ and that for some $A \subseteq E$ the oriented matroid $_{-A}M$ does not satisfy property $(C2)$. Then there is an oriented matroid $M_1 \neq {_{-A}M}$ with the same family, $\mathcal{C}_2$, of rank two circuits then $_{-A}M$. But then $_{-A}M_1$ and M would be different oriented matroids with the same family, $_{-A}\mathcal{C}_2$, of rank 2 circuits, a contradiction. $\square$

2. Main theorem

Consider $B(m,n) := \{(p,q) \in \mathbb{N}^2 : 1 \leq p \leq m \ \ 1 \leq q \leq n\}$ and $\mathcal{L}(m,n)$ the oriented matroid of affine dependencies of $B(m,n)$ over $\mathbb{R}$.

Denote by $\mathcal{C}(m,n)$ the family of signed circuits of $\mathcal{L}(m,n)$ and by $\mathcal{C}_2(m,n)$ its subfamily of rank 2 signed circuits.

Theorem 2.1. *For every $m,n \in \mathbb{N}$, such that $m,n \geq 3$ the oriented matroid $\mathcal{L}(m,n)$ can be reconstructed from the family $\mathcal{C}_2(m,n)$ of its rank 2 circuits.*

The theorem is proven by induction in the next two Lemmas.

Lemma 2.2 establishes the result for $m = n = 3$. Lemma 2.3 is the induction step. It says that given $m \geq 3, n \geq 4$ and assuming the theorem is true for (m,n') with $n' < n$ then the theorem is true for (m,n). This is enough to prove the theorem since if the theorem is true for $\mathcal{L}(m,n)$, $m,n \geq 3$, then, by symmetry, it is also true for $\mathcal{L}(n,m)$.

Lemma 2.2. *The oriented matroid $\mathcal{L}(3,3)$ can be reconstructed from the family $\mathcal{C}_2(3,3)$ of its rank 2 circuits.*

As pointed out in the previous section the proof of this lemma is the result of "playing" Game 1 for the 20 lines of the set E.

Lemma 2.3. *Consider $m, n \in \mathbb{N}$, $m \geq 3$ and $n \geq 4$.*

Assume that for all $n' \in \mathbb{N}$, $3 \leq n' < n$ the oriented matroid $\mathcal{L}(m, n')$ can be reconstructed from the family $\mathcal{C}_2(m, n')$ of its rank 2 signed circuits. Then $\mathcal{L}(m,n)$ can be reconstructed from $\mathcal{C}_2(m,n)$.

Proof. Consider $m, n \in \mathbb{N}$, $m \geq 3$ and $n \geq 4$ in the conditions of the Lemma. To simplify denote by $\mathcal{L}$ the oriented matroid $\mathcal{L}(m,n)$ and put $\mathcal{C}_2 := \mathcal{C}_2(m,n)$. Denote by $\mathcal{D}$ the family of signed cocircuits of $\mathcal{L}$.

The proof consists in describing a procedure to reconstruct $\mathcal{D}$ from $\mathcal{C}_2$.

Since $\mathcal{C}_2$ determines the underlying matroid we know the lines (flats of rank 2) and the unsigned cocircuits of $\mathcal{L}$. Given a line $\mathbf{s}$ of $\mathcal{L}$ we denote by $\underline{X}_\mathbf{s}$ the unsigned cocircuit complementary of $\mathbf{s}$.

In order to use the induction assumption, it is convenient to define the following subsets of $B(m,n)$:

$$A := \{1, \ldots, m\} \times \{1, \ldots, n-1\} \text{ and } A' := \{1, \ldots, m\} \times \{2, \ldots, n\}.$$

Clearly the restrictions $\mathcal{L}(A)$ and $\mathcal{L}(A')$ of the oriented matroid $\mathcal{L}$ to A and A' are isomorphic to $\mathcal{L}(m, n-1)$.

Given a line $\mathbf{s}$ of $\mathcal{L}$ if $\mathbf{s} \cap A$ is a line of $\mathcal{L}(A)$ then the restriction $\underline{X}_\mathbf{s}(A)$ of the cocircuit $\underline{X}_\mathbf{s}$ to A is a cocircuit of $\mathcal{L}(A)$.

We use the same letter to denote the line $\mathbf{s}$ of the matroid and the (straight) line $\mathbf{s}$ of $\mathbb{R}^2$. The lines $\mathbf{h_i} := \{(p, i) : 1 \leq p \leq m\}$ are called horizontal lines and the lines $\mathbf{v_j} := \{(j, q) : 1 \leq q \leq n\}$ vertical lines.

In order to prove the lemma we describe a procedure to determine, for every line $\mathbf{s}$ of $\mathcal{L}$, the unique pair $X_\mathbf{s}, -X_\mathbf{s}$ of signed sets with support $\underline{X}_\mathbf{s}$. This procedure combines essentially orthogonality between the circuits and elimination for pairs of modular cocircuits of an oriented matroid and is divided in 3 cases:

Case 1) The line $\mathbf{s}$ satisfies one of the following conditions:

  i) $\mathbf{s} \cap A$ is a line of $\mathcal{L}(A)$ and $\mathbf{s} \cap A'$ is a line of $\mathcal{L}(A')$ or
  ii) $\mathbf{s}$ is one of the horizontal lines $\mathbf{h_1}$ or $\mathbf{h_n}$.

Note that this case includes all the lines that contain three or more points of $\mathcal{L}$.

Case 2) $\mathbf{s}$ contains exactly two points and satisfies one of the following conditions:

  i) $\mathbf{s} \cap A$ is a line of $\mathcal{L}(A)$ and $\mathbf{s} \cap A'$ is not a line of $\mathcal{L}(A')$ or
  ii) $\mathbf{s} \cap A'$ is a line of $\mathcal{L}(A')$ and $\mathbf{s} \cap A$ is not a line of $\mathcal{L}(A)$.

Note that by symmetry of $B(m,n)$ (and $\mathcal{L}$) we only have to consider one of these two conditions.

Case 3) **s** *contains exactly two points and* $\mathbf{s} \cap A$ *is not a line of* $\mathcal{L}(A)$ *and* $\mathbf{s} \cap A'$ *is not a line of* $\mathcal{L}(A')$

Case 1-i) $\mathbf{s} \cap A$ *is a line of* $\mathcal{L}(A)$ *and* $\mathbf{s} \cap A'$ *is a line of* $\mathcal{L}(A')$.

In this case the restriction $\underline{X}_\mathbf{s}(A)$ of the cocircuit $\underline{X}_\mathbf{s}$ to A is a cocircuit of $\mathcal{L}(A)$. By the induction assumption $\mathcal{C}_2$ determines the unique pair of signed cocircuits, $X_A, -X_A$, of $\mathcal{L}(A)$ with support $\underline{X}_\mathbf{s}(A)$. Similarly, $\mathcal{C}_2$ determines the unique pair of signed cocircuits, $X_{A'}, -X_{A'}$, of $\mathcal{L}(A')$ with support $\underline{X}_\mathbf{s}(A')$.

Since $n \geq 4$ we have $\underline{X}_\mathbf{s}(A) \cap \underline{X}_\mathbf{s}(A') \neq \emptyset$. Let x be an element in $\underline{X}_\mathbf{s}(A) \cap \underline{X}_\mathbf{s}(A')$ and assume, without loss of generality, that x has the same sign in X_A and $X_{A'}$. The restriction of a signed cocircuit X_s of $\mathcal{L}$ with support $\underline{X}_s$ to A (resp. A') is either $X_A, -X_A$ (resp. $X_{A'}, -X_{A'}$). Since x has the same sign on X_A and on $X_{A'}$ we conclude that the signed cocircuits complementary of **s** in $\mathcal{L}$ are: $X_s = (X_A^+ \cup X_{A'}^+, X_A^- \cup X_{A'}^-)$ and $-X_\mathbf{s}$.

Case 1-ii) **s** *is the horizontal line* $\mathbf{h_1} := \{(i, 1) : 1 \leq i \leq m\}$. Let $X_{\mathbf{h_1}}$ be a signed cocircuit of $\mathcal{L}$ complementary of $\mathbf{h_1}$. Since $\mathbf{h_1}$ is a line of $\mathcal{L}(A)$. By definition of $\mathcal{L}(A)$ and by the induction assumption, we may assume that the restriction of X_s to A is the positive cocircuit $X_s(A) = (A \setminus \mathbf{h_1}, \emptyset)$. Then, orthogonality with the circuits of $\mathcal{C}_2$ of the form $(i, 1)^+(i, j)^-(i, n)^+$ with $1 < j < n$ implies that X_s is a positive cocircuit as well.

The case of the horizontal line $\mathbf{h_n}$ is similar.

From now on we only consider lines with two points. Since $B(m, n)$ has an horizontal axe of symmetry we may restrict our attention to lines with a "positive slope", i.e., to lines of $\mathcal{L}$ of the form:

$$\mathbf{s} = \{(p, q), (p', q')\} \quad \text{with} \quad 1 \leq p < p' \leq m \quad \text{and} \quad 1 < q < q' \leq n$$

Case 2) **s** *contains exactly two points,* $\mathbf{s} \cap A$ *is a line of* $\mathcal{L}(A)$ *and* $\mathbf{s} \cap A'$ *is not a line of* $\mathcal{L}(A')$. Note that in this case we have:

$$\mathbf{s} = \{(p, 1), (p', q')\} \quad \text{with} \quad 1 \leq p < p' \leq m \quad \text{and} \quad 1 < q' \leq n - 1\}.$$

We consider separately two subcases:

A) the straight line of $\mathbb{R}^2$ spanned by **s** crosses the horizontal straight line (spanned by) $\mathbf{h_n}$ between (i, n) and $(i + 1, n)$ for some $1 < i < m$ (Figure 3.A) and

B) the straight line of $\mathbb{R}^2$ spanned by **s** crosses the vertical straight line (spanned by) $\mathbf{v_m}$ between (m, j) and $(m, j + 1)$ (eventually in (m, j)) for some $1 < j \leq n - 1$ (Figure 3.B)

Case 2A) The line $\mathbf{s} = \{(p, 1), (p', q')\}$ with $1 \leq p < p' \leq m$ and $1 < q' \leq n - 1$ crosses the horizontal straight line $\mathbf{h_n}$ between (i, n) and $(i + 1, n)$. Let $X_\mathbf{s}$ be the cocircuit of $\mathcal{L}$ complementary of **s** and X_A the restriction of X_s to A.

Consider the points $P_i := (p', q') + (p' - i, q' - n)$ and $P_{i+1} := (p', q') + (p' - (i+1), q' - n)$ (see Figure 3.A). Since the line **s** contains exactly two points of $B(m, n)$, P_i and P_{i+1} belong to the grey rectangle $R \subseteq A$ of Figure 3.A, therefore $\mathcal{C}_2$ contains

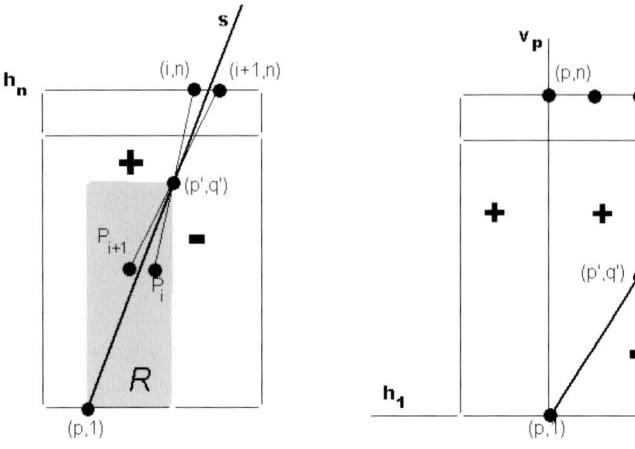

Figure 3.A　　　　　　　　　　Figure 3.B

the signed circuits $(i,n)^+(p',q')^- P_i^+$ and $(i,n)^+(p',q')^- P_{i+1}^+$. Orthogonality with these circuits implies that $X_{\mathbf{s}}((i,n)) = -X_A(P_i)$ and $X_{\mathbf{s}}((i+1,n)) = -X_A(P_{i+1})$.

By definition of $\mathcal{L}(A)$ we have $X_A(P_i) = -X_A(P_{i+1})$ implying that

$$X_{\mathbf{s}}((i,n)) = -X_{\mathbf{s}}((i+1,n)).$$

Then, orthogonality with the signed circuits of $\mathcal{C}_2$ contained in the horizontal line $\mathbf{h_n}$ determines X_s.

Case 2B) The line $\mathbf{s} = \{(p,1), (p',q')\}$ *with* $1 \leq p < p' \leq m$ *and* $1 < q' \leq n-1$ *crosses the vertical straight line* $\mathbf{v_m}$ *between* (m,j) *and* $(m,j+1)$ *(eventually in* (m,j)*) for some* $1 < j \leq n-1$.

Let $X_{\mathbf{s}}$ be a signed cocircuit of $\mathcal{L}$ complementary of $\mathbf{s}$ and X_A the restriction X_s to A. (See Figure 3.B.)

Orthogonality with the circuits of $\mathcal{C}_2$ contained in the vertical lines $\mathbf{v_p}, \ldots, \mathbf{v_m}$ determines the sign of the elements $(p,n), \ldots, (m,n)$ in $X_{\mathbf{s}}$. The sign of $X_{\mathbf{s}}$ in the remaining elements of the horizontal line $\mathbf{h_n}$ is determined using elimination between one of the modular pair of signed cocircuits of $\mathcal{L}$ determined by the horizontal line $\mathbf{h_1}$ and the vertical line $\mathbf{v_p}$ (these two pairs of modular signed cocircuits of $\mathcal{L}$ are known from Case 1).

Case 3) $\mathbf{s}$ *contains exactly two points and* $\mathbf{s} \cap A$ *is not line of* $\mathcal{L}(A)$ *and* $\mathbf{s} \cap A'$ *is not a line of* $\mathcal{L}(A')$. In this case we have:

$$\mathbf{s} = \{(p,1),(p',n)\} \quad \text{with} \quad 1 \leq p < p' \leq m\}.$$

We consider separately the two subcases: A) $p' < m$ (Figure 4.A) and B) $p' = m$ (Figure 4.B).

Case 3A) $p' < m$. Consider the triangles T_1, T_2 of $\mathbb{R}^2$ with vertices, respectively, $((p,1), (p',n), (p'-1,n))$ and $((p,1), (p',n), (p'+1,n))$ (Figure 4.A).

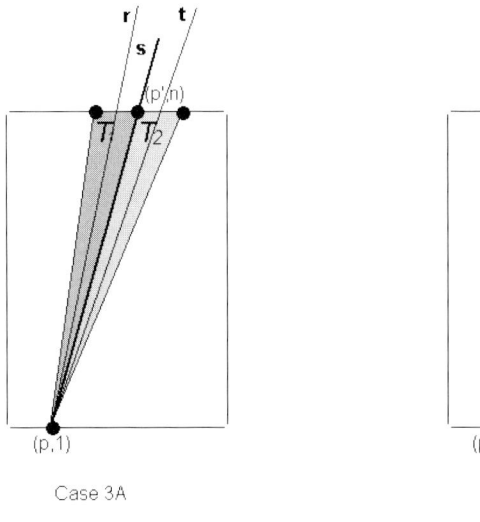
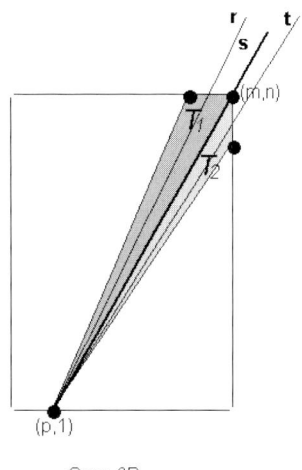

Case 3A   Case 3B

Figure 4.A   Figure 4.B

Since area(T_1) = area$(T_2) = \frac{n-1}{2} > \frac{1}{2}$ $(n \geq 4)$ by Pick's Theorem both these triangles contain (in their closure) points of $B(m,n)$ other then the vertices. Such points are not contained neither in the line **s** nor in the horizontal line $\mathbf{h_n}$.

For all points $P_i \in T_1$ (resp. $Q_j \in T_2$), different from the vertices of the triangle denote by $\mathbf{r_i}$ (resp. $\mathbf{t_j}$) the line of $\mathcal{L}$ containing $(p,1), P_i$ (resp. $(p,1), Q_j$). Note that all lines $\mathbf{r_i}, \mathbf{t_j}$ fall into a previous case and therefore we know the signed complementary cocircuits of $\mathcal{L}$.

Denote by **r** the line $\mathbf{r_i}$ that minimizes the angle between the straight lines $\mathbf{r_i}$ and **s**. Similarly denote by **t** the line $\mathbf{t_j}$ that minimizes the angle between the straight lines $\mathbf{t_j}$ and **s**.

By definition of $\mathcal{L}$ the signed cocircuits complementary of **s** are obtained by elimination between the two modular pairs of cocircuits $\mathbf{X_r}, \mathbf{X_t}$ such that $\mathbf{X_r}^+ \cap \mathbf{X_t}^- = \{(p', n)\}$.

Case 3B) $p' = m$. A similar reasoning applies to the triangles T_1 with vertices $(p,1), (m,n), (m-1,n)$ and T_2 with vertices $(p,1), (m,n), (m-1,n)$. In this case area$(T_1) = \frac{n-1}{2} > \frac{1}{2}$ $(n \geq 4)$ and area$(T_2) = \frac{m-p}{2} \geq \frac{1}{2}$. If $p < m-1$ then we proceed as in the previous case. If $p = m-1$ then the triangle T_2 contains no other points of $B(m,n)$ but the vertices. In this case the line $\mathbf{t} = \{(p,1),(m,n-1)\}$ falls in case 2A. $\square$

As immediate consequences of the results of Section 1.3. we have:

Corollary 2.4. *For every* $m, n \in \mathbb{N}$, *such that* $m, n \geq 3$ *and every* $A \subseteq B(m,n)$ *the oriented matroid* $_{-A}\mathcal{L}(m,n)$ *can be reconstructed from the family* $\mathcal{C}_2(m,n)$ *of its rank 2 circuits.*

Corollary 2.5. *The oriented matroid* $\mathcal{L}(3,3)$ *has more than one reorientation class.*

Note that the oriented matroids $Aff(E)$, $Aff(E')$ and $Aff(E^*)$ where E, E' are the sets of points in the plane of Game 1, Game 2 and E^* described in Section 1.4. represent, in fact, three different orientation classes of $\mathcal{L}(3,3)$.

3. Final remarks

1. We would like to point out that although $\mathcal{L}(3,3)$ has more then one class of orientations it is not clear whether $\mathcal{L}(m,n)$ will have more then one class of orientations $\forall m, n \geq 3$ or if there is (m_0, n_0) such that for $m \geq m_0, n \geq n_0$ $\mathcal{L}(m,n)$ will have exactly one class of orientations.
2. Questions 1 and 2 considered in this paper (Section 1), generalize to the following problem: *characterize those oriented matroids of rank r whose orientation can be reconstructed from the subfamily $\mathcal{C}_k$ of its rank k signed circuits.*
3. In this paper we have restricted our study to rank 3 simple oriented matroids whose orientation can be recovered from its subfamily of rank 2 circuits. In this case it is obvious that not only the orientation but also the underlying matroid can be recovered form $\mathcal{C}_2$. For different values of r and k it is not clear *whether or not the oriented property:* "the orientation is defined by its rank k signed circuits" *implies the non oriented property:* "the underlying matroid is defined by its rank k circuits".

Conclusion. Perhaps the best conclusion was given by Claude Berge, in 1968, [2]:

"Si l'on regarde tous les aspects de la Combinatoire que nous avons essayé d'énumerer ici, on est frappé par la prolifération récente des problèmes qui peuvent se poser à propos de configurations, et de la diversité des outils pour les résoudre."

References

[1] M. Aigner, G.M. Ziegler, *Proofs from THE BOOK*, Springer 1998
[2] C. Berge, "Principes de Combinatoire", Dunod 1968
[3] A. Björner, M. Las Vergnas, B. Sturmfels, N. White, G. Ziegler, *Oriented Matroids*, Encycl. of Maths. and Appl. 46, Cambridge University Press, 1999 (2nd Edition).
[4] R. Bland, M. Las Vergnas, "Orientability of matroids", *J. Combinatorial Theory*, Ser.B, **24** (1978), 94–123.
[5] J.E. Goodman, R. Pollack, "Multidimensional sorting", SIAM J. Computing, **12** (1983), 484–503.

Ilda P.F. da Silva
CELC/University of Lisbon
Faculdade de Ciências – Dep. Matemática
Edifício C6- Piso 2
P-1749-016 Lisboa, Portugal
e-mail: `isilva@cii.fc.ul.pt`

The Normal Graph Conjecture is True for Circulants

Annegret K. Wagler

Abstract. Normal graphs are defined in terms of cross-intersecting set families and turned out to be "weaker" perfect graphs w.r.t. several aspects, e.g., by means of co-normal products (Körner [5]) and graph entropy (Cziszár et al. [4]). Perfect graphs have been recently characterized as those graphs without odd holes and odd antiholes as induced subgraphs (Chudnovsky et al. [2]). In analogy, Körner and de Simone [7] conjectured that every $(C_5, C_7, \overline{C}_7)$-free graph is normal (Normal Graph Conjecture). We prove this conjecture for a first class of graphs that generalize both odd holes and odd antiholes: the circulants. For that, we characterize all the normal circulants by explicitly constructing the required set families for all normal circulants and showing that the remaining ones are not $(C_5, C_7, \overline{C}_7)$-free.

Mathematics Subject Classification (2000). Primary 05C17; Secondary 05C69.

Keywords. Perfect graphs, normal graphs, circulants.

1. Introduction

Normal graphs come up in a natural way in an information theoretic context [6]. A graph G is called *normal* if G admits a clique cover $\mathcal{Q}$ and a stable set cover $\mathcal{S}$ such that every clique in $\mathcal{Q}$ intersects every stable set in $\mathcal{S}$. (A set is a clique (resp. stable set) if its nodes are mutually adjacent (resp. non-adjacent).) In Figure 1, the last two graphs are normal (the clique covers consist of the shaded cliques and the cross-intersecting stable set covers consist of 3 (resp. 6) stable sets of the form indicated by the black nodes).

The interest in normal graphs is caused by the fact that they form "weaker" perfect graphs w.r.t several concepts, e.g. by co-normal products [5] and graph entropy [4, 9]. Berge [1] introduced the latter class in 1960 and conjectured that a graph is perfect if and only if it does not contain chordless odd cycles C_{2k+1}

This work was supported by the Deutsche Forschungsgemeinschaft (Gr 883/9–2).

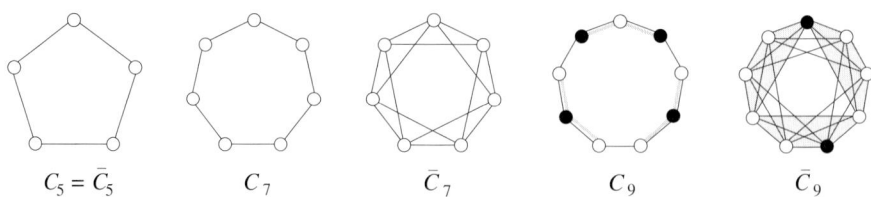

Figure 1. Small odd holes and odd antiholes

with $k \geq 2$, called *odd holes*, or their complements, the *odd antiholes* $\overline{C}_{2k+1}$, as induced subgraphs (see Figure 1 for small examples). The Strong Perfect Graph Conjecture stimulated the study of perfect graphs; it turned out that they have many fascinating properties and interesting relationships to other fields of scientific enquiry, see e.g. [8]. Recently Chudnovsky et al. [2] verified this conjecture.

Körner and de Simone [7] asked whether the similarity of perfect and normal graphs is also reflected in terms of forbidden subgraphs. Körner [5] showed that an odd (anti)hole is normal iff its length is at least 9. In particular, the three smallest odd (anti)holes C_5, C_7, and $\overline{C}_7$ are *not* normal. These three graphs are even *minimally* not normal since all of their proper induced subgraphs are perfect and, hence, normal. This led Körner and de Simone conjecture:

Conjecture 1.1 (Normal Graph Conjecture [7]). Graphs without any C_5, C_7, or $\overline{C}_7$ as induced subgraph are normal.

Our aim is to verify the conjecture for a first graph class. We consider graphs with circular symmetry of their maximum cliques and stable sets, introduced in [3] as generalization of odd holes and odd antiholes: A *circulant* C_n^k is a graph with nodes $1, \ldots, n$ where ij is an edge if i and j differ by at most k (mod n) and $i \neq j$. We assume $k \geq 1$ in order to exclude the degenerated case when C_n^k is a stable set. Circulants include all holes $C_n = C_n^1$ and odd antiholes $\overline{C}_{2k+1} = C_{2k+1}^{k-1}$.

The main result is the characterization of all normal circulants in Section 2. As all the non-normal circulants contain one of the graphs C_5, C_7, and $\overline{C}_7$ as induced subgraph, this implies that the Normal Graph Conjecture is true for circulants, see Section 3.

2. The normal circulants

The circulants C_n^1 are holes, thus C_n^1 is normal iff $n \neq 5, 7$. For any $k \geq 2$, C_n^k is a clique if $n < 2(k+1)$, a perfect graph, namely a clique minus a perfect matching, if $n = 2(k+1)$, and an odd antihole if $n = 2(k+1)+1$. Thus, we have to consider circulants C_n^k with $k \geq 2$ and $n \geq 2(k+1)+2$ only. Note that the size of a maximum clique of C_n^k is $k+1$, and the size of a maximum stable set, called stability number α, is $\lfloor \frac{n}{k+1} \rfloor$. Unless stated otherwise, arithmetic is always performed modulo the number of nodes of the circulant involved in the computation.

Our goal is to characterize all the normal circulants. We explicitly construct the required set families in Section 2.1: According to the circular structure of circulants, we introduce cyclic clique covers $\mathcal{Q}$ of odd size $2t-1$ and construct the corresponding cross-intersecting covers $\mathcal{S}$ consisting of stable t-sets. We show that such a pair $(\mathcal{Q},\mathcal{S})$ exists for each circulant C_n^k satisfying $t(k+1) \leq n \leq (2t-1)k$. In Section 2.2, we figure out for which circulants C_n^k such an appropriate parameter t exists and for which circulants not. Proving that the latter circulants are indeed not normal finishes the characterization of all normal circulants.

2.1. Cyclic clique covers and cross-intersecting stable set covers

According to the circular symmetry of the maximum cliques in a circulant, we shall construct clique covers with an appropriate structure. Let $\mathcal{Q} = \{Q_1, \ldots, Q_l\}$ be a clique cover of C_n^k consisting of maximum cliques only. We call $\mathcal{Q}$ *cyclic* if each clique Q_i has a non-empty intersection with precisely the cliques Q_{i-1} and Q_{i+1} (where the indices are taken modulo l), see Figure 2 for three examples.

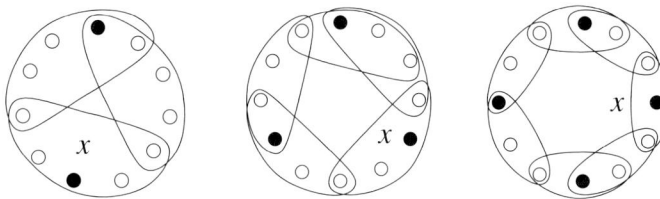

FIGURE 2. Cyclic clique covers of different size in C_{11}^4, C_{12}^3, and C_{12}^2

The next lemma answers the question when a circulant C_n^k admits a cyclic clique cover of a certain size, thereby establishing a 1-1-correspondence between cyclic clique covers of size l and induced holes C_l^1 in C_n^k.

Lemma 2.1. C_n^k with $k \geq 2$ admits a cyclic clique cover of size l if and only if
$$\tfrac{1}{2}(k+1)l \ \leq \ n \ \leq \ kl$$
holds.

Proof. Consider a cyclic clique cover $\mathcal{Q} = \{Q_1, \ldots, Q_l\}$ and denote by q_i the first node in Q_i. Then q_i is adjacent to q_{i-1} and q_{i+1} but not to q_j with $i+1 < j < i-1 \pmod{l}$ by definition; thus $q_1, \ldots, q_l$ induce an l-hole in C_n^k. On the other hand, consider $C_l^1 \subseteq C_n^k$ with nodes $q_1, \ldots, q_l$ and the maximum cliques $Q(q_i) = \{q_i, \ldots, q_i + k\}$ of C_n^k starting in q_i. A result of Trotter [10] shows that $Q(q_i)$ contains precisely two nodes of C_l^1, namely q_i and q_{i+1}; thus $\mathcal{Q} = \{Q(q_1), \ldots, Q(q_l)\}$ is a cyclic clique cover. Hence cyclic clique covers of size l and holes $C_l^1 \subseteq C_n^k$ correspond to each other. Furthermore, Trotter [10] shows that
$$C_l^1 \subseteq C_n^k \text{ iff } \tfrac{(k+1)}{2} l \ \leq \ n \ \leq \ \tfrac{k}{1} l \tag{2.1}$$
holds, as required. □

Note that the assertion of the above lemma remains true if $l = 3$. In this case, inequality (2.1) shows that C_n^k contains a triangle C_3^1 consisting of *non*-consecutive nodes only if $n \leq 3k$ holds. This implies in particular:

Lemma 2.2. *A circulant C_n^k has maximal cliques consisting of non-consecutive nodes only if $n \leq 3k$.*

Proof. Consider such a maximal clique $Q \subseteq C_n^k$. Then Q can clearly neither be a single node nor a single edge, hence Q must contain a triangle C_3^1 consisting of non-consecutive nodes. By inequality (2.1), this is possible only if $n \leq 3k$ holds. □

Let $q(x, \mathcal{Q})$ stand for the number of cliques in $\mathcal{Q}$ containing node x. By the definition of $\mathcal{Q}$, we have $q(x, \mathcal{Q}) \in \{1, 2\}$ for all nodes x (since $\mathcal{Q}$ covers all nodes but no three cliques intersect). We call x a *1-node* (resp. *2-node*) w.r.t. $\mathcal{Q}$ if $q(x, \mathcal{Q}) = 1$ (resp. $q(x, \mathcal{Q}) = 2$) holds.

For a cyclic clique cover $\mathcal{Q}$ consisting of l cliques of size $k+1$ each, the lower bound $\frac{1}{2}(k+1)l \leq n$ in inequality (2.1) is attained if there are 2-nodes only, whereas the upper bound $n \leq kl$ guarantees that there is at least one 2-node in the intersection of two consecutive cliques (if $l = 3$, the 2-nodes induce a clique).

For our purpose, we are interested in cyclic clique covers $\mathcal{Q}$ of *odd* size $2t - 1$ due to the following reason. If a 1-node x belongs to $Q \in \mathcal{Q}$, then $\mathcal{Q} - Q$ consists of $2t - 2$ cliques or, in other words, of $t - 1$ pairs of intersecting cliques. We denote by $S(x, \mathcal{Q})$ a t-set containing x and one node from the intersection of the $t - 1$ pairs of cliques (see the black nodes in Figure 2 and Figure 3). Thus $S(x, \mathcal{Q})$ intersects all cliques in $\mathcal{Q}$ by construction; we shall show that there exist *stable* sets $S(x, \mathcal{Q})$ whose union *covers* all nodes.

Lemma 2.3. *If $t(k+1) \leq n \leq (2t-1)k$ and $k, t \geq 2$, then C_n^k has a cyclic clique cover $\mathcal{Q}$ of size $2t - 1$, and for each 1-node x w.r.t. $\mathcal{Q}$ of C_n^k there is a stable set $S(x, \mathcal{Q})$ of size t in C_n^k.*

Proof. By $\frac{(2t-1)(k+1)}{2} < t(k+1)$, C_n^k has a cyclic clique cover $\mathcal{Q} = \{Q_1, \ldots, Q_{2t-1}\}$ due to Lemma 2.1. Furthermore, $t(k+1) \leq n$ guarantees that C_n^k contains stable sets of size t by $t \leq \alpha(C_n^k) = \lfloor \frac{n}{k+1} \rfloor$.

Consider a 1-node x of C_n^k and assume w.l.o.g. that x belongs to $Q_1 \in \mathcal{Q}$. We construct a stable set $S(x, \mathcal{Q}) = \{x, x_1, \ldots, x_{t-1}\}$ s.t. $x_i \in Q_{2i} \cap Q_{2i+1}$ for $1 \leq i \leq t - 1$, see Figure 3.

Since $x \in Q_1 - Q_2$, there is a non-neighbor of x in $Q_2 \cap Q_3$ (at least the last node in Q_2 is not adjacent to x but belongs to Q_3). We choose $x_1 = x + (k+1) + d_1 \in Q_2 \cap Q_3$ with $d_1 \in \mathbb{N} \cup \{0\}$ minimal.

In order to construct x_i from x_{i-1} for $2 \leq i \leq t - 1$, notice that we have $x_{i-1} \in Q_{2i-2} \cap Q_{2i-1}$, in particular $x_{i-1} \in Q_{2i-1} - Q_{2i}$. As before, there is a non-neighbor of x_{i-1} in $Q_{2i} \cap Q_{2i+1}$ and we choose $x_i = x_{i-1} + (k+1) + d_i \in Q_{2i} \cap Q_{2i+1}$ with $d_i \in \mathbb{N} \cup \{0\}$ minimal. Then $S(x, \mathcal{Q})$ is a stable set if x_{t-1} and x are non-adjacent (all other nodes are non-adjacent by construction).

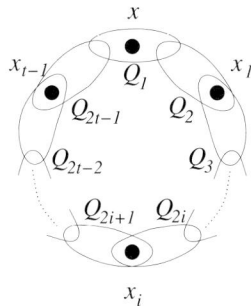

FIGURE 3. Constructing the stable set $S(x, \mathcal{Q})$ for $x \in Q_1$

If $d_i = 0$ for $1 \leq i \leq t-1$, then $x_{t-1} = x + (t-1)(k+1)$. Hence, there are at least $k+1$ nodes between x_{t-1} and x (in increasing order modulo n) due to $n \geq t(k+1)$ and we are done. Otherwise, let j be the smallest index s.t. $d_j > 0$. Then x_j is the first node in Q_{2j+1} since we choose d_j minimal: By $x_{j-1} \notin Q_{2j}$, we have $x_{j-1} + (k+1) \in Q_{2j}$. The only reason for choosing $d_j > 0$ was, therefore, $x_{j-1} + (k+1) + d'_j \notin Q_{2j+1}$ for all $0 \leq d'_j < d_j$ by the minimality of d_j. Hence, x_j is indeed the first node in Q_{2j+1}. This implies that its first non-neighbor is the node $x_j + (k+1)$ belonging to $Q_{2j+2} - Q_{2j+1}$ and $x_{j+1} = x_j + (k+1) + d_{j+1} \in Q_{2j+2} \cap Q_{2j+3}$ is, by the minimality of d_{j+1}, the first node of Q_{2j+3}. The same argumentation shows that *every* further x_i with $i > j+1$ is the first node in Q_{2i+1}; in particular, x_{t-1} is the first node of Q_{2t-1}. Hence, $x \in Q_1 - Q_{2t-1}$ shows that x_{t-1} and x are non-adjacent. Thus $S(x, \mathcal{Q}) = \{x, x_1, \ldots, x_{t-1}\}$ is a stable set of size t and intersects all cliques of $\mathcal{Q}$ by $x \in Q_1$ and $x_i \in Q_{2i} \cap Q_{2i+1}$ for $1 \leq i \leq t-1$. □

Lemma 2.3 implies that there is, for each 1-node x, *at least* one stable set $S(x, \mathcal{Q})$. There can be several such sets; for the cyclic clique cover $\mathcal{Q}$ of C_{11}^4 in Figure 2, e.g., there are *two* stable sets $S(x, \mathcal{Q})$ for the 1-node x.

It is left to show that the union of *all* stable sets $S(x, \mathcal{Q})$ covers the circulant.

Lemma 2.4. *Consider a cyclic clique cover $\mathcal{Q}$ of C_n^k of size $2t-1$ where $t(k+1) \leq n \leq (2t-1)k$ and $k, t \geq 2$. Then the union $\mathcal{S}$ of the stable sets $S(x, \mathcal{Q})$, where x is a 1-node of C_n^k w.r.t $\mathcal{Q}$, covers all nodes of C_n^k.*

Proof. Assume to the contrary that there is a node y in C_n^k not covered by $\mathcal{S}$. Then there is no stable set $S(x, \mathcal{Q})$ with $y \in S(x, \mathcal{Q})$. In particular, y is a 2-node w.r.t. $\mathcal{Q}$ by Lemma 2.3. W.l.o.g. let $y \in Q_1 \cap Q_{2t-1}$. We first show $y_l = y + l(k+1) \in Q_{2l+1}$ for $0 \leq l \leq t-2$. Clearly, we have $y = y_0 = y + 0(k+1) \in Q_1$ by assumption and prove that $y_{i-1} = y + (i-1)(k+1) \in Q_{2i-1}$ implies $y_i = y + i(k+1) \in Q_{2i+1}$ for $1 \leq i \leq t-2$.

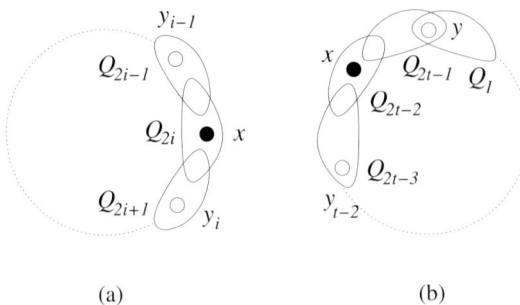

FIGURE 4. Constructing the nodes $y_i \in Q_{2i+1}$

If there is a 1-node x in $Q_{2i} \setminus (Q_{2i-1} \cup Q_{2i+1})$, then x is adjacent to $y_{i-1} = y + (i-1)(k+1)$, see Figure 4(a) (otherwise, there is a stable set $S(x, \mathcal{Q})$ containing x and $y_0, \ldots, y_{i-1}$ in contradiction to our assumption) and $x < y + i(k+1)$ yields $y_i = y + i(k+1) \in Q_{2i+1}$.

If $Q_{2i} \setminus (Q_{2i-1} \cup Q_{2i+1}) = \emptyset$, then $y_i = y + i(k+1)$ clearly belongs to Q_{2i+1} (since we have $y_{i-1} = y + (i-1)(k+1) \in Q_{2i-1}$ and $|Q_{2i-1}| = k+1$).

In particular, we have $y_{t-2} = y + (t-2)(k+1) \in Q_{2t-3}$. Any 1-node x in $Q_{2t-2} \setminus (Q_{2t-3} \cup Q_{2t-1})$ is adjacent to y_{t-2} or to y, see Figure 4(b) (otherwise, x together with $y_0, \ldots, y_{t-2}$ would be a set $S(x, \mathcal{Q}) \in \mathcal{S}$ in contradiction to our assumption). We distinguish three cases:

If x is adjacent to y, then $x > y - (k+1) = y_{-1}$ follows and y_{-1} is adjacent to y_{t-2}: either y_{-1} belongs to Q_{2t-3} or is as 1-node adjacent to y_{t-2}; thus, $y_{-1} = y - (k+1) \leq y + (t-2)(k+1) + k = y_{t-2} + k$ implies $y \leq y + (t-1)(k+1) + k$.

If x is adjacent to y_{t-2}, we obtain $x < y_{t-2} + (k+1) = y_{t-1}$ and y_{t-1} either belongs to Q_{2t-1} or is a 1-node adjacent to y; here, $y_{t-1} \geq y - k$ and, therefore, $y + (t-1)(k+1) \geq y - k$ holds.

The non-existence of a 1-node in Q_{2t-2} implies $y_{t-1} \in Q_{2t-2}$ and, therefore, $y_{t-1} \geq y - k$ follows again.

All three cases imply $n \leq (t-1)(k+1) + k$. By $n \geq t(k+1)$, we obtain

$$t(k+1) \leq (t-1)(k+1) + k$$

yielding the final contradiction. Hence the union $\mathcal{S}$ of the stable sets $S(x, \mathcal{Q})$, where x is a 1-node of C_n^k w.r.t $\mathcal{Q}$, covers all nodes of C_n^k. □

Since each $S(x, \mathcal{Q})$ meets all cliques in $\mathcal{Q}$ by construction, $\mathcal{S}$ is the required stable set cover. Thus Lemma 2.1, Lemma 2.3, and Lemma 2.4 together imply:

Theorem 2.5. *A circulant C_n^k with $k \geq 2$ admits, for $t \geq 2$,*
- *a cyclic clique cover $\mathcal{Q}$ of size $2t - 1$ and*
- *a cross-intersecting stable set cover $\mathcal{S}$ of stable t-sets*

if $t(k+1) \leq n \leq (2t-1)k$ holds.

2.2. Characterizing the normal circulants

Theorem 2.5 shows that a circulant C_n^k with $k \geq 2$ is normal if there is a $t \geq 2$ with $t(k+1) \leq n \leq (2t-1)k$. It is left to figure out for which C_n^k such a t exists.

Lemma 2.6. *A circulant C_n^k with $k \geq 2$ and $n \geq 2(k+1)+2$ is normal if*

- $k = 2$ and $n \neq 8, 11$,
- $k \geq 3$ and $n \neq 3k+1, 3k+2$.

Proof. We shall ensure, for such a C_n^k, the existence of a $t \geq 2$ with $t(k+1) \leq n \leq (2t-1)k$. For that we check, for fixed k, whether there are gaps between the ranges $t(k+1) \leq n \leq (2t-1)k$ and $(t+1)(k+1) \leq n \leq (2t+1)k$ for two consecutive values of $t \geq 2$. There is *no* gap between the two ranges if

$$(t+1)(k+1) \leq (2t-1)k + 1$$

which is true for $k = 2$ if $t \geq 4$ and for $k \geq 3$ if $t \geq 3$ (in Figure 5 the dotted (resp. solid, resp. dashed) line indicates the range for $t = 2$ (resp. $t = 3$, resp. $t = 4$). Thus Theorem 2.5 shows normality for all circulants C_n^k with $k \geq 2$ *except* the cases $n = 3k+1, 3k+2$ (gap between the ranges for $t = 2$ and $t = 3$) and C_{11}^2 (gap between the ranges for $t = 3$ and $t = 4$), see Figure 5. □

It remains to prove that C_{11}^2 and C_{3k+1}^k, C_{3k+2}^k are *not* normal for all $k \geq 2$.

Lemma 2.7. *The circulant C_{11}^2 is not normal.*

Proof. Suppose conversely that C_{11}^2 admits a clique cover $\mathcal{Q}$ and a cross-intersecting stable set cover $\mathcal{S}$. Lemma 2.2 implies that all maximal cliques of C_{11}^2 are of the form $Q(i) = \{i, i+1, i+2\}$.

First, $\mathcal{Q}$ must not contain three *disjoint* cliques Q, Q', Q'': These three cliques cannot cover all nodes of C_{11}^2 by $11 > 9 = |Q \cup Q' \cup Q''|$. Hence, there is a node $x \notin Q \cup Q' \cup Q''$ but due to $\alpha(C_{11}^2) = 3$, there is no stable set in $\mathcal{S}$ containing x and meeting all three cliques Q, Q', Q''.

Thus $\mathcal{Q}$ contains at most two disjoint cliques Q and Q'. They must not be consecutive (see Figure 6(a)): If $Q = \{1, 2, 3\}$ and $Q' = \{4, 5, 6\}$ we need a clique Q'' among $\{7, 8, 9\}, \{8, 9, 10\}, \{9, 10, 11\}$ in order to cover node 9, but Q, Q', Q'' would be disjoint in any case.

There is no possibility for avoiding two consecutive disjoint cliques: Suppose $Q = \{1, 2, 3\}$. If $Q' = \{5, 6, 7\}$ (Figure 6(b)), then $\{8, 9, 10\}, \{9, 10, 11\} \notin \mathcal{Q}$ follows, but we need the two consecutive cliques $\{7, 8, 9\}$ and $\{10, 11, 1\}$ to cover the nodes 9 and 10. If $Q' = \{6, 7, 8\}$ (Figure 6(c)), then $\{4, 5, 6\}, \{3, 4, 5\} \notin \mathcal{Q}$ follows. Thus we need the two consecutive cliques $\{2, 3, 4\}$ and $\{5, 6, 7\}$ to cover the nodes 4 and 5, which yields the final contradiction. □

Lemma 2.8. *For all $k \geq 2$, the circulants C_{3k+1}^k and C_{3k+2}^k are not normal.*

Proof. Suppose in contrary, C_n^k with $n \in \{3k+1, 3k+2\}$ has a clique cover $\mathcal{Q}$ and a cross-intersecting stable set cover $\mathcal{S}$. Lemma 2.2 implies that all maximal

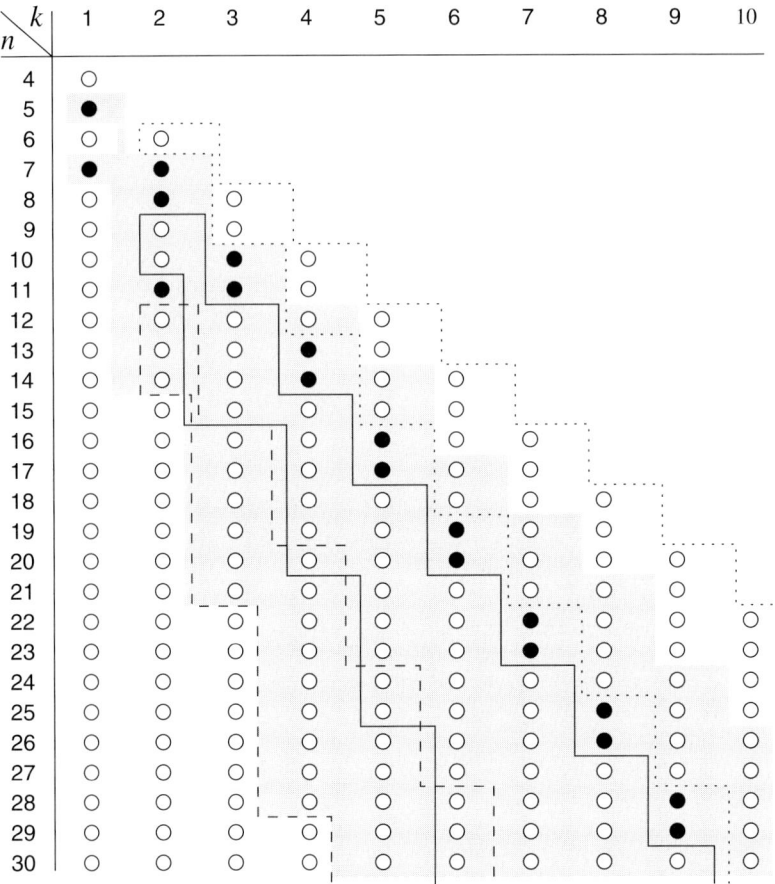

FIGURE 5. The ranges of normal circulants for $t = 2, 3, 4$

cliques of C_n^k have maximum size $k+1$ as $n > 3k$. Hence, $\mathcal{Q}$ consists of cliques $Q(i) = \{i, \ldots, i+k\}$ only.

In particular, there are two *disjoint* cliques Q and Q' in $\mathcal{Q}$: Otherwise, every two cliques of $\mathcal{Q}$ would intersect. For $|\mathcal{Q}| = 2$, we would obtain $n \leq 2k+1$, a contradiction to $k \geq 2$ and $n > 3k$. For $|\mathcal{Q}| \geq 3$, the cliques intersect pairwise only if $n \leq 3k$, a contradiction to $n > 3k$ again.

The two disjoint cliques Q and Q' cannot cover all nodes of C_n^k with $k \geq 2$ by $n \geq 3k+1 > 2(k+1) = |Q \cup Q'|$. Hence, there is a node $x \notin Q \cup Q'$. Due to $\alpha(C_n^k) = 2$, there is no stable set in $\mathcal{S}$ which contains x and meets both cliques Q and Q' (such a stable set had to contain three nodes). Thus, C_n^k with $n = 3k+1, 3k+2$ and $k \geq 2$ does not admit a clique cover and a cross-intersecting stable set cover and is, therefore, not normal. □

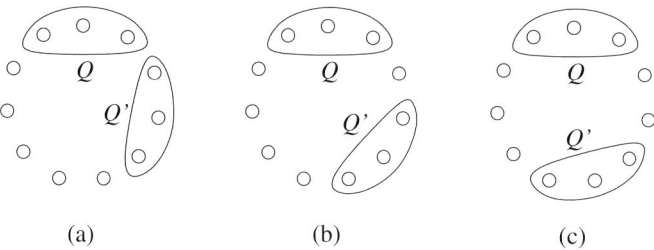

FIGURE 6. Two disjoint cliques Q and Q' in C_{11}^2

As a consequence, we are able to characterize all the normal circulants:

Theorem 2.9. *A circulant C_n^k is normal if and only if*
- *$k = 1$ and $n \neq 5, 7$,*
- *$k = 2$ and $n \neq 7, 8, 11$,*
- *$k \geq 3$ and $n \neq 3k+1, 3k+2$.*

3. Conclusions

Theorem 2.9 does not only characterize all the normal circulants, but also which circulants are *not* normal (see the black circles in Figure 5). With the help of inequality (2.1) it is a routine to check that all the non-normal circulants C_n^k different from $C_5, C_7, \overline{C}_7$ contain either a C_5 (if $n = 3k+1, 3k+2$) or a C_7 (for C_{11}^2) as induced subgraph. Thus, all the non-normal circulants are not $(C_5, C_7, \overline{C}_7)$-free which finally verifies the Normal Graph Conjecture for circulants:

Corollary 3.1. *The Normal Graph Conjecture is true for circulants.*

Since the class of normal graphs is closed under taking complements (by definition), we obtain the same assertion for the complementary class. Thus, we verified the Normal Graph Conjecture for the first two graph classes. However, even proving the Normal Graph Conjecture in general would not yield a *characterization* of normal graphs, as induced subgraphs of normal graphs are not necessarily normal. In order to get a better analogy with perfect graphs, Körner and de Simone [7] introduced a hereditary property by defining *strongly normal graphs* as those normal graphs whose induced subgraphs are all normal. In terms of strongly normal graphs the Normal Graph Conjecture is equivalent to the following:

Conjecture 3.2 (Strongly Normal Graph Conjecture [7]). A graph G is strongly normal if and only if neither G nor its complement $\overline{G}$ contain any C_5 or C_7 as induced subgraph.

The interest of this conjecture lies in the fact that it would immediately lead to a polynomial time recognition algorithm for strongly normal graphs.

Acknowledgement

The author is grateful to an anonymous referee whose suggestions improved the presentation of the results.

References

[1] C. Berge, *Färbungen von Graphen, deren sämtliche bzw. deren ungerade Kreise starr sind*. Wiss. Zeitschrift der Martin-Luther-Universität Halle-Wittenberg 10 (1961) 114–115.

[2] M. Chudnovsky, N. Robertson, P. Seymour, and R. Thomas, *The Strong Perfect Graph Theorem*, to appear in: Annals of Mathematics.

[3] V. Chvátal, *On the Strong Perfect Graph Conjecture*. J. Combin. Theory B 20 (1976) 139–141.

[4] I. Cziszár, J. Körner, L. Lovász, K. Marton, and G. Simonyi, *Entropy splitting for antiblocking corners and perfect graphs*. Combinatorica 10 (1990) 27–40.

[5] J. Körner, *An Extension of the Class of Perfect Graphs*. Studia Math. Hung. 8 (1973) 405–409.

[6] J. Körner and G. Longo, *Two-step encoding of finite memoryless sources*. IEEE Trans. Inform. Theory 19 (1973) 778–782.

[7] J. Körner and C. de Simone, *On the Odd Cycles of Normal Graphs*. Discrete Appl. Math. 94 (1999) 161–169.

[8] J. Ramirez-Alfonsin and B. Reed, *Perfect Graphs*. Wiley (2001).

[9] G. Simonyi, Perfect Graphs and Graph Entropy, In: *Perfect Graphs*, J.L. Ramirez-Alfonsin and B.A. Reed (eds.). Wiley, pages 293–328, 2001.

[10] L.E. Trotter, jr., *A Class of Facet Producing Graphs for Vertex Packing Polyhedra*. Discrete Math. 12 (1975) 373–388.

Annegret K. Wagler
Otto-von-Guericke-University of Magdeburg
Faculty of Mathematics / IMO
Universitätsplatz 2
D-39106 Magdeburg, Germany
e-mail: `wagler@imo.math.uni-magdeburg.de`

Graph Theory
Trends in Mathematics, 375–380
© 2006 Birkhäuser Verlag Basel/Switzerland

Two-arc Transitive Near-polygonal Graphs

Sanming Zhou

In memory of Claude Berge

Abstract. For an integer $m \geq 3$, a near m-gonal graph is a pair $(\Sigma, \mathbf{E})$ consisting of a connected graph Σ and a set $\mathbf{E}$ of m-cycles of Σ such that each 2-arc of Σ is contained in exactly one member of $\mathbf{E}$, where a 2-arc of Σ is an ordered triple $(\sigma, \tau, \varepsilon)$ of distinct vertices such that τ is adjacent to both σ and ε. The graph Σ is called $(G, 2)$-arc transitive, where $G \leq \mathrm{Aut}(\Sigma)$, if G is transitive on the vertex set and on the set of 2-arcs of Σ. From a previous study it arises the question of when a $(G, 2)$-arc transitive graph is a near m-gonal graph with respect to a G-orbit on m-cycles. In this paper we answer this question by providing necessary and sufficient conditions in terms of the stabiliser of a 2-arc.

Mathematics Subject Classification (2000). Primary 05C25; Secondary 20B25.

Keywords. Symmetric graph, 2-arc transitive graph, near-polygonal graph, 3-arc graph construction, group action.

1. Introduction

We consider finite, undirected and simple graphs only. For a graph $\Sigma = (V(\Sigma), E(\Sigma))$ and an integer $s \geq 1$, an s-arc of Σ is a sequence $(\sigma_0, \sigma_1, \ldots, \sigma_s)$ of $s+1$ vertices of Σ such that σ_{i-1} and σ_i are adjacent for $1 \leq i \leq s$ and $\sigma_{i-1} \neq \sigma_{i+1}$ for $1 \leq i \leq s-1$. For an integer $m \geq 3$, a *near m-gonal graph* [13] is a pair $(\Sigma, \mathbf{E})$, where Σ is a connected graph and $\mathbf{E}$ is a set of m-cycles of Σ, such that each 2-arc of Σ is contained in a unique member of $\mathbf{E}$. Here and in the following by an *m-cycle* we mean an undirected cycle of length m. In this case we also say that Σ is a near m-gonal graph with respect to $\mathbf{E}$, and we call cycles in $\mathbf{E}$ *basic cycles* of $(\Sigma, \mathbf{E})$. From the definition it follows that near m-gonal graphs are associated with the Buekenhout geometries [3, 13] of the following diagram:

Supported by a Discovery Project Grant (DP0558677) from the Australian Research Council and a Melbourne Early Career Researcher Grant from The University of Melbourne.

In such a geometry associated with $(\Sigma, \mathbf{E})$, the maximal flags are those triples (σ, e, C) such that $\sigma \in V(\Sigma)$, $e \in E(\Sigma)$ is incident with σ in Σ, and C is a member of $\mathbf{E}$ containing e. A near m-gonal graph with girth m is called an *m-gonal graph* [7]. (The *girth* of a graph Σ is the length of a shortest cycle of Σ if Σ contains cycles, and is defined to be ∞ otherwise.) In fact, the concept of a near-polygonal graph was introduced [13] as a generalisation of that of a polygonal graph. As a simple example, the (3-dimensional) cube together with its faces (taking as 4-cycles) is a 4-gonal graph. There are exactly four 6-cycles in the cube with the property that no three consecutive edges on the cycle belong to the same face; the cube together with these four 6-cycles is a near 6-gonal graph. Another example is the well-known embedding of the Petersen graph on the projective plane as the dual of K_6, which together with the six faces (taking as 5-cycles) is a near 5-gonal graph. The reader is referred to [7, 8, 9, 10, 11, 12, 16, 17] and [13, 14] respectively for results, constructions and more examples on polygonal graphs and near-polygonal graphs. For group-theoretic notation and terminology used in the paper, the reader may consult [1, 2].

This paper was motivated by a recent study [19] where the author found an intimate connection between near-polygonal graphs and a class of imprimitive symmetric graphs with 2-arc transitive quotients. Let Γ be a graph and G a group. If G acts on $V(\Gamma)$ as a group of automorphisms of Γ such that G is transitive on $V(\Gamma)$ and, in its induced action, transitive on the set of s-arcs of Γ, then Γ is said [1, 18] to be (G, s)-*arc transitive*. Usually, a 1-arc is called an *arc* and a $(G, 1)$-arc transitive graph is called a *G-symmetric graph*. A G-symmetric graph Γ is said to be *imprimitive* if G is imprimitive on $V(\Gamma)$, that is, $V(\Gamma)$ admits a partition $\mathbf{B}$ such that $1 < |B| < |V(\Gamma)|$ and $B^g \in \mathbf{B}$ for any *block* $B \in \mathbf{B}$ and element $g \in G$, where $B^g := \{\sigma^g : \sigma \in B\}$. In this case the *quotient graph* $\Gamma_{\mathbf{B}}$ of Γ with respect to this *G-invariant partition* $\mathbf{B}$ is defined to be the graph with vertex set $\mathbf{B}$ such that two blocks $B, C \in \mathbf{B}$ are adjacent if and only if there exists at least one edge of Γ between B and C. Denote by $\Gamma(B)$ the set of vertices of Γ adjacent to at least one vertex in B. In [19, Theorem 1.1] we proved that, if $(\Gamma, \mathbf{B})$ is an imprimitive G-symmetric graph with connected but non-complete $\Gamma_{\mathbf{B}}$ such that the subgraph (without including isolated vertices) induced by two adjacent blocks B, C of $\mathbf{B}$ is a matching of $|B| - 1 \geq 2$ edges and that $\Gamma(C) \cap B \neq \Gamma(D) \cap B$ for different blocks C, D of $\mathbf{B}$ adjacent to B, then $\Gamma_{\mathbf{B}}$ must be a $(G, 2)$-arc transitive near m-gonal graph with respect to a certain G-orbit on m-cycles of $\Gamma_{\mathbf{B}}$, where $m \geq 4$ is an even integer. Moreover, any $(G, 2)$-arc transitive near m-gonal graph (where $m \geq 4$ is even) with respect to a G-orbit on m-cycles can occur as such a quotient $\Gamma_{\mathbf{B}}$. Furthermore, the graph Γ can be reconstructed from $\Gamma_{\mathbf{B}}$ by using the 3-arc graph construction introduced in [6] by Li, Praeger and the author. For more information about this construction, its extension and applications, see [6, 20], [21, 22] and [4, 5, 19, 21], respectively.

The result above motivated us to ask when a $(G, 2)$-arc transitive graph is a near m-gonal graph with respect to a G-orbit on m-cycles. In this paper we answer this question by giving necessary and sufficient conditions in terms of the stabiliser of a 2-arc.

2. Main result

For a G-symmetric graph Σ and $\sigma, \tau, \varepsilon \in V(\Sigma)$, denote by $G_{\sigma\tau\varepsilon}$ the pointwise stabiliser of $\{\sigma, \tau, \varepsilon\}$ in G, that is, the subgroup of G consisting of those elements of G which fix each of σ, τ and ε. Denote by $\Sigma(\sigma)$ the subset of vertices of Σ which are adjacent to σ in Σ. For a subgroup H of G, let $N_G(H)$ denote the normalizer of H in G. For a near m-gonal graph $(\Sigma, \mathbf{E})$, define [13] $\mathrm{Aut}(\Sigma, \mathbf{E})$ to be the subgroup of $\mathrm{Aut}(\Sigma)$ consisting of those elements g of $\mathrm{Aut}(\Sigma)$ which leave $\mathbf{E}$ invariant, that is, $\mathbf{E}^g = \mathbf{E}$ under the induced action of $\mathrm{Aut}(\Sigma)$ on the set of m-cycles of Σ. Note that, for a near m-gonal graph $(\Sigma, \mathbf{E})$ such that Σ is $(G, 2)$-arc transitive, $G \leq \mathrm{Aut}(\Sigma, \mathbf{E})$ holds if and only if $\mathbf{E}$ is a G-orbit on m-cycles of Σ [19, Lemma 2.6].

Theorem 1. *Suppose that Σ is a connected $(G, 2)$-arc transitive graph, where $G \leq \mathrm{Aut}(\Sigma)$. Let $(\sigma, \tau, \varepsilon)$ be a 2-arc of Σ and set $H = G_{\sigma\tau\varepsilon}$. Then the following conditions (a)-(c) are equivalent:*

(a) *there exist an integer $m \geq 3$ and a G-orbit $\mathbf{E}$ on m-cycles of Σ such that $(\Sigma, \mathbf{E})$ is a near m-gonal graph;*
(b) *H fixes at least one vertex in $\Sigma(\varepsilon) \setminus \{\tau\}$;*
(c) *there exists $g \in N_G(H)$ such that $(\sigma, \tau)^g = (\tau, \varepsilon)$.*

Moreover, if one of these conditions is satisfied, then $G \leq \mathrm{Aut}(\Sigma, \mathbf{E})$ and G is transitive on the maximal flags of the Buekenhout geometry associated with $(\Sigma, \mathbf{E})$.

Proof. (a) $\Rightarrow$ (b) Suppose that $(\Sigma, \mathbf{E})$ is a near m-gonal graph for a G-orbit $\mathbf{E}$ on m-cycles of Σ, where $m \geq 3$. Let $C(\sigma, \tau, \varepsilon) = (\sigma, \tau, \varepsilon, \eta, \ldots, \sigma)$ be the basic cycle containing the 2-arc $(\sigma, \tau, \varepsilon)$. Then we have $\eta \in \Sigma(\varepsilon) \setminus \{\tau\}$. (Note that η coincides with σ when $m = 3$.) We claim that η is fixed by H. Suppose otherwise and let $\eta^g \neq \eta$ for some $g \in H$. Then, since $\mathbf{E}$ is a G-orbit on m-cycles of Σ, $(C(\sigma, \tau, \varepsilon))^g = (\sigma, \tau, \varepsilon, \eta^g, \ldots, \sigma)$ is a basic cycle containing $(\sigma, \tau, \varepsilon)$ which is different from $C(\sigma, \tau, \varepsilon)$. This contradicts with the uniqueness of the basic cycle containing a given 2-arc, and hence (b) holds.

(b) $\Rightarrow$ (c) Suppose H fixes $\eta \in \Sigma(\varepsilon) \setminus \{\tau\}$. Then we have $H \leq G_{\tau\varepsilon\eta}$. Since Σ is $(G, 2)$-arc transitive, there exists $g \in G$ such that $(\sigma, \tau, \varepsilon)^g = (\tau, \varepsilon, \eta)$ and hence $G_{\tau\varepsilon\eta} = H^g$. Therefore, $H^g = H$ and $g \in N_G(H)$.

(c) $\Rightarrow$ (a) Suppose that there exists $g \in N_G(H)$ such that $(\sigma, \tau)^g = (\tau, \varepsilon)$. Set $\eta := \varepsilon^g$. Then $\eta \in \Sigma(\varepsilon) \setminus \{\tau\}$, $(\sigma, \tau, \varepsilon)^g = (\tau, \varepsilon, \eta)$ and hence $G_{\tau\varepsilon\eta} = H^g = H$. Set $\sigma_0 = \sigma, \sigma_1 = \tau, \sigma_2 = \varepsilon$ and $\sigma_3 = \eta$, and set $\sigma_4 = \sigma_3^g$. Then $\sigma_4 \in \Sigma(\sigma_3) \setminus \{\sigma_2\}$ and $G_{\sigma_2\sigma_3\sigma_4} = (G_{\sigma_1\sigma_2\sigma_3})^g = H^g = H$. Now set $\sigma_5 = \sigma_4^g$, then similarly $\sigma_5 \in \Sigma(\sigma_4) \setminus \{\sigma_3\}$ and $G_{\sigma_3\sigma_4\sigma_5} = (G_{\sigma_2\sigma_3\sigma_4})^g = H^g = H$. Continuing this process, we obtain inductively a sequence $\sigma_0, \sigma_1, \sigma_2, \sigma_3, \sigma_4, \sigma_5, \ldots$ of vertices of Σ with the following properties:

(1) $\sigma_i = \sigma_{i-1}^g$ for all $i \geq 1$, and hence $\sigma_{i+1} \in \Sigma(\sigma_i) \setminus \{\sigma_{i-1}\}$ for $i \geq 1$ and $\sigma_i = \sigma_0^{g^i}$ for $i \geq 0$; and
(2) $G_{\sigma_{i-1}\sigma_i\sigma_{i+1}} = H$ for all $i \geq 1$.

Since we have finitely many vertices in Σ, this sequence will eventually contain repeated terms. Suppose σ_m is the first vertex in this sequence which coincides with one of the preceding vertices. Without loss of generality we may suppose that σ_m coincides with σ_0 for if $\sigma_m = \sigma_i$ for some $i \geq 1$ then we can begin with σ_i and relabel the vertices in the sequence. Thus, we obtain an m-cycle

$$J := (\sigma_0, \sigma_1, \sigma_2, \sigma_3, \sigma_4, \ldots, \sigma_{m-1}, \sigma_0)$$

of Σ. (It may happen that $m = 3$ if the girth of Σ is 3.) Let $\mathbf{E}$ denote the G-orbit on m-cycles of Σ containing J. In the following we will prove that each 2-arc of Σ is contained in exactly one of the "basic cycles" in $\mathbf{E}$ and hence $(\Sigma, \mathbf{E})$ is indeed a near m-gonal graph.

By the $(G,2)$-arc transitivity of Σ, it is clear that each 2-arc $(\sigma', \tau', \varepsilon')$ of Σ is contained in at least one member J^x of $\mathbf{E}$, where $x \in G$ is such that $(\sigma', \tau', \varepsilon') = (\sigma, \tau, \varepsilon)^x$. So it suffices to show that if two members of $\mathbf{E}$ have a 2-arc in common then they are identical; or, equivalently, if J^x and J have a 2-arc in common then they are identical.

Suppose then that J^x and J have a 2-arc in common for some $x \in G$. Note that, for each $i \geq 0$, g^i maps each vertex σ_j to σ_{j+i} and so $\langle g \rangle$ leaves J invariant (subscripts modulo m here and in the rest of this proof). So, replacing J^x by J^{xg^i} for some i if necessary, we may suppose without loss of generality that $(\sigma_0, \sigma_1, \sigma_2)$ is a common 2-arc of J^x and J. Then $(\sigma_0, \sigma_1, \sigma_2) \in J^x$ implies that $(\sigma_0, \sigma_1, \sigma_2) = (\sigma_{i-1}, \sigma_i, \sigma_{i+1})^x$ for some $1 \leq i \leq m$. Thus, $(\sigma_0, \sigma_1, \sigma_2) = (\sigma_0, \sigma_1, \sigma_2)^{g^{i-1}x}$ and hence $g^{i-1}x \in H$. From the properties (1)-(2) above, we then have $\sigma_{j+i-1}^x = \sigma_j^{g^{i-1}x} = \sigma_j$ for each vertex σ_j on J. That is, $\sigma_j^x = \sigma_{j-i+1}$ for each j and hence $J^x = J$. Thus, we have proved that each 2-arc of Σ is contained in exactly one member of $\mathbf{E}$, and so $(\Sigma, \mathbf{E})$ is a near m-gonal graph.

So far we have proved the equivalence of (a), (b) and (c). Now assume that one of these conditions is satisfied, so that $(\Sigma, \mathbf{E})$ is a near m-gonal graph for a G-orbit $\mathbf{E}$ on m-cycles of Σ, where $m \geq 3$. Clearly, we have $G \leq \text{Aut}(\Sigma, \mathbf{E})$. Let (α, e, C), (α', e', C') be maximal flags of the Buekenhout geometry associated with $(\Sigma, \mathbf{E})$. Denote $e = \{\alpha, \beta\}$, $e' = \{\alpha', \beta'\}$, $C = (\alpha, \beta, \gamma, \ldots, \alpha)$ and $C' = (\alpha', \beta', \gamma', \ldots, \alpha')$. Since Σ is $(G,2)$-arc transitive there exists $h \in G$ such that $(\alpha, \beta, \gamma)^h = (\alpha', \beta', \gamma')$. Hence $\alpha^h = \alpha'$, $e^h = e'$ and $C^h = C'$. That is, $(\alpha, e, C)^h = (\alpha', e', C')$, and thus G is transitive on the maximal flags of the Buekenhout geometry associated with $(\Sigma, \mathbf{E})$. $\square$

3. Remarks

The proof above gives a procedure for generating the near m-gonal graph $(\Sigma, \mathbf{E})$ guaranteed by Theorem 1. Unfortunately, it does not tell us any information about the relationship between m and the girth of Σ. Moreover, the basic cycles of $(\Sigma, \mathbf{E})$ are not necessarily induced cycles of Σ, that is, they may have chords. (See [19, Example 3.3, Proposition 3.4] for an example of such graphs. A *chord* of a cycle is an edge joining two non-consecutive vertices on the cycle.) Furthermore, from [19, Lemma 2.6(e)] such basic cycles contain chords only when either G_τ is sharply

2-transitive on $\Sigma(\tau)$ or $G_{\sigma\tau}$ is imprimitive on $\Sigma(\tau) \setminus \{\sigma\}$, where σ, τ are adjacent vertices of Σ, G_τ is the stabiliser of τ in G and $G_{\sigma\tau}$ is the pointwise stabiliser of $\{\sigma, \tau\}$ in G.

It is hoped that Theorem 1 would be useful in constructing 2-arc transitive near-polygonal graphs. In view of the 3-arc graph construction [6] and [19, Theorem 1.1], it would also be helpful in studying imprimitive G-symmetric graphs $(\Gamma, \mathbf{B})$ such that the subgraph (excluding isolated vertices) induced by two adjacent blocks B, C of $\mathbf{B}$ is a matching of $|B| - 1 \geq 2$ edges and that $\Gamma(C) \cap B \neq \Gamma(D) \cap B$ for different blocks C, D of $\mathbf{B}$ adjacent to B. A sufficient condition was given in [19] for a connected, non-complete, $(G, 2)$-arc transitive graph Σ of valency at least 3 to be a near m-gonal graph with respect to a G-orbit on m-cycles, where $m \geq 4$ is even. It was shown in [19, Corollary 4.1] that this is the case if G_σ is sharply 2-transitive on $\Sigma(\sigma)$ and one of the G-orbits on 3-arcs of Σ is self-paired. (A set A of 3-arcs is called *self-paired* if $(\sigma, \tau, \varepsilon, \delta) \in A$ implies $(\delta, \varepsilon, \tau, \sigma) \in A$.) Another sufficient condition was given in [13, Theorem 2.2] for a connected, non-complete, $(G, 2)$-arc transitive graph Σ to be a near m-gonal graph with respect to a G-orbit on m-cycles, where $m \geq 4$ is not necessarily even. Note that a near m-gonal graph $(\Sigma, \mathbf{E})$ is $(G, 2)$-arc transitive if and only if G is transitive on the maximal flags of the Buekenhout geometry associated with $(\Sigma, \mathbf{E})$. The "only if" part of this statement was proved in the last paragraph of the proof of Theorem 1, and the "if" part was part of [13, Theorem 1.8] and can be verified easily.

Finally, in the original definition [13] of a near m-gonal graph Σ, it was required that the girth of Σ be at least 4 and subsequently $m \geq 4$. In the definition given at the beginning of the introduction, we removed this requirement since the case of girth 3 is not entirely uninteresting when the graph is not 2-arc transitive. Of course for 2-arc transitive graphs this case is not so interesting, because a connected 2-arc transitive graph has girth 3 if and only if it is a complete graph (see e.g. [19, Lemma 2.5]). This is perhaps the main reason [15] for requiring girth ≥ 4 in a near-polygonal graph in [13], since the research in the area is focused on 2-arc transitive near-polygonal graphs. Recently, the author showed [23] that every connected trivalent $(G, 2)$-arc transitive graph (other than K_4) of type G_2^1 are near polygonal with respect to two G-orbits on even cycles.

Acknowledgment

The author thanks an anonymous referee for his/her helpful comments and interesting examples.

References

[1] N.L. Biggs, *Algebraic Graph Theory*. 2nd edition, Cambridge University Press, Cambridge, 1993.

[2] J.D. Dixon and B. Mortimer, *Permutation Groups*. Springer, New York, 1996.

[3] F. Buekenhout, *Diagrams for geometries and groups*. J. Combinatorial Theory Ser. A **27** (1979), 121–151.

[4] A. Gardiner, C.E. Praeger and S. Zhou, *Cross ratio graphs*. J. London Math. Soc. (2) **64** (2001), no. 2, 257–272.

[5] M.A. Iranmanesh, C.E. Praeger and S. Zhou, *Finite symmetric graphs with two-arc transitive quotients*. J. Combinatorial Theory Ser. B **94** (2005), 79–99.

[6] C.H. Li, C.E. Praeger and S. Zhou, *A class of finite symmetric graphs with 2-arc transitive quotients*. Math. Proc. Camb. Phil. Soc. **129** (2000), no. 1, 19–34.

[7] M. Perkel, *Bounding the valency of polygonal graphs with odd girth*. Canad. J. Math. **31** (1979), 1307–1321.

[8] M. Perkel, *A characterization of J_1 in terms of its geometry*. Geom. Dedicata **9** (1980), no. 3, 291–298.

[9] M. Perkel, *A characterization of* PSL(2, 31) *and its geometry*. Canad. J. Math. **32** (1980), no. 1, 155–164.

[10] M. Perkel, *Polygonal graphs of valency four*. Congr. Numer. **35** (1982), 387–400.

[11] M. Perkel, *Trivalent polygonal graphs*. Congr. Numer. **45** (1984), 45–70.

[12] M. Perkel, *Trivalent polygonal graphs of girth* 6 *and* 7. Congr. Numer. **49** (1985), 129–138.

[13] M. Perkel, *Near-polygonal graphs*. Ars Combinatoria **26(A)** (1988), 149–170.

[14] M. Perkel, *Some new examples of polygonal and near-polygonal graphs with large girth*. Bull. Inst. Combin. Appl. **10** (1994), 23–25.

[15] M. Perkel, *Personal communication*, 2005.

[16] M. Perkel, C.E. Praeger, *Polygonal graphs: new families and an approach to their analysis*. Congr. Numer. **124** (1997), 161–173.

[17] M. Perkel, C.E. Praeger, R. Weiss, *On narrow hexagonal graphs with a 3-homogeneous suborbit*. J. Algebraic Combin. **13** (2001), 257–273.

[18] C.E. Praeger, *Finite transitive permutation groups and finite vertex transitive graphs*, in: G. Hahn and G. Sabidussi eds., *Graph Symmetry* (Montreal, 1996, NATO Adv. Sci. Inst. Ser. C, Math. Phys. Sci., **497**), Kluwer Academic Publishing, Dordrecht, 1997, pp. 277–318.

[19] S. Zhou, *Almost covers of* 2*-arc transitive graphs*. Combinatorica **24** (4) (2004), 731–745.

[20] S. Zhou, *Imprimitive symmetric graphs,* 3*-arc graphs and* 1*-designs*. Discrete Math. **244** (2002), 521–537.

[21] S. Zhou, *Constructing a class of symmetric graphs*. European J. Combinatorics **23** (2002), 741–760.

[22] S. Zhou, *Symmetric graphs and flag graphs*. Monatshefte für Mathematik **139** (2003), 69–81.

[23] S. Zhou, *Trivalent 2-arc transitive graphs of type G_2^1 are near polygonal*. reprint, 2006.

Sanming Zhou
Department of Mathematics and Statistics
The University of Melbourne
Parkville, VIC 3010, Australia
e-mail: smzhou@ms.unimelb.edu.au

Open Problems

Edited by U.S.R. Murty

This chapter contains open problems that were presented at the problem session of the GT04 conference, complemented by several submitted later. Comments and questions of a technical nature should be addressed to the poser of the problem.

Problem GT04-1: Superstrongly perfect graphs

B.D. Acharya
Department of Science and technology,
Government of India, New Mehrauli Road,
New Delhi – 110 016, India

e-mail: bdacharya@yahoo.com

A graph G is *superstrongly perfect* if every induced subgraph H possesses a minimal dominating set that meets all the maximal complete subgraphs of H. Clearly, every strongly perfect graph is superstrongly perfect, but not conversely.

Problem. Characterize superstrongly perfect graphs.

Problem GT04-2: Eulerian Steinhaus graphs

M. Augier
EPFL,
Lausanne, Switzerland
e-mail: maxime.augier@epfl.ch

S. Eliahou
LMPA-ULCO,
B.P. 699, F-62228 Calais cedex, France
e-mail: eliahou@lmpa.univ-littoral.fr

To every binary string $s = x_1 x_2 \ldots x_{n-1} \in \mathbb{F}_2^{n-1}$ is associated a simple graph $G(s)$ on the vertex set $\{0, 1, \ldots, n-1\}$, whose adjacency matrix $M = (a_{i,j})_{0 \le i,j \le n-1} \in \mathcal{M}_n(\mathbb{F}_2)$ satisfies $a_{i,i} = 0$ for $i \ge 0$, $a_{0,i} = x_i$ for $i \ge 1$, and $a_{i,j} = a_{i-1,j-1} + a_{i-1,j}$ for $1 \le i < j \le n-1$. The graph $G(s)$ is called the *Steinhaus graph* associated to s.

Problem. Is it true that an Eulerian Steinhaus graph is completely determined by its vertex degree sequence?

It is known that $G(s)$ is connected unless s is the zero string $0\ldots 0$. Denote $d_i = d_i(s)$ the degree of vertex i in $G(s)$, and

$$d(s) = (d_1, d_2, \ldots, d_n)$$

the vertex degree sequence of $G(s)$.

The example $s_1 = 010$, $s_2 = 100$ with $d(s_1) = d(s_2) = (1, 2, 2, 1)$ shows that $d(s)$ alone does not determine s or $G(s)$ in general.

A conjecture of Dymacek states that the only *regular* Steinhaus graphs are those corresponding to the binary strings $1, 0\ldots 0$, and $110110\ldots 110$, see [6]. In [2], we investigate *parity-regular* Steinhaus graphs $G(s)$, where the degrees $d_i(s)$ all have the same parity, even or odd. The even case corresponds to Eulerian Steinhaus graphs, except for $s = 0\ldots 0$. Dymacek has shown that there are exactly $2^{\lfloor \frac{n-1}{3} \rfloor} - 1$ Eulerian Steinhaus graphs on n vertices [6].

In our study of parity-regular Steinhaus graphs, we came upon the observation that it is much harder in this context to find *collisions*, i.e., binary strings $s_1 \neq s_2 \in \mathbb{F}_2^{n-1}$ with $d(s_1) = d(s_2)$. The smallest collision in the parity-regular case occurs at $n = 26$ and is unique in this size:

$s_1 = 0010101001111110010101001$
$s_2 = 1101010110000001101010111 = s_1 + 1111111111111111111111110$

giving rise to the common degree sequence

$$d(s_1) = d(s_2)$$
$$= (13, 15, 9, 9, 13, 13, 17, 11, 9, 19, 9, 9, 11, 11, 9, 9, 19, 9, 11, 17, 13, 13, 9, 9, 15, 13).$$

Up to $n \leq 50$ vertices, we have found a total of 29 collisions in the parity-regular case. They only occur if $n \equiv 2 \bmod 4$, namely at $n = 26, 34, 38, 42, 46$ and 50 (yes, 30 is missing), and they all satisfy $s_1 + s_2 = 11\ldots 10$. We conjecture in [2] that these properties always hold.

Finally, in these 29 instances, all vertex degrees turn out to be odd. Thus, *up to $n \leq 50$ vertices, there are no collisions $s_1 \neq s_2$ where all degrees in $d(s_1) = d(s_2)$ are even*.

In other words, Eulerian Steinhaus graphs $G(s)$ are completely determined by their degree sequence $d(s) \in \mathbb{N}^n$ for $n \leq 50$. Does this remain true for $n \geq 51$?

Problem GT04-3: Edge-disjoint paths in planar graphs with a fixed number of terminal pairs

C. Bentz
CEDRIC, CNAM, Paris, France

e-mail: `cedric.bentz@cnam.fr`

Input: An undirected planar graph G, a list of k pairs of terminal vertices (source s_i, sink t_i), k being a fixed integer.

Problem. Find in $\bigcup_i P_i$ a maximum number of edge-disjoint paths (i.e., edge-disjoint in G), where, for each i, P_i is the set of elementary paths linking s_i to t_i in G.

If the graph is not planar, the problem is $\mathcal{NP}$-hard, even for $k = 2$ [7]. If k is not fixed, the problem is $\mathcal{NP}$-hard even in outerplanar graphs [9]. Moreover, in general graphs, the problem is tractable if the maximum degree is bounded or if we allow only one path between s_i and t_i for each i (in this case, one can solve the problem by solving a constant number of instances of the *edge-disjoint paths problem* and using the algorithm given in [13]). When $k = 2$ and adding the 2 edges (s_1, t_1) and (s_2, t_2) to G does not destroy planarity, the problem is polynomial-time solvable [11].

Problem GT04-4: Shortest alternating cycle

M.-C. Costa,
CEDRIC, CNAM, Paris, France
e-mail: `costa@cnam.fr`

D. de Werra,
EPFL, Lausanne, Suisse
e-mail: `dewerra.ima@epfl.ch`

C. Picouleau
CEDRIC, CNAM, Paris, France
e-mail: `chp@cnam.fr`

B. Ries
EPFL, Lausanne, Suisse
e-mail: `bernard.ries@epfl.ch`

The decision problem SAC (*Shortest Alternating Cycle*) is formally defined as follows:

Instance: A graph $G = (V, E)$ and a positive integer $L \leq |V|$.
Question: Is there a maximum matching M and an even cycle C with $|C| \leq L$ and $|C \cap M| = \frac{1}{2}|C|$?
Problem. Determine the complexity status of SAC.

The complexity status of SAC is unknown even if G is a 3-regular bipartite graph. Notice that the problem SAC becomes solvable in polynomial time if either a cycle C or a perfect matching M is given.

Problem GT04-5: Edge 3-coloration of K_{mn} with pre-specified colored degrees

M.-C. Costa,
CEDRIC, CNAM, Paris, France
e-mail: `costa@cnam.fr`

D. de Werra,
EPFL, Lausanne, Suisse
e-mail: `dewerra.ima@epfl.ch`

C. Picouleau
CEDRIC, CNAM, Paris, France
e-mail: `chp@cnam.fr`

B. Ries
EPFL, Lausanne, Suisse
e-mail: `bernard.ries@epfl.ch`

Let $G = (X, Y, E) = K_{mn}$ be the complete bipartite graph with $X = \{x_1, \ldots, x_m\}$ and $Y = \{y_1, \ldots, y_n\}$. Let $L_1 = (a_1, \ldots, a_m)$, $L_2 = (b_1, \ldots, b_m)$, $R_1 = (c_1, \ldots, c_n)$ and $R_2 = (d_1, \ldots, d_n)$ be four sequences of nonnegative integers such that $a_i + b_i \leq n, i = 1, \ldots, m$ and $c_i + d_i \leq m, i = 1, \ldots, n$.

The goal is to find a partition E_1, E_2, E_3 of the edge set E such that:
- $\delta_{G_1}(x_i) = a_i, i = 1,\ldots,m$ and $\delta_{G_1}(y_i) = c_i, i = 1,\ldots,n$
- $\delta_{G_2}(x_i) = b_i, i = 1,\ldots,m$ and $\delta_{G_2}(y_i) = d_i, i = 1,\ldots,n$
- $\delta_{G_3}(x_i) = n - a_i - b_i, i = 1,\ldots,m$ and $\delta_{G_3}(y_i) = m - c_i - d_i, i = 1,\ldots,n$

Here $\delta_{G_i}(z)$ denotes the degree of the vertex z in the partial graph
$$G_i = (X, Y, E_i), \quad i = 1, 2, 3.$$

Problem. Determine the complexity status of this problem.

This problem can be seen as a special case of a two-commodity integral flow problem in a complete bipartite network, for which the complexity status is also unknown. The problem can also be seen as the three-colored matrix reconstruction problem well known in discrete tomography.

Problem GT04-6: List matrix partition

T. Feder
268 Waverley St. Palo Alto,
CA 94301, USA
e-mail: `tomas@theory.stanford.edu`

P. Hell
Simon Fraser University,
Burnaby, B.C., Canada, V5A 1S6
e-mail: `pavol@cs.sfu.ca`

Problem. Find a polynomial time algorithm for the following decision problem: Given a complete graph with edges labeled by 1,2,3, can the vertices be labeled, also by 1,2,3, so that there is no monochromatic edge (i.e., an edge labeled by i whose both ends are also labeled by i)?

The best-known algorithm, due to T. Feder, P. Hell, D. Kral and J. Sgall [8] has the (subexponential) complexity $O(n^{\log n / \log \log n})$, indicating the problem is unlikely to be $\mathcal{NP}$-complete.

In [5], the authors define a problem of partitioning a given graph into four parts with certain constraints. That problem also turned out to be difficult to classify, and the authors have dubbed it the 'stubborn' problem. The problem described here is a relative of the stubborn problem (they can be shown to be polynomially equivalent). In particular, the algorithm of Feder, Hell, Kral and Sgall mentioned above also solves the stubborn problem in time $O(n^{\log n / \log \log n})$.

Problem GT04-7: A point-line configuration

Harald Gropp
Mühlingstr. 19, D-69121 Heidelberg, Germany
e-mail: `d12@ix.urz.uni-heidelberg.de`

A *configuration* v_6 is a finite incidence structure of v points and v lines such that each line consists of 6 points, there are 6 lines through each point, and two different points are connected by at most one line.

Problem. Does a configuration 33_6 exist ?

This is the last unsettled case after the construction of a configuration 34_6 by Krčadinac in 2004. The configuration 31_6 is the projective plane of order 5; a configuration 32_6 does not exist; for all other values of $v \geq 34$ a configuration v_6 exists.

Problem GT04-8: Graph partitioning

M. Haviv
Department of Statistics,

The Hebrew University of Jerusalem,
Israel
e-mail: `haviv@mscc.huji.ac.il`

E. Korach
Department of Industrial Engineering
and Management,
Ben-Gurion University of the Negev,
Israel
e-mail: `korach@bgumail.bgu.ac.il`

Problem. Given an undirected graph $G = (V, E)$, a constant δ and a positive integral weight function ω on E where, $\omega_{s,t}$ is the weight of the edge (s, t), find a partition $V = (J(1), \ldots, J(q))$ of the vertices of G into q subsets such that q is as large as possible and

$$\max_{1 \leq i \leq q} \max_{s \in J(i)} \sum_{\substack{1 \leq j \leq q \\ j \neq i}} \sum_{t \in J(j)} \omega_{s,t} \leq \delta,$$

i.e., a partition such that for every vertex v in G, the sum of the weights of the edges connecting v with vertices that are not in the same part of the partition as v, is at most δ.

It is known that the analogous problem for directed graph is $\mathcal{NP}$-hard [10]. A motivation for this problem comes from the subject of partitioning Markov chains. A related problem was defined and solved polynomially in [10].

Problem GT04-9: Number of independent sets in multicolorings

Dániel Marx
Dept. of Computer Science and Information Theory
Budapest University of Technology and Economics
H-1521 Budapest, Hungary
e-mail: `dmarx@cs.bme.hu`

Multicoloring is a generalization of ordinary vertex coloring. A graph $G(V, E)$ is given with integer weights $w : V \to \mathbb{N}$ on the vertices. A *multicoloring* is a function $\Psi : V \to 2^{\mathbb{N}}$ that assigns $w(v)$ colors to each vertex v in such a way that adjacent vertices receive disjoint sets of colors. The goal is to minimize the number of colors used. Clearly, ordinary vertex coloring is the special case where every weight is 1.

Another way to look at a multicoloring is to consider it as a collection of independent sets. This implies an integer linear programming formulation. Let A be a matrix where the rows correspond to the vertices of G and the columns are the incidence vectors of the maximal independent sets. Let **w** be a vector whose elements correspond to the weights of the vertices. The optimum of the following integer linear program determines the minimum number of colors required by a multicoloring:

$$\min \mathbf{1}^\top \cdot \mathbf{x}$$
$$A\mathbf{x} \geq \mathbf{w}$$

The length of the vector $\mathbf{x}$ can be exponential in the size of the graph. However, if the graph is perfect, then it is well known that there is an optimum solution $\mathbf{x}$ with at most $|V|$ non-zero components. But what can we say about the number of non-zero components if the graph is not perfect?

Problem. Is it true that for every multicoloring problem there is an optimum solution with a polynomial (or even linear) number of independent sets?

The following example shows that if the graph is not perfect, then we may need more than $|V|$ independent sets for a solution. Consider the *complement* of the following graph G:

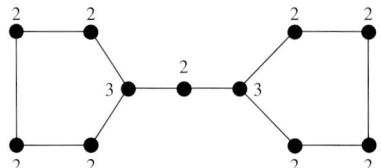

The graph $\overline{G}$ has 11 vertices and 12 maximal independent sets, each of size 2. It is not difficult to show that the only way to satisfy the weights shown on the figure is to select each of these 12 independent sets once.

The significance of the problem comes from the fact that it is not known whether the multicoloring problem is in the complexity class NP. The multicoloring itself is not a good certificate, since the weights can be exponentially large, and the color sets cannot be described in polynomial size. However, if we can prove that a polynomial number of independent sets is always sufficient for the multicoloring, then this implies that the problem is in NP.

Paul Seymour observed that if G is the complement of a triangle-free graph, then a linear number of independent sets are always sufficient. In this case the maximal independent sets of G are the edges of $\overline{G}$ (as in the example). If the independent sets of a solution form an even cycle in $\overline{G}$, then the solution can be simplified by alternately increasing and decreasing the multiplicities of the independent sets in the cycle. Thus the solution corresponds to an even-cycle-free subgraph of $\overline{G}$, which has a linear number of edges.

Problem GT04-10: Geodesic convexity

Ignacio M. Pelayo
Departament de Matemàtica Aplicada III,
Universitat Politècnica de Catalunya,
Barcelona, Spain

e-mail: `ignacio.m.pelayo@upc.edu`

Given vertices u, v in a connected graph G, the *geodetically closed interval* $I[u, v]$ is the set of vertices of all $u-v$ shortest paths. For $S \subseteq V$, the *geodetic closure* $I[S]$ of S is the union of all geodetically closed intervals $I[u, v]$ over all pairs $u, v \in S$, i.e., $I[S] = \bigcup_{u,v \in S} I[u, v]$. A vertex set S is called *convex* if $I[S] = S$. The smallest convex set containing S is denoted $[S]$ and is called the *convex hull* of S. A non-empty set $A \subseteq V$ is called a *hull set* if $[A] = V$, and it is said to be *geodetic* if moreover $I[A] = V$.

The *eccentricity* of a vertex $u \in V(G)$ is defined as $ecc_G(u) = \max\{d(u,v) \mid v \in V(G)\}$. A vertex $v \in V$ is called a *contour vertex* of G if no neighbor vertex of v has an eccentricity greater than $ecc(v)$. The *contour* $Ct(G)$ of G is the set all of its contour vertices:

$$Ct(G) = \{v \in V \mid ecc(u) \leq ecc(v), \forall u \in N(v)\}.$$

A set of vertices S is called *redundant* if $[S] = \bigcup_{a \in S}[S - a]$, and *irredundant* otherwise. The *Carathéodory number* $c = c(G)$ is the maximum cardinality of an irredundant set. A set of vertices S is called *exchange-dependent* if $[S - x] \subseteq \bigcup_{a \in S-x}[S - a]$ for all $x \in S$, and *exchange independent* otherwise. The *exchange number*, $e = e(G)$, is the maximum cardinality of an exchange independent set [15].

Problem 1. Is it true that $I^2[Ct(G)] = V(G)$ for every connected graph G?

It was proved in [3] that the contour of every graph is a hull set. It was shown in [4], firstly, that the contour of every chordal graph is geodetic, and secondly, that this statement is not true for every graph. We know of no example of a graph G having a contour whose geodetic closure is not geodetic.

Problem 2. Is it true that $I[Ct(G)] = V(G)$ for every bipartite graph G?

It was shown in [4], that the contour of every chordal graph is geodetic, that the contour of every distance hereditary graph is geodetic, and that this statement is not true for every perfect graph. We know of no example of a bipartite graph G whose contour be not geodetic.

Problem 3. Is it true that $e - 1 \leq c \leq e$ hold in general for every connected graph G?

In [14], G. Sierksma proved that, for every convexity space, $e - 1 \leq c$. We know of no example of a graph G for which $c(G) > e(G)$ when considering the geodesic convexity.

Problem GT04-11: Partition into closed trails

J.L. Ramírez Alfonsín
Université Pierre et Marie Curie, Paris 6,
Equipe Combinatoire et Optimisation – Case 189,
4 Place Jussieu Paris 75252 Cedex 05, France

e-mail: `ramirez@math.jussieu.fr`

Let $\mathcal{H}$ be a family of graphs consisting of m_i graphs H_i for $i = 1, \ldots, l$. A *decomposition* of a graph G into $\mathcal{H}$ is a partition of the edges of G into $\sum_{i=1}^{l} m_i$ edge-disjoint subgraphs such that m_i of which are isomorphic to H_i for each $i = 1, \ldots, l$.

Problem 1. Let $a_1, \ldots, a_m, n$ be positive integers. If n is odd and $a_1 + \cdots + a_m = \frac{n(n-1)}{2}$ (respectively, if n is even and $a_1 + \cdots + a_m = \frac{n(n-2)}{2}$), $3 \leq a_i \leq n$, does there exist a decomposition of K_n into $T_{a_1}, \ldots, T_{a_m}$ (respectively, a decomposition of $K_n - F$ into $T_{a_1}, \ldots, T_{a_m}$) where $K_n - F$ denote the complete graph on n vertices from which a 1-factor has been removed and T_{a_i} is any closed trail of length a_i [12]?

The above question is a weaker version of Alspach' conjecture [1] in which each closed trail T_{a_i} is replaced by a cycle of length a_i.

Consider the following two questions, see [12]:

Input: Paths $P_{a_1}, \ldots, P_{a_m}$ with $\sum_{i=1}^{m} a_i = \frac{n(n-1)}{2}$.
Question: Does there exist a decomposition of K_n into $P_{a_1}, \ldots, P_{a_m}$?

Input: Rooted trees $T_{a_1}, \ldots, T_{a_m}$ with $\sum_{i=1}^{m} a_i = \frac{n(n-1)}{2}$ and $m \leq n$.
Question: Does there exist a rooted decomposition of K_n into $T_{a_1}, \ldots, T_{a_m}$ (that is, the root of each tree T_{a_i} starts at different vertex of K_n)?

Problem 2. Are these questions decidable in polynomial time?

References

[1] B. Alspach, Research problems, Problem 3, *Discrete Mathematics* **36** (1981), 333.

[2] M. Augier and S. Eliahou, Parity-regular Steinhaus graphs, in preparation.

[3] J. Cáceres, A. Márquez, O.R. Oellerman, M.L. Puertas, Rebuilding convex sets in graphs, *Discrete Math.* **297**(1-3) (2005), 26–37.

[4] J. Cáceres, C. Hernando, M. Mora, I.M. Pelayo, M.L. Puertas and C. Seara. Geodeticity of the contour of chordal graphs, Submitted to *Disc. Appl. Math.*

[5] K. Cameron, E.E. Eschen, C.T. Hoang and R. Sritharan, The list partition problem for graphs, *Proc. 15th Annual ACM-SIAM Symposium on Discrete Algorithms (SODA)* (2004), 391–399.

[6] W. Dymacek, Steinhaus graphs, Proc. 10th southeast. Conf. Combinatorics, Graph Theory and Computing, Boca Raton 1979, Vol. I, *Congr. Numerantium* **23** (1979) 399–412.

[7] S. Even, A. Itai and A. Shamir, On the complexity of timetable and multicommodity flow problems, *SIAM J. Comp.* **5** (1976) 691–703.

[8] T. Feder, P. Hell, D. Kral and J. Sgall, Two algorithms for list matrix partitions, *Proc. 16th Annual ACM-SIAM Symposium on Discrete Algorithms (SODA)*, (2005), 870–876.

[9] N. Garg, V.V. Vazirani and M. Yannakakis, Primal-dual approximation algorithms for integral flow and multicut in trees, *Algorithmica* **18** (1997) 3–20.

[10] M. Haviv and E. Korach, On graph partitioning and determining near uncoupling in Markov chains, in preparation.

[11] E. Korach and M. Penn, A fast algorithm for maximum integral two-commodity flow in planar graphs, *Discrete Applied Mathematics* **47** (1993) 77–83.

[12] J.L. Ramírez Alfonsín, Topics in Combinatorics and Computational Complexity, D.Phil. Thesis, University of Oxford, U.K., (1993).

[13] N. Robertson and P.D. Seymour, Graphs minors XIII: The disjoint paths problem, *J. Comb. Theory, Ser. B*, **63** (1995) 65–110.

[14] G. Sierksma, Relationships between Carathéodory, Helly, Radon and exchange numbers of convexity spaces, *Nieuw Archief voor Wisk.*, **25** (2) (1977) 115–132.

[15] M. Van de Vel, Theory of convex structures, *North-Holland*, Amsterdam, MA (1993).

Edited by U.S.R. Murty
University of Waterloo, Canada
e-mail: `usrmurty@math.uwaterloo.ca`